APPROXIMATION THEORY

PURE AND APPLIED MATHEMATICS

A Program of Monographs, Textbooks, and Lecture Notes

MONOGRAPHS AND TEXTBOOKS IN PURE AND APPLIED MATHEMATICS

1. *K. Yano,* Integral Formulas in Riemannian Geometry (1970)
2. *S. Kobayashi,* Hyperbolic Manifolds and Holomorphic Mappings (1970)
3. *V. S. Vladimirov,* Equations of Mathematical Physics (A. Jeffrey, ed.; A. Littlewood, trans.) (1970)
4. *B. N. Pshenichnyi,* Necessary Conditions for an Extremum (L. Neustadt, translation ed.; K. Makowski, trans.) (1971)
5. *L. Narici et al.,* Functional Analysis and Valuation Theory (1971)
6. *S. S. Passman,* Infinite Group Rings (1971)
7. *L. Dornhoff,* Group Representation Theory. Part A: Ordinary Representation Theory. Part B: Modular Representation Theory (1971, 1972)
8. *W. Boothby and G. L. Weiss, eds.,* Symmetric Spaces (1972)
9. *Y. Matsushima,* Differentiable Manifolds (E. T. Kobayashi, trans.) (1972)
10. *L. E. Ward, Jr.,* Topology (1972)
11. *A. Babakhanian,* Cohomological Methods in Group Theory (1972)
12. *R. Gilmer,* Multiplicative Ideal Theory (1972)
13. *J. Yeh,* Stochastic Processes and the Wiener Integral (1973)
14. *J. Barros-Neto,* Introduction to the Theory of Distributions (1973)
15. *R. Larsen,* Functional Analysis (1973)
16. *K. Yano and S. Ishihara,* Tangent and Cotangent Bundles (1973)
17. *C. Procesi,* Rings with Polynomial Identities (1973)
18. *R. Hermann,* Geometry, Physics, and Systems (1973)
19. *N. R. Wallach,* Harmonic Analysis on Homogeneous Spaces (1973)
20. *J. Dieudonné,* Introduction to the Theory of Formal Groups (1973)
21. *I. Vaisman,* Cohomology and Differential Forms (1973)
22. *B.-Y. Chen,* Geometry of Submanifolds (1973)
23. *M. Marcus,* Finite Dimensional Multilinear Algebra (in two parts) (1973, 1975)
24. *R. Larsen,* Banach Algebras (1973)
25. *R. O. Kujala and A. L. Vitter, eds.,* Value Distribution Theory: Part A; Part B: Deficit and Bezout Estimates by Wilhelm Stoll (1973)
26. *K. B. Stolarsky,* Algebraic Numbers and Diophantine Approximation (1974)
27. *A. R. Magid,* The Separable Galois Theory of Commutative Rings (1974)
28. *B. R. McDonald,* Finite Rings with Identity (1974)
29. *J. Satake,* Linear Algebra (S. Koh et al., trans.) (1975)
30. *J. S. Golan,* Localization of Noncommutative Rings (1975)
31. *G. Klambauer,* Mathematical Analysis (1975)
32. *M. K. Agoston,* Algebraic Topology (1976)
33. *K. R. Goodearl,* Ring Theory (1976)
34. *L. E. Mansfield,* Linear Algebra with Geometric Applications (1976)
35. *N. J. Pullman,* Matrix Theory and Its Applications (1976)
36. *B. R. McDonald,* Geometric Algebra Over Local Rings (1976)
37. *C. W. Groetsch,* Generalized Inverses of Linear Operators (1977)
38. *J. E. Kuczkowski and J. L. Gersting,* Abstract Algebra (1977)
39. *C. O. Christenson and W. L. Voxman,* Aspects of Topology (1977)
40. *M. Nagata,* Field Theory (1977)
41. *R. L. Long,* Algebraic Number Theory (1977)
42. *W. F. Pfeffer,* Integrals and Measures (1977)
43. *R. L. Wheeden and A. Zygmund,* Measure and Integral (1977)
44. *J. H. Curtiss,* Introduction to Functions of a Complex Variable (1978)
45. *K. Hrbacek and T. Jech,* Introduction to Set Theory (1978)
46. *W. S. Massey,* Homology and Cohomology Theory (1978)
47. *M. Marcus,* Introduction to Modern Algebra (1978)
48. *E. C. Young,* Vector and Tensor Analysis (1978)
49. *S. B. Nadler, Jr.,* Hyperspaces of Sets (1978)
50. *S. K. Segal,* Topics in Group Kings (1978)
51. *A. C. M. van Rooij,* Non-Archimedean Functional Analysis (1978)
52. *L. Corwin and R. Szczarba,* Calculus in Vector Spaces (1979)
53. *C. Sadosky,* Interpolation of Operators and Singular Integrals (1979)

54. *J. Cronin,* Differential Equations (1980)
55. *C. W. Groetsch,* Elements of Applicable Functional Analysis (1980)
56. *I. Vaisman,* Foundations of Three-Dimensional Euclidean Geometry (1980)
57. *H. I. Freedan,* Deterministic Mathematical Models in Population Ecology (1980)
58. *S. B. Chae,* Lebesgue Integration (1980)
59. *C. S. Rees et al.,* Theory and Applications of Fourier Analysis (1981)
60. *L. Nachbin,* Introduction to Functional Analysis (R. M. Aron, trans.) (1981)
61. *G. Orzech and M. Orzech,* Plane Algebraic Curves (1981)
62. *R. Johnsonbaugh and W. E. Pfaffenberger,* Foundations of Mathematical Analysis (1981)
63. *W. L. Voxman and R. H. Goetschel,* Advanced Calculus (1981)
64. *L. J. Corwin and R. H. Szczarba,* Multivariable Calculus (1982)
65. *V. I. Istrățescu,* Introduction to Linear Operator Theory (1981)
66. *R. D. Järvinen,* Finite and Infinite Dimensional Linear Spaces (1981)
67. *J. K. Beem and P. E. Ehrlich,* Global Lorentzian Geometry (1981)
68. *D. L. Armacost,* The Structure of Locally Compact Abelian Groups (1981)
69. *J. W. Brewer and M. K. Smith, eds.,* Emmy Noether: A Tribute (1981)
70. *K. H. Kim,* Boolean Matrix Theory and Applications (1982)
71. *T. W. Wieting,* The Mathematical Theory of Chromatic Plane Ornaments (1982)
72. *D. B.Gauld,* Differential Topology (1982)
73. *R. L. Faber,* Foundations of Euclidean and Non-Euclidean Geometry (1983)
74. *M. Carmeli,* Statistical Theory and Random Matrices (1983)
75. *J. H. Carruth et al.,* The Theory of Topological Semigroups (1983)
76. *R. L. Faber,* Differential Geometry and Relativity Theory (1983)
77. *S. Barnett,* Polynomials and Linear Control Systems (1983)
78. *G. Karpilovsky,* Commutative Group Algebras (1983)
79. *F. Van Oystaeyen and A. Verschoren,* Relative Invariants of Rings (1983)
80. *I. Vaisman,* A First Course in Differential Geometry (1984)
81. *G. W. Swan,* Applications of Optimal Control Theory in Biomedicine (1984)
82. *T. Petrie and J. D. Randall,* Transformation Groups on Manifolds (1984)
83. *K. Goebel and S. Reich,* Uniform Convexity, Hyperbolic Geometry, and Nonexpansive Mappings (1984)
84. *T. Albu and C. Năstăsescu,* Relative Finiteness in Module Theory (1984)
85. *K. Hrbacek and T. Jech,* Introduction to Set Theory: Second Edition (1984)
86. *F. Van Oystaeyen and A. Verschoren,* Relative Invariants of Rings (1984)
87. *B. R. McDonald,* Linear Algebra Over Commutative Rings (1984)
88. *M. Namba,* Geometry of Projective Algebraic Curves (1984)
89. *G. F. Webb,* Theory of Nonlinear Age-Dependent Population Dynamics (1985)
90. *M. R. Bremner et al.,* Tables of Dominant Weight Multiplicities for Representations of Simple Lie Algebras (1985)
91. *A. E. Fekete,* Real Linear Algebra (1985)
92. *S. B. Chae,* Holomorphy and Calculus in Normed Spaces (1985)
93. *A. J. Jerri,* Introduction to Integral Equations with Applications (1985)
94. *G. Karpilovsky,* Projective Representations of Finite Groups (1985)
95. *L. Narici and E. Beckenstein,* Topological Vector Spaces (1985)
96. *J. Weeks,* The Shape of Space (1985)
97. *P. R. Gribik and K. O. Kortanek,* Extremal Methods of Operations Research (1985)
98. *J.-A. Chao and W. A. Woyczynski, eds.,* Probability Theory and Harmonic Analysis (1986)
99. *G. D. Crown et al.,* Abstract Algebra (1986)
100. *J. H. Carruth et al.,* The Theory of Topological Semigroups, Volume 2 (1986)
101. *R. S. Doran and V. A. Belfi,* Characterizations of C*-Algebras (1986)
102. *M. W. Jeter,* Mathematical Programming (1986)
103. *M. Altman,* A Unified Theory of Nonlinear Operator and Evolution Equations with Applications (1986)
104. *A. Verschoren,* Relative Invariants of Sheaves (1987)
105. *R. A. Usmani,* Applied Linear Algebra (1987)
106. *P. Blass and J. Lang,* Zariski Surfaces and Differential Equations in Characteristic $p > 0$ (1987)
107. *J. A. Reneke et al.,* Structured Hereditary Systems (1987)
108. *H. Busemann and B. B. Phadke,* Spaces with Distinguished Geodesics (1987)
109. *R. Harte,* Invertibility and Singularity for Bounded Linear Operators (1988)

110. *G. S. Ladde et al.*, Oscillation Theory of Differential Equations with Deviating Arguments (1987)
111. *L. Dudkin et al.*, Iterative Aggregation Theory (1987)
112. *T. Okubo*, Differential Geometry (1987)
113. *D. L. Stancl and M. L. Stancl*, Real Analysis with Point-Set Topology (1987)
114. *T. C. Gard*, Introduction to Stochastic Differential Equations (1988)
115. *S. S. Abhyankar*, Enumerative Combinatorics of Young Tableaux (1988)
116. *H. Strade and R. Farnsteiner*, Modular Lie Algebras and Their Representations (1988)
117. *J. A. Huckaba*, Commutative Rings with Zero Divisors (1988)
118. *W. D. Wallis*, Combinatorial Designs (1988)
119. *W. Więsław*, Topological Fields (1988)
120. *G. Karpilovsky*, Field Theory (1988)
121. *S. Caenepeel and F. Van Oystaeyen*, Brauer Groups and the Cohomology of Graded Rings (1989)
122. *W. Kozlowski*, Modular Function Spaces (1988)
123. *E. Lowen-Colebunders*, Function Classes of Cauchy Continuous Maps (1989)
124. *M. Pavel*, Fundamentals of Pattern Recognition (1989)
125. *V. Lakshmikantham et al.*, Stability Analysis of Nonlinear Systems (1989)
126. *R. Sivaramakrishnan*, The Classical Theory of Arithmetic Functions (1989)
127. *N. A. Watson*, Parabolic Equations on an Infinite Strip (1989)
128. *K. J. Hastings*, Introduction to the Mathematics of Operations Research (1989)
129. *B. Fine*, Algebraic Theory of the Bianchi Groups (1989)
130. *D. N. Dikranjan et al.*, Topological Groups (1989)
131. *J. C. Morgan II*, Point Set Theory (1990)
132. *P. Biler and A. Witkowski*, Problems in Mathematical Analysis (1990)
133. *H. J. Sussmann*, Nonlinear Controllability and Optimal Control (1990)
134. *J.-P. Florens et al.*, Elements of Bayesian Statistics (1990)
135. *N. Shell*, Topological Fields and Near Valuations (1990)
136. *B. F. Doolin and C. F. Martin*, Introduction to Differential Geometry for Engineers (1990)
137. *S. S. Holland, Jr.*, Applied Analysis by the Hilbert Space Method (1990)
138. *J. Okniński*, Semigroup Algebras (1990)
139. *K. Zhu*, Operator Theory in Function Spaces (1990)
140. *G. B. Price*, An Introduction to Multicomplex Spaces and Functions (1991)
141. *R. B. Darst*, Introduction to Linear Programming (1991)
142. *P. L. Sachdev*, Nonlinear Ordinary Differential Equations and Their Applications (1991)
143. *T. Husain*, Orthogonal Schauder Bases (1991)
144. *J. Foran*, Fundamentals of Real Analysis (1991)
145. *W. C. Brown*, Matrices and Vector Spaces (1991)
146. *M. M. Rao and Z. D. Ren*, Theory of Orlicz Spaces (1991)
147. *J. S. Golan and T. Head*, Modules and the Structures of Rings (1991)
148. *C. Small*, Arithmetic of Finite Fields (1991)
149. *K. Yang*, Complex Algebraic Geometry (1991)
150. *D. G. Hoffman et al.*, Coding Theory (1991)
151. *M. O. González*, Classical Complex Analysis (1992)
152. *M. O. González*, Complex Analysis (1992)
153. *L. W. Baggett*, Functional Analysis (1992)
154. *M. Sniedovich*, Dynamic Programming (1992)
155. *R. P. Agarwal*, Difference Equations and Inequalities (1992)
156. *C. Brezinski*, Biorthogonality and Its Applications to Numerical Analysis (1992)
157. *C. Swartz*, An Introduction to Functional Analysis (1992)
158. *S. B. Nadler, Jr.*, Continuum Theory (1992)
159. *M. A. Al-Gwaiz*, Theory of Distributions (1992)
160. *E. Perry*, Geometry: Axiomatic Developments with Problem Solving (1992)
161. *E. Castillo and M. R. Ruiz-Cobo*, Functional Equations and Modelling in Science and Engineering (1992)
162. *A. J. Jerri*, Integral and Discrete Transforms with Applications and Error Analysis (1992)
163. *A. Charlier et al.*, Tensors and the Clifford Algebra (1992)
164. *P. Biler and T. Nadzieja*, Problems and Examples in Differential Equations (1992)
165. *E. Hansen*, Global Optimization Using Interval Analysis (1992)
166. *S. Guerre-Delabrière*, Classical Sequences in Banach Spaces (1992)

167. *Y. C. Wong*, Introductory Theory of Topological Vector Spaces (1992)
168. *S. H. Kulkarni and B. V. Limaye*, Real Function Algebras (1992)
169. *W. C. Brown*, Matrices Over Commutative Rings (1993)
170. *J. Loustau and M. Dillon*, Linear Geometry with Computer Graphics (1993)
171. *W. V. Petryshyn*, Approximation-Solvability of Nonlinear Functional and Differential Equations (1993)
172. *E. C. Young*, Vector and Tensor Analysis: Second Edition (1993)
173. *T. A. Bick*, Elementary Boundary Value Problems (1993)
174. *M. Pavel*, Fundamentals of Pattern Recognition: Second Edition (1993)
175. *S. A. Albeverio et al.*, Noncommutative Distributions (1993)
176. *W. Fulks*, Complex Variables (1993)
177. *M. M. Rao*, Conditional Measures and Applications (1993)
178. *A. Janicki and A. Weron*, Simulation and Chaotic Behavior of α-Stable Stochastic Processes (1994)
179. *P. Neittaanmäki and D. Tiba*, Optimal Control of Nonlinear Parabolic Systems (1994)
180. *J. Cronin*, Differential Equations: Introduction and Qualitative Theory, Second Edition (1994)
181. *S. Heikkilä and V. Lakshmikantham*, Monotone Iterative Techniques for Discontinuous Nonlinear Differential Equations (1994)
182. *X. Mao*, Exponential Stability of Stochastic Differential Equations (1994)
183. *B. S. Thomson*, Symmetric Properties of Real Functions (1994)
184. *J. E. Rubio*, Optimization and Nonstandard Analysis (1994)
185. *J. L. Bueso et al.*, Compatibility, Stability, and Sheaves (1995)
186. *A. N. Michel and K. Wang*, Qualitative Theory of Dynamical Systems (1995)
187. *M. R. Darnel*, Theory of Lattice-Ordered Groups (1995)
188. *Z. Naniewicz and P. D. Panagiotopoulos*, Mathematical Theory of Hemivariational Inequalities and Applications (1995)
189. *L. J. Corwin and R. H. Szczarba*, Calculus in Vector Spaces: Second Edition (1995)
190. *L. H. Erbe et al.*, Oscillation Theory for Functional Differential Equations (1995)
191. *S. Agaian et al.*, Binary Polynomial Transforms and Nonlinear Digital Filters (1995)
192. *M. I. Gil'*, Norm Estimations for Operation-Valued Functions and Applications (1995)
193. *P. A. Grillet*, Semigroups: An Introduction to the Structure Theory (1995)
194. *S. Kichenassamy*, Nonlinear Wave Equations (1996)
195. *V. F. Krotov*, Global Methods in Optimal Control Theory (1996)
196. *K. I. Beidar et al.*, Rings with Generalized Identities (1996)
197. *V. I. Arnautov et al.*, Introduction to the Theory of Topological Rings and Modules (1996)
198. *G. Sierksma*, Linear and Integer Programming (1996)
199. *R. Lasser*, Introduction to Fourier Series (1996)
200. *V. Sima*, Algorithms for Linear-Quadratic Optimization (1996)
201. *D. Redmond*, Number Theory (1996)
202. *J. K. Beem et al.*, Global Lorentzian Geometry: Second Edition (1996)
203. *M. Fontana et al.*, Prüfer Domains (1997)
204. *H. Tanabe*, Functional Analytic Methods for Partial Differential Equations (1997)
205. *C. Q. Zhang*, Integer Flows and Cycle Covers of Graphs (1997)
206. *E. Spiegel and C. J. O'Donnell*, Incidence Algebras (1997)
207. *B. Jakubczyk and W. Respondek*, Geometry of Feedback and Optimal Control (1998)
208. *T. W. Haynes et al.*, Fundamentals of Domination in Graphs (1998)
209. *T. W. Haynes et al.*, Domination in Graphs: Advanced Topics (1998)
210. *L. A. D'Alotto et al.*, A Unified Signal Algebra Approach to Two-Dimensional Parallel Digital Signal Processing (1998)
211. *F. Halter-Koch*, Ideal Systems (1998)
212. *N. K. Govil et al.*, Approximation Theory (1998)

Additional Volumes in Preparation

APPROXIMATION THEORY

In Memory of A. K. Varma

edited by

N. K. Govil
Auburn University
Auburn, Alabama

R. N. Mohapatra
University of Central Florida
Orlando, Florida

Z. Nashed
University of Delaware
Newark, Delaware

A. Sharma
University of Alberta
Edmonton, Alberta, Canada

J. Szabados
Hungarian Academy of Sciences
Budapest, Hungary

MARCEL DEKKER, INC. NEW YORK • BASEL • HONG KONG

Library of Congress Cataloging-in-Publication Data

Approximation theory: in memory of A. K. Varma / edited by N. K. Govil
. . . [et al.].
p. cm.—(Monographs and textbooks in pure and applied mathematics; 212)
Includes bibliographical references and index.
ISBN 0-8247-0185-2 (alk. paper)
1. Approximation theory. I. Govil, N. K. (Narendra Kumar) II. Series.
QA221.A657 1998
511'.4—dc21 98-16700
CIP

The publisher offers discounts on this book when ordered in bulk quantities. For more information, write to Special Sales/Professional Marketing at the address below.

This book is printed on acid-free paper.

MARCEL DEKKER, INC.
270 Madison Avenue, New York, New York 10016
http://www.dekker.com

Current printing (last digit):
10 9 8 7 6 5 4 3 2 1

PRINTED IN THE UNITED STATES OF AMERICA

Foreword

The volume of papers assembled here is a memorial to Professor A Kumar Varma, and it reflects quite accurately his interests in approximation theory and the related branches of mathematics. This Foreword presents a brief summary of the contents of the volume. There is an astonishing breadth of research reported here, and the contributors are to be thanked for making possible this outstanding memorial to Professor Varma.

The first paper, by Agarwal and Wong, studies Lidstone interpolation, which is Hermite-Birkhoff interpolation at two points utilizing only even-ordered derivatives. The polynomial interpolant can be written down at once using Lidstone polynomials. Error bounds are given, and applications to Lidstone boundary value problems are investigated. The authors also introduce Lidstone spline interpolation and establish error bounds for it.

The contribution by Anastassiou studies four linear approximation operators. In each, the basis functions for the approximating function are translates and dilates of a single function. Thus, they are similar to wavelet approximations. He gives error bounds on the approximation in terms of derivatives (or moduli of continuity) of the function being approximated.

The paper by Balázs studies the interpolation problem of finding a polynomial p of least degree such that $p(x_1), \ldots, p(x_n)$, $p'(x_n)$, and $(wp)''(x_1)$, $\ldots, (wp)''(x_{n-1})$ take prescribed values. Here, w is a weight function. The classical weight functions are of special interest here.

In his paper, Baran gives a new proof of the Markov Inequality that relates the norm of a polynomial to the norm of its derivative. The L^p norm is considered, and the constants occurring in the inequality are improvements over those previously known.

The contribution by Beaucoup and Carraro treats a problem in approximation that arose in reliability theory. Namely, can a continuous monotone function ϕ on $[0,1]$ satisfying $\phi(0) = 0$ and $\phi(1) = 1$ be approximated uniformly with arbitrary precision by functions in the family generated from $x \mapsto x^\lambda$ by using only the operations $f \times g = fg$ and $f * g = f + g - fg$?

In the paper by Bojanov, five disjoint topics are considered, of which we single out the following one: For each n, find the polynomials P and Q of degree at most n which minimize the supremum of $|Q(x) - |P(x)||$ on $[-1,1]$, subject to the constraints $P(-1) = -1$, $P(1) = 1$, and $P'(x) \geq 0$ on $[-1,1]$.

In their contribution, Borwein and Erdélyi prove some Remez-type inequalities for series $\sum a_i x^{\lambda_i}$. These enable them to prove new results about the density of spaces of such functions in weighted L^p-spaces on arbitrary subsets of the real line. There are Müntz-type theorems and theorems akin to the Clarkson-Erdős results.

The next paper, by Casazza and Christensen, discusses the notions of "Schauder basis", "Riesz basis", "unconditional basis", and "frame" in the setting of Hilbert space. These concepts have come to the fore in recent years because of their role in wavelet theory, and the relationships among them are still not completely understood. The authors give an example to show the unexpected result that a frame need not contain a Schauder basis.

The paper by Cheney and Sun discusses the problem of interpolation of data on the m−dimensional sphere S^m by functions that are strictly positive definite on the sphere. Such a function f is defined on the interval $[-1, 1]$ and has the property that when distinct nodes $x_1, \ldots, x_n$ are chosen on the sphere the matrix $A_{ij} = f(\langle x_i, x_j\rangle)$ is nonsingular. Hence, the functions $x \mapsto f(\langle x, x_j\rangle)$ provide an ideal base for interpolation at the given nodes.

In his paper, Ciesielski shows how Bernstein polynomials and B-splines can be used to carry out non-parametric density estimation on finite intervals. He employs the Bernstein-Durrmeyer operators and the de Casteljau algorithm in the polynomial case.

The paper of de Bruin, Sharma, and Szabados addresses the Birkhoff interpolation problem of type $(0, 1, \ldots, r-2, r)$. Their principal theorem establishes that this problem is regular when $r \geq 2$, $0 < \alpha < 1$ and the nodes are given by $z_k = (\omega_k - \alpha)/(1 - \alpha\omega_k)$, where ω_k are the n−th roots of unity.

The contribution by Dechevsky, Dryanov, and Rahman studies a theorem of Clunie. This theorem asserts that if a polynomial f has only real zeros and if -1 and 1 are consecutive zeros, the integral of $|1 - f|$ over $[-1, 1]$ is at least $1/10$. The interesting questions (which they answer) are whether the inequality can be improved and generalized to include other functions and other integrands, such as $|1 - f|^p$.

The first paper by Floater and Micchelli introduces a general concept of a "mean": it is a function m of three variables with the properties $m(0, x, y) = x$, $m(1, x, y) = y$ and $m(r, m(s, x, y), m(t, x, y)) = m((1 - r)s + rt, x, y)$. In these equations, x, y, and $m(t, x, y)$ can be in $\mathbf{R}^d$. The purpose of this general notion of a mean is to extend the algorithms of geometric modeling beyond the familiar case where m is the function that maps (t, x, y) to $(1 - t)x + ty$. In particular, the generalized means need not be linear.

In their second paper, Floater and Micchelli study a class of nonlinear subdivision schemes that includes schemes based on local rational interpolation. In particular, they discuss convergence and convexity–preserving properties of these algorithms.

The paper by Fournier, Ma, and Ruscheweyh studies convex univalent functions f that are defined on the unit disk $\mathbf{D}$ and omit certain values. For example, if f is such a function having the form $f(z) = z + az^2 + bz^3 + \cdots$ then the real part of $af(z) + 1/2$ is positive for each z in $\mathbf{D}$.

In his contribution, Goodman investigates curves produced from control polygons by totally positive bases or by totally positive subdivision matrices. In general, the total angle turned through by the tangent to the curve is no greater than the total angle turned through by the polygon, but if, in this assertion, the tangent is replaced by the binormal the relationship is reversed: the total angle turned through by the binormal is greater for the curve than for the polygon.

In the paper by Govil and Mohapatra, they study the family of rational functions f having numerator of degree at most n and denominator of degree n, and having all poles exterior to the unit disk and zeros on or outside the unit circle. They seek bounds on the ratio of the maximum modulus of f on the circle $|z| = R$ to the maximum modulus of f on the unit disk. For a polynomial, this ratio is bounded above by $(R^n + 1)/2$, when $R \geq 1$, according to a theorem of Ankeny and Rivlin. For the rational functions described, a corresponding inequality is proved, in which R^n is replaced by the maximum modulus of $\Pi(1 - \bar{a_k}z)/(z - a_k)$ on the circle $|z| = R$. Here, $a_1, \ldots, a_n$ are the poles of the function f.

In his paper, Don Hong provides a survey of recent results in the area of multivariate splines. Three topics are emphasized: B-nets and smoothness conditions of multivariate splines; dimension and local bases of multivariate spline spaces; approximation powers of multivariate spline spaces. This extensive survey, with its many references, will bring the reader up-to-date in the area of multivariate splines.

The paper by Jakimovski and Sharma begins by establishing a formula for the Hermite interpolation of a function at the zeros of $(T_m(x))^p$, thus extending work of Rivlin, who had already done the "Lagrange" case, $p = 1$. In this work, the function being interpolated is assumed to be analytic inside the ellipse whose foci are ± 1 and whose semi-axes add up to R. The authors then obtain the region of Walsh equiconvergence of the interpolant as $m \to \infty$. The description of this region is sharp for functions having a singularity on the given ellipse.

In the contribution by Johnson, the question of what is meant by saying that a function changes sign at a point is carefully examined. (This issue arises in several situations of classical approximation theory.) The author shows that for any perfect subset S of $[a, b]$ there exists a continuous function that "changes sign around each point" of S, but never "properly crosses the axis".

In the paper of Kilgore, $Y(n, \alpha, \beta)$ is the space of polynomials of degree at most n, each multiplied by the Jacobi weight, $(1 - x)^\alpha (1 + x)^\beta$, where $\alpha > 0$ and $\beta > 0$. The minimal–norm interpolating operator that projects continuous functions into $Y(n, \alpha, \beta)$ is characterized by the fact that its nodes satisfy the Bernstein-Erdős conditions. That is, the Lebesgue function should exhibit local maxima of equal magnitude. The author proves this and gives the corresponding results when one of the parameters α and β is zero.

The paper of Li establishes a new inequality that generalizes those of Chebyshev and Turán. It states that if $z_1, \ldots, z_n$ are arbitrary complex numbers and if $p(z) = (z - z_1) \cdots (z - z_n)$ then the supremum norm of p on the interval $[-1, 1]$ is at least $2^{-n}(A + A^{-1})$, where A is defined to be $\Pi_{k=1}^{n} |z_k + (z_k^2 - 1)^{1/2}|$. A number of related results are given.

In his contribution, Lubinsky explores the question of how "smoothness" of the coefficient-sequence in $f(z) = \sum_{j=0}^{\infty} a_j z^j$ is reflected in nice behavior of the Padé approximants of f. For example, he proves that if $a_j \neq 0$ and $\lim a_{j-1} a_{j+1} a_j^{-2} = q$, then for each $n \geq 1$, $\lim_m D(m/n) a_m^n$ exists and equals $\Pi_1^{n-1}(1 - q^j)^{n-j}$. Here, $D(m/n)$ is the determinant of the matrix $(a_{m-j+k})_{j,k=1}^{n}$.

The paper by Mhaskar and Prestin considers the Marcinkiewicz-Zygmund Inequality, expressed as $||S||_{p,\mu} \leq c_1 ||S||_{p,\nu} \leq c_2 ||S||_{p,\mu}$, where S is a trigonometric polynomial, μ is Lebesgue measure on $[-\pi, \pi]$, and ν is the discrete measure that assigns measure $1/(2n+1)$ to each point $2\pi j/(2n+1)$ for $1 \leq j \leq 2n + 1$. They generalize this by considering any measurable space, two σ-finite measures, and a general kernel, $K(x, t)$, which leads to two linear integral operators, one using the $\mu-$integral and the other the $\nu-$integral. The theorems they obtain are then applied to certain classical situations, such as the integral operator of de La Vallée Poussin.

The work of Milovanović considers extremal problems typified by the classical Markov Inequality. Thus, one specifies a family of polynomials $\mathcal{F}$ and a norm $|| \; ||$, and asks for the value of $\sup ||p'||/||p||$ when p runs over $\mathcal{F}$. The author surveys this field of research, emphasizing weighted integral norms, pairs of different norms, the class of nonnegative polynomials on $[0, \infty]$, and many other interesting variations on this theme. A valuable bibliography of over 50 items is included.

The contribution by Milovanović and Rassias surveys extremal problems of the type $\inf ||p^{(m)}||/||p||$, where p runs over some prescribed set of polynomials. The resulting inequalities are known as "inequalities of Turán type" after Turán's inequality $||p'|| \geq B_n ||p||$, for polynomials of degree n having all zeros in $[-1, 1]$. Here, the norm is the sup-norm on $[-1, 1]$, and $B_n = \sqrt{n}/6$. This field of research was one in which Professor Varma made many contributions.

The paper by Pleśniak gives a survey of multivariable Markov inequalities and their applications. A first such example is that $||\nabla(p)||_E \leq M n^{2k} ||p||_E$ when E is the set of points (x, y) such that $0 \leq x \leq 1$ and

$0 \le y \le x^k$ and p is any polynomial in two variables of degree at most n. Applications to the problem of C^∞–extension of functions are described. The bibliography contains nearly 60 items.

In his contribution, Schmeisser considers an orthogonal system of polynomials, ϕ_n. If a given polynomial, p, is expressed as $p = \sum_{k=0}^{n} a_k \phi_k$, can the location of the zeros of p be determined roughly from the coefficients a_k? Papers by Turán, Specht, and others contributed to this problem. Schmeisser gives a sufficient condition for p to have a specified number of variations in sign.

The paper by Singh concerns the family $M_{2,\mu}$ consisting of all functions f analytic in the open unit disk, satisfying $f(0) = 0$, $f'(0) = 1$ and $|zf''(z)| < \mu$ in the disk. He shows, for example, that $M_{2,\mu}$ is convex if and only if $0 \le \mu \le 1/2$ and that $M_{2,\mu}$ is star-like if and only if $0 \le \mu \le 1$.

In his contribution, Vértesi investigates weighted Lagrange interpolation. This refers to interpolation by a function of the form pw, where p is a polynomial and w is a weight function. By taking w to be a "generalized Jacobi weight" and by taking a set of nodes that are especially computed for the situation at hand, the author gets a sequence of Lebesgue constants of order $\log n$, which is optimal.

Zalik's paper concerns generalized extended Chebyshev systems and their linear spans. The functions arising are expected to mimic the behavior of polynomials to some extent, and here the question is whether theorems such as Descartes' Rule of Signs are valid.

E. W. Cheney
University of Texas
Austin

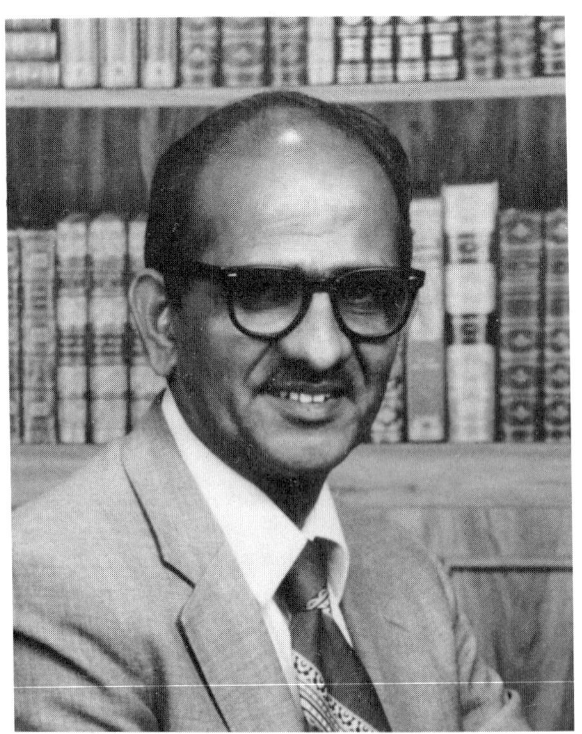

Arum Kumar Varma (1934–1994)

Preface

With inspiration from Professor Turán, Professor Paul Erdös, and other approximation theorists of the Hungarian school, Dr. Arun Kumar Varma made substantial contributions in the fields of interpolation and approximation of functions, including Birkhoff interpolation and approximation by spline functions. In the fall of 1994 he and his wife Manjuka visited Professor Sharma at Edmonton, Canada and later went to Hungary to spend about two weeks with Professor Szabados. Professor Varma was taken ill and was in constant pain. He had to undergo an emergency surgery in a hospital in Budapest. Although his health needed further attention, he insisted on going to India to visit friends and relatives. This turned out to be his last visit. After his return from India, he was seriously ill. One of the editors went to see him in Gainesville, Florida. In a frail voice he discussed some of the problems that he had solved during his visit. He also mentioned that after his recovery he would write a book on interpolation and approximation. Unfortunately, his untimely death left his desire to write a book unfulfilled.

This volume, Approximation Theory: In Memory of A. K. Varma, is a token of the appreciation of the contributors and the editors for his devotion to the subject and for the admirable manner in which he conducted his professional activities until his last breath.

Professor Erdös had kindly accepted our invitation to contribute a paper entitled "Some of My Favourite Problems and Results on Polynomials and Interpolation", but he was unable to complete his paper before his death. We are deeply saddened by his sudden demise.

This volume has thirty-one papers covering a wide area of approximation theory. Most of the authors took the trouble to provide us with camera ready copy, for which we are extremely grateful. Each paper was read by an expert in the field and by one of the editors prior to its acceptance. In editing the papers, we have borne in mind the importance of minimizing the time interval between the contribution of the paper and the publication of the volume. Nevertheless we have endeavored by repeated scrutiny to eliminate errors of all sorts and we trust that not too many still remain.

With so many authors from different countries, there is bound to be some variation in notation. We have not attempted to standardize this, but we have tried to impose, as far as possible, a uniform layout style. The papers are not grouped according to topic, but are arranged alphabetically according to the surnames of the authors.

It is a pleasure to acknowledge our gratitude to the authors for contributing to this memorial volume and to the many referees who gave their valuable time in scrutinizing the papers. We are also immensely grateful to Professors R. P. Agarwal, G. A. Anastassiou, M. Datta, Zeev Ditzian, C. Frappier, G. Harris, M. Ismail, O. Jenda, Rong-Qing Jia, Peter D. Johnson, I. Joo, T. Kilgore, G. A. Kozlowski, S. L. Lee, Xin Li, D. S. Lubinsky, A. Menezes, H. N. Mhaskar, J. Prasad, C. Rodger, S. Ruscheweyh, G. Schmeisser, Edward Slaminka, M. Smith, H. M. Srivastava, X. H. Sun, and P. Vértesi who helped us during the preparation of this volume. In addition, we most gratefully acknowledge the contributions of Professors D. Hankerson, Q. I. Rahman and R. A. Zalik who helped us immensely in improving the presentation of this volume, by providing timely guidance, technical help, and countless suggestions. Their hard work contributed greatly to the quality of this volume, and it was truly a pleasure working with them. Our special thanks are due to Professor E. W. Cheney, who took upon himself the task of reading all the papers and writing the Foreword. We are also thankful to Ms. Rosie Torbert and Ms. Cecilia Price for their help in typing some of the papers and in putting together the manuscripts in a uniform style. Finally, we sincerely thank the editorial staff of Marcel Dekker, Inc., for all their help and support.

N. K. Govil
R. N. Mohapatra
Z. Nashed
A. Sharma
J. Szabados

Contents

Contributors

Ravi P. Agarwal National University of Singapore, Singapore

George A. Anastassiou The University of Memphis, Memphis, Tennessee

J. Balázs Eötvös Loránd University, Budapest, Hungary

Mirosław Baran Institute of Mathematics, Jagiellonian University of Cracow, Cracow, Poland

Franck Beaucoup School of Mines of Saint-Etienne, Saint-Etienne, France

Borislav Bojanov University of Sofia, Sofia, Bulgaria

Peter Borwein Simon Fraser University, Burnaby, British Columbia, Canada

M. G. de Bruin Delft University of Technology, Delft, The Netherlands

Laurent Carraro School of Mines of Saint-Etienne, Saint-Etienne, France

Peter G. Casazza University of Missouri, Columbia, Missouri

E. W. Cheney University of Texas, Austin, Texas

Ole Christensen Technical University of Denmark, Lyngby, Denmark

Z. Ciesielski Instytut Matematyczny Pan, Sopot, Poland

L. T. Dechevsky University of Montreal, Montreal, Canada

D. P. Dryanov University of Montreal, Montreal, Canada

Tamás Erdélyi Texas A&M University, College Station, Texas

Michael S. Floater SINTEF, Oslo, Norway

Richard Fournier University of Montreal, Montreal, Canada

T. N. T. Goodman University of Dundee, Dundee, Scotland

N. K. Govil Auburn University, Auburn, Alabama

Don Hong East Tennessee State University, Johnson City, Tennessee

A. Jakimovski School of Mathematical Sciences, Tel-Aviv University, Tel-Aviv, Israel

Peter D. Johnson, Jr. Auburn University, Auburn, Alabama

Theodore Kilgore Auburn University, Auburn, Alabama

Xin Li University of Central Florida, Orlando, Florida

D. S. Lubinsky University of the Witwatersrand, Wits, South Africa

Jinxi Ma Mathematics Institute, University of Würzburg, Würzburg, Germany

H. N. Mhaskar California State University, Los Angeles, California

Charles A. Micchelli T. J. Watson Research Center, IBM Corporation, Yorktown Heights, New York

Gradimir V. Milovanović Faculty of Electronic Engineering, University of Niš, Niš, Yugoslavia

R. N. Mohapatra University of Central Florida, Orlando, Florida

W. Pleśniak Institute of Mathematics, Jagiellonian University of Cracow, Cracow, Poland

J. Prestin University of Rostock, Rostock, Germany

Q. I. Rahman University of Montreal, Montreal, Canada

Themistocles M. Rassias National Technical University of Athens, Athens, Greece

Stephan Ruscheweyh Mathematics Institute, University of Würzburg, Würzburg, Germany

Gerhard Schmeisser Mathematics Institute, University of Erlangen–Nürnberg, Erlangen, Germany

A. Sharma University of Alberta, Edmonton, Alberta, Canada

Vikramaditya Singh Kanpur, India

Xingping Sun Southwestern Missouri State University, Springfield, Missouri

J. Szabados Mathematical Institute, Hungarian Academy of Sciences, Budapest, Hungary

P. Vértesi Hungarian Academy of Sciences, Budapest, Hungary

Patricia J. Y. Wong Nanyang Technological University, Singapore

R. A. Zalik Auburn University, Auburn, Alabama

Arum Kumar Varma: Some Reminiscences

My first meeting with Arun (or Varmaji, as I have ever since addressed him) was perhaps on a late August afternoon in 1958 when the sun was still high but the hot summer winds (called "loo") had subsided. I then lived with my family in a small house called "Noor Manzil" in New Hyderabad, Lucknow, State of Uttar Pradesh, India. We were getting ready for the afternoon tea after the noon siesta, when we heard a gentle knocking on the door. I opened the door to find a suave young diminutive person in simple dress who expressed his desire to see me. On being invited to come in and be seated, he introduced himself as Arun Kumar Varma and told me that he was the nephew of the late Acharaya Narendra Deoji who had been the Vice-Chancellor of the Lucknow University from 1947 to 1952. We talked for some time about how, as Vice-Chancellor of the Lucknow University, Acharyaji lived in a bungalow just half a block from Noor Manzil where we lived. His family still lived in the same bungalow. I knew that Acharyaji was an erudite scholar and a very powerful orator. Acharyaji was the leader of the Socialist Party of India and all throughout his life he was passionately involved in the fight for the freedom of India. I had heard him once when he gave an extemporaneous address to the staff and students at a convocation. On February 10, 1956 he died at Erodé in the South India and the body was brought next day to Lucknow where he was cremated on the banks of the River Gomati near the University which he served so well. A memorial stands today to mark the spot.

Varmaji continued his story and told me that he had received his B.Sc. at the Benares Hindu University in 1955, that he took his M.Sc. in mathematics at the Lucknow University in 1958 and that he was then teaching in a small elementary school in Shivgarh, a small village in District Rae Bareilley. He told me that he wanted to do research in mathematics and asked if I would agree to accept him as a candidate for the Ph.D. degree for students who wanted to pursue research. I explained to him that there was no financial support I could offer him. But he assured me that he could support himself as he was working in a school in Shivgarh and was single. Noticing a spirit of determination in his speech and observing his cheerful and polite behavior, I decided to encourage him. I told him not to worry about formalities or registration, and to start reading and studying. I gave him a copy of the Ph.D. thesis of Dr. R. B. Saxena on lacunary interpolation and asked him to see me in two/three weeks.

During the next few months Varmaji visited us often and stayed for a cup of tea. We were able to get some information about him during casual conversation. We learnd that Arun Varma was born on October 20, 1934 at

Faizabad in the state of Uttar Pradesh (India) where his father, Dr. Yogendra Deo, was a general practitioner. He ran a clinic and was well known for his surgical skill. He was the younger brother of Acharaya Narendra Deoji and was born in 1899. In his reminiscences Acharyaji wrote that he was born at Sitapur in the state of Uttar Pradesh, India on October 30, 1889 where his father, Mr. Baldev Prasad, was practicing law, although their ancestral home was at Faizabad. His grandfather, Babu Sohan Lal, was a professor in the old Canning College, Lucknow. Acharyaji writes: "After passing Intermediate from Canning College, Lucknow, my father took his degree in law. Owing to eye trouble, he could not pass B.A. My grandfather read to him books on law and that was how he prepared for the examination. After passing law, my father started practice under the guidance of Munshi Murlidhar, a student of my grandfather. Munshiji had no children. He loved his nephew and my elder brother like his sons. About two years after my birth, as my grandfather was no more, my father was forced to leave Sitapur. He started practicing in Faizabad".

Varmaji's mother, Krishna Devi, was a pious lady. Varmaji had two older sisters, Shanti and Gayatri, and two older brothers, Harish Kumar and Vinay Kumar. He had another sister, Kumari Vidya (nicknamed Tikkho), who died at the tender age of 14 of meningitis. He had a younger sister, Indira and a younger brother, Krishna Kumar. His sisters are all married and settled. His brother, Harish Kumar, who is a doctor, is now retired from his U.P. State Government job and is practicing medicine in Faizabad. Vinay Kumar received a law degree but because of poor health and circumstances (family litigation and related pressures) he was unable to practice law. The family clinic at Faizabad which was run by his father is run by his two nephews, their wives and Dr. Harish Kumar. Krishna Kumar came to Calgary, Canada in 1965, married a Canadian girl and is happily settled in Washington, D.C. as a civil engineer. He is a structural bridge welding engineer for the Federal Highway Administration.

Krishna Kumar Varma recalled that their home in Faizabad is well known as "Baldev Nivas" and that it was a large, beautiful home with fountains and beautiful gardens and two courtyards inside the house. The family owned several acres of land and a few buildings in the Faizabad and Ayodhya area. The younger brothers, Yogendra Deo and Acharyaji, fully entrusted the management of the property to their eldest brother, Mahendra Deo, who was a lawyer. Most of the land was rented or leased to farmers. Later most of the land was lost because of new Zamindari regulations. Today the house Baldev Nivas stands but the property is divided between the families of the three brothers.

Varmaji, Indira and Krishna Kumar were small when their father died on December 19, 1944. The family income plummeted after Yogendra Deoji's death. Krishna Kumar recalls that he and Arun Bhaiya used to run the pharmacy with the help of friends and neighbors so that they could retain the clientele. The job of feeding the four or five milch cows and looking after them fell on the young and tender shoulders of Arun and Krishna Kumar, since the elder brother Harish was studying medicine at Agra and his younger brother Vinay Kumar was studying law at Lucknow. Krishna Kumar tells me that they maintained four or five milch cows with their calves. A male calf was either given away or sold to a farmer or to a family friend while a female calf was kept at home. His mother tried to make sure that the person who took the calf did not kill it. The cows were tethered to separate posts in the large courtyard. Arun and K.K. Varma used to buy fodder by the cartload from farmers and store it in the house in a shed. Sometimes one of the cattle would break loose and land in the Kanjee House for stray cattle. Then the two brothers paid the fine and brought the cow through the city to their home.

Luckily for Varmaji our library at Lucknow had a fairly good collection of books and journals thanks to the foresight of the late Dr. A. N. Singh, who moved the library of Benares Mathematical Society from Benares to Lucknow. Thus we received many European journals in exchange for the Proceedings of the Benares Mathematical Society (now called "Ganita"). Somehow the first paper of Professor P. Turán and J. Balázs on $(0, 2)$ interpolation on zeros of $\Pi_n(x)$ had come to our library and this led Dr. R. B. Saxena to complete his Ph.D. thesis "On Lacunary Interpolation...". While working with Dr. Saxena, I learnt a good deal myself as I got interested in the problem. When Varmaji came back with Saxena's thesis which I had lent him, he told me that he felt interested in the work. So we agreed that he should try the problem of $(0, 2)$ interpolation on the zeros of Chebyshev polynomials $T_n(x)$.

At this point the wheels of fortune began to turn for me (and also for Varmaji). I received an invitation from the Vice-Chancellor of the University of Rajasthan, Jaipur, to join as Reader in the Mathematics Department. Since Jaipur is my home town, my father and my wife were both in favor of my accepting the offer. I made a request to the Vice-Chancellor at Jaipur that my research student Arun Varma be supported as well, since he would be coming with me. The then Vice-Chancellor, Dr. Mohan Singh Mehta, was a dynamic person with vision and he arranged to take Arun Varma as a Research Scholar at a stipend of Rs. 75 (equivalent to about $15.00 in those days) per month. Varmaji joined me in October 1960. We rented two rooms within walking distance from the University. Here we cooked our breakfast together on a kerosene stove and took our meals at a nearby restaurant or at the university cafeteria. Since they needed graduate assistants, the Chairman, Dr. G. C. Patni, promoted

Varmaji to Tutor at a salary of Rs. 150 (equivalent to about \$30.00 in those days) per month. From September 1, 1961, he was given a salary of Rs. 200 (in those days equivalent to about \$40.00) a month.

When in 1961, I received the offer of a Fellowship at Harvard University, I decided to accept. The Vice-Chancellor refused to grant me leave within such a short time of my appointment. So I resigned my job at the University of Rajasthan. Since I could not get leave from the University of Lucknow, I resigned from Lucknow also and left for the U.S. early in September 1961.

After my stay at Harvard for the year I decided to look for a position in the U.S. or Canada since I had burnt my boats. I was able to get an invitation from the University of California, Los Angeles, for the summer months. There I got to know Professor T. S. Motzkin, Professor E. G. Strauss and Professor E. W. Cheney. When I got an offer from the University of Alberta, Calgary, I was in a fix. On the advice of my friend Professor E. W. Cheney I accepted the offer and early in September 1962, I came to Calgary. After arriving at Calgary and settling down, I began to make plans to get my family to join me there. I also told the Chairman about Varmaji, who had been working for the Ph.D. degree with me at Lucknow and Jaipur. The chairman offered him a teaching assistantship at Calgary and so Varmaji left his job at Jaipur, India, on August 31, 1962 and came to Calgary. When he came there, he was advised to take some courses and continue the work he had been doing for his research.

During my stay abroad, my father would often visit Jaipur. He always stayed with Varmaji, who looked after him like his own son. My wife and I recall all his kindness to my father with gratitude. A notable characteristic of Varmaji was his saintly nature and his general modesty about himself, his family and his family mansion.

When Varmaji was at Lucknow, he had copied (in his own hand) papers of P. Turán and J. Balázs and other papers of interest to him on interpolation. It was hard and tedious work but it paid dividends when he moved to Jaipur with me. Since he had the copies of the relevant papers with him, he completed his first paper on $(0, 2)$ interpolation at Jaipur. He had the lucky idea of modifying the problem slightly with two extra conditions at the end points which enabled him to obtain simpler expressions for the fundamental polynomials. This idea occurs like a refrain in many of his other papers on lacunary interpolation.

Before he came to Calgary, he had completed the paper on $(0, 1, 2)$ interpolation on the zeros of the Chebychev polynomial $T_n(x)$. At Calgary, both Varmaji and myself were lucky because Professor A. Meir joined the faculty at Calgary soon after my coming there. It was here that Varmaji began his paper on trigonometric interpolation which was motivated by a paper of O. Kis on the $(0, 2)$ trigonometric case. I had done the $(0, 3)$

case and when I was at Chicago one summer, I succeeded in settling the $(0,4)$ case. When I showed these special cases to Professor A. Zygmund he remarked with his characteristic smile that in this way I could get a large number of papers to my credit. So when I returned to Calgary, Varmaji and I started to look at the $(0,m)$ case of trigonometric interpolation. We had useful discussions with Professor Meir in connection with some lemmas and the paper was completed in due course. Since in my opinion Varmaji had enough material for a Ph.D. thesis, I advised him to get the work typed by a secretary and submit it to the Registrar. In keeping with his parsimonious habits, he got different chapters typed by different secretaries and this saved him some money. When I saw this, it was too late and so the thesis was sent to the examiners. The external examiners were Professor P. Turán and Professor G. Szegö. Professor G. Szegö sent a favorable report, but Professor Turán was able to come in person for the oral defense. After Varmaji's oral presentation, Professor Turán as the external examiner spoke first and explained that the results obtained were significant and useful. The thesis was accepted subject to certain minor corrections, involving a few grammatical flaws and typographical errors, which were pointed out by the other members of the committee. I learnt soon after that Varmaji had already booked his flight to India after 5 or 6 days. I tried to explain to him that it would not be possible to make corrections in such a short period of time. But his mind was made up and so it fell to Professor Meir and myself to get the corrections done later. Varmaji always recalled the help of Dr. Meir with gratitude.

After going to India, Varmaji joined his duties at the University of Rajasthan, Jaipur, from September 1, 1964. His relatives arranged for his marriage and he was married to Manjuka on December 13, 1965. He stayed at Jaipur with Manjuka until August 1966 when he came to Edmonton as a postdoctoral fellow. It was perhaps in this period that we completed our joint paper on the $(0,2,3)$ case of trigonometric interpolation.

From here he joined the University of Florida, Gainesville, in October 1967. Their first son, Rajeev, was born on December 19, 1967. Nitin was born on April 20, 1969 and their daughter, Latika, was born on January 29, 1973. Varmaji was a loving parent and a devoted husband. On June 22, 1976, he acquired U.S. citizenship.

Varmaji very much loved his younger sister Indira and his younger brother Krishna Kumar. His sister tells a story about his childlike simplicity and naivete. She had come to Delhi to see him off before he went abroad. During the day, she had to go out on some errand and was rather late in returning home. When she came, she found him in tears. When asked to explain, he said that he was worried about her. "What would I say to your husband if something had happened?" Varmaji loved his mother dearly and invited her to come and visit him and his family at Gainesville. She did come but did not want to stay in the U.S. too long. She returned to Faizabad, where she died on March 3, 1988. His elder brother Dr. Harish Kumar, also visited him at Gainesville and recalls his courtesy and kindness in a personal letter to me.

Since January 1994, Varmaji was not feeling well. He and Manjuka visited us in May 1994 for two or three weeks. He went to Israel to attend a conference at the end of May 1994 and returned to Gainesville. In September 1994, he went to Hungary, where he had to be hospitalized. In spite of the advice of his wife and friends, he went to India from Hungary. He returned soon after to Gainesville, where the doctors diagnosed him with cancer of the pancreas. He passed away peacefully on December 8, 1994. It was his desire that after his death his ashes be immersed in the holy confluence of the Ganges and Yamuna rivers. This was done by his sons Nitin and Rajeev later according to the Hindu custom.

The story of his career at Gainesville, the details of the five students he supervised for Master's and Ph.D. degrees and a survey of his research has been very well described by Professor J. Szabados in his article published in the Journal of Approximation Theory, Vol. 84 (1), 1996, 1-11. The late Professor P. Erdös used to visit Gainesville whenever he was in the U.S. and he invariably stayed with Varmaji. Professors P. Vértesi and J. Szabados have also been his guests and have enjoyed his hospitality. Varmaji was a very friendly and amiable person, always willing to help and oblige. Varmaji was a pragmatist by nature and upbringing, and his genial nature made it possible for him to do joint research work with many mathematicians. He had a passion for work. He liked the style of writing and speaking of Professor P. Turán and he tried to follow the same style. He was deeply influenced by the work of the Hungarian school in approximation. Although he did not have the flair for languages which his famous uncle Acharya Narendra Deoji possessed, he had in him an indomitable spirit and a strong will which stayed with him until the end.

A. Sharma

APPROXIMATION THEORY

Error Bounds for the Derivatives of Lidstone Interpolation and Applications

Ravi P. Agarwal
Department of Mathematics
National University of Singapore, Singapore

Patricia J. Y. Wong
Division of Mathematics
Nanyang Technological University, Singapore

Abstract

We introduce Lidstone polynomials and establish several identities and inequalities involving these polynomials. Then, we construct Lidstone interpolating polynomial, provide explicit representation of the error, and obtain best possible error bounds for the derivatives of Lidstone interpolation. These errorbounds are used to study the existence and uniqueness of Lidstone boundary value problems. This is followed piecewise-Lidstone and Lidstone-spline interpolation. For these interpolating polynomials we offer explicit error bounds for their derivatives. These error bounds play a fundamental role in approximating the solutions of differential and integral equations.

1 Introduction

In the year 1929 Lidstone [27] introduced a generalization of Taylor's series, it approximates a given function in the neighborhood of two points instead of one. From the practical point of view such a development is very useful; and in terms of completely continuous functions it has been characterized in the works of Boas [12,13], Poritsky [32], Schoenberg [33], Whittaker [43,44], Widder [45,46], and others [8,11,30]. In the field of approximation theory [3,4,6,7,17,19,20,23,25,26,42,44], the *Lidstone interpolating polynomial* $P(t)$ of degree $(2m-1)$ satisfies the *Lidstone conditions*

$$P^{(2i)}(0) = \alpha_i, \quad P^{(2i)}(1) = \beta_i, \ \ 0 \le i \le m-1. \tag{1.1}$$

Further, boundary value problems consisting of $2m$-th order ordinary differential equation

$$(-1)^m x^{(2m)} = f(t, \mathbf{x}) \tag{1.2}$$

where $\mathbf{x} = (x, x', ..., x^{(q)})$, $0 \le q \le 2m-1$ but fixed, f is continuous at least in the interior of the domain of interest; and the *Lidstone boundary conditions*

$$x^{(2i)}(0) = \alpha_i, \quad x^{(2i)}(1) = \beta_i, \ 0 \le i \le m-1 \tag{1.3}$$

and several of its particular cases, have been the subject matter of recent investigations [1-3,5,9,10,14-16,18,21,31,35-41].

In addition, *Lidstone-spline interpolation* has attracted considerable attention and has proved to be very useful in approximating the solutions of integral and boundary value problems [34,47-51]. In this paper we shall survey some of our recent work on Lidstone interpolation, Lidstone boundary value problems and Lidstone-spline interpolation. The motivation of our investigations comes from the contribution of Professor *A.K. Varma* in the year 1983.

2 Equalities and Inequalities for Lidstone Polynomials

Definition 2.1 *The unique polynomial $\Lambda_n(t)$ of degree $(2n+1)$ recursively defined by: $\Lambda_0(t) = t$, and for $n \ge 1$,*

$$\Lambda_n''(t) = \Lambda_{n-1}(t), \quad \Lambda_n(0) = \Lambda_n(1) = 0, \tag{2.1}$$

is called Lidstone polynomial.

Lemma 2.1 *The Lidstone polynomial $\Lambda_n(t)$ can be expressed as*

$$\Lambda_n(t) = \int_0^1 g_n(t,s) s\, ds, \quad n \ge 1 \tag{2.2}$$

where

$$g_1(t,s) = \begin{cases} (t-1)s, & s \le t \\ (s-1)t, & t \le s \end{cases} \tag{2.3}$$

and

$$g_n(t,s) = \int_0^1 g_1(t,t_1) g_{n-1}(t_1,s)\, dt_1, \quad n \ge 2. \tag{2.4}$$

Proof. It is clear that the solution $\Lambda_1(t)$ of the boundary value problem $\Lambda_1''(t) = t$, $\Lambda_1(0) = \Lambda_1(1) = 0$ can be written as (2.2) with $n = 1$ ($g_1(t,s)$ is the Green's function). If (2.2) is true for $n \ge 1$, then the solution $\Lambda_{n+1}(t)$ of the boundary value problem $\Lambda''_{n+1}(t) = \int_0^1 g_n(t,s)s\,ds,\ \Lambda_{n+1}(0) = \Lambda_{n+1}(1) = 0$ can be written as

$$\begin{aligned}\Lambda_{n+1}(t) &= \int_0^1 \left[g_1(t,t_1)\int_0^1 g_n(t_1,s)s\,ds\right] dt_1 \\ &= \int_0^1 \left[\int_0^1 g_1(t,t_1)g_n(t_1,s)\,dt_1\right] s\,ds = \int_0^1 g_{n+1}(t,s)s\,ds. \quad \blacksquare\end{aligned}$$

Remark 2.1 *From (2.3), $g_1(t,s) \le 0,\ 0 \le s,t \le 1$. Thus, from (2.4) it follows that $0 \le (-1)^n g_n(t,s) = |g_n(t,s)|,\ 0 \le s,t \le 1$. Hence, in view of (2.2) we have $(-1)^n \Lambda_n(t) \ge 0,\ 0 \le t \le 1$.*

Lemma 2.2 *The following equalities hold*

$$\int_0^1 g_n(t,s)\sin k\pi s\,ds = (-1)^n \frac{1}{(k\pi)^{2n}} \sin k\pi t, \quad 0 \le t \le 1 \quad (k \ge 1) \tag{2.5}$$

$$\int_0^1 |g_n(t,s)| \sin \pi s\,ds = \frac{1}{\pi^{2n}} \sin \pi t, \quad 0 \le t \le 1. \tag{2.6}$$

Proof. Since

$$\begin{aligned}\int_0^1 g_1(t,s)\sin k\pi s ds &= -(1-t)\int_0^t s\sin k\pi s ds - t\int_t^1 (1-s)\sin k\pi s ds \\ &= -\frac{1}{(k\pi)^2}\sin k\pi t\end{aligned}$$

the equality (2.5) follows from (2.4) by using an inductive argument. Equality (2.6) is a consequence of (2.5) and Remark 2.1. ■

Lemma 2.3 *The Lidstone polynomial $\Lambda_n(t)$ can be expressed as*

$$\Lambda_n(t) = (-1)^n \frac{2}{\pi^{2n+1}} \sum_{k=1}^{\infty} \frac{(-1)^{k+1}}{k^{2n+1}} \sin k\pi t, \quad n \ge 1. \tag{2.7}$$

Proof. We multiply the Fourier series

$$s = \frac{2}{\pi}\sum_{k=1}^{\infty} \frac{(-1)^{k+1}}{k} \sin k\pi s, \quad 0 < s < 1$$

by $g_n(t,s)$ and then integrate with respect to s from 0 to 1. The relation (2.7) now follows from (2.2) and (2.5). ■

Remark 2.2 *Lidstone and Bernoulli polynomials are connected by the relation [43]*

$$\Lambda_n(t) = \frac{2^{2n+1}}{(2n+1)!} B_{2n+1}\left(\frac{1+t}{2}\right).$$

Remark 2.3 *An explicit representation of Lidstone polynomial* $\Lambda_n(t)$, $n \geq 1$ *which follows by inductive arguments, is*

$$\Lambda_n(t) = \frac{1}{6}\left[\frac{6t^{2n+1}}{(2n+1)!} - \frac{t^{2n-1}}{(2n-1)!}\right] - \sum_{k=0}^{n-2} \frac{2(2^{2k+3}-1)}{(2k+4)!} B_{2k+4} \frac{t^{2n-2k-3}}{(2n-2k-3)!},$$

where B_{2k+4} *is the* $(2k+4)$*th Bernoulli number.*

Lemma 2.4 *The following hold*

$$\int_0^1 g_n(t,s)\,ds = (-1)^n \frac{4}{\pi^{2n+1}} \sum_{k=0}^{\infty} \frac{\sin(2k+1)\pi t}{(2k+1)^{2n+1}} = E_{2n}(t), \quad n \geq 1 \tag{2.8}$$

where $E_{2n}(t)$ *is the Euler polynomial.*

Proof. The proof of the first equality is similar to that of Lemma 2.3 except that it uses the Fourier series

$$1 = \frac{4}{\pi}\sum_{k=0}^{\infty} \frac{\sin(2k+1)\pi s}{(2k+1)}, \quad 0 < s < 1.$$

The second equality is in fact the Fourier series expansion of $E_{2n}(t)$ due to Lindelöf (see [22,24,29]). ∎

Remark 2.4 *As in Remark 2.3, by induction it follows that for* $n \geq 1$,

$$\int_0^1 g_n(t,s)ds = \frac{1}{2}\left[\frac{2t^{2n}}{(2n)!} - \frac{t^{2n-1}}{(2n-1)!}\right] - \sum_{k=0}^{n-2} \frac{2(2^{2k+4}-1)}{(2k+4)!} B_{2k+4} \frac{t^{2n-2k-3}}{(2n-2k-3)!}.$$

Corollary 2.5 *The following hold*

$$\int_0^1 |g_n(t,s)|\,ds = (-1)^n E_{2n}(t) \leq (-1)^n E_{2n}\left(\frac{1}{2}\right) = \frac{(-1)^n E_{2n}}{2^{2n}(2n)!}, \tag{2.9}$$

where E_{2n} *is the* $2n$ *th Euler number.*

Proof. Equality holds from (2.8) and Remark 2.1. Further, since the extrema of the function $E_{2n}(t)$ is at $t = \frac{1}{2}$ (see [22,24,29]) the inequality in (2.9) is obvious. ∎

Lemma 2.6 *The following equalities hold*

$$\Lambda_n(1-t) = \int_0^1 g_n(t,s)(1-s)\,ds = (-1)^n \frac{2}{\pi^{2n+1}} \sum_{k=1}^{\infty} \frac{\sin k\pi t}{k^{2n+1}}, \quad n \geq 1. \tag{2.10}$$

Proof. From (2.2), (2.7) and (2.8), we have

$$\begin{aligned}
&\int_0^1 g_n(t,s)(1-s)\,ds \\
&= (-1)^n \frac{4}{\pi^{2n+1}} \sum_{k=0}^{\infty} \frac{\sin(2k+1)\pi t}{(2k+1)^{2n+1}} \\
&\quad -(-1)^n \frac{2}{\pi^{2n+1}} \left\{ \sum_{k=0}^{\infty} \frac{\sin(2k+1)\pi t}{(2k+1)^{2n+1}} - \sum_{k=1}^{\infty} \frac{\sin 2k\pi t}{(2k)^{2n+1}} \right\} \\
&= (-1)^n \frac{2}{\pi^{2n+1}} \left\{ \sum_{k=0}^{\infty} \frac{\sin(2k+1)\pi t}{(2k+1)^{2n+1}} + \sum_{k=1}^{\infty} \frac{\sin 2k\pi t}{(2k)^{2n+1}} \right\} \\
&= (-1)^n \frac{2}{\pi^{2n+1}} \sum_{k=1}^{\infty} \frac{\sin k\pi t}{k^{2n+1}} \\
&= (-1)^n \frac{2}{\pi^{2n+1}} \sum_{k=1}^{\infty} \frac{(-1)^{k+1} \sin k\pi(1-t)}{k^{2n+1}} \quad = \quad \Lambda_n(1-t). \qquad \blacksquare
\end{aligned}$$

Lemma 2.7 *The following hold*

$$\begin{aligned}
\int_0^1 |g_n'(t,s)|\,ds &= (-1)^n \left[2E_{2n}(t) + (1-2t)E_{2n-1}(t)\right] \\
&\le (-1)^{n+1} \frac{2(2^{2n}-1)}{(2n)!} B_{2n}.
\end{aligned} \tag{2.11}$$

Proof. From (2.4) we have

$$\begin{aligned}
\int_0^1 |g_n'(t,s)|\,ds &= \int_0^1 \int_0^t t_1 (-1)^{n-1} g_{n-1}(t_1,s)\,dt_1\,ds \\
&\quad + \int_0^1 \int_t^1 (1-t_1)(-1)^{n-1} g_{n-1}(t_1,s)\,dt_1\,ds,
\end{aligned}$$

which on changing the order of integration and using the equality (2.9) gives

$$\int_0^1 |g_n'(t,s)|ds = (-1)^{n-1} \left[\int_0^t t_1 E_{2n-2}(t_1)dt_1 + \int_t^1 (1-t_1)E_{2n-2}(t_1)dt_1 \right]. \tag{2.12}$$

Now in (2.12), we use the relation $E_k'(t) = E_{k-1}(t)$ (see [22,24,29]) and the conditions $E_{2n}(0) = E_{2n}(1) = 0$, to obtain

$$\int_0^1 |g_n'(t,s)|ds = (-1)^n \left[2E_{2n}(t) + (1-2t)E_{2n-1}(t)\right].$$

Next, since the derivative of the right side of the equality (2.11) is $-(1-2t)(-1)^{n-1}E_{2n-2}(t)$ and $(-1)^{n-1}E_{2n-2}(t) \geq 0$ for all $t \in [0,1]$, it follows that

$$\int_0^1 |g_n'(t,s)|\,ds \;\leq\; (-1)^n \max\{E_{2n-1}(0), -E_{2n-1}(1)\}.$$

However, since $E_{2n-1}(0) + E_{2n-1}(1) = 0$ (see [22,24,29]), we get

$$\int_0^1 |g_n'(t,s)|\,ds \;\leq\; (-1)^n E_{2n-1}(0) \;=\; (-1)^n E_{2n}'(0). \tag{2.13}$$

Finally, from (2.8) and Remark 2.4, it is easy to find

$$E_{2n}'(0) \;=\; -\frac{2(2^{2n}-1)}{(2n)!}B_{2n}.$$

Hence, the inequality (2.13) is the same as in (2.11). ∎

Lemma 2.8 *For $t \in [0,1]$ the following inequalities hold*

$$\int_0^1 |g_n(t,s)|\,ds \;\leq\; \frac{1}{\pi^{2n-2}}\left(\frac{1}{2\pi}\right)\sin\pi t, \tag{2.14}$$

$$\int_0^1 |g_n'(t,s)|\,ds \;\leq\; \frac{1}{\pi^{2n-2}}\left(\frac{1}{2\pi}\right)[2\sin\pi t + \pi(1-2t)\cos\pi t], \tag{2.15}$$

$$\int_0^1 |g_n'(t,s)|\sin\pi s\,ds \leq \frac{1}{\pi^{2n}}[2\sin\pi t + \pi(1-2t)\cos\pi t], \tag{2.16}$$

$$\int_0^1 |g_n(t,s)|[2\sin\pi s + \pi(1-2s)\cos\pi s]\,ds \;\leq\; \frac{1}{\pi^{2n-2}}\left(\frac{4}{\pi^2}\right)\sin\pi t, \tag{2.17}$$

$$\int_0^1 |g_n'(t,s)|[2\sin\pi s + \pi(1-2s)\cos\pi s]\,ds$$

$$\leq \frac{1}{\pi^{2n-2}}\left(\frac{4}{\pi^2}\right)[2\sin\pi t + \pi(1-2t)\cos\pi t]. \tag{2.18}$$

Proof. We shall prove only (2.14). Since

$$\int_0^1 |g_1(t,s)|\,ds \;=\; \frac{1}{2}t(1-t) \;\leq\; \frac{1}{2\pi}\sin\pi t$$

(2.14) is true for $n = 1$.

Next, if (2.14) is true for $k \geq 1$, then from (2.4) we have

$$
\begin{aligned}
&\int_0^1 |g_{k+1}(t,s)|\,ds \\
&\leq (1-t)\int_0^1\int_0^t |g_k(t_1,s)|t_1\,dt_1\,ds + t\int_0^1\int_t^1 |g_k(t_1,s)|(1-t_1)\,dt_1\,ds \\
&\leq \left[(1-t)\int_0^t t_1\sin\pi t_1\,dt_1 + t\int_t^1 (1-t_1)\sin\pi t_1\,dt_1\right]\frac{1}{\pi^{2k-2}}\left(\frac{1}{2\pi}\right) \\
&= \frac{1}{\pi^{2k}}\left(\frac{1}{2\pi}\right)\sin\pi t. \qquad \blacksquare
\end{aligned}
$$

3 Lidstone Interpolation and Error Inequalities

Theorem 3.1 *The Lidstone interpolating polynomial $P(t)$ can be expressed as*

$$P(t) = \sum_{k=0}^{m-1}[\alpha_k\Lambda_k(1-t) + \beta_k\Lambda_k(t)]. \tag{3.1}$$

Proof. It is clear that $P(t)$ in (3.1) is a polynomial of degree at most $(2m-1)$. Further, since in view of (2.1)

$$
\begin{aligned}
P^{(2i)}(t) &= \sum_{k=i}^{m-1}\left[\alpha_k\Lambda_k^{(2i)}(1-t) + \beta_k\Lambda_k^{(2i)}(t)\right] \\
&= \sum_{k=i}^{m-1}[\alpha_k\Lambda_{k-i}(1-t) + \beta_k\Lambda_{k-i}(t)] \\
&= \sum_{k=0}^{m-i-1}[\alpha_{k+i}\Lambda_k(1-t) + \beta_{k+i}\Lambda_k(t)], \quad 0 \leq i \leq m-1
\end{aligned}
$$

for $0 \leq i \leq m-1$, it follows that

$$P^{(2i)}(0) = \alpha_i\Lambda_0(1-t)|_{t=0} = \alpha_i, \quad \text{and} \quad P^{(2i)}(1) = \beta_i\Lambda_0(t)|_{t=1} = \beta_i,$$

The uniqueness of $P(t)$ is obvious. $\blacksquare$

. Let $\alpha_i = x^{(2i)}(0)$, $\beta_i = x^{(2i)}(1)$, $0 \leq i \leq m-1$ where the function $x(t)$ is assumed to be $2m$ times continuously differentiable on $[0,1]$. In such a case $P(t)$ is called the Lidstone interpolating polynomial of the function $x(t)$. For the associated error $e(t) = x(t) - P(t)$ we have the following representation.

Theorem 3.2 *If $x(t) \in C^{(2m)}[0,1]$, then*

$$e(t) = \int_0^1 g_m(t,s)x^{(2m)}(s)\,ds. \tag{3.2}$$

Proof. Since $\partial^{2n} g_m(t,s)/\partial t^{2n} = g_{m-n}(t,s)$, and $g_{m-n}(0,s) = g_{m-n}(1,s) = 0,\ 0 \le n \le m-1$ it is clear that $e^{(2i)}(0) = e^{(2i)}(1) = 0,\ 0 \le i \le m-1$. Thus, it suffices to prove that $e^{(2m)}(t) = x^{(2m)}(t)$. But, this is immediate from

$$e^{(2m-2)}(t) = \int_0^1 g_1(t,s) x^{(2m)}(s)\, ds. \qquad \blacksquare$$

Theorem 3.3 *If $x(t) \in C^{(2m)}[0,1]$, then for $0 \le i \le m-1$ the following hold*

$$\begin{aligned} |e^{(2i)}(t)| &= \left| x^{(2i)}(t) - \sum_{k=0}^{m-i-1} \left[x^{(2k+2i)}(0)\Lambda_k(1-t) + x^{(2k+2i)}(1)\Lambda_k(t) \right] \right| \\ &\le (-1)^{m-i} E_{2m-2i}(t) M_{2m} \le \frac{(-1)^{m-i} E_{2m-2i}}{2^{2m-2i}(2m-2i)!} M_{2m} \qquad (3.3) \end{aligned}$$

and

$$\begin{aligned} |e^{(2i+1)}(t)| &= \left| x^{(2i+1)}(t) - \sum_{k=0}^{m-i-1} \left[x^{(2k+2i)}(0)\Lambda'_k(1-t) + x^{(2k+2i)}(1)\Lambda'_k(t) \right] \right| \\ &\le (-1)^{m-i} \left[2E_{2m-2i}(t) + (1-2t) E_{2m-2i-1}(t) \right] M_{2m} \\ &\le (-1)^{m-i+1} \frac{2(2^{2m-2i}-1)}{(2m-2i)!} B_{2m-2i} M_{2m}, \qquad (3.4) \end{aligned}$$

where $M_{2m} = \max_{0 \le t \le 1} |x^{(2m)}(t)|$.

Proof. Following as in Theorem 3.1, the relation (3.2) for each $0 \le i \le m-1$ successively leads to

$$\begin{aligned} x^{(2i)}(t) &= \sum_{k=0}^{m-i-1} \left[x^{(2k+2i)}(0)\Lambda_k(1-t) + x^{(2k+2i)}(1)\Lambda_k(t) \right] \\ &\quad + \int_0^1 g_{m-i}(t,s) x^{(2m)}(s)\, ds \qquad (3.5) \end{aligned}$$

from which it is clear that

$$|e^{(2i)}(t)| \le M_{2m} \int_0^1 |g_{m-i}(t,s)|\, ds, \quad 0 \le i \le m-1.$$

Inequalities (3.3) are now immediate from (2.9).

From (3.5) it also follows that

$$|e^{(2i+1)}(t)| \le M_{2m} \int_0^1 |g'_{m-i}(t,s)|\, ds, \quad 0 \le i \le m-1.$$

Inequalities (3.4) are now obvious from (2.11). ∎

Remark 3.1 *Inequalities (3.3) and (3.4) are the best possible, as throughout the equality holds for the function $x(t) = E_{2m}(t)$ whose Lidstone interpolating polynomial $P(t) \equiv 0$, and only for this function up to a constant factor.*

4 Lidstone Boundary Value Problems

If $f = 0$ then the Lidstone boundary value problem (1.2), (1.3) obviously has a unique solution $x(t) = P(t)$; and if f is linear, i.e., $f = \sum_{i=0}^{q} a_i(t)x^{(i)}$ then (1.2), (1.3) gives the possibility of interpolation by the solutions of the differential equation (1.2).

Theorem 4.1 *Suppose that*

(i) $K_i > 0,\ 0 \le i \le q$ *are given real numbers and let* Q *be the maximum of* $|f(t, x_0, x_1, ..., x_q)|$ *on the compact set* $[0,1] \times D_0$, *where*

$$D_0 = \{(x_0, x_1, ..., x_q) \ : \ |x_i| \le 2K_i,\ 0 \le i \le q\};$$

(ii) $Q\, \dfrac{(-1)^{m-i}E_{2m-2i}}{2^{2m-2i}(2m-2i)!} \ \le\ K_{2i},\ 0 \le i \le \left[\dfrac{q}{2}\right];$

(iii) $Q\, \dfrac{(-1)^{m-i+1}2\left(2^{2m-2i}-1\right)B_{2m-2i}}{(2m-2i)!} \ \le\ K_{2i+1},\ 0 \le i \le \left[\dfrac{q-1}{2}\right];$

(iv) $max\{|\alpha_i|, |\beta_i|\} + \left[\displaystyle\sum_{k=1}^{m-i-1} max\{|\alpha_{k+i}|, |\beta_{k+i}|\}\, \frac{(-1)^k E_{2k}}{2^{2k}(2k)!}\right] = C_{2i}$
$\le K_{2i},\ 0 \le i \le \left[\dfrac{q}{2}\right];$

(v) $|\alpha_i - \beta_i| + \left[\displaystyle\sum_{k=1}^{m-i-1} max\{|\alpha_{k+i}|, |\beta_{k+i}|\}\, \frac{(-1)^{k+1}2\left(2^{2k}-1\right)B_{2k}}{(2k)!}\right]$
$= C_{2i+1} \ \le\ K_{2i+1}, 0 \le i \le \left[\dfrac{q-1}{2}\right].$

Then, the boundary value problem (1.2), (1.3) has a solution in D_0.

Proof. The set

$$B[0,1] \ = \ \left\{x(t) \in C^{(q)}[0,1] \ : \ \|x^{(i)}\| = \max_{0 \le t \le 1} |x^{(i)}(t)| \le 2K_i,\ 0 \le i \le q\right\}$$

is a closed convex subset of the Banach space $C^{(q)}[0,1]$. We define an operator $T : C^{(q)}[0,1] \to C^{(2m)}[0,1]$ as follows

$$(Tx)(t) = \sum_{k=0}^{m-1} \left[\alpha_k \Lambda_k(1-t) + \beta_k \Lambda_k(t)\right] + \int_0^1 |g_m(t,s)| f\left(s, \mathbf{x}(s)\right) ds. \quad (4.1)$$

In view of Theorem 3.2, it is clear that any fixed point of (4.1) is a solution of the boundary value problem (1.2), (1.3).

Let $x(t) \in B[0,1]$. Then from (4.1), (2.10), (2.9) and hypotheses (i), (ii), (iv), we find

$$\begin{aligned}
&|(Tx)^{(2i)}(t)| \\
&\le \sum_{k=0}^{m-i-1} \left[|\alpha_{k+i}||\Lambda_k(1-t)| + |\beta_{k+i}||\Lambda_k(t)|\right] + Q\int_0^1 |g_{m-i}(t,s)|ds \\
&\le \max_{0\le t\le 1} \left[|\alpha_i|(1-t) + |\beta_i|t\right] + \sum_{k=1}^{m-i-1} \max\left\{|\alpha_{k+i}|, |\beta_{k+i}|\right\} \\
&\quad \times \int_0^1 |g_k(t,s)|(s+1-s)\, ds + Q\int_0^1 |g_{m-i}(t,s)|ds \\
&\le \max\left\{|\alpha_i|, |\beta_i|\right\} + \sum_{k=1}^{m-i-1} \max\left\{|\alpha_{k+i}|, |\beta_{k+i}|\right\} \frac{(-1)^k E_{2k}}{2^{2k}(2k)!} \\
&\quad + Q\frac{(-1)^{m-i}E_{2m-2i}}{2^{2m-2i}(2m-2i)!} \\
&\le K_{2i} + K_{2i} \ = \ 2K_{2i}, \quad 0 \le i \le \left[\frac{q}{2}\right].
\end{aligned} \tag{4.2}$$

Similarly, from (4.1), (2.10), (2.11) and hypotheses (i), (iii), (v), we get

$$|(Tx)^{(2i+1)}(t)| \ \le \ 2K_{2i+1}, \quad 0 \le i \le \left[\frac{q-1}{2}\right]. \tag{4.3}$$

This shows that $TB[0,1] \subseteq B[0,1]$. The inequalities (4.2) and (4.3) imply that the sets $\left\{(Tx)^{(i)}(t) \ : \ x(t) \in B[0,1]\right\}$, $0 \le i \le q$ are uniformly bounded and equicontinuous in $[0,1]$. Hence, $\overline{TB}[0,1]$ is compact follows from the Ascoli-Arzela theorem. The Schauder fixed point theorem is applicable and a fixed point of T in D_0 exists. ∎

Corollary 4.2 *Assume that the function $f(t,x_0,x_1,...,x_q)$ on $[0,1]\times \mathbb{R}^{q+1}$ satisfies the following condition*

$$|f(t,x_0,x_1,...,x_q)| \ \le \ L + \sum_{i=0}^{q} L_i |x_i|^{\alpha_i},$$

where L, L_i, $0 \le i \le q$ are nonnegative constants, and $0 \le \alpha_i < 1$, $0 \le i \le q$. Then, the boundary value problem (1.2), (1.3) has a solution.

Theorem 4.3 *Suppose that the function $f(t,x_0,x_1,...,x_q)$ on $[0,1] \times D_1$ satisfies the following condition*

$$|f(t,x_0,x_1,...,x_q)| \ \le \ L + \sum_{i=0}^{q} L_i |x_i|, \tag{4.4}$$

where

$$D_1 = \Big\{(x_0, x_1, ..., x_q) \ : \ |x_{2i}| \le (1-\theta)^{-1}\frac{1}{\pi^{2m-2i-2}}\left(\frac{1}{2\pi}\right) C\sin\pi t$$
$$+C_{2i}, \ 0 \le i \le \left[\frac{q}{2}\right]; \quad |x_{2i+1}| \le (1-\theta)^{-1}\frac{1}{\pi^{2m-2i-2}}\left(\frac{1}{2\pi}\right) \times$$
$$C[2\sin\pi t + \pi(1-2t)\cos\pi t] + C_{2i+1}, \ \ 0 \le i \le \left[\frac{q-1}{2}\right]\Big\}$$

and

$$C = L + \sum_{i=0}^{[\frac{q}{2}]} L_{2i}C_{2i} + \sum_{i=0}^{[\frac{q-1}{2}]} L_{2i+1}C_{2i+1},$$

$$\theta = \sum_{i=0}^{[\frac{q}{2}]} L_{2i}\frac{1}{\pi^{2m-2i}} + \sum_{i=0}^{[\frac{q-1}{2}]} L_{2i+1}\frac{1}{\pi^{2m-2i-2}}\left(\frac{4}{\pi^2}\right) < 1. \tag{4.5}$$

Then, the boundary value problem (1.2), (1.3) has a solution in D_1.

Proof. Let $y(t) = x(t) - P(t)$, so that (1.2), (1.3) is the same as

$$(-1)^m y^{(2m)}(t) = f\left(t, y(t)+P(t), y'(t)+P'(t), ..., y^{(q)}(t)+P^{(q)}(t)\right)$$

$$y^{(2i)}(0) = y^{(2i)}(1) = 0, \ \ 0 \le i \le m-1. \tag{4.6}$$

Define $M[0,1]$ as the space of q times continuously differentiable functions satisfying (4.6). If we introduce in $M[0,1]$ the finite norm

$$\|y\| = \max\Big\{ \sup_{0\le t\le 1}\frac{\pi^{-2i}|y^{(2i)}(t)|}{\sin\pi t}, \ 0 \le i \le \left[\frac{q}{2}\right];$$
$$\sup_{0\le t\le 1}\frac{\pi^{-2i}|y^{(2i+1)}(t)|}{2\sin\pi t + \pi(1-2t)\cos\pi t}, \ 0 \le i \le \left[\frac{q-1}{2}\right]\Big\} \tag{4.7}$$

then it becomes a Banach space. As in Theorem 4.1 it suffices to show that $T : M[0,1] \to M[0,1]$ defined by

$$(Ty)(t) = \int_0^1 |g_m(t,s)| f\left(s, y(s)+P(s), ..., y^{(q)}(s)+P^{(q)}(s)\right) ds$$

maps the set

$$B_1[0,1] = \left\{y(t) \in M[0,1] \ : \ \|y\| \le (1-\theta)^{-1}\frac{1}{\pi^{2m-2}}\left(\frac{1}{2\pi}\right) C\right\}$$

into itself. For this, in addition to the condition (4.4) in the even case (2.14), (2.6) and (2.17), and in the odd case (2.15), (2.16) and (2.18) are needed. ∎

Remark 4.1 *In Theorem 4.3 the inequality (4.5) for $q = 0$ is the best possible, i.e., θ cannot be replaced by a smaller number. Indeed, in case of* *equality* $L_0 = \pi^{2m}$ *the boundary value problem* $(-1)^m x^{(2m)} = L_0 x$; $x(0) = \epsilon(\neq 0)$, $x^{(2i)}(0) = 0$, $1 \leq i \leq m-1$; $x^{(2i)}(1) = 0$, $0 \leq i \leq m-1$ *has no solution.*

Definition 4.1 *A function* $\bar{x}(t) \in C^{(2m)}[0,1]$ *is called an approximate solution of (1.2), (1.3) if there exist nonnegative constants* δ *and* ϵ *such that*

$$\max_{0\leq t\leq 1} |(-1)^m \bar{x}^{(2m)}(t) - f(t, \bar{\mathbf{x}}(t))| \leq \delta \tag{4.8}$$

and

$$|P^{(2i)}(t) - \bar{P}^{(2i)}(t)| \leq \epsilon \frac{1}{\pi^{2m-2i-2}} \left(\frac{1}{2\pi}\right) \sin \pi t, \quad 0 \leq i \leq \left[\frac{q}{2}\right]$$

$$|P^{(2i+1)}(t) - \bar{P}^{(2i+1)}(t)| \leq \epsilon \frac{1}{\pi^{2m-2i-2}} \left(\frac{1}{2\pi}\right) [2\sin \pi t + \pi(1-2t)\cos \pi t],$$

$$0 \leq i \leq \left[\tfrac{q-1}{2}\right]$$

where $P(t)$ *and* $\bar{P}(t)$ *are polynomials of degree* $(2m-1)$ *satisfying (1.1), and*

$$\bar{P}^{(2i)}(0) = \bar{x}^{(2i)}(0), \quad \bar{P}^{(2i)}(1) = \bar{x}^{(2i)}(1); \quad 0 \leq i \leq m-1$$

respectively.

The inequality (4.8) means that there exists a continuous function $\eta(t)$ such that $(-1)^m \times \bar{x}^{(2m)}(t) = f(t, \bar{\mathbf{x}}(t)) + \eta(t)$ and $\max_{0\leq t\leq 1} |\eta(t)| \leq \delta$. Thus, from Theorem 3.2 the approximate solution $\bar{x}(t)$ can be expressed as

$$\bar{x}(t) = \bar{P}(t) + \int_0^1 |g_m(t,s)| [f(s, \bar{\mathbf{x}}(s)) + \eta(s)]\, ds.$$

In the following three results also we shall consider the Banach space $C^{(q)}[0,1]$ and for $y(t) \in C^{(q)}[0,1]$ the norm $\|y\|$ is defined in (4.7).

Theorem 4.4 *(Picard's Iteration) With respect to the boundary value problem (1.2), (1.3) we assume that there exists an approximate solution* $\bar{x}(t)$ *and*

(i) the function $f(t, x_0, x_1, ..., x_q)$ *satisfies the Lipschitz condition*

$$|f(t, x_0, x_1, \cdots, x_q) - f(t, \bar{x}_0, \bar{x}_1, \cdots, \bar{x}_q)| \leq \sum_{i=0}^{q} L_i |x_i - \bar{x}_i|$$

on $[0,1] \times D_2$, *where*

$$D_2 = \Big\{(x_0, \cdots, x_q) : |x_{2i} - \bar{x}^{(2i)}(t)| \leq \pi^{2i} N \sin \pi t, \quad 0 \leq i \leq \left[\frac{q}{2}\right];$$
$$|x_{2i+1} - \bar{x}^{(2i+1)}(t)| \leq \pi^{2i} N[2\sin \pi t + \pi(1-2t)\cos \pi t], \ 0 \leq i \leq \left[\frac{q-1}{2}\right]\Big\};$$

(ii) $N_0 = (1-\theta)^{-1}(\epsilon+\delta)\dfrac{1}{\pi^{2m-2}}\left(\dfrac{1}{2\pi}\right) \le N.$

Then, the following hold

(1) there exists a solution $x^*(t)$ *of (1.2), (1.3) in* $\bar{S}(\bar{x}, N_0) = \{x \in C^{(q)}[0,1] : \|x-\bar{x}\| \le N_0\}$;

(2) $x^*(t)$ *is the unique solution of (1.2), (1.3) in* $\bar{S}(\bar{x}, N)$;

(3) the Picard iterative sequence $\{x_n(t)\}$, *defined by*

$$x_{n+1}(t) = P(t) + \int_0^1 |g_m(t,s)| f(s, \mathbf{x}_n(s))\, ds; \quad n = 0,\ 1, \cdots$$

where $x_0(t) = \bar{x}(t)$ *converges to* $x^*(t)$ *with* $\|x^* - x_n\| \le \theta^n N_0$, *and*

$$\|x^* - x_n\| \le \theta(1-\theta)^{-1}\|x_n - x_{n-1}\|.$$

Proof. For the operator $T : \bar{S}(\bar{x}, N) \to C^{(q)}[0,1]$ defined in (4.1) it suffices to show that the conditions of Contraction mapping theorem are satisfied. ∎

The quasilinear iterative scheme for (1.2), (1.3) is defined as

$$(-1)^m x_{n+1}^{(2m)}(t) = f(t, \mathbf{x}_n(t)) + \sum_{i=0}^{q}\left(x_{n+1}^{(i)}(t) - x_n^{(i)}(t)\right)\frac{\partial}{\partial x_n^{(i)}(t)} f(t, \mathbf{x}_n(t)) \tag{4.9}$$

$$x_{n+1}^{(2i)}(0) = \alpha_i, \quad x_{n+1}^{(2i)}(1) = \beta_i; \quad 0 \le i \le m-1, \quad n = 0, 1, \ldots \tag{4.10}$$

where $x_0(t) = \bar{x}(t)$ is an approximate solution of (1.2), (1.3).

Theorem 4.5 *(Newton's Iteration) With respect to the boundary value problem (1.2), (1.3) we assume that there exists an approximate solution* $\bar{x}(t)$, *and*

(i) the function $f(t, x_0, x_1, \ldots, x_q)$ *is continuously differentiable with respect to all* x_i, $0 \le i \le q$ *on* $[0,1] \times D_2$;

(ii) there exist nonnegative constants L_i, $0 \le i \le q$ *such that for all* $(t, x_0, x_1, \ldots, x_q) \in [0,1] \times D_2$, $\left|\dfrac{\partial}{\partial x_i} f(t, x_0, x_1, \ldots, x_q)\right| \le L_i$;

(iii) $3\theta < 1$;

(iv) $N_3 = (1-3\theta)^{-1}(\epsilon+\delta)\dfrac{1}{\pi^{2m-2}}\left(\dfrac{1}{2\pi}\right) \le N.$

Then, the following hold

(1) the sequence $\{x_n(t)\}$ *generated by the iterative scheme (4.9), (4.10) remains in* $\bar{S}(\bar{x}, N_3)$;

(2) the sequence $\{x_n(t)\}$ converges to the unique solution $x^*(t)$ of the boundary value problem (1.2), (1.3);

(3) a bound on the error is given by

$$\|x_n - x^*\| \leq \left(\frac{2\theta}{1-\theta}\right)^n \left(1 - \frac{2\theta}{1-\theta}\right)^{-1} (1-\theta)^{-1}(\epsilon+\delta)\frac{1}{\pi^{2m-2}}\left(\frac{1}{2\pi}\right).$$

Proof. The proof requires the equalities and the inequalities established in Section 2, and is based on inductive arguments. ∎

Finally, in this section we state the following result which provides the quadratic convergence of the iterative scheme (4.9), (4.10).

Theorem 4.6 *Let the conditions of Theorem 4.5 be satisfied. Further, let $f(t, x_0, x_1, ..., x_q)$ be twice continuously differentiable with respect to all x_i, $0 \leq i \leq q$ on $[0,1] \times D_2$, and*

$$\left|\frac{\partial^2}{\partial x_i \partial x_j} f(t, x_0, x_1, ..., x_q)\right| \leq L_i L_j K, \quad 0 \leq i, j \leq q.$$

Then,

$$\begin{aligned} \|x_{n+1} - x_n\| &\leq \alpha\|x_n - x_{n-1}\|^2 \leq \frac{1}{\alpha}\left(\alpha\|x_1 - x_0\|\right)^{2^n} \\ &\leq \frac{1}{\alpha}\left[\frac{(\epsilon+\delta)}{\pi^{2m-1}}\frac{\theta}{(1-\theta)^2}\frac{K}{4}max\{\pi\psi, \phi+2\psi\}\right]^{2^n}, \end{aligned}$$

where

$$\alpha = \frac{\theta}{1-\theta}\frac{K}{2}max\{\pi\psi, \phi+2\psi\}, \quad \phi = \sum_{i=0}^{[\frac{q}{2}]} \pi^{2i} L_{2i}, \quad \psi = \sum_{i=0}^{[\frac{q-1}{2}]} \pi^{2i} L_{2i+1}.$$

Thus, the convergence is quadratic if

$$\frac{(\epsilon+\delta)}{\pi^{2m-1}}\frac{\theta}{(1-\theta)^2}\frac{K}{4}max\{\pi\psi, \phi+2\psi\} < 1.$$

5 Piecewise-Lidstone Interpolation and Error Inequalities

Let $-\infty < a < b < \infty$. For the interval $[a, b]$ we let $\Delta : a = t_0 < t_1 < \cdots < t_{N+1} = b$ denote a uniform partition of $[a, b]$ with stepsize $h = (b-a)/(N+1)$. For each positive integer r, we will let $PC^{r,\infty}[a, b]$ be the set of all real-valued functions $x(t)$ such that : (i) $x(t)$ is $(r-1)$ times continuously differentiable on $[a,b]$, (ii) there exist s_i, $0 \le i \le L+1$ with $a = s_0 < s_1 < \cdots < s_{L+1} = b$ such that on each open subinterval (s_i, s_{i+1}), $0 \le i \le L$, $D^{r-1}x$ $(D = d/dt)$ is continuously differentiable, and (iii) the L_∞- norm of $D^r x$ is finite, i.e., $\|D^r x\|_\infty = \max_{0 \le i \le L} \sup_{t \in (s_i, s_{i+1})} |D^r x(t)| < \infty$.

For a fixed Δ, we define the set $L_m(\Delta) = \{h(t) \in C[a,b] : h(t)$ is a polynomial of degree at most $(2m-1)$ in each subinterval $[t_i, t_{i+1}]$, $0 \le i \le N\}$. It is clear that $L_m(\Delta)$ is of dimension $[2m(N+1) - N]$.

Definition 5.1 *For a given function $x(t) \in C^{(2m-2)}[a,b]$ we say $L_m^\Delta x(t)$ is the $L_m(\Delta)$ - interpolate of $x(t)$, also known as Lidstone interpolate of $x(t)$ if $L_m^\Delta x(t) \in L_m(\Delta)$ with $D^{2k} L_m^\Delta x(t_i) = x^{(2k)}(t_i) = x_i^{(2k)}$; $0 \le k \le m-1$, $0 \le i \le N+1$.*

In view of (2.1) – (2.4), (2.10) it follows that for $x(t) \in C^{(2m-2)}[a,b]$, $L_m^\Delta x(t)$ uniquely exists and in the subinterval $[t_i, t_{i+1}]$ can be explicitly expressed as

$$L_m^\Delta x(t) = \sum_{k=0}^{m-1} \left[x_i^{(2k)} \Lambda_k \left(\frac{t_{i+1} - t}{h} \right) + x_{i+1}^{(2k)} \Lambda_k \left(\frac{t - t_i}{h} \right) \right] h^{2k}. \quad (5.1)$$

Therefore, it follows that

$$L_m^\Delta x(t) = \sum_{i=0}^{N+1} \sum_{j=0}^{m-1} r_{m,i,j}(t)\, x_i^{(2j)}, \quad (5.2)$$

where $r_{m,i,j}(t)$, $0 \le i \le N+1$, $0 \le j \le m-1$ are the basic elements of $L_m(\Delta)$ satisfying

$$D^{2\nu} r_{m,i,j}(t_\mu) = \delta_{i\mu}\, \delta_{2\nu,j} \ ; \quad 0 \le \nu \le m-1, \ 0 \le \mu \le N+1$$

and appear as

$$\begin{aligned} r_{m,i,j}(t) &= \Lambda_j \left(\frac{t_{i+1} - t}{h} \right) h^{2j}, \quad t_i \le t \le t_{i+1}, \ 0 \le i \le N \\ &= \Lambda_j \left(\frac{t - t_{i-1}}{h} \right) h^{2j}, \quad t_{i-1} \le t \le t_i, \ 1 \le i \le N+1 \\ &= 0, \quad \text{otherwise.} \end{aligned} \quad (5.3)$$

Theorem 5.1 *Let $x(t) \in PC^{2m,\infty}[a,b]$. Then,*

$$\|D^k(x - L_m^\Delta x)\|_\infty \le d_{2m,k}\, h^{2m-k} \|D^{2m} x\|_\infty \ , \quad 0 \le k \le 2m-1 \quad (5.4)$$

where

$$d_{2n,k} = \begin{cases} \dfrac{(-1)^{n-i}E_{2n-2i}}{2^{2n-2i}(2n-2i)!}, \quad k = 2i,\ 0 \le i \le n \\ (-1)^{n-i+1}\dfrac{2(2^{2n-2i}-1)}{(2n-2i)!}B_{2n-2i}, \quad k = 2i+1, \quad 0 \le i \le n-1 \\ 2, \quad k = 2n+1. \end{cases} \tag{5.5}$$

Proof. The proof is clear from Theorem 3.3. ∎

Theorem 5.2 *Let $x(t) \in PC^{2m-2,\infty}[a,b]$. Then,*

$$\begin{aligned} &\|D^k(x - L^{\Delta}_m x)\|_\infty \\ &\le d_{2m-2,k}h^{2m-2-k} \max_{0\le i\le N} \sup_{t\in(t_i,t_{i+1})} \Bigg| D^{2m-2}x(t) - \left(\frac{t_{i+1}-t}{h}\right)x_i^{(2m-2)} \\ &\qquad - \left(\frac{t-t_i}{h}\right)x_{i+1}^{(2m-2)}\Bigg| \qquad (5.6) \\ &\le 2\ d_{2m-2,k}\ h^{2m-2-k}\|D^{2m-2}x\|_\infty, \quad 0 \le k \le 2m-2. \qquad (5.7) \end{aligned}$$

Proof. Without loss of generality we assume $a = 0, b = 1$ and $h = 1$, so that

$$L^{\Delta}_m x(t) = \sum_{k=0}^{m-1}[x^{(2k)}(0)\Lambda_k(1-t) + x^{(2k)}(1)\Lambda_k(t)]$$

and hence, in view of (2.2) – (2.4) and (2.10) directly it follows that

$$x(t) - L^{\Delta}_m x(t) = \int_0^1 g_{m-1}(t,s)\left[D^{2m-2}x(s) - (1-s)x^{(2m-2)}(0) - sx^{(2m-2)}(1)\right]ds. \tag{5.8}$$

Now for each $0 \le k \le 2m-2$ from (2.4), Corollary 2.5 and Lemma 2.7, we find

$$\|D^k(x - L^{\Delta}_m x)\|_\infty \le d_{2m-2,k} \sup_{t\in(0,1)} |D^{2m-2}x(t) - (1-t)x^{(2m-2)}(0) - tx^{(2m-2)}(1)|. \quad ∎$$

Remark 5.1 *For each k the inequality (5.6) is the best possible, as equality holds for the function $x(t) = E_{2m-2}(t)$ for $a = 0, b = 1$ and $h = 1$, whose Lidstone interpolate is $L^{\Delta}_m E_{2m-2}(t) = E_{2m-2}(t)$, and only for this function up to a constant factor. We also note that the inequality connecting the* *right sides of (5.6) and (5.7) is also the best possible. For this, it suffices to note that for the continuous function*

$$x(t) = \begin{cases} -1+4t, & 0 \le t \le 1/2 \\ 3-4t, & 1/2 \le t \le 1 \end{cases}$$

the equality $\max_{0\le t\le 1} |x(t) - (1-t)x(0) - t\, x(1)| \ = 2 \ \max_{0\le t\le 1} |x(t)|$ *holds. However, as such the sharpness of (5.7) remains undecided.*

Theorem 5.3 *Let* $x(t) \in PC^{2m-1,\infty}[a,b],\ 1 \le m \le 3$. *Then,*

$$\|D^k(x - L_m^{\Delta} x)\|_\infty \ \le \ e_{2m-1,k}\, h^{2m-1-k}\|D^{2m-1}x\|_\infty, \ \ 0 \le k \le 2m-2 \quad (5.9)$$

where the constants $e_{2m-1,k}$ *are given in the following:*

Table 5.1.

k \ m	1	2	3
0	$\frac{1}{2}$	$\frac{1}{24}$	$\frac{1}{240}$
1		$\frac{2}{9}\sqrt{\frac{1}{3}}$	$\frac{1}{225}\left(1-\sqrt{\frac{8}{15}}\right)^{\frac{1}{2}}\left(2+5\sqrt{\frac{8}{15}}\right)$
2		$\frac{1}{2}$	$\frac{1}{24}$
3			$\frac{2}{9}\sqrt{\frac{1}{3}}$
4			$\frac{1}{2}$

Proof. Without loss of generality we assume $a = 0, b = 1$ and $h = 1$. Then, from Peano's Kernel Theorem it follows that

$$x(t) - L_m^{\Delta} x(t) \ = \ \int_0^1 \bar{g}_m(t,s) D^{2m-1}x(s)\, ds, \qquad (5.10)$$

where

$$\bar{g}_m(t,s) \ = \ -\frac{\partial g_m(t,s)}{\partial s} \ = \ \begin{cases} -\sum_{k=0}^{m-1} \Lambda_k(t)\, \frac{(1-s)^{2m-2k-2}}{(2m-2k-2)!}, & t \le s \\ \sum_{k=0}^{m-1} \Lambda_k(1-t)\, \frac{s^{2m-2k-2}}{(2m-2k-2)!}, & s \le t. \end{cases} \qquad (5.11)$$

From (5.10), we have

$$|D^k(x - L_m^\Delta x)(t)| \le \left(\int_0^1 \left|\frac{\partial^k \bar{g}_m(t,s)}{\partial t^k}\right| ds\right) \|D^{2m-1}x\|_\infty, \quad 0 \le k \le 2m-2. \tag{5.12}$$

Hence, for $m = 3$ it suffices to show that

$$\max_{0\le t\le 1} \int_0^1 \left|\frac{\partial^k \bar{g}_3(t,s)}{\partial t^k}\right| ds = e_{5,k}, \quad 0 \le k \le 4. \tag{5.13}$$

Further, since $\frac{\partial^2 \bar{g}_m(t,s)}{\partial t^2} = \bar{g}_{m-1}(t,s)$ the proof of (5.13) will automatically include the proof for $m = 2$ and 1. Finally, we remark that the proof of (5.13) is by a direct computation. ∎

6 Lidstone-Spline Interpolation and Error Inequalities I

We begin with the remark that in Definition 5.1 as well as in the representation (5.2) the function $x(t)$ need not be in $C^{(2m-2)}[a,b]$, rather it is sufficient (which we shall assume throughout) that for the function $x(t)$, $D^{2k}L_m^\Delta x(t_i) = x^{(2k)}(t_i) = x_i^{(2k)}$; $0 \le k \le m-1$, $0 \le i \le N+1$ exist.

For a fixed Δ we define the spline space $S_m(\Delta) = \{s(t) \in C^{(2m-2)}[a,b] :$ $s(t)$ is a polynomial of degree at most $(2m-1)$ in each subinterval $[t_i, t_{i+1}]$, $0 \le i \le N\}$. It is clear that $S_m(\Delta)$ is of dimension $(N+2m)$.

Definition 6.1 *For a given $x(t) \in C^{(2m-2)}[a,b]$ we say $LS_m^\Delta x(t)$ is the Lidstone $S_m(\Delta)$-interpolate of $x(t)$, also known as Lidstone-spline interpolate of $x(t)$ if $LS_m^\Delta x(t) \in S_m(\Delta)$ with*

$$\begin{aligned} LS_m^\Delta x(t_i) &= x(t_i) = x_i, \quad 1 \le i \le N \\ D^{2k} LS_m^\Delta x(t_i) &= x^{(2k)}(t_i) = x_i^{(2k)}; \quad i = 0, N+1, \quad 0 \le k \le m-1. \end{aligned} \tag{6.1}$$

Since $S_m(\Delta) \subset L_m(\Delta)$ we can represent $LS_m^\Delta x(t)$ in terms of the basic elements $r_{m,i,j}(t)$; $0 \le i \le N+1$, $0 \le j \le m-1$ of $L_m(\Delta)$ even though these functions may not belong to $S_m(\Delta)$. In fact, we have

$$LS_m^\Delta x(t) = \sum_{i=0}^{N+1} \sum_{j=0}^{m-1} r_{m,i,j}(t)\, D^{2j} LS_m^\Delta x(t_i). \tag{6.2}$$

For $m = 3$ we shall show that the unknown constants $D^{2j}LS_m^\Delta x(t_i)$, $1 \le i \le N$, $1 \le j \le 2$ are the solutions of diagonally dominant system of linear algebraic equations, and hence can be obtained explicitly in terms of the known quantities.

Lemma 6.1 *Let $1 \le i \le N$ but fixed, and $p(t)$, $q(t)$ be two quintic polynomials in $[t_{i-1}, t_i]$ and $[t_i, t_{i+1}]$, respectively. Suppose $D^{2k}p(t_i) = D^{2k}q(t_i) = x_i^{(2k)}$, $0 \le k \le 2$ then $Dp(t_i) = Dq(t_i)$, if and only if*

$$\frac{h^2}{6}\left[D^2p(t_{i-1}) + 4\ x_i'' + D^2q(t_{i+1})\right] = p(t_{i-1}) - 2\ x_i + q(t_{i+1}) + \frac{h^4}{360}\left[7D^4p(t_{i-1}) + 16\ x_i^{(4)} + 7D^4q(t_{i+1})\right], \quad (6.3)$$

and $D^3p(t_i) = D^3q(t_i)$ if and only if

$$D^2p(t_{i-1}) - 2\ x_i'' + D^2q(t_{i+1}) = \frac{h^2}{6}[D^4p(t_{i-1}) + 4\ x_i^{(4)} + D^4q(t_{i+1})]. \quad (6.4)$$

Proof. The proof is by construction. ∎

Lemma 6.2 *Let $h(t) \in L_3(\Delta)$ be a function for which, say, $c_i = h(t_i)$, $c_i'' = D^2h(t_i)$ and $c_i^{(4)} = D^4h(t_i)$, $0 \le i \le N+1$ exist. Then, $h(t) \in S_3(\Delta)$ if and only if c_i'', $c_i^{(4)}$, $1 \le i \le N$ satisfy the following relations*

$$65\ c_1'' + 26\ c_2'' + c_3'' = -8\ c_0'' + \frac{2}{3}\ h^2\ c_0^{(4)} + \frac{20}{h^2}(4\ c_0 - 7\ c_1 + 2\ c_2 + c_3) \quad (6.5)$$

$$c_{i-2}'' + 26\ c_{i-1}'' + 66\ c_i'' + 26\ c_{i+1}'' + c_{i+2}'' = \frac{20}{h^2}(c_{i-2} + 2\ c_{i-1} - 6\ c_i + 2\ c_{i+1} + c_{i+2}), \quad 2 \le i \le N-1 \quad (6.6)$$

$$c_{N-2}'' + 26\,c_{N-1}'' + 65\,c_N'' = -8\,c_{N+1}'' + \frac{2}{3}\,h^2\ c_{N+1}^{(4)} + \frac{20}{h^2}(c_{N-2} + 2\,c_{N-1} - 7\,c_N + 4\,c_{N+1}) \quad (6.7)$$

$$65\ c_1^{(4)} + 26\ c_2^{(4)} + c_3^{(4)} = \frac{120}{h^2}\ c_0'' - 18\ c_0^{(4)} + \frac{120}{h^4}(-2\ c_0 + 5\ c_1 - 4\ c_2 + c_3) \quad (6.8)$$

$$c_{i-2}^{(4)} + 26\ c_{i-1}^{(4)} + 66\ c_i^{(4)} + 26\ c_{i+1}^{(4)} + c_{i+2}^{(4)} = \frac{120}{h^4}(c_{i-2} - 4\ c_{i-1} + 6\ c_i - 4\ c_{i+1} + c_{i+2}), \quad 2 \le i \le N-1 \quad (6.9)$$

$$c_{N-2}^{(4)} + 26\,c_{N-1}^{(4)} + 65\,c_N^{(4)} = \frac{120}{h^2}\,c_{N+1}'' - 18\,c_{N+1}^{(4)} + \frac{120}{h^4}(c_{N-2} - 4\,c_{N-1} + 5\,c_N - 2\,c_{N+1}). \quad (6.10)$$

Moreover, from the systems (6.5) – (6.7) and (6.8) – (6.10) the unknowns c_i'' and $c_i^{(4)}$, $1 \le i \le N$ can be obtained uniquely in terms of c_i, $0 \le i \le N+1$, c_0'', c_{N+1}'', $c_0^{(4)}$, and $c_{N+1}^{(4)}$.

Proof. Relations (6.5) – (6.10) follow directly from (6.1) and (6.2) and the continuity requirements. In matrix form systems (6.5) – (6.7) and (6.8) –

(6.10), respectively, can be written as

$$B_1 c^2 \;=\; k_1 \tag{6.11}$$

and

$$B_1 c^4 \;=\; k_2, \tag{6.12}$$

where $c^2 = [c_i'']$, $c^4 = [c_i^{(4)}]$,

$$B_1 \;=\; [b_{ij}^1] \;=\; \begin{cases} 26, & |i-j| = 1 \\ 65, & i = j = 1, N \\ 66, & i = j = 2, \cdots, N-1 \\ 1, & |i-j| = 2 \\ 0, & \text{otherwise} \end{cases} \tag{6.13}$$

and $k_1 = [k_i^1]$, $k_2 = [k_i^2]$, where

$$k_i^1 = \begin{cases} -8c_0'' + \frac{2}{3}h^2 c_0^{(4)} + \frac{20}{h^2}(4c_0 - 7c_1 + 2c_2 + c_3), \; i = 1 \\ -c_0'' + \frac{20}{h^2}(c_0 + 2c_1 - 6c_2 + 2c_3 + c_4), \; i = 2 \\ \frac{20}{h^2}(c_{i-2} + 2c_{i-1} - 6c_i + 2c_{i+1} + c_{i+2}), \; 3 \le i \le N-2 \\ -c_{N+1}'' + \frac{20}{h^2}(c_{N-3} + 2c_{N-2} - 6c_{N-1} + 2c_N + c_{N+1}), \; i = N-1 \\ -8c_{N+1}'' + \frac{2}{3}h^2 c_{N+1}^{(4)} + \frac{20}{h^2}(c_{N-2} + 2c_{N-1} - 7c_N + 4c_{N+1}), \; i = N \end{cases} \tag{6.14}$$

$$k_i^2 = \begin{cases} \frac{120}{h^2}c_0'' - 18c_0^{(4)} + \frac{120}{h^4}(-2c_0 + 5c_1 - 4c_2 + c_3), \; i = 1 \\ -c_0^{(4)} + \frac{120}{h^4}(c_0 - 4c_1 + 6c_2 - 4c_3 + c_4), \; i = 2 \\ \frac{120}{h^4}(c_{i-2} - 4c_{i-1} + 6c_i - 4c_{i+1} + c_{i+2}), \; 3 \le i \le N-2 \\ -c_{N+1}^{(4)} + \frac{120}{h^4}(c_{N-3} - 4c_{N-2} + 6c_{N-1} - 4c_N + c_{N+1}), \; i = N-1 \\ \frac{120}{h^2}c_{N+1}'' - 18c_{N+1}^{(4)} + \frac{120}{h^4}(c_{N-2} - 4c_{N-1} + 5c_N - 2c_{N+1}), \; i = N. \end{cases} \tag{6.15}$$

Lemma 6.3 *For a given $x(t) \in C^{(4)}[a,b]$, $LS_3^{\Delta}x(t)$ exists and is unique.*

Proof. For a given $g(t) \in C^{(4)}[a,b]$, $L_3^{\Delta}g(t)$ exists and is unique. Further, by Lemma 6.2 for the given set of numbers $c_i = x_i$, $1 \le i \le N+1$, $c_i'' = x_i''$, $c_i^{(4)} = x_i^{(4)}$, $i = 0, N+1$ there exist unique c_i'', $c_i^{(4)}$, $1 \le i \le N$ satisfying (6.5) – (6.10). Now, let $g(t) \in C^{(4)}[a,b]$ be such that $g^{(2k)}(t_i) = c_i^{(2k)}$, $0 \le i \le N+1$, $0 \le k \le 2$. Then, again by Lemma 6.2, $L_3^{\Delta}g(t) \in S_3(\Delta)$. However, from the definition, this $L_3^{(\Delta)}g(t)$ is the same as $LS_3^{\Delta}x(t)$. ∎

Remark 6.1 *From Lemma 6.3 and (6.2) it is clear that $LS_3^{\Delta}x(t)$ can be expressed as*

$$LS_3^{\Delta}x(t) = \sum_{i=0}^{N+1} r_{3,i,0}(t)x_i + r_{3,0,1}(t)x_0'' + r_{3,N+1,1}(t)x_{N+1}'' + \sum_{i=1}^{N} r_{3,i,1}(t)c_i''$$

$$+ r_{3,0,2}(t)x_0^{(4)} + r_{3,N+1,2}(t)x_{N+1}^{(4)} + \sum_{i=1}^{N} r_{3,i,2}(t)c_i^{(4)}, \quad (6.16)$$

where c_i'' and $c_i^{(4)}$, $1 \le i \le N$ satisfy (6.5) - (6.10).

Now let $x(t) \in PC^{n,\infty}[a,b]$, $2m-2 \le n \le 2m$. To obtain upper bounds for $\|D^k(x - LS_m^{\Delta}x)\|$, $0 \le k \le n-1$ in terms of $\|D^n x\|$ we begin with the equality

$$x(t) - LS_m^{\Delta}x(t) = \left(x(t) - L_m^{\Delta}x(t)\right) + \left(L_m^{\Delta}x(t) - LS_m^{\Delta}x(t)\right). \quad (6.17)$$

In (6.17) the term $\left(L_m^{\Delta}x(t) - LS_m^{\Delta}x(t)\right)$ belongs to $L_m(\Delta)$, and

$$D^{2k}\left(L_m^{\Delta}x(t_i) - LS_m^{\Delta}x(t_i)\right) = 0;$$
$$1 \le i \le N, \quad k = 0, \quad i = 0, N+1, \quad 0 \le k \le m-1.$$

Hence, from (5.2) and (6.2) it follows that

$$L_m^{\Delta}x(t) - LS_m^{\Delta}x(t) = \sum_{i=1}^{N}\sum_{j=1}^{m-1} r_{m,i,j}(t)\, e_{i,m}^{2j}, \quad (6.18)$$

where $e_{i,m}^{2j} = x_i^{(2j)} - D^{2j}LS_m^{\Delta}x(t_i)$. Thus, on substituting (6.18) into (6.17) and differentiating the resulting relation k times, $0 \le k \le n-1$ at $t \in [t_i, t_{i+1}]$, $0 \le i \le N$, we obtain

$$D^k\left(x(t) - LS_m^{\Delta}x(t)\right) = D^k\left(x(t) - L_m^{\Delta}x(t)\right) + \sum_{i=1}^{N}\sum_{j=1}^{m-1} r_{m,i,j}^{(k)}(t)\, e_{i,m}^{2j}.$$

Let the vector $\left[e_{i,m}^{2j}\right]$ be denoted as e_m^{2j}. Then, from the triangle inequality, we find for $0 \le k \le n-1$,

$$\|D^k(x - LS_m^{\Delta}x)\| \le \|D^k(x - L_m^{\Delta}x)\|$$

$$+ \sum_{j=1}^{m-1} \|e_m^{2j}\| \max_{0\le i\le N} \max_{t_i\le t\le t_{i+1}} \left[|r_{m,i,j}^{(k)}(t)| + |r_{m,i+1,j}^{(k)}(t)|\right], \quad (6.19)$$

where we have used the fact that $r_{m,i,j}^{(k)}(t)$ is nonzero only in the interval $(t_{i-1}, t_i)\bigcup(t_i, t_{i+1})$.

In the right side of (6.19) for the first term the inequalities obtained in Theorems 5.1 – 5.3 can be used, whereas for the second term, we need to prove the following:

Lemma 6.4 *For $0 \le i \le N$ and $0 \le k \le 2j+1$, $0 \le j \le m-1$ the following equalities hold*

$$\max_{t_i \le t \le t_{i+1}} \left[|D^k r_{m,i,j}(t)| + |D^k r_{m,i+1,j}(t)| \right] \;=\; d_{2j,k}\, h^{2j-k}. \qquad (6.20)$$

Proof. For $t_i \le t \le t_{i+1}$, from (5.3), (2.10) and (2.2), we find that

$$\begin{aligned} |D^k r_{m,i,j}(t)| &= h^{2j} \left| D^k \Lambda_j \left(1 - \frac{t-t_i}{h}\right)\right| \\ &= h^{2j-k} \left| \frac{d^k}{d\theta^k} \int_0^1 g_j(\theta,\phi)(1-\phi)d\phi \right|, \quad \theta = \frac{t-t_i}{h} \\ &\le h^{2j-k} \int_0^1 \left| \frac{\partial^k g_j(\theta,\phi)}{\partial \theta^k} \right| (1-\phi)d\phi, \quad 0 \le k \le 2j-1 \end{aligned}$$

and

$$|D^k r_{m,i+1,j}(t)| \;\le\; h^{2j-k} \int_0^1 \left| \frac{\partial^k g_j(\theta,\phi)}{\partial \theta^k} \right| \phi\, d\phi, \quad 0 \le k \le 2j-1.$$

Therefore, it follows that

$$\begin{aligned} &\max_{t_i \le t \le t_{i+1}} \left[|D^k r_{m,i,j}(t)| + |D^k r_{m,i+1,j}(t)| \right] \\ &\qquad \le h^{2j-k} \max_{0 \le \theta \le 1} \int_0^1 \left| \frac{\partial^k g_j(\theta,\phi)}{\partial \theta^k} \right| d\phi, \quad 0 \le k \le 2j-1. \end{aligned}$$

Now, (6.20) for $0 \le k \le 2j-1$ follows from (2.1), (2.9) and (2.11).

For $k = 2j$, from (5.3) and (2.1), we have

$$\begin{aligned} &\max_{t_i \le t \le t_{i+1}} \left[|D^{2j} r_{m,i,j}(t)| + |D^{2j} r_{m,i+1,j}(t)\,| \right] \\ &\qquad = \max_{t_i \le t \le t_{i+1}} \left[\left| \frac{t_{i+1}-x}{h} \right| + \left| \frac{x-t_i}{h} \right| \right] = 1 = d_{2j,2j}. \end{aligned}$$

Finally, for $k = 2j+1$ it suffices to note that $D^{2j+1} r_{m,i,j}(t) = -1/h$ and $D^{2j+1} r_{m,i+1,j}(t) = 1/h$. ∎

Lemma 6.5 *Let $x(t) \in PC^{n,\infty}[a,b]$. Then,*

(i) if $4 \le n \le 6$

$$\|e_3^2\| \;\le\; \bar{\alpha}_n\, h^{n-2} \|D^n x\|, \qquad (6.21)$$

where $\bar{\alpha}_4 = 0.38179896+0$, $\bar{\alpha}_5 = 0.67806574-1$ and $\bar{\alpha}_6 = 0.18518519-1$; and

(ii) if $4 \le n \le 6$

$$\|e_3^4\| \;\le\; \beta_n\; h^{n-4}\|D^n x\|, \tag{6.22}$$

where $\beta_4 = 20$, $\beta_5 = 0.25465679 + 1$ *and* $\beta_6 = 0.83333333 + 0$.

Proof. We shall prove only (6.21) for $n = 4$. The proof of the other cases is similar. Let $r = [r_i(x)]$ be an $N \times 1$ vector defined by

$$r \;=\; B_1\; e_3^2, \tag{6.23}$$

where the matrix B_1 is given in (6.13). Then, it follows that

$$B_1\; x^2 \;=\; k_1 + r, \tag{6.24}$$

where k_1 is defined in (6.14) and $x^2 = [x_i'']$ is an $N \times 1$ vector.

For $i = 1$, from (6.24), (6.13) and (6.14), we have

$$r_1(x) \;=\; 8\, x_0'' + 65\, x_1'' + 26\, x_2'' + x_3'' - \frac{2}{3}h^2\, x_0^{(4)} - \frac{20}{h^2}(4\, x_0 - 7\, x_1 + 2\, x_2 + x_3),$$

which in view of Peano's kernel theorem can be written as

$$r_1(x) \;=\; \frac{1}{3!}\int_{t_0}^{t_3} (r_1)_t(t-s)_+^3\; D^4x(s)\, ds - \frac{2}{3}h^2\, x_0^{(4)}, \tag{6.25}$$

where

$$\begin{aligned} |(r_1)_t(t-s)_+^3| \;=\; & \Big| 6\,[8(t_0-s)_+ + 65(t_1-s)_+ + 26(t_2-s)_+ + (t_3-s)_+] \\ & - \frac{20}{h^2}\left[4(t_0-s)_+^3 - 7(t_1-s)_+^3 + 2(t_2-s)_+^3 + (t_3-s)_+^3\right] \Big|. \end{aligned}$$

Therefore, from (6.25) it follows that

$$|r_1(x)| \;\le\; (4.5815875)\; h^2\|D^4x\|. \tag{6.26}$$

For $2 \le i \le N-1$, again from (6.24), (6.13) and (6.14), we have

$$\begin{aligned} r_i(x) \;=\; & x_{i-2}'' + 26\, x_{i-1}'' + 66\, x_i'' + 26\, x_{i+1}'' + x_{i+2}'' \\ & - \frac{20}{h^2}(x_{i-2} + 2\, x_{i-1} - 6\, x_i + 2\, x_{i+1} + x_{i+2}) \end{aligned}$$

and again by Peano's kernel theorem, we have

$$r_i(x) \;=\; \frac{1}{3!}\int_{t_{i-2}}^{t_{i+2}} (r_i)_t(t-s)_+^3\; D^4x(s)\, ds.$$

Thus, as in the case $i = 1$, we find that

$$|r_i(x)| \;\le\; (4.36317497)\; h^2\|D^4x\|, \quad 2 \le i \le N-1. \tag{6.27}$$

Finally, for $i = N$ we have

$$r_N(x) = x''_{N-2} + 26\, x''_{N-1} + 65\, x''_N + 8\, x''_{N+1} - \frac{2}{3}h^2\, x^{(4)}_{N+1} - \frac{20}{h^2}(x_{N-2} + 2x_{N-1} - 7\, x_N + 4\, x_{N+1}),$$

and hence as earlier in the case $i = 1$, we obtain

$$|r_N(x)| \le (4.5815875)\, h^2\|D^4x\|. \tag{6.28}$$

Now multiplying both sides of (6.23) by the diagonal matrix $\Box = [d_{ij}]$, where $d_{ii} = 1/a,\ a \in \mathbb{R}^+,\ 1 \le i \le N$ to obtain $\Box\, B_1\, e_3^2 = \Box\, r$, which gives that

$$\|e_3^2\| \le \|(\Box\, B_1)^{-1}\|\ \|\Box\, r\|. \tag{6.29}$$

Writing $\Box\, B_1 = I + A$, where A is an $N \times N$ matrix with the property that $\|A\| < 1$, it follows from (6.29) that

$$\|e_3^2\| \le \frac{1}{1-\|A\|}\ \frac{1}{a}\ \max_{1\le i\le N}|r_i(x)|. \tag{6.30}$$

To obtain the smallest bound in (6.30), we need to maximize $(1-\|A\|)\,a$ over $a \in \mathbb{R}^+$. For this, from (6.13) we have

$$\begin{aligned}\|A\| &= \max\left\{\frac{|65-a|+27}{a},\ \frac{|66-a|+54}{a}\right\}\\ &= \begin{cases}\max\left\{\dfrac{92}{a}-1,\ \dfrac{120}{a}-1\right\}, & 0 < a \le 65\\[2mm] \max\left\{1-\dfrac{38}{a},\ \dfrac{120}{a}-1\right\}, & 65 \le a \le 66\\[2mm] \max\left\{1-\dfrac{38}{a},\ 1-\dfrac{12}{a}\right\}, & a \ge 66\end{cases}\\ &= \begin{cases}\dfrac{120}{a}-1, & 0 < a \le 66\\[2mm] 1-\dfrac{12}{a}, & a \ge 66.\end{cases}\end{aligned} \tag{6.31}$$

Thus, the condition $\|A\| < 1$ is equivalent to $(120/a)-1 < 1$ $(1-(12/a) < 1$ for all $a \ge 66)$, which gives that $a > 60$. Hence, from (6.31) we find

$$\max_{a>60}(1-\|A\|)\,a = \max\left\{\max_{60<a\le 66}(2a-120),\ \max_{a\ge 66}(12)\right\} = 12. \tag{6.32}$$

Using (6.26) – (6.28) and (6.32) in (6.30), we obtain

$$\|e_3^2\| \le (0.38179896 + 0)h^2\|D^4x\|. \quad \blacksquare$$

Theorem 6.6 *Let $x(t) \in PC^{n,\infty}[a,b]$, $4 \le n \le 6$. Then,*

$$\|D^k(x - LS_3^{\Delta}x)\| \le \gamma_{3,n,k}\, h^{n-k}\|D^n x\|, \quad 0 \le k \le n-1 \tag{6.33}$$

where the constants $\gamma_{3,n,k}$ are given in the following table.

Table 6.1.

k \ n	4	5	6
0	0.33418320 + 0	0.45800925 − 1	0.14489293 − 1
1	0.11075662 + 1	0.15305465 + 0	0.48148148 − 1
2	0.31317990 + 1	0.42779423 + 0	0.13570602 + 0
3	0.11763598 + 2	0.15371972 + 1	0.49537037 + 0
4		0.30465679 + 1	0.95833333 + 0
5			0.21666667 + 1

Proof. Using Theorems 5.1 – 5.3 and Lemmas 6.4 and 6.5 in (6.19) the inequalities (6.33) are immediate. ∎

Now for a given function $x(t) \in PC^{6,\infty}[a,b]$ and a fixed partition Δ we shall construct approximates for the quintic Lidstone-spline interpolate $LS_3^{\Delta}x(t)$ when

1. the values of x_i'', $i = 0, N+1$ are unknown;
2. the values of $x_i^{(4)}$, $i = 0, N+1$ are unknown;
3. the values of x_i'', $x_i^{(4)}$, $i = 0, N+1$ are unknown;
4. the values of x_i, $1 \le i \le N$ are unknown;
5. the values of x_i, $1 \le i \le N$, $x_i^{(4)}$, $i = 0, N+1$ are unknown; and
6. the values of x_i, $0 \le i \le N+1$, x_i'', $x_i^{(4)}$, $i = 0, N+1$ are unknown.

Case 1 We approximate x_i'', $i = 0, N+1$ by the following relations:

$$x_0'' \simeq \tilde{x}_0'' = \frac{1}{h^2}\left(\frac{15}{4}x_0 - \frac{77}{6}x_1 + \frac{107}{6}x_2 - 13x_3 + \frac{61}{12}x_4 - \frac{5}{6}x_5\right), \tag{6.34}$$

$$x_{N+1}'' \simeq \tilde{x}_{N+1}'' = \frac{1}{h^2}\left(-\frac{5}{6}x_{N-4} + \frac{61}{12}x_{N-3} - 13x_{N-2} + \frac{107}{6}x_{N-1} - \frac{77}{6}x_N + \frac{15}{4}x_{N+1}\right). \tag{6.35}$$

It is easy to see that each of these relations has $\mathcal{O}(h^6)$ truncation error.

Definition 6.2 *We say $S_1x(t)$ is an approximate for $LS_3^{\Delta}x(t)$ if $S_1x(t) \in S_3(\Delta)$ with $S_1x(t_i) = x_i$, $0 \le i \le N+1$, and $D^2S_1x(t_i) = \tilde{x}_i''$, $D^{(4)}S_1x(t_i) = x_i^{(4)}$, $i = 0, N+1$.*

We use (6.34) and (6.35) to replace x_i'', $i = 0, N+1$ in the systems (6.5) – (6.7) and (6.8) – (6.10) and observe that the resulting unknowns c_{1i}'', $c_{1i}^{(4)}$, $1 \le i \le N$, say, can be obtained uniquely in terms of x_i, $0 \le i \le N+1$, and $x_0^{(4)}$ and $x_{N+1}^{(4)}$. Further, by Remark 6.1, $S_1x(t)$ can be explicitly expressed as

$$S_1x(t) = \sum_{i=0}^{N+1} r_{3,i,0}(t)x_i + r_{3,0,1}(t)\tilde{x}_0'' + r_{3,N+1,1}(t)\tilde{x}_{N+1}'' + \sum_{i=1}^{N} r_{3,i,1}(t)c_{1i}''$$

$$+ r_{3,0,2}(t)x_0^{(4)} + r_{3,N+1,2}(t)x_{N+1}^{(4)} + \sum_{i=1}^{N} r_{3,i,2}(t)c_{1i}^{(4)}. \quad (6.36)$$

To obtain a priori bound for $\|D^k(x - S_1x)\|$, $0 \le k \le 5$ we use the inequality

$$\|D^k(x - S_1x)\| \le \|D^k(x - LS_3^{\Delta}x)\| + \|D^k(LS_3^{\Delta}x - S_1x)\| \quad (6.37)$$

in which the first term of the right side can be estimated by Theorem 6.6, whereas for the second term we proceed as follows: from Remark 6.1 and (6.36), we have

$$(S_1x - LS_3^{\Delta}x)(t)$$

$$= r_{3,0,1}(t)\theta_0'' + r_{3,N+1,1}(t)\theta_{N+1}'' + \sum_{i=1}^{N} r_{3,i,1}(t)\ \theta_i'' + \sum_{i=1}^{N} r_{3,i,2}(t)\ \theta_i^{(4)}, \quad (6.38)$$

where $\theta_0'' = \tilde{x}_0'' - x_0''$, $\theta_{N+1}'' = \tilde{x}_{N+1}'' - x_{N+1}''$, and $\theta_i'' = c_{1i}'' - c_i''$, $\theta_i^{(4)} = c_{1i}^{(4)} - c_i^{(4)}$. Thus, for $0 \le k \le 5$ it follows that

$$\begin{aligned} &\|D^k(S_1x - LS_3^{\Delta}x)\| \\ &\le \left(|\theta_0''| + |\theta_{N+1}''|\right) \max_{0\le i\le N}\ \max_{t_i\le t\le t_{i+1}} |D^k r_{3,i,1}(t)| \\ &\quad +\|\theta^2\| \max_{0\le i\le N}\ \max_{t_i\le t\le t_{i+1}} \left[|D^k r_{3,i,1}(t)| + |D^k r_{3,i+1,1}(t)|\right] \\ &\quad +\|\theta^4\| \max_{0\le i\le N}\ \max_{t_i\le t\le t_{i+1}} \left[|D^k r_{3,i,2}(t)| + |D^k r_{3,i+1,2}(t)|\right], \quad (6.39) \end{aligned}$$

where θ^2 and θ^4 are the vectors $[\theta_i'']$ and $[\theta_i^{(4)}]$, respectively. In (6.39) we have used the fact that $D^k r_{3,i,1}(t)$ is symmetrical about the line $t = t_i$, and that $D^k r_{3,i,1}(t)$ as well as $D^k r_{3,i,2}(t)$ is nonzero only in the interval $(t_{i-1}, t_i) \bigcup (t_i, t_{i+1})$.

Lemma 6.7 *For $0 \le i \le N$ and $0 \le k \le 5$ the following equalities hold*

$$\max_{t_i\le t\le t_{i+1}} |D^k r_{3,i,j}(t)| = b_{2j,k}\ h^{2j-k}, \quad j = 1, 2 \quad (6.40)$$

where the constants $b_{2j,k}$ are given in the following table.

Table 6.2.

j \ k	0	1	2	3	4	5
1	$\frac{1}{9}\sqrt{\frac{1}{3}}$	$\frac{1}{3}$	1	1	0	0
2	$0.65221842-2$	$\frac{1}{45}$	$\frac{1}{9}\sqrt{\frac{1}{3}}$	$\frac{1}{3}$	1	1

Proof. The proof is by direct computation, and we convince by noting that

$$\begin{aligned}
&\max_{t_i\le t\le t_{i+1}} |r_{3,i,2}(t)| \\
&= \max_{t_i\le t\le t_{i+1}} \left|\frac{1}{120}\left(\frac{t_{i+1}-t}{h}\right)^5 - \frac{1}{36}\left(\frac{t_{i+1}-t}{h}\right)^3 + \frac{7}{360}\left(\frac{t_{i+1}-t}{h}\right)\right| h^4 \\
&= \max_{0\le t\le 1} \left[\frac{1}{120}(1-t)^5 - \frac{1}{36}(1-t)^3 + \frac{7}{360}(1-t)\right] h^4 \\
&= \left(1-\sqrt{\frac{8}{15}}\right)^{1/2}\left(\frac{1}{225}+\frac{1}{90}\sqrt{\frac{8}{15}}\right) h^4 \;=\; (0.65221842-2)h^4. \qquad \blacksquare
\end{aligned}$$

Lemma 6.8 *If $x(t) \in PC^{6,\infty}[a,b]$, then*

$$|\theta_i''| \le \frac{137}{180}\, h^4 \|D^6 x\|, \quad i = 0, N+1. \tag{6.41}$$

Proof. We shall prove (6.41) only for $i = 0$. Denoting θ_0'' by $r(x)$, we have

$$r(x) = \frac{1}{h^2}\left(\frac{15}{4}x_0 - \frac{77}{6}x_1 + \frac{107}{6}x_2 - 13x_3 + \frac{61}{12}x_4 - \frac{5}{6}x_5\right) - x_0'',$$

which in view of Peano's kernel theorem can be written as

$$r(x) = \frac{1}{5!}\int_{t_0}^{t_5} (r)_t (t-s)_+^5\, D^6 x(s)\, ds, \tag{6.42}$$

where

$$\begin{aligned}
(r)_t(t-s)_+^5 &= \frac{1}{12\,h^2}\left[45(t_0-s)_+^5 - 154(t_1-s)_+^5 + 214(t_2-s)_+^5\right. \\
&\qquad \left. -156(t_3-s)_+^5 + 61(t_4-s)_+^5 - 10(t_5-s)_+^5 - 240h^2(t_0-s)_+^3\right].
\end{aligned}$$

It is easy to see that $(r)_t(t-s)_+^5$ is nonpositive for all $t \in [t_0, t_5]$. Thus, from (6.42) it follows that

$$|r(x)| \le -\frac{1}{5!}\int_{t_0}^{t_5} (r)_t(t-s)_+^5\, |D^6 x(s)|\, ds = \frac{137}{180}\, h^4 \|D^6 x\|. \qquad \blacksquare$$

Lemma 6.9 *If $x(t) \in PC^{6,\infty}[a,b]$, then*

$$\|\theta^2\| \ \le \ \frac{137}{270}\, h^4 \|D^6 x\|. \tag{6.43}$$

Proof. Letting $c^2 = [c_i'']$ and $c_1^2 = [c_{1i}'']$, we have (6.11) and

$$B_1 c_1^2 \ = \ \tilde{k}_1, \tag{6.44}$$

where $\tilde{k}_1$ is obtained from k_1 by replacing x_i'' with $\tilde{x}_i''$, $i = 0, N+1$. Thus, from (6.11) and (6.44) it follows that

$$B_1(c^2 - c_1^2) = k_1 - \tilde{k}_1. \tag{6.45}$$

Multiplying both sides of (6.45) by the diagonal matrix $\Box = [d_{ij}]$, where $d_{ii} = 1/a,\ a \in \mathbb{R}^+,\ 1 \le i \le N$, we get

$$\Box B_1(c^2 - c_1^2) = \Box(k_1 - \tilde{k}_1),$$

which implies that

$$\|\theta^2\| \ = \ \|c^2 - c_1^2\| \ \le \ \|(\Box\, B_1)^{-1}\|\ \|\Box\,(k_1 - \tilde{k}_1)\|. \tag{6.46}$$

Now as in Lemma 6.5 writing $\Box\, B_1 = I + A$, where A is an $N \times N$ matrix with the property that $\|A\| < 1$, it follows from (6.46) and Lemma 6.8 that

$$\begin{aligned} \|\theta^2\| \ &\le \ \frac{1}{1 - \|A\|}\ \frac{1}{a}\ \max\left\{8|\theta_0''|,\ |\theta_0''|,\ |\theta_{N+1}''|,\ 8|\theta_{N+1}''|\right\} \\ &\le \ \frac{1}{1 - \|A\|}\ \frac{1}{a}\ 8\left[\frac{137}{180} h^4 \|D^6 x\|\right]. \end{aligned} \tag{6.47}$$

Since in Lemma 6.5 we have noted that the maximum of $(1 - \|A\|)a$, $a \in \mathbb{R}^+$ subject to the condition $\|A\| < 1$ is 12, from (6.47) the inequality (6.43) is immediate. ∎

Lemma 6.10 *If $x(t) \in PC^{6,\infty}[a,b]$, then*

$$\|\theta^4\| \ \le \ \frac{137}{18}\, h^2 \|D^6 x\|. \tag{6.48}$$

Proof. Letting $c^4 = [c_i^{(4)}]$ and $c_1^4 = [c_{1i}^{(4)}]$, we have (6.12) and

$$B_1 c_1^4 \ = \ \tilde{k}_2, \tag{6.49}$$

where $\tilde{k}_2$ is obtained from k_2 by replacing x_i'' with $\tilde{x}_i''$, $i = 0, N+1$. Thus, from (6.12) and (6.49) as in Lemma 6.9 it follows that

$$\begin{aligned} \|\theta^4\| = \|c^4 - c_1^4\| &\le \|(\Box\, B_1)^{-1}\|\, \|\Box\,(k_2 - \tilde{k}_2)\| \\ &\le \frac{1}{1-\|A\|}\,\frac{1}{a}\,\max\left\{\frac{120}{h^2}|\theta_0''|,\ \frac{120}{h^2}|\theta_{N+1}''|\right\} \\ &\le \frac{1}{1-\|A\|}\,\frac{1}{a}\,\frac{120}{h^2}\left[\frac{137}{180}h^4\|D^6x\|\right] \\ &= \frac{1}{12}\frac{120}{h^2}\left[\frac{137}{180}h^4\|D^6x\|\right] = \frac{137}{18}h^2\|D^6x\|. \quad \blacksquare \end{aligned}$$

Theorem 6.11 *If $x(t) \in PC^{6,\infty}[a,b]$, then*

$$\|D^k(x - S_1x)\| \le \beta_{1,k}\, h^{6-k}\|D^6x\|, \quad 0 \le k \le 5 \tag{6.50}$$

where $\beta_{1,0} = 0.27466883+0$, $\beta_{1,1} = 0.11263889+1$, $\beta_{1,2} = 0.31167245+1$, $\beta_{1,3} = 0.68379630+1$, $\beta_{1,4} = 0.85694444+1$, *and* $\beta_{1,5} = 0.17388889+2$.

Proof. We use Lemmas 6.4, 6.7 – 6.10 in (6.39) to obtain an upper estimate for $\|D^k(LS_3^{\Delta}x - S_1x)\|$, this estimate together with Theorem 6.6 in (6.37) then gives (6.50). $\blacksquare$

Case 2 As in Case 1 we approximate $x_0^{(4)}$ and $x_{N+1}^{(4)}$ by the formulae

$$x_0^{(4)} \simeq \tilde{x}_0^{(4)} = \frac{1}{h^4}\left(3\,x_0 - 14\,x_1 + 26\,x_2 - 24\,x_3 + 11\,x_4 - 2\,x_5\right), \tag{6.51}$$

$$\begin{aligned} x_{N+1}^{(4)} \simeq \tilde{x}_{N+1}^{(4)} = \frac{1}{h^4}&\left(-2\,x_{N-4} + 11\,x_{N-3} - 24\,x_{N-2}\right. \\ &\left. +26\,x_{N-1} - 14\,x_N + 3\,x_{N+1}\right), \end{aligned} \tag{6.52}$$

which have $\mathcal{O}(h^6)$ truncation error.

Definition 6.3 *We say $S_2x(t)$ is an approximate for $LS_3^{\Delta}x(t)$ if $S_2x(t) \in S_3(\Delta)$ with $S_2x(t_i) = x_i$, $0 \le i \le N+1$ and $D^2S_2x(t_i) = x_i''$, $D^4S_2x(t_i) = \tilde{x}_i^{(4)}$, $i = 0, N+1$.*

As in Case 1 we use (6.51) and (6.52) to replace $x_i^{(4)}$, $i = 0, N+1$ in the systems (6.5) – (6.7) and (6.8) – (6.10) and note that the resulting unknowns c_{2i}'', $c_{2i}^{(4)}$, $1 \le i \le N$, say, can be obtained uniquely in terms of x_i, $0 \le i \le N+1$, and x_0'' and x_{N+1}''. Further, by Remark 6.1, for $S_2x(t)$ an explicit representation similar to (6.36) holds. Moreover, on following as for the Case 1, we obtain the following result.

Theorem 6.12 *If $x(t) \in PC^{6,\infty}[a,b]$, then*

$$\|D^k(x - S_2x)\| \le \beta_{2,k}\, h^{6-k}\|D^6x\|, \quad 0 \le k \le 5$$

where $\beta_{2,0} = 0.12646280 + 0$, $\beta_{2,1} = 0.42986111 + 0$, $\beta_{2,2} = 0.11878803 + 1$, $\beta_{2,3} = 0.48240741 + 1$, $\beta_{2,4} = 0.10875000 + 2$, *and* $\beta_{2,5} = 0.16333333 + 2$.

Case 3

Definition 6.4 *We say* $S_3x(t)$ *is an approximate for* $LS_3^{\Delta}x(t)$ *if* $S_3x(t) \in S_3(\Delta)$ *with* $S_3x(t_i) = x_i$, $0 \le i \le N+1$ *and* $D^2S_3x(t_i) = \tilde{x}_i''$, $D^4S_3x(t_i) = \tilde{x}_i^{(4)}$, $i = 0, N+1$.

We use (6,34), (6.35), (6.51) and (6.52) to replace x_i'', $x_i^{(4)}$, $i = 0, N+1$ in the systems (6.5) – (6.7) and (6.8) – (6.10) and note that the resulting unknowns c_{3i}'', $c_{3i}^{(4)}$, $1 \le i \le N$, say, can be obtained uniquely in terms of x_i, $0 \le i \le N+1$. Further, by Remark 6.1, for $S_3x(t)$ an explicit representation similar to (6.36) holds. Moreover, the following result holds.

Theorem 6.13 *If* $x(t) \in PC^{6,\infty}[a,b]$, *then*

$$||D^k(x - S_3x)|| \ \le \ \beta_{3,k}\ h^{6-k}||D^6x||, \quad 0 \le k \le 5$$

where $\beta_{3,0} = 0.38664234 + 0$, $\beta_{3,1} = 0.15081019 + 1$, $\beta_{3,2} = 0..41688988 + 1$, $\beta_{3,3} = 0.11166667 + 2$, $\beta_{3,4} = 0.18486111 + 2$, *and* $\beta_{3,5} = 0.31555556 + 2$.

Case 4

Definition 6.5 *We say* $S_4x(t)$ *is an approximate for* $LS_3^{\Delta}x(t)$ *if* $S_4x(t) \in S_3(\Delta)$ *with* $D^{2k}S_4x(t_i) = x_i^{(2k)}$, $0 \le k \le 2$, $i = 0, N+1$ *and* $S_4x(t_i) = g_i$, $1 \le i \le N$, *where the given* g_i, $1 \le i \le N$ *are such that* $\max_{1\le i\le N} |x_i - g_i| = \xi$.

In systems (6.5) – (6.7) and (6.8) – (6.10) we replace x_i by g_i, $1 \le i \le N$ and note that the resulting unknowns c_{4i}'', $c_{4i}^{(4)}$, $1 \le i \le N$, say, can be obtained uniquely in terms of $x_i^{(2k)}$, $0 \le k \le 2$, $i = 0, N+1$ and g_i, $1 \le i \le N$. Further, by Remark 6.1, for $S_4x(t)$ an explicit representation similar to (6.36) holds. Moreover, the following result holds.

Theorem 6.14 *If* $x(t) \in PC^{6,\infty}[a,b]$, *then*

$$||D^k(x - S_4x)|| \ \le \ \gamma_{3,6,k}\ h^{6-k}||D^6x|| + \ \tau_k\ h^{-k}\ \xi, \quad 0 \le k \le 5$$

where $\gamma_{3,6,k}$ *are defined in Table 6.1 and* $\tau_0 = 0.55833333{+}1$, $\tau_1 = 0.16666667 + 2$, $\tau_2 = 0.40000000 + 2$, $\tau_3 = 0.12000000 + 3$, $\tau_4 = 0.16000000 + 3$, *and* $\tau_5 = 0.32000000 + 3$.

Case 5

Definition 6.6 *We say* $S_5x(t)$ *is an approximate for* $LS_3^{\Delta}x(t)$ *if* $S_5x(t) \in S_3(\Delta)$ *with* $D^{2k}S_5x(t_i) = x_i^{(2k)}$, $k = 0, 1$, $i = 0, N+1$, $S_5x(t_i) = g_i$, $1 \le i \le N$, *where the given* g_i, $1 \le i \le N$ *are such that* $\max_{1\le i\le N} |x_i - g_i| = \xi^*$, *and* $D^{(4)}S_5x(t_i) = x_i^{*(4)}$, $i = 0, N+1$ *where* $x_i^{*(4)}$ *is obtained from* $\tilde{x}_i^{(4)}$ *by replacing* x_i *with* g_i, $1 \le i \le N$.

In systems (6.5) – (6.7) and (6.8) – (6.10) we replace x_i by g_i, $1 \le i \le N$ and $x_i^{(4)}$ by $x_i^{*(4)}$, $i = 0, N+1$ and note that the resulting unknowns c''_{5i}, $c_{5i}^{(4)}$, $1 \le i \le N$, say, can be obtained uniquely in terms of $x_i^{(2k)}$, $k = 0, 1$, $i = 0, N+1$ and g_i, $1 \le i \le N$. Further, by Remark 6.1, for $S_5x(t)$ an explicit representation similar to (6.36) holds. Moreover, the following result can be proved.

Theorem 6.15 *If* $x(t) \in PC^{6,\infty}[a,b]$, *then*

$$\|D^k(x - S_5x)\| \le \beta_{2,k}\, h^{6-k}\|D^6x\| + \tau_k^*\, h^{-k}\, \xi^*, \quad 0 \le k \le 5$$

where $\beta_{2,k}$ *are defined in Theorem 6.12 and* $\tau_0^* = 0.73104060+1$, $\tau_1^* = 0.22401389+2$, $\tau_2^* = 0.56816605+2$, $\tau_3^* = 0.19908333+3$, $\tau_4^* = 0.36950000+3$, *and* $\tau_5^* = 0.58500000+3$.

Case 6

Definition 6.7 *We say* $S_6x(t)$ *is an approximate for* $LS_3^{\Delta}x(t)$ *if* $S_6x(t) \in S_3(\Delta)$ *with* $S_6x(t_i) = g_i$, $0 \le i \le N+1$, *where the given* g_i, $0 \le i \le N+1$ *are such that* $\max_{0 \le i \le N+1} |x_i - g_i| = \hat{\xi}$, *and* $D^{(2k)}S_6x(t_i) = \hat{x}_i^{(2k)}$, $k = 1, 2$, $i = 0, N+1$ *where* $\hat{x}_i^{(2k)}$ *is obtained from* $\tilde{x}_i^{(2k)}$ *by replacing* x_i *with* g_i, $0 \le i \le N+1$.

In systems (6.5) – (6.7) and (6.8) – (6.10) we replace x_i by g_i, $0 \le i \le N+1$ and $x_i^{(2k)}$ by $\hat{x}_i^{(2k)}$, $k = 1, 2$, $i = 0, N+1$ and note that the resulting unknowns c''_{6i}, $c_{6i}^{(4)}$, $1 \le i \le N$, say, can be obtained uniquely in terms of g_i, $0 \le i \le N+1$. Further, by Remark 6.1, for $S_6x(t)$ an explicit representation similar to (6.36) holds. Moreover, the following result can be proved.

Theorem 6.16 *If* $x(t) \in PC^{6,\infty}[a,b]$, *then*

$$\|D^k(x - S_6x)\| \le \beta_{3,k}\, h^{6-k}\|D^6x\| + \hat{\tau}_k\, h^{-k}\, \hat{\xi}, \quad 0 \le k \le 5$$

where $\beta_{3,k}$ *are defined in Theorem 6.13 and* $\hat{\tau}_0 = 0.16150108+2$, $\hat{\tau}_1 = 0.65111111+2$, $\hat{\tau}_2 = 0.18115289+3$, $\hat{\tau}_3 = 0.36177778+3$, $\hat{\tau}_4 = 0.45333333+3$, *and* $\hat{\tau}_5 = 0.74666667+3$.

7 Lidstone-Spline Interpolation and Error Inequalities II

For a fixed Δ we define the spline space $S_m(\Delta) = \{s(t) \in C^{(2m-3)}[a,b] : s(t)$ is a polynomial of degree at most $(2m-1)$ in each subinterval $[t_i, t_{i+1}]$, $0 \le i \le N\}$. It is clear that $S_m(\Delta)$ is of dimension $2(N+m)$.

Definition 7.1 *For a given $x(t) \in C^{(2m-2)}[a,b]$ we say $LS_m^\Delta x(t)$ is the Lidstone $S_m(\Delta)$-interpolate of $x(t)$, also known as Lidstone-spline interpolate of $x(t)$ if $LS_m^\Delta x(t) \in S_m(\Delta)$ with*

$$\begin{aligned} D^{2k}LS_m^\Delta x(t_i) &= x^{(2k)}(t_i) = x_i^{(2k)};\ 1 \le i \le N,\ 0 \le k \le m-2 \\ D^{2k}LS_m^\Delta x(t_i) &= x^{(2k)}(t_i) = x_i^{(2k)};\ i = 0, N+1,\ 0 \le k \le m-1. \end{aligned} \tag{7.1}$$

Remark 7.1 *It is clear that for this $LS_m^\Delta x(t)$ also the representation (6.2) holds. Further, from Lemma 6.1 we have the following analog of Lemma 6.2.*

Lemma 7.1 *Let $h(t) \in L_3(\Delta)$ be a function for which, say, $c_i = h(t_i)$, $c_i'' = D^2h(t_i)$ and $c_i^{(4)} = D^4h(t_i)$, $0 \le i \le N+1$ exist. Then, $h(t) \in S_3(\Delta)$ if and only if $c_i^{(4)}$, $1 \le i \le N$ satisfy the relations (6.8) – (6.10). Moreover, from the system (6.8) – (6.10) the unknowns $c_i^{(4)}$, $1 \le i \le N$ can be obtained uniquely in terms of c_i, $0 \le i \le N+1$, c_0'', c_{N+1}'', $c_0^{(4)}$, and $c_{N+1}^{(4)}$.*

Remark 7.2 *In system form (6.8) – (6.10) is the same as (6.12) where again $c^4 = [c_i^{(4)}]$, and B_1 and k_i^2 are defined in (6.13) and (6.15), respectively. We also note that a result corresponding to Lemma 6.3 holds for this $LS_3^\Delta x(t)$ also.*

Remark 7.3 *As in Remark 6.1 this $LS_3^\Delta x(t)$ can be expressed as*

$$\begin{aligned} LS_3^\Delta x(t) &= \sum_{i=0}^{N+1} [r_{3,i,0}(t)x_i + r_{3,i,1}(t)x_i''] + r_{3,0,2}(t)x_0^{(4)} \\ &\quad + r_{3,N+1,2}(t)x_{N+1}^{(4)} + \sum_{i=1}^{N} r_{3,i,2}(t)c_i^{(4)} \end{aligned} \tag{7.2}$$

where $c_i^{(4)}$, $1 \le i \le N$ satisfy (6.8) – (6.10).

Now let $x(t) \in PC^{n,\infty}[a,b]$, $2m-2 \le n \le 2m$. To obtain upper bounds for $\|D^k(x - LS_m^\Delta x)\|$, $0 \le k \le n-1$ in terms of $\|D^n x\|$ we begin with the equality (6.17). As noted in Section 6 in (6.17) the term $(L_m^\Delta x(t) - LS_m^\Delta x(t))$ belongs to $L_m(\Delta)$, and

$$\begin{aligned} D^{2k}(L_m^\Delta x(t_i) - LS_m^\Delta x(t_i)) = 0;\quad & 1 \le i \le N,\ 0 \le k \le m-2 \\ & i = 0, N+1,\ 0 \le k \le m-1. \end{aligned}$$

Hence, from (5.2) and (6.2) it follows that

$$L_m^\Delta x(t) - LS_m^\Delta x(t) = \sum_{i=1}^{N} r_{m,i,m-1}(t)\ e_{i,m}^{2m-2}, \tag{7.3}$$

where $e_{i,m}^{2m-2} = x_i^{(2m-2)} - D^{2m-2}LS_m^{\Delta}x(t_i)$. Thus, on substituting (7.3) into (6.17) and differentiating the resulting relation k times, $0 \le k \le n-1$ at $t \in [t_i, t_{i+1}]$, $0 \le i \le N$, we obtain

$$D^k(x(t) - LS_m^{\Delta}x(t)) = D^k(x(t) - L_m^{\Delta}x(t)) + \sum_{i=1}^{N} r_{m,i,m-1}^{(k)}(t)\, e_{i,m}^{2m-2}. \quad (7.4)$$

Let the vector $[e_{i,m}^{2m-2}]$ be denoted as e_m^{2m-2}. Then, from the triangle inequality, we find that

$$\|D^k(x - LS_m^{\Delta}x)\| \le \|D^k(x - L_m^{\Delta}x)\| + \|e_m^{2m-2}\| \times$$
$$\max_{0\le i\le N}\ \max_{t_i\le t\le t_{i+1}} \left[|r_{m,i,m-1}^{(k)}(t)| + |r_{m,i+1,m-1}^{(k)}(t)|\right],\ 0 \le k \le n-1 \quad (7.5)$$

where we have used the fact that $r_{m,i,m-1}^{(k)}(t)$ is nonzero only in the interval $(t_{i-1}, t_i) \bigcup (t_i, t_{i+1})$.

In the right side of (7.5) the equalities and the inequalities obtained in Theorems 5.1 – 5.3, and Lemma 6.7 can be used, and hence we only need to compute an upper estimate for $\|e_m^{2m-2}\|$. However, for $m = 3$ this has been done in Lemma 6.5(ii). Thus, the following result follows:

Theorem 7.2 *Let $x(t) \in PC^{n,\infty}[a,b]$, $4 \le n \le 6$. Then,*

$$\|D^k(x - LS_3^{\Delta}x)\| \le \gamma_{n,k}\, h^{n-k}\|D^n x\|, \quad 0 \le k \le n-1 \quad (7.6)$$

where the constants $\gamma_{n,k}$ are given in the following table.

Table 7.1.

k \ n	4	5	6
0	0.28645833 + 0	0.37325103 − 1	0.12174479 − 1
1	0.91666667 + 0	0.11915136 + 0	0.38888889 − 1
2	0.27500000 + 1	0.35998765 + 0	0.11718750 + 0
3	0.11000000 + 2	0.14015840 + 1	0.45833333 + 0
4		0.30465679 + 1	0.95833333 + 0
5			0.21666667 + 1

Now for a given function $x(t) \in PC^{6,\infty}[a,b]$ and a fixed partition Δ we shall construct approximates for the quintic Lidstone-spline interpolate $LS_3^{\Delta}x(t)$ when

1. the values of $x_i^{(4)}$, $i = 0, N+1$ are unknown;
2. the values of x_i'', $0 \le i \le N+1$ are unknown;
3. the values of x_i'', $0 \le i \le N+1$, $x_i^{(4)}$, $i = 0, N+1$ are unknown;
4. the values of x_i'', $1 \le i \le N$, $x_i^{(4)}$, $i = 0, N+1$ are unknown;
5. the values of x_i, x_i'', $0 \le i \le N+1$, $x_i^{(4)}$, $i = 0, N+1$ are unknown;

6. the values of $x_i,\ x_i'',\ 1 \le i \le N,\ x_i^{(4)},\ i = 0, N+1$ are unknown; and
7. the values of $x_i,\ x_i'',\ 1 \le i \le N$ are unknown.

Case 1 We approximate $x_0^{(4)}$ and $x_{N+1}^{(4)}$ by the formulae (6.51) and (6..52).

Definition 7.2 *We say $S_1x(t)$ is an approximate for $LS_3^{\Delta}x(t)$ if $S_1x(t) \in S_3(\Delta)$ with $S_1x(t_i) = x_i,\ D^2S_1x(t_i) = x_i'',\ 0 \le i \le N+1$ and $D^4S_1x(t_i) = \tilde{x}_i^{(4)},\ i = 0, N+1$.*

We use (6.51) and (6.52) to replace $x_i^{(4)},\ i = 0, N+1$ in the system (6.8) – (6.10) and note that the resulting unknowns $c_{1i}^{(4)},\ 1 \le i \le N$, say, can be obtained uniquely in terms of $x_i,\ 0 \le i \le N+1,\ x_0''$ and x_{N+1}''. Further, by Remark 7.1, $S_1x(t)$ can be explicitly expressed as

$$S_1x(t) = \sum_{i=0}^{N+1} [r_{3,i,0}(t)x_i + r_{3,i,1}(t)x_i''] + r_{3,0,2}(t)\tilde{x}_0^{(4)} + r_{3,N+1,2}(t)\tilde{x}_{N+1}^{(4)} + \sum_{i=1}^{N} r_{3,i,2}(t)c_{1i}^{(4)}. \quad (7.7)$$

Moreover, the following result holds.

Theorem 7.3 *If $x(t) \in PC^{6,\infty}[a,b]$, then*

$$\|D^k(x - S_1x)\| \le \beta_{1,k}\, h^{6-k}\|D^6x\|,\ 0 \le k \le 5$$

where $\beta_{1,0} = 0.10447206+0$, $\beta_{1,1} = 0.34189815+0$, $\beta_{1,2} = 0.10119543+1$, $\beta_{1,3} = 0.44722222+1$, $\beta_{1,4} = 0.10875000+2$, *and* $\beta_{1,5} = 0.16333333+2$.

Case 2 We approximate x_i'' by the relations (6.34) and (6.35) for $i = 0$ and $N+1$, respectively, and for $1 \le i \le N$ by

$$x_1'' \simeq \tilde{x}_1'' = \frac{1}{h^2}\left(\frac{5}{6}x_0 - \frac{5}{4}x_1 - \frac{1}{3}x_2 + \frac{7}{6}x_3 - \frac{1}{2}x_4 + \frac{1}{12}x_5\right), \quad (7.8)$$

$$x_i'' \simeq \tilde{x}_i'' = \frac{1}{h^2}\left(-\frac{1}{12}x_{i-2} + \frac{4}{3}x_{i-1} - \frac{5}{2}x_i + \frac{4}{3}x_{i+1} - \frac{1}{12}x_{i+2}\right), \quad 2 \le i \le N-1 \quad (7.9)$$

$$x_N'' \simeq \tilde{x}_N'' = \frac{1}{h^2}\left(\frac{1}{12}x_{N-4} - \frac{1}{2}x_{N-3} + \frac{7}{6}x_{N-2} - \frac{1}{3}x_{N-1} - \frac{5}{4}x_N + \frac{5}{6}x_{N+1}\right). \quad (7.10)$$

It is easy to see that each of these relations has $\mathcal{O}(h^6)$ truncation error.

Definition 7.3 *We say $S_2x(t)$ is an approximate for $LS_3^{\Delta}x(t)$ if $S_2x(t) \in S_3(\Delta)$ with $S_2x(t_i) = x_i$, $D^2S_2x(t_i) = \tilde{x}_i''$, $0 \le i \le N+1$, and $D^4S_2x(t_i) = x_i^{(4)}$, $i = 0, N+1$.*

As in Case 1 we use (6.34) and (6.35) to replace x_i'', $i = 0, N+1$ in the system (6.8) – (6.10) and note that the resulting unknowns $c_{2i}^{(4)}$, $1 \le i \le N$, say, can be obtained uniquely in terms of x_i, $0 \le i \le N+1$, $x_0^{(4)}$ and $x_{N+1}^{(4)}$. Further, by Remark 7.1, for $S_2x(t)$ an explicit representation similar to (7.7) holds. Moreover, the following result can be proved.

Theorem 7.4 *If $x(t) \in PC^{6,\infty}[a,b]$, then*

$$\|D^k(x - S_2x)\| \;\le\; \beta_{2,k}\, h^{6-k}\|D^6x\|, \quad 0 \le k \le 5$$

where $\beta_{2,0} = 0.20641638+0$, $\beta_{2,1} = 0.73657407+0$, $\beta_{2,2} = 0.18296875+1$, $\beta_{2,3} = 0.57861111+1$, $\beta_{2,4} = 0.85694444+1$, *and* $\beta_{2,5} = 0.17388889+2$.

Case 3

Definition 7.4 *We say $S_3x(t)$ is an approximate for $LS_3^{\Delta}x(t)$ if $S_3x(t) \in S_3(\Delta)$ with $S_3x(t_i) = x_i$, $D^2S_3x(t_i) = \tilde{x}_i''$, $0 \le i \le N+1$ and $D^4S_3x(t_i) = \tilde{x}_i^{(4)}$, $i = 0, N+1$.*

We use (6.34), (6.35), (6.51) and (6.52) to replace x_i'', $x_i^{(4)}$, $i = 0, N+1$ in the system (6.8) – (6.10) and note that the resulting unknowns $c_{3i}^{(4)}$, $1 \le i \le N$, say, can be obtained uniquely in terms of x_i, $0 \le i \le N+1$. Further, by Remark 7.1, for $S_2x(t)$ an explicit representation similar to (7.7) holds. Moreover, the following result can be proved.

Theorem 7.5 *If $x(t) \in PC^{6,\infty}[a,b]$, then*

$$\|D^k(x - S_3x)\| \;\le\; \beta_{3,k}\, h^{6-k}\|D^6x\|, \quad 0 \le k \le 5$$

where $\beta_{3,0} = 0.29871396+0$, $\beta_{3,1} = 0.10395833+1$, $\beta_{3,2} = 0.27244543+1$, $\beta_{3,3} = 0.98000000+1$, $\beta_{3,4} = 0.18486111+2$, *and* $\beta_{3,5} = 0.31555556+2$.

Case 4

Definition 7.5 *We say $S_4x(t)$ is an approximate for $LS_3^{\Delta}x(t)$ if $S_4x(t) \in S_3(\Delta)$ with $S_4x(t_i) = x_i$, $0 \le i \le N+1$, $D^2S_4x(t_i) = \tilde{x}_i''$, $1 \le i \le N$, $D^2S_4x(t_i) = x_i''$, $i = 0, N+1$ and $D^4S_4x(t_i) = \tilde{x}_i^{(4)}$, $i = 0, N+1$.*

We use (6.51) and (6.52) to replace $x_i^{(4)}$, $i = 0, N+1$ in the system (6.8) – (6.10) and note that the resulting unknowns $c_{1i}^{(4)}$, $1 \le i \le N$ (Case 1) can be obtained uniquely in terms of x_i, $0 \le i \le N+1$, x_0'' and x_{N+1}''. Further, by Remark 7.1, for $S_4x(t)$ an explicit representation similar to (7.7) holds. Moreover, the following result can be proved.

Theorem 7.6 *If $x(t) \in PC^{6,\infty}[a,b]$, then*

$$\|D^k(x - S_4x)\| \;\le\; \beta_{4,k}\; h^{6-k}\|D^6 x\|, \quad 0 \le k \le 5$$

where $\beta_{4,0} = 0.11349984 + 0$, $\beta_{4,1} = 0.37800926 + 0$, $\beta_{4,2} = 0.10841766 + 1$, $\beta_{4,3} = 0.46166667 + 1$, $\beta_{4,4} = 0.10875000 + 2$, *and* $\beta_{4,5} = 0.16333333 + 2$.

Case 5

Definition 7.6 *We say $S_5x(t)$ is an approximate for $LS_3^{\Delta}x(t)$ if $S_5x(t) \in S_3(\Delta)$ with $S_5x(t_i) = g_i$, $0 \le i \le N+1$, where the given g_i, $0 \le i \le N+1$ are such that $\max_{0\le i\le N+1}|x_i - g_i| = \xi$, and $D^2S_5x(t_i) = \bar{x}_i''$, $0 \le i \le N+1$, $D^4S_5x(t_i) = \bar{x}_i^{(4)}$, $i = 0, N+1$, where $\bar{x}_i''$ and $\bar{x}_i^{(4)}$ are obtained from $\tilde{x}_i''$ and $\tilde{x}_i^{(4)}$ by replacing x_i with g_i, $0 \le i \le N+1$.*

In system (6.8) – (6.10) we replace x_i by g_i, $0 \le i \le N+1$ and $x_i^{(2k)}$ by $\bar{x}_i^{(2k)}$, $k = 1, 2$, $i = 0, N+1$, and note that the resulting unknowns $c_{5i}^{(4)}$, $1 \le i \le N$, say, can be obtained uniquely in terms of g_i, $0 \le i \le N+1$. Further, by Remark 7.1, for $S_5x(t)$ an explicit representation similar to (7.7) holds. Moreover, the following result can be proved.

Theorem 7.7 *If $x(t) \in PC^{6,\infty}[a,b]$, then*

$$\|D^k(x - S_5x)\| \;\le\; \beta_{3,k}\; h^{6-k}\|D^6 x\| + \;\tau_k\; h^{-k}\;\xi, \quad 0 \le k \le 5$$

where $\beta_{3,k}$ are defined in Theorem 7.5 and $\tau_0 = 0.12529661 + 2$, $\tau_1 = 0.42444444+2$, $\tau_2 = 0.10026400+3$, $\tau_3 = 0.30666667+3$, $\tau_4 = 0.45333333+3$, *and* $\tau_5 = 0.74666667 + 3$.

Case 6

Definition 7.7 *We say $S_6x(t)$ is an approximate for $LS_3^{\Delta}x(t)$ if $S_6x(t) \in S_3(\Delta)$ with $D^{2k}S_6x(t_i) = x_i^{(2k)}$, $k = 0, 1$, $i = 0, N+1$, $S_6x(t_i) = g_i$, $1 \le i \le N$, where the given g_i, $1 \le i \le N$ are such that*

$$\max_{1\le i\le N}|x_i - g_i| \;= \xi^*, \tag{7.11}$$

and $D^2S_6x(t_i) = \hat{x}_i''$, $1 \le i \le N$, $D^4S_6x(t_i) = \hat{x}_i^{(4)}$, $i = 0, N+1$ where $\hat{x}_i''$ and $\hat{x}_i^{(4)}$ are obtained from $\tilde{x}_i''$ and $\tilde{x}_i^{(4)}$ by replacing x_i with g_i, $1 \le i \le N$.

In system (6.8) – (6.10) we replace x_i by g_i, $1 \le i \le N$ and $x_i^{(4)}$ by $\hat{x}_i^{(4)}$, $i = 0, N+1$ and note that the resulting unknowns $c_{6i}^{(4)}$, $1 \le i \le N$, say, can be obtained uniquely in terms of $x_i^{(2k)}$, $k = 0, 1$, $i = 0, N+1$ and g_i, $1 \le i \le N$. Further, by Remark 7.1, for $S_6x(t)$ an explicit representation similar to (7.7) holds. Moreover, the following result can be proved.

Theorem 7.8 *If* $x(t) \in PC^{6,\infty}[a,b]$, *then*

$$\|D^k(x - S_6x)\| \ \le \ \beta_{4,k}\ h^{6-k}\|D^6x\| + \ \tau_k^*\ h^{-k}\ \xi^* \ \equiv \ M_k, \ \ 0 \le k \le 5$$

where $\beta_{4,k}$ *are defined in Theorem 7.6 and* $\tau_0^* = 0.54770726 + 1$, $\tau_1^* = 0.15068056 + 2$, $\tau_2^* = 0.42149938 + 2$, $\tau_3^* = 0.16975000 + 3$, $\tau_4^* = 0.36950000 + 3$, *and* $\tau_5^* = 0.58500000 + 3$.

Case 7

Definition 7.8 *We say* $S_7x(t)$ *is an approximate for* $LS_3^{\Delta}x(t)$ *if* $S_7x(t) \in S_3(\Delta)$ *with* $D^{2k}S_7x(t_i) = x_i^{(2k)}$, $0 \le k \le 2$, $i = 0, N+1$, $S_7x(t_i) = g_i$, $1 \le i \le N$, *where the given* g_i, $1 \le i \le N$ *are such that (7.11) holds, and* $D^2S_7x(t_i) = \hat{x}_i''$, $1 \le i \le N$ *where* $\hat{x}_i''$ *are obtained from* $\tilde{x}_i''$ *by replacing* x_i *with* g_i, $1 \le i \le N$.

In system (6.8) – (6.11) we replace x_i by g_i, $1 \le i \le N$ and note that the resulting unknowns $c_{7i}^{(4)}$, $1 \le i \le N$, say, can be obtained uniquely in terms of $x_i^{(2k)}$, $0 \le k \le 2$, $i = 0, N+1$ and g_i, $1 \le i \le N$. Further, by Remark 7.1, for $S_7x(t)$ an explicit representation similar to (7.7) holds. Moreover, the following result can be proved.

Theorem 7.9 *If* $x(t) \in PC^{6,\infty}[a,b]$, *then*

$$\|D^k(x - S_7x)\| \ \le \ M_k + \beta_{7,k}\ h^{6-k}\|D^6x\| + \ \tau_k'\ h^{-k}\ \xi^*, \ 0 \le k \le 5$$

where M_k *are defined in Theorem 7.8, and* $\beta_{7,0} = 0.92297585 - 1$, $\beta_{7,1} = 0.30300926 + 0$, $\beta_{7,2} = 0.89476684 + 0$, $\beta_{7,3} = 0.40138889 + 1$, $\beta_{7,4} = 0.99166667 + 1$, $\beta_{7,5} = 0.14166667 + 2$, $\tau_0' = 0.25083226 + 1$, $\tau_1' = 0.82347222 + 1$, $\tau_2' = 0.24316605 + 2$, $\tau_3' = 0.10908333 + 3$, $\tau_4' = 0.26950000 + 3$, *and* $\tau_5' = 0.38500000 + 3$.

8 Problems and Remarks

1. Results of Sections 5 – 7 have been extended to functions of two independent variables over rectangular domains, respectively, in [7], [50] and [51], however over other domains, e.g., triangles, not much seems to have been done.

2. Sharp inequalities similar to (5.4), (5.7) and (5.9) in L_p- norm ($p \ge 1$) are available in [7], however, reasonable analogs of the inequalities discussed in Sections 6 and 7 in L_p- norm need more attention.

3. It will be interesting to find error inequalities similar to those obtained in Sections 5 –7 for functions which are not 'very' smooth.

4. The results of Sections 5 –7 have been used to construct approximate solutions of differential and integral equations [47-51]. We expect many more applications in future.

References

[1] R.P. Agarwal and G. Akrivis, Boundary value problems occurring in plate deflection theory, J. Comp. Appl. Math. 8(1982), 145-154.

[2] R.P. Agarwal, *Boundary Value Problems for Higher Order Differential Equations,* World Scientific, Singapore, 1986.

[3] R.P. Agarwal and P.J.Y. Wong, Lidstone polynomials and boundary value problems, Computers Math. Applic. 17(1989), 1397-1421.

[4] R.P. Agarwal, Sharp inequalities in polynomial interpolation, in General Inequalities 6, Ed. W. Walter, International Series of Numerical Mathematics, Birkhäuser Verlag 103(1992), 73-92.

[5] R.P. Agarwal and P.J.Y. Wong, Quasilinearization and approximate quasilinearization for Lidstone boundary value problems, Intern. J. Computer Math. 42(1992), 99-116.

[6] R.P. Agarwal and P.J.Y. Wong, *Error Inequalities in Polynomial Interpolation and their Applications,* Kluwer Academic Publishers, Dordrecht, 1993.

[7] R.P. Agarwal and P.J.Y. Wong, Explicit error bounds for the derivatives of piecewise-Lidstone interpolation, J. Comp. Appl. Math. 58(1995), 67-81.

[8] D. Amir and Z. Ziegler, Expansion of generalized completely convex functions, SIAM J. Math. Anal. 10(1979), 643-654.

[9] P. Baldwin, Asymptotic estimates of the eigenvalues of a sixth- order boundary-value problem obtained by using global phase-integral method, Phil. Trans. R. Soc. London A322(1987), 281-305.

[10] P. Baldwin, A localized instability in a Bénard layer, Applicable Analysis 24(1987), 117-156.

[11] C. Berg, Representation of completely convex functions by the extreme-point method, Enseignement Math. (2), 23(1977), 181-190.

[12] R.P. Boas, A note on functions of exponential type, Bull. Amer. Math. Soc. 47(1941), 750-754.

[13] R.P. Boas, Representation of functions by Lidstone series, Duke Math. J. 10(1943), 239-245.

[14] A. Boutayeb and E.H. Twizell, Finite-difference methods for twelfth-order boundary value problems, J. Comp. Appl. Math. 35(1991), 133-138.

[15] A. Boutayeb and E.H. Twizell, Numerical methods for the solution of special sixth-order boundary-value problems, Intern. J. Computer Math. 45(1992), 207-233.

[16] A. Boutayeb and E.H. Twizell, Finite difference methods for the solution of eighth-order boundary-value problems, Intern. J. Computer Math. 48(1993), 63-75.

[17] J.D. Buckholtz and J.K. Shaw, On functions expandable in Lidstone series, J. Math. Anal. Appl. 47(1974), 626-632.

[18] M.M. Chawla and C.P. Katti, Finite difference methods for two - point boundary value problems involving higher order differential equations, BIT 19(1979), 27-33.

[19] P.J. Davis, *Interpolation and Approximation,* Blaisdell Publishing Co., Boston, 1961.

[20] I.H. Dimovski, Lidstone-type formulas and nonharmonic sine expansions, in Constructive Function Theory, Bulgar. Acad. Sci., Sofia, 1983, 279-287.

[21] P. Forster, Existenzaussagen und Fehlerabschätzungen bei gewissen nichtlinearen Randwertaufgaben mit gewöhnlichen Differentialgleichungen, Numerishe Mathematik 10(1967), 410-422.

[22] T. Fort, *Finite Differences and Difference Equations in the Real Domain,* Oxford University Press, London, 1948.

[23] G.O. Golightly, Absolutely convergent Lidstone series, J. Math. Anal. Appl. 125(1987), 72-80.

[24] C. Jordan, *Calculus of Finite Differences,* Chelsea Pub. Co., New York, 1960.

[25] Ju.A. Kazmin, Lidstone's problem and certain of its generalizations, Vestnik Moskov. Univ. Ser. I Mat. Meh. 21(1966), 40-51.

[26] D.J. Leeming, Representation of functions by generalized Lidstone series, J. Approximation Theory 5(1972), 123-136.

[27] G.J. Lidstone, Notes on the extension of Aitken's theorem (for polynomial interpolation) to the Everett types, Proc. Edinburgh Math. Soc. (2), 2(1929), 16-19.

[28] Y.L.Luke, *The Special Functions and their Approximations,* Academic Press, New York, 1969.

[29] L.M. Milne-Thomson, *The Calculus of Finite Differences,* MacMillan, London, 1960.

[30] D.H. Mugler, Completely convex functions and convergence, SIAM J. Math. Anal. 10(1979), 292-296.

[31] Muhammad Aslam Noor and S.I. Tirmizi, Numerical methods for unilateral problems, J. Comp. Appl. Math. 16(1986), 387-395.

[32] H. Poritsky, On certain polynomial and other approximations to analytic functions, Trans. Amer. Math. Soc. 34(1932), 274-331.

[33] I.J. Schoenberg, On certain two-point expansions of integral functions of exponential type, Bull. Amer. Math. Soc. 42(1936), 284-288.

[34] H. Spath, The numerical calculation of high degree Lidstone splines with equidistant knots by blockunderrelaxation, Computing 7(1971), 65-74.

[35] J. Toomre, J.R. Jahn, J. Latour and E.A. Spiegel, Stellar convection theory II : single - mode study of the second convection zone in an A - type star, Astrophys. J. 207(1976), 545-563.

[36] E.H. Twizell and S.I.A. Tirmizi, A sixth order multiderivative method for two beam problems, Int. J. Numer. Methods Engg. 23(1986), 2089-2102.

[37] E.H. Twizell, Numerical methods for sixth - order boundary value problems, International Series of Numerical Mathematics 86(1988), 495-506.

[38] E.H. Twizell and S.I.A. Tirmizi, Multiderivative methods for nonlinear beam problems, Comm. Appl. Numer. Methods 4(1988), 43-50.

[39] E.H. Twizell and A. Boutayeb, Numerical methods for the solution of special and general sixth-order boundary value problems, with applications to Bénard layer eigenvalue problems, Proc. R. Soc. London A431(1990), 433-450.

[40] E.H. Twizell, A. Boutayeb and K. Djidjli, Numerical methods for eighth-, tenth-, and twelfth-order eigenvalue problems arising in thermal instability, Advances in Computational Mathematics, 2(1994), 407-436.

[41] R.A. Usmani, Solving boundary value problems in plate deflection theory, Simulation, December(1981), 195-206.

[42] A.K. Varma and G. Howell, Best error bounds for derivatives in two point Birkhoff interpolation problem, J. Approximation Theory 38(1983), 258-268.

[43] J.M. Whittaker, On Lidstone's series and two-point expansions of analytic functions, Proc. London Math. Soc. (2), 36(1933-34), 451-459.

[44] J.M. Whittaker, *Interpolatory Function Theory,* Cambridge, 1935.

[45] D.V. Widder, Functions whose even derivatives have a prescribed sign, Proc. National Acad. Sciences 26(1940), 657-659.

[46] D.V. Widder, Completely convex functions and Lidstone series, Trans. Amer. Math. Soc. 51(1942), 387-398.

[47] P.J.Y. Wong and R.P. Agarwal, Explicit error estimates for quintic and biquintic spline interpolation, Computers Math. Applic. 18(1989), 701-722.

[48] P.J.Y. Wong and R.P. Agarwal, Quintic spline solutions of Fredholm integral equations of the second kind, Intern. J. Computer Math. 33(1990), 237-249.

[49] P.J.Y. Wong and R.P. Agarwal, Explicit error estimates for quintic and biquintic spline interpolation II, Computers Math. Applic., 28, No. 7 (1994), 51-69.

[50] P.J.Y. Wong and R.P. Agarwal, Sharp error bounds for the derivatives of Lidstone-spline interpolation, Computers Math. Applic., 28, No. 9 (1994), 23-53.

[51] P.J.Y. Wong and R.P. Agarwal, Sharp error bounds for the derivatives of Lidstone-spline interpolation II, Computers Math. Applic., 31, No. 3 (1996), 61-90.

Higher Order Univariate Wavelet Type Approximation

George A. Anastassiou
Department of Mathematical Sciences
The University of Memphis, Memphis, TN 38152

Abstract

Higher order differentiable univariate functions are approximated by wavelet type operators, old and new. The higher order of this approximation is estimated by establishing some Jackson type inequalities. Sharpness of some of these inequalities over convex functions is achieved nontrivally.

1 Introduction

Recently there has been great interest in the wavelet type approximations. In [1], [2] the author and X.M. Yu introduced and studied extensively the wavelet type operators A_k, B_k over continuous functions on $\mathbf{R}$. They studied their approximation properties, as well as their shape and probabilistic preserving capabilities. In this paper among others two new interesting wavelet type operators C_k, D_k are introduced. These arise in a natural well-justified way. We study their approximation to the unit over $f \in C^N(\mathbf{R})$, $N \geq 0$. We produce related Jackson type inequalities, which give very close upper bounds to the error of this higher order approximation. When $N = 0$, sharpness over convexity is established for C_k, D_k operators. The same was established for B_k operator earlier in [2].

Also we give some interesting applications. All the produced inequalities involve the first modulus of continuity ω_1 of $f^{(N)}$, and the scale function of compact support φ is left without any assumption about its orthogonality.

The presented results are totally new and hopefully will open new avenues in the research of wavelet type approximations.

2 Results

Next we give our first main result

Theorem 1 *Let $f \in C^N(\mathbf{R})$, $N \geq 1$, $x \in \mathbf{R}$ and $k \in \mathbf{Z}$. Let φ be a bounded function of compact support $\subseteq [-a, a]$, $a > 0$ such that $\sum_{j=-\infty}^{\infty} \varphi(x-j) = 1$ all $x \in \mathbf{R}$. Assume $\varphi \geq 0$. Call*

$$(B_k(f))(x) := \sum_{j=-\infty}^{\infty} f\left(\frac{j}{2^k}\right) \varphi(2^k x - j).$$

Then

$$|(B_k(f))(x) - f(x)| \leq \sum_{i=1}^{N} \frac{|f^{(i)}(x)|}{i!} \frac{a^i}{2^{ki}} + \frac{a^N}{2^{kN} N!} \omega_1\left(f^{(N)}, \frac{a}{2^k}\right), \tag{1}$$

which is attained by constant function.

Remark 1 i) Given that $f^{(N)}$ is continuous and bounded or uniformly continuous on $\mathbf{R}$, we get that $\omega_1\left(f^{(N)}, \frac{a}{2^k}\right) < \infty$ and $B_k f \to f$, pointwise over $\mathbf{R}$, as $k \to \infty$.

ii) Given that $f \in C_b^N(\mathbf{R})$ (i.e., $f, f', \ldots, f^{(N)}$ are continuous and bounded) we obtain

$$\|B_k f - f\|_\infty \leq \sum_{i=1}^{N} \frac{\|f^{(i)}\|_\infty}{i!} \frac{a^i}{2^{ki}} + \frac{a^N}{2^{kN} N!} \omega_1\left(f^{(N)}, \frac{a}{2^k}\right). \tag{2}$$

That is, $B_k f \to f$, uniformly over $\mathbf{R}$, as $k \to \infty$.

Proof. (Of Theorem 1). Since $f \in C^N(\mathbf{R})$, $N \geq 1$ we get

$$f\left(\frac{j}{2^k}\right) = f(x) + \sum_{i=1}^{N} \frac{f^{(i)}(x)}{i!} \left(\frac{j}{2^k} - x\right)^i + \int_x^{j/2^k} (f^{(N)}(t) - f^{(N)}(x)) \frac{(\frac{j}{2^k} - t)^{N-1}}{(N-1)!} dt.$$

Thus

$$f\left(\frac{j}{2^k}\right) \varphi(2^k x - j) = f(x)\varphi(2^k x - j) + \sum_{i=1}^{N} \frac{f^{(i)}(x)}{i!} \left(\frac{j}{2^k} - x\right)^i \varphi(2^k x - j) + \varphi(2^k x - j) \int_x^{j/2^k} (f^{(N)}(t) - f^{(N)}(x)) \frac{(\frac{j}{2^k} - t)^{N-1}}{(N-1)!} dt.$$

and

$$\sum_{j=-\infty}^{\infty} f\left(\frac{j}{2^k}\right)\varphi(2^k x - j) = f(x)\sum_{j=-\infty}^{\infty}\varphi(2^k x - j)$$
$$+\sum_{j=-\infty}^{\infty}\sum_{i=1}^{N}\frac{f^{(i)}(x)}{i!}\left(\frac{j}{2^k}-x\right)^i\varphi(2^k x - j)$$
$$+\sum_{j=-\infty}^{\infty}\varphi(2^k x - j)\int_x^{j/2^k}(f^{(N)}(t)-f^{(N)}(x))\frac{(\frac{j}{2^k}-t)^{N-1}}{(N-1)!}dt.$$

Thus

$$\begin{aligned}\mathcal{E}_k(x) &:= (B_k(f))(x) - f(x)\\ &= \sum_{i=1}^{N}\frac{f^{(i)}(x)}{i!}\left(\sum_{j=-\infty}^{\infty}\left(\frac{j}{2^k}-x\right)^i\varphi(2^k x - j)\right)+\mathcal{R},\end{aligned}$$

where

$$\mathcal{R} := \sum_{j=-\infty}^{\infty}\varphi(2^k x - j)\int_x^{j/2^k}(f^{(N)}(t)-f^{(N)}(x))\frac{(\frac{j}{2^k}-t)^{N-1}}{(N-1)!}dt.$$

Hence

$$|\mathcal{E}_k(x)| \le \sum_{i=1}^{N}\frac{|f^{(i)}(x)|}{i!}\left|\sum_{j=-\infty}^{\infty}\left(\frac{j}{2^k}-x\right)^i\varphi(2^k x - j)\right| + |\mathcal{R}|.$$

Since φ is of compact support $\subseteq [-a, a]$ in order to have nonzero terms in the last inequality we need

$$-a \le 2^k x - j \le a \quad \text{i.e.,} \quad \left|x - \frac{j}{2^k}\right| \le \frac{a}{2^k}.$$

Furthermore

$$2^k x - a \le j \le 2^k x + a,$$

collapsing

$$\sum_{j=-\infty}^{\infty} \cdots = \sum_{j=\lceil 2^k x - a\rceil}^{[2^k x + a]} \cdot,$$

where $[\cdot]$, $\lceil\cdot\rceil$ denote the integral part and ceiling of the number, respectively. Consequently,

$$|\mathcal{E}_k(x)| \le \sum_{i=1}^{N}\frac{|f^{(i)}(x)|}{i!}\frac{a^i}{2^{ki}} + |\mathcal{R}|. \tag{3}$$

Call

$$\Gamma_j(x) := \left| \int_x^{j/2^k} (f^{(N)}(t) - f^{(N)}(x)) \frac{(\frac{j}{2^k} - t)^{N-1}}{(N-1)!} dt \right| .$$

Here we have

$$|\mathcal{R}| \leq \sum_{j=-\infty}^{\infty} \varphi(2^k x - j) \Gamma_j(x).$$

Next we would like to estimate $\Gamma_j(x)$.

i) If $x \leq j/2^k$, then

$$\begin{aligned}
\Gamma_j(x) &\leq \int_x^{j/2^k} |f^{(N)}(t) - f^{(N)}(x)| \frac{(\frac{j}{2^k} - t)^{N-1}}{(N-1)!} dt \\
&\leq \int_x^{j/2^k} \omega_1(f^{(N)}, |t-x|) \frac{(\frac{j}{2^k} - t)^{N-1}}{(N-1)!} dt \\
&\leq \omega_1 \left(f^{(N)}, \left| x - \frac{j}{2^k} \right| \right) \int_x^{j/2^k} \frac{(\frac{j}{2^k} - t)^{N-1}}{(N-1)!} dt \\
&\leq \omega_1 \left(f^{(N)}, \frac{a}{2^k} \right) \frac{(\frac{j}{2^k} - x)^N}{N!} \leq \omega_1 \left(f^{(N)}, \frac{a}{2^k} \right) \frac{a^N}{N! 2^{kN}} .
\end{aligned}$$

I.e., if $x \leq j/2^k$ we get that

$$\Gamma_j(x) \leq \omega_1 \left(f^{(N)}, \frac{a}{2^k} \right) \frac{a^N}{2^{kN} N!} .$$

ii) If $x \geq j/2^k$, then

$$\begin{aligned}
\Gamma_j(x) &= \left| \int_{j/2^k}^x (f^{(N)}(t) - f^{(N)}(x)) \frac{(\frac{j}{2^k} - t)^{N-1}}{(N-1)!} dt \right| \\
&\leq \int_{j/2^k}^x |f^{(N)}(t) - f^{(N)}(x)| \frac{(t - \frac{j}{2^k})^{N-1}}{(N-1)!} dt \\
&\leq \int_{j/2^k}^x \omega_1(f^{(N)}, |t-x|) \frac{(t - \frac{j}{2^k})^{N-1}}{(N-1)!} dt \\
&\leq \omega_1 \left(f^{(N)}, \left| x - \frac{j}{2^k} \right| \right) \int_{j/2^k}^x \frac{(t - \frac{j}{2^k})^{N-1}}{(N-1)!} dt \\
&\leq \omega_1 \left(f^{(N)}, \frac{a}{2^k} \right) \frac{(x - \frac{j}{2^k})^N}{N!} \leq \omega_1 \left(f^{(N)}, \frac{a}{2^k} \right) \frac{a^N}{2^{kN} N!} .
\end{aligned}$$

I.e., if $x \geq j/2^k$ then

$$\Gamma_j(x) \leq \omega_1 \left(f^{(N)}, \frac{a}{2^k} \right) \frac{a^N}{2^{kN} N!} .$$

So we have established that

$$\Gamma_j(x) \le \Gamma^*, \quad \text{any } x \in \mathbf{R},$$

where

$$\Gamma^* := \omega_1\left(f^{(N)}, \frac{a}{2^k}\right)\frac{a^N}{2^{kN}N!}\,.$$

And

$$|\mathcal{R}| \le \left(\sum_{j=-\infty}^{\infty} \varphi(2^k x - j)\right)\Gamma^* = \Gamma^*.$$

I.e.,

$$|\mathcal{R}| \le \frac{a^N}{2^{kN}N!}\omega_1\left(f^{(N)}, \frac{a}{2^k}\right). \tag{4}$$

That is from (3) and (4) we get

$$|\mathcal{E}_k(x)| \le \sum_{i=1}^{N} \frac{|f^{(i)}(x)|}{i!}\,\frac{a^i}{2^{ki}} + \frac{a^N}{2^{kN}N!}\omega_1\left(f^{(N)}, \frac{a}{2^k}\right),$$

proving (1). □

Our second main result follows

Theorem 2 *Same assumptions as in Theorem* 1. *Additionally assume that* φ *is Lebesgue measurable (then* $\int_{-\infty}^{+\infty} \varphi(x)dx = 1$*). Define*

$$\varphi_{kj}(x) := 2^{k/2}\varphi(2^k x - j) \quad \text{all } k, j \in \mathbf{Z},$$

$$\langle f, \varphi_{kj}\rangle := \int_{-\infty}^{\infty} f(t)\varphi_{kj}(t)dt,$$

and

$$\begin{aligned}(A_k(f))(x) &:= \sum_{j=-\infty}^{\infty} \langle f, \varphi_{kj}\rangle \varphi_{kj}(x)\\ &= \sum_{j=-\infty}^{\infty}\left(\int_{-\infty}^{\infty} f\left(\frac{u}{2^k}\right)\varphi(u-j)du\right)\varphi(2^k x - j).\end{aligned}$$

Then

$$|(A_k(f))(x) - f(x)| \le \sum_{i=1}^{N} \frac{|f^{(i)}(x)|}{i!}\,\frac{a^i}{2^{i(k-1)}} + \frac{a^N}{N!2^{N(k-1)}}\omega_1\left(f^{(N)}, \frac{a}{2^{k-1}}\right), \tag{5}$$

which is attained by constants.

Remark 2 i) Given that $f^{(N)}$ is continuous and bounded or uniformly continuous on $\mathbf{R}$, we get that $\omega_1(f^{(N)}, \frac{a}{2^{k-1}}) < \infty$ and $A_k f \to f$, pointwise over $\mathbf{R}$, as $k \to \infty$.

ii) Given that $f \in C_b^N(\mathbf{R})$ we obtain

$$\|A_k f - f\|_\infty \leq \sum_{i=1}^{N} \frac{\|f^{(i)}\|_\infty}{i!} \frac{a^i}{2^{i(k-1)}} + \frac{a^N}{N! 2^{N(k-1)}} \omega_1\left(f^{(N)}, \frac{a}{2^{k-1}}\right). \quad (6)$$

That is $A_k f \to f$, uniformly over $\mathbf{R}$, as $k \to \infty$.

Corollary 3 *Additionally assume that φ is right continuous. Here $f \in C^1(\mathbf{R})$ is a probability distribution function such that $\lambda := f'$ is the density function. Then, $k \in \mathbf{Z}$,*

$$|B_k(f)(x) - f(x)| \leq \frac{a}{2^k}\left(\lambda(x) + \omega_1\left(\lambda, \frac{a}{2^k}\right)\right) \quad (7)$$

and

$$|A_k(f)(x) - f(x)| \leq \frac{a}{2^{k-1}}\left(\lambda(x) + \omega_1\left(\lambda, \frac{a}{2^{k-1}}\right)\right).$$

From [1] here $A_k(f)$, $B_k(f)$ are distribution functions, given that φ is bell-shaped.

Proof. (Of Theorem 2.) By $\sum_{j=-\infty}^{\infty} \varphi(2^k x - j) = 1$ we get that

$$\int_{-\infty}^{\infty} \varphi(u - j) du = 1.$$

Notice that

$$A_k(f)(x) - f(x) = 2^{k/2} \sum_{j=-\infty}^{\infty} [\langle f, \varphi_{k_j}\rangle - 2^{-k/2} f(x)] \varphi(2^k x - j). \quad (8)$$

Also we observe that

$$\begin{aligned} \langle f, \varphi_{k_j}\rangle - 2^{-k/2} f(x) &= 2^{k/2} \int_{-\infty}^{\infty} f(t)\varphi(2^k t - j) dt - 2^{-k/2} f(x) \\ &= 2^{-k/2} \int_{-\infty}^{\infty} f\left(\frac{u}{2^k}\right) \varphi(u - j) du - 2^{-k/2} f(x) \\ &= 2^{-k/2} \int_{-\infty}^{\infty} \left[f\left(\frac{u}{2^k}\right) - f(x)\right] \varphi(u - j) du. \end{aligned}$$

By supp $\varphi \subseteq [-a, a]$ we have that $2^{-k}(-a + j) \leq x \leq 2^{-k}(a + j)$. Hence

$$|\langle f, \varphi_{k_j}\rangle - 2^{-k/2} f(x)| = |2^{-k/2} \Lambda|, \quad (9)$$

where

$$\Lambda := \int_{-a+j}^{a+j} \left[f\left(\frac{u}{2^k}\right) - f(x) \right] \varphi(u-j)du.$$

Next we have that

$$f\left(\frac{u}{2^k}\right) - f(x) = \sum_{i=1}^{N} \frac{f^{(i)}(x)}{i!} \left(\frac{u}{2^k} - x\right)^i + \int_x^{u/2^k} (f^{(N)}(t) - f^{(N)}(x)) \frac{(\frac{u}{2^k} - t)^{N-1}}{(N-1)!} dt.$$

Thus

$$\left(f\left(\frac{u}{2^k}\right) - f(x)\right) \varphi(u-j) = \left(\sum_{i=1}^{N} \frac{f^{(i)}(x)}{i!} \left(\frac{u}{2^k} - x\right)^i \right) \varphi(u-j) + \varphi(u-j) \int_x^{u/2^k} (f^{(N)}(t) - f^{(N)}(x)) \frac{(\frac{u}{2^k} - t)^{N-1}}{(N-1)!} dt.$$

Now we notice that $|\frac{u}{2^k} - x| \le \frac{a}{2^{k-1}}$ and

$$\Lambda = \sum_{i=1}^{N} \frac{f^{(i)}(x)}{i!} \int_{-a+j}^{a+j} \left(\frac{u}{2^k} - x\right)^i \varphi(u-j)du + \mathcal{R}^*, \tag{10}$$

where

$$\mathcal{R}^* := \left\{ \int_{-a+j}^{a+j} \varphi(u-j) \left(\int_x^{u/2^k} (f^{(N)}(t) - f^{(N)}(x)) \frac{(\frac{u}{2^k} - t)^{N-1}}{(N-1)!} dt \right) du \right\}. \tag{11}$$

Therefore

$$|\Lambda| \le \sum_{i=1}^{N} \frac{|f^{(i)}(x)|}{i!} \frac{a^i}{2^{i(k-1)}} + |\mathcal{R}^*|. \tag{12}$$

Furthermore

$$|\mathcal{R}^*| \le \int_{-a+j}^{a+j} \varphi(u-j)\rho(u)du, \tag{13}$$

where

$$\rho(u) := \left| \int_x^{u/2^k} (f^{(N)}(t) - f^{(N)}(x)) \frac{(\frac{u}{2^k} - t)^{N-1}}{(N-1)!} dt \right|. \tag{14}$$

Next we estimate $\rho(u)$: i) If $x \le \frac{u}{2^k}$, then

$$\begin{aligned}\rho(u) &\le \int_x^{u/2^k} |f^{(N)}(t) - f^{(N)}(x)| \cdot \frac{(\frac{u}{2^k} - t)^{N-1}}{(N-1)!} dt \\ &\le \int_x^{u/2^k} \omega_1(f^{(N)}, |t-x|) \frac{(\frac{u}{2^k} - t)^{N-1}}{(N-1)!} dt \\ &\le \omega_1\left(f^{(N)}, \left|x - \frac{u}{2^k}\right|\right) \int_x^{u/2^k} \frac{(\frac{u}{2^k} - t)^{N-1}}{(N-1)!} dt \\ &\le \omega_1\left(f^{(N)}, \frac{a}{2^{k-1}}\right) \frac{(\frac{u}{2^k} - x)^N}{N!} \\ &\le \omega_1\left(f^{(N)}, \frac{a}{2^{k-1}}\right) \frac{a^N}{2^{N(k-1)}N!}.\end{aligned}$$

So if $x \le \frac{u}{2^k}$ we find that

$$\rho(u) \le \omega_1\left(f^{(N)}, \frac{a}{2^{k-1}}\right) \frac{a^N}{N!2^{N(k-1)}}.$$

ii) If $x \ge \frac{u}{2^k}$ then

$$\begin{aligned}\rho(u) &\le \int_{u/2^k}^{x} |f^{(N)}(t) - f^{(N)}(x)| \frac{(t - \frac{u}{2^k})^{N-1}}{(N-1)!} dt \\ &\le \omega_1\left(f^{(N)}, \frac{a}{2^{k-1}}\right) \frac{(x - \frac{u}{2^k})^N}{N!} \\ &\le \omega_1\left(f^{(N)}, \frac{a}{2^{k-1}}\right) \frac{a^N}{N!2^{N(k-1)}}.\end{aligned}$$

So in both cases we obtain that

$$\rho(u) \le \omega_1\left(f^{(N)}, \frac{a}{2^{k-1}}\right) \frac{a^N}{N!2^{N(k-1)}} =: T. \tag{15}$$

Consequently we have (cf. (13), (15))

$$|\mathcal{R}^*| \le \left(\int_{-a+j}^{a+j} \varphi(u-j)du\right) T = 1T.$$

I.e.,

$$|\mathcal{R}^*| \le \omega_1\left(f^{(N)}, \frac{a}{2^{k-1}}\right) \frac{a^N}{N!2^{N(k-1)}}. \tag{16}$$

Hence (cf. (12), (16))

$$|\Lambda| \le \sum_{i=1}^{N} \frac{|f^{(i)}(x)|}{i!} \frac{a^i}{2^{i(k-1)}} + \frac{a^N}{N!2^{N(k-1)}} \omega_1\left(f^{(N)}, \frac{a}{2^{k-1}}\right) =: \theta, \tag{17}$$

and

$$2^{-k/2}|\Lambda| \le 2^{-k/2}\theta.$$

I.e., (cf. (9))

$$|\langle f, \varphi_{kj}\rangle - 2^{-k/2} f(x)| \le 2^{-k/2}\theta. \tag{18}$$

Finally from (8), (17), and (18) we have

$$|(A_k(f))(x) - f(x)| \le 2^{k/2} \sum_{j=-\infty}^{\infty} |[\langle f, \varphi_{kj}\rangle - 2^{-k/2} f(x)]|\varphi(2^k x - j)$$

$$\le 2^{k/2} \sum_{j=-\infty}^{\infty} 2^{-k/2}\theta\varphi(2^k x - j)$$

$$= \theta 2^{k/2} 2^{-k/2} \left(\sum_{j=-\infty}^{\infty} \varphi(2^k x - j) \right) = \theta 1,$$

proving (5). □

The next wavelet type operator is studied for the first time.

Theorem 4 *Same assumptions as in Theorem* 1. *Define*

$$C_k(f)(x) := \sum_{j=-\infty}^{\infty} \gamma_{kj}(f)\varphi(2^k x - j)$$

$$:= \sum_{j=-\infty}^{\infty} \left(2^k \int_{2^{-k}j}^{2^{-k}(j+1)} f(t)dt \right) \varphi(2^k x - j),$$

i.e.,

$$\gamma_{kj}(f) := 2^k \int_{2^{-k}j}^{2^{-k}(j+1)} f(t)dt = 2^k \int_0^{2^{-k}} f\left(t + \frac{j}{2^k}\right) dt.$$

Then ($x \in \mathbf{R}$, $k \in \mathbf{Z}$)

$$|(C_k(f))(x) - f(x)| \le \sum_{i=1}^{N} \frac{|f^{(i)}(x)|}{i!} \frac{(a+1)^i}{2^{ki}} + \frac{(a+1)^N}{2^{kN} N!} \omega_1 \left(f^{(N)}, \frac{a+1}{2^k} \right), \tag{19}$$

which is attained by constants.

Remark 3 i) Given that $f^{(N)}$ is continuous and bounded or uniformly continuous we get that $\omega_1(f^{(N)}, \frac{a+1}{2^k}) < \infty$ and thus $(C_k f)(x) \to f(x)$,

pointwise over $\mathbf{R}$, as $k \to \infty$.
ii) Given also that $f \in C_b^N(\mathbf{R})$ we get that

$$\|C_k f - f\|_\infty \leq \sum_{i=1}^{N} \frac{\|f^{(i)}\|_\infty}{i!} \frac{(a+1)^i}{2^{ki}} + \frac{(a+1)^N}{2^{kN} N!} \omega_1 \left(f^{(N)}, \frac{a+1}{2^k} \right), \quad (20)$$

and hence $C_k f \to f$, uniformly on $\mathbf{R}$, as $k \to \infty$.

iii) ($N = 1$ case). We have that

$$|(C_k f)(x) - f(x)| \leq \left(\frac{a+1}{2^k} \right) \left(|f'(x)| + \omega_1 \left(f', \frac{a+1}{2^k} \right) \right). \quad (21)$$

Proof. (Of Theorem 4.) We have that

$$f\left(t + \frac{j}{2^k}\right) = f(x) + \sum_{i=1}^{N} \frac{f^{(i)}(x)}{i!} \left(t + \frac{j}{2^k} - x \right)^i + \int_x^{t+\frac{j}{2^k}} (f^{(N)}(\tau) - f^{(N)}(x)) \frac{(t + \frac{j}{2^k} - \tau)^{N-1}}{(N-1)!} d\tau.$$

Hence

$$\begin{aligned}\gamma_{kj}(f) &= 2^k \int_0^{2^{-k}} f\left(t + \frac{j}{2^k}\right) dt \\ &= f(x) + \sum_{i=1}^{N} \frac{f^{(i)}(x)}{i!} 2^k \left(\int_0^{2^{-k}} \left(t + \frac{j}{2^k} - x \right)^i dt \right) \\ &\quad + 2^k \int_0^{2^{-k}} \left(\int_x^{t+\frac{j}{2^k}} (f^{(N)}(\tau) - f^{(N)}(x)) \frac{(t + \frac{j}{2^k} - \tau)^{N-1}}{(N-1)!} d\tau \right) dt.\end{aligned}$$

Therefore

$$(C_k(f))(x) - f(x) = \sum_{i=1}^{N} \frac{f^{(i)}(x)}{i!} \sum_{j=-\infty}^{\infty} \varphi(2^k x - j) 2^k \int_0^{2^{-k}} \left(t + \frac{j}{2^k} - x \right)^i dt + \tilde{\mathcal{R}}, \quad (22)$$

where

$$\tilde{\mathcal{R}} := \sum_{j=-\infty}^{\infty} \varphi(2^k x - j) 2^k \int_0^{2^{-k}} \chi(t) dt, \quad (23)$$

where

$$\chi(t) := \int_x^{t+\frac{j}{2^k}} (f^{(N)}(\tau) - f^{(N)}(x)) \frac{(t + \frac{j}{2^k} - \tau)^{N-1}}{(N-1)!} d\tau. \quad (24)$$

Here $0 \le t \le 2^{-k}$, $|x - \frac{j}{2^k}| \le \frac{a}{2^k}$. We would like to estimate $\chi(t)$:

i) If $x \le t + \frac{j}{2^k}$, then

$$|\chi(t)| \le \int_x^{t+\frac{j}{2^k}} \omega_1(f^{(N)}, |\tau - x|) \frac{(t + \frac{j}{2^k} - \tau)^{N-1}}{(N-1)!} d\tau \le \text{(as earlier)}$$

$$\le \omega_1\left(f^{(N)}, \frac{a+1}{2^k}\right) \frac{(a+1)^N}{2^{kN} N!}.$$

ii) If $x \ge t + \frac{j}{2^k}$, then

$$|\chi(t)| \le \int_{t+\frac{j}{2^k}}^{x} \omega_1(f^{(N)}, |\tau - x|) \frac{(\tau - (t + \frac{j}{2^k}))^{N-1}}{(N-1)!} d\tau \le \text{(as earlier)}$$

$$\le \omega_1\left(f^{(N)}, \frac{a+1}{2^k}\right) \frac{(a+1)^N}{2^{kN} N!}.$$

So always it holds that

$$|\chi(t)| \le \omega_1\left(f^{(N)}, \frac{a+1}{2^k}\right) \frac{(a+1)^N}{2^{kN} N!} =: \lambda. \tag{25}$$

Thus (cf. (25))

$$\begin{aligned} |\tilde{\mathcal{R}}| &\le \sum_{j=-\infty}^{\infty} \varphi(2^k x - j) \left(2^k \int_0^{2^{-k}} |\chi(t)| dt\right) \\ &\le \sum_{j=-\infty}^{\infty} \varphi(2^k x - j) 2^k \int_0^{2^{-k}} \lambda\, dt \\ &= \lambda \left(\sum_{j=-\infty}^{\infty} \varphi(2^k x - j)\right) = \lambda 1. \end{aligned}$$

I.e.,

$$|\tilde{\mathcal{R}}| \le \omega_1\left(f^{(N)}, \frac{a+1}{2^k}\right) \frac{(a+1)^N}{2^{kN} N!}. \tag{26}$$

Furthermore notice that

$$\begin{aligned} \left|2^k \int_0^{2^{-k}} \left(t + \frac{j}{2^k} - x\right)^i dt\right| &\le 2^k \int_0^{2^{-k}} \left(|t| + \left|x - \frac{j}{2^k}\right|\right)^i dt \\ &\le 2^k \int_0^{2^{-k}} \left(\frac{1}{2^k} + \frac{a}{2^k}\right)^i dt = \frac{(a+1)^i}{2^{ki}}. \end{aligned} \tag{27}$$

Therefore (cf. (27))

$$\left| \sum_{j=-\infty}^{\infty} \varphi(2^k x - j) 2^k \int_0^{2^{-k}} \left(t + \frac{j}{2^k} - x \right)^i dt \right|$$

$$\leq \left(\sum_{j=-\infty}^{\infty} \varphi(2^k x - j) \right) \left(\frac{(a+1)^i}{2^{ki}} \right) = \frac{(a+1)^i}{2^{ki}}. \tag{28}$$

Finally from (22), (26), and (28) we find that

$$|(C_k f)(x) - f(x)| \leq \sum_{i=1}^{N} \frac{|f^{(i)}(x)|}{i!} \frac{(a+1)^i}{2^{ki}} + \omega_1 \left(f^{(N)}, \frac{a+1}{2^k} \right) \frac{(a+1)^N}{2^{kN} N!},$$

proving (19). □

Also the next quadrature wavelet type operator is studied for the first time.

Theorem 5 *Same assumptions as in Theorem* 1. *Define* $(k, j \in \mathbf{Z},\ x \in \mathbf{R})$

$$(D_k f)(x) := \sum_{j=-\infty}^{\infty} \delta_{kj}(f) \varphi(2^k x - j),$$

where

$$\delta_{kj}(f) := \sum_{r=0}^{n} w_r f \left(\frac{j}{2^k} + \frac{r}{2^k n} \right),$$

$n \in \mathbf{N}$, $w_r \geq 0$, $\sum_{r=0}^{n} w_r = 1$. *Then*

$$|(D_k f)(x) - f(x)| \leq \sum_{i=1}^{N} \frac{|f^{(i)}(x)|}{i!} \frac{(a+1)^i}{2^{ki}} + \frac{(a+1)^N}{2^{kN} N!} \omega_1 \left(f^{(N)}, \frac{(a+1)}{2^k} \right), \tag{29}$$

which is attained by constants.

Remark 4 i) Given that $f^{(N)}$ is continuous and bounded or uniformly continuous we get that $\omega_1(f^{(N)}, \frac{a+1}{2^k}) < \infty$ and thus $D_k f \to f$, pointwise over $\mathbf{R}$, as $k \to \infty$.

ii) Given also that $f \in C_b^N(\mathbf{R})$ we have that

$$\|D_k f - f\|_\infty \leq \sum_{i=1}^{N} \frac{\|f^{(i)}\|_\infty}{i!} \frac{(a+1)^i}{2^{ki}} + \frac{(a+1)^N}{2^{kN} N!} \omega_1 \left(f^{(N)}, \frac{(a+1)}{2^k} \right), \tag{30}$$

and so $D_k f \to f$, uniformly on $\mathbf{R}$, as $k \to \infty$.

iii) ($N = 1$ case). We get that

$$|(D_k f)(x) - f(x)| \leq \frac{(a+1)}{2^k}\left(|f'(x)| + \omega_1\left(f', \frac{a+1}{2^k}\right)\right). \tag{31}$$

Proof. (Of Theorem 5.) Again we have that

$$f\left(\frac{j}{2^k} + \frac{r}{2^k n}\right) = f(x) + \sum_{i=1}^{N} \frac{f^{(i)}(x)}{i!}\left(\frac{j}{2^k} + \frac{r}{2^k n} - x\right)^i$$
$$+ \int_x^{\frac{j}{2^k}+\frac{r}{2^k n}} (f^{(N)}(t) - f^{(N)}(x)) \frac{(\frac{j}{2^k} + \frac{r}{2^k n} - t)^{N-1}}{(N-1)!} dt.$$

Then

$$\delta_{kj}(f) = \sum_{r=0}^{N} w_r f\left(\frac{j}{2^k} + \frac{r}{2^k n}\right)$$
$$= f(x) + \sum_{i=1}^{N} \frac{f^{(i)}(x)}{i!}\left(\sum_{r=0}^{n} w_r \left(\frac{j}{2^k} + \frac{r}{2^k n} - x\right)^i\right) + \sum_{r=0}^{n} w_r \psi_{jr},$$

where

$$\psi_{jr} := \int_x^{\frac{j}{2^k}+\frac{r}{2^k n}} (f^{(N)}(t) - f^{(N)}(x)) \frac{(\frac{j}{2^k} + \frac{r}{2^k n} - t)^{N-1}}{(N-1)!} dt. \tag{32}$$

Therefore

$$(\mathcal{E}_k(f))(x) := (D_k f)(x) - f(x) = \sum_{i=1}^{N} \frac{f^{(i)}(x)}{i!}\left(\sum_{j=-\infty}^{\infty} \varphi(2^k x - j)\right) \cdot$$
$$\cdot \left(\sum_{r=0}^{n} w_r \left(\frac{j}{2^k} + \frac{r}{2^k n} - x\right)^i\right) + \hat{\mathcal{R}}, \tag{33}$$

where

$$\hat{\mathcal{R}} := \sum_{j=-\infty}^{\infty} \varphi(2^k x - j)\left(\sum_{r=0}^{n} w_r \psi_{jr}\right). \tag{34}$$

Using a similar method as in earlier theorems we find that

$$|\psi_{jr}| \leq \omega_1\left(f^{(N)}, \frac{a+1}{2^k}\right) \frac{(a+1)^N}{2^{kN} N!}. \tag{35}$$

And

$$|\hat{\mathcal{R}}| \leq \omega_1\left(f^{(N)}, \frac{a+1}{2^k}\right)\frac{(a+1)^N}{2^{kN}N!}. \tag{36}$$

Finally from (33) and (36) we obtain

$$|(\mathcal{E}_k f)(x)| \leq \sum_{i=1}^{N}\frac{|f^{(i)}(x)|}{i!}\left\{\left(\sum_{j=-\infty}^{\infty}\varphi(2^k x - j)\right)\left(\sum_{r=0}^{n} w_r\left|\frac{j}{2^k}+\frac{r}{2^k n}-x\right|^i\right)\right\}$$

$$+ |\hat{\mathcal{R}}| \leq \sum_{i=1}^{N}\frac{|f^{(i)}(x)|}{i!}\left\{\left(\sum_{j=-\infty}^{\infty}\varphi(2^k x - j)\right)\left(\sum_{r=0}^{n} w_r\right)\frac{(a+1)^i}{2^{ki}}\right\}$$

$$+ |\hat{\mathcal{R}}| \leq \sum_{i=1}^{N}\frac{|f^{(i)}(x)|}{i!}\frac{(a+1)^i}{2^{ki}} + \omega_1\left(f^{(N)}, \frac{a+1}{2^k}\right)\frac{(a+1)^N}{2^{kN}N!}$$

proving (29). □

Related results are presented in

Proposition 6 *Here φ is as in Theorem 1, $f \in C(\mathbf{R})$, $x \in \mathbf{R}$, $k \in \mathbf{Z}$. Operators C_k, D_k are as in Theorem 4, 5 respectively. Then*

$$|(C_k(f))(x) - f(x)| \leq \omega_1\left(f, \frac{a+1}{2^k}\right), \tag{37}$$

and

$$|(D_k(f))(x) - f(x)| \leq \omega_1\left(f, \frac{a+1}{2^k}\right). \tag{38}$$

Proof. Notice that

$$(C_k(f))(x) - f(x) = \sum_{j=-\infty}^{\infty}\left(2^k\int_0^{2^{-k}}\left(f\left(t+\frac{j}{2^k}\right) - f(x)\right)dt\right)\varphi(2^k x - j).$$

So that

$$\begin{aligned} &|(C_k(f))(x) - f(x)| \\ &\quad \leq \sum_{j=-\infty}^{\infty}\left(2^k\int_0^{2^{-k}}\left|f\left(t+\frac{j}{2^k}\right) - f(x)\right|dt\right)\varphi(2^k x - j) \\ &\quad \leq \sum_{j=-\infty}^{\infty}\left(2^k\int_0^{2^{-k}}\omega_1\left(f, \left|x - t - \frac{j}{2^k}\right|\right)dt\right)\varphi(2^k x - j) \end{aligned}$$

$$\leq \sum_{j=-\infty}^{\infty} \left(2^k \int_0^{2^{-k}} \omega_1 \left(f, |t| + \left| x - \frac{j}{2^k} \right| \right) dt \right) \varphi(2^k x - j)$$

$$\leq \sum_{j=-\infty}^{\infty} \left(2^k \int_0^{2^{-k}} \omega_1 \left(f, \frac{1+a}{2^k} \right) dt \right) \varphi(2^k - j) = \omega_1 \left(f, \frac{1+a}{2^k} \right),$$

which proves (37).

Next we observe that

$$(D_k(f))(x) - f(x) = \sum_{j=-\infty}^{\infty} \left\{ \sum_{r=0}^{n} w_r \left(f \left(\frac{j}{2^k} + \frac{r}{2^k n} \right) - f(x) \right) \right\} \varphi(2^k x - j).$$

Hence

$$|(D_k f)(x) - f(x)| \leq \sum_{j=-\infty}^{\infty} \left(\sum_{r=0}^{n} w_r \left| f \left(\frac{j}{2^k} + \frac{r}{2^k n} \right) - f(x) \right| \right) \varphi(2^k x - j)$$

$$\leq \sum_{j=-\infty}^{\infty} \left(\sum_{r=0}^{n} w_r \omega_1 \left(f, \left| \frac{j}{2^k} + \frac{r}{2^k n} - x \right| \right) \right) \varphi(2^k x - j)$$

$$\leq \sum_{j=-\infty}^{\infty} \left(\sum_{r=0}^{n} w_r \omega_1 \left(f, \frac{a+1}{2^k} \right) \right) \varphi(2^k x - j) = \omega_1 \left(f, \frac{a+1}{2^k} \right),$$

proving (38). □

Observe that

$$(C_k(f))(x) = C_0(f(2^{-k}\cdot))(2^k x),$$

and (39)

$$(D_k f)(x) = D_0(f(2^{-k}\cdot))(2^k x).$$

Optimality is established in

Proposition 7 *Inequalities* (37), (38) *are sharp over convex* $f \in C(\mathbf{R})$.

Proof. Enough to prove it for $k = 0$, for any other $k \neq 0$ it follows similarly.

i) We would like to prove that

$$|(C_0(f))(x) - f(x)| \leq \omega_1(f, a+1), \tag{40}$$

is sharp over convex $f \in C(\mathbf{R})$. Assume that there exists $0 < c < 1$ such that

$$|(C_0(f))(x) - f(x)| \leq c\omega_1(f, a+1). \tag{41}$$

Pick $f(x) := g(x) := (x-1)_+$ which is a convex continuous function on $\mathbf{R}$. Here $\omega_1(g,h) = h$, any $h > 0$. Choose ($\lceil\cdot\rceil$ is the ceiling of number)

$$\beta := \left\lceil \frac{|2c - \frac{1}{2}|}{1-c} \right\rceil + 1 \left(> \frac{2c - \frac{1}{2}}{1-c} \right).$$

Therefore

$$\beta + \frac{1}{2} > c(\beta + 2). \tag{42}$$

Define

$$\varphi_1(x) := \begin{cases} 1+x, & -1 \le x \le 0, \\ 1-x, & 0 \le x \le 1, \\ 0, & \text{elsewhere}, \end{cases}$$

and

$$\varphi_3(x) := \varphi_1(x+\beta) := \begin{cases} x+\beta+1, & -\beta-1 \le x \le -\beta, \\ 1-x-\beta, & -\beta \le x \le 1-\beta, \\ 0, & \text{elsewhere}, \end{cases}$$

which is continuous and bounded on $\mathbf{R}$. Observe that φ_3 fulfills

1) $\sum_{j=-\infty}^{+\infty} \varphi_3(x-j) = 1$ on $\mathbf{R}$,

2) $\sum_{j=-\infty}^{\infty} j\varphi_3(x-j) = x+\beta$ on $\mathbf{R}$,

3) φ_3 is convex over $(-\infty, -1-\beta]$, and $[1-\beta, +\infty)$, also φ_3 is concave over $[-1-\beta, 1-\beta]$,

4) φ_3 is nondecreasing for $x \le -\beta$ and nonincreasing for $x \ge -\beta$, i.e., $\varphi_3 \ge 0$.

I.e., φ_3 has all the properties of φ in Theorem 1. Here

$$\varphi_3(x) \neq 0 \quad \text{iff} \quad -1-\beta < x < 1-\beta.$$

Thus for $j \in \mathbf{Z}$ we get that

$$\varphi_3(1-j) \neq 0 \quad \text{iff} \quad \beta < j < \beta + 2.$$

Hence

$$\varphi_3(1-(\beta+1)) = \varphi_3(-\beta) = \varphi_1(0) = 1.$$

I.e., $\varphi_3(1-j) = 0$, all $j \in \mathbf{Z} - \{\beta+1\}$. And

$$\ell_{0,\beta+1}(g) := \int_0^1 g(t+\beta+1)dt = \int_0^1 (t+\beta)dt = \frac{1}{2} + \beta.$$

So that

$$\ell_{0,\beta+1}(g) = \beta + \frac{1}{2}. \tag{43}$$

Here $\ell_{0,j}(g) := \gamma_{0,j}(g)$, all $j \in \mathbf{Z}$. Thus

$$C_0(g)(1) = \sum_{j=-\infty}^{\infty} \ell_{0,j}(g)\varphi_3(1-j) \overset{(43)}{=} \ell_{0,\beta+1}(g)\varphi_3(1-(\beta+1)) = \beta + \frac{1}{2}.$$

Also $g(1) = 0$. Therefore by assumption (41) we get that (here $a = \beta + 1$, by supp $\varphi_3 \subseteq [-\beta - 1, \beta + 1]$)

$$\begin{aligned} \beta + \frac{1}{2} = |(C_0(g))(1) - g(1)| &\le c\omega_1(g, a+1) \\ &= c(a+1) \\ &= c(\beta+2). \end{aligned}$$

I.e.,

$$\left(\beta + \frac{1}{2}\right) \le c(\beta + 2),$$

which contradicts (42).

ii) We would like to prove that

$$|(D_0(f))(x) - f(x)| \le \omega_1(f, a+1) \tag{44}$$

is sharp over convex $f \in C(\mathbf{R})$. The proof is similar to (i).

Let $0 < c < 1$ be such that

$$|(D_0 f)(x) - f(x)| \le c\omega_1(f, a+1). \tag{45}$$

Here pick

$$0 < \gamma := \frac{\sum_{r=0}^{n} r w_r}{n} \le 1$$

and

$$\beta := \left\lceil \frac{|2c-\gamma|}{1-c} \right\rceil + 1 > \frac{2c-\gamma}{1-c}.$$

Thus

$$\beta + \gamma > c(\beta + 2). \tag{46}$$

Notice that (g as in (i))

$$\delta_{0,\beta+1}(g) = \sum_{r=0}^{n} w_r g\left(\beta + 1 + \frac{r}{n}\right) = \sum_{r=0}^{n} w_r \left(\beta + \frac{r}{n}\right) = \beta + \gamma.$$

I.e.,

$$\delta_{0,\beta+1}(g) = \beta + \gamma. \tag{47}$$

Hence (cf. (47))

$$(D_0(g))(1) = \sum_{j=-\infty}^{\infty} \delta_{0,j}(g)\varphi_3(1-j) = \delta_{0,\beta+1}(g)\varphi_3(1-(\beta+1))$$

$$= (\beta+\gamma)\varphi_3(-\beta) = \beta+\gamma.$$

I.e.,

$$(D_0(g))(1) = \beta+\gamma.$$

Finally (cf. (45))

$$\beta+\gamma = |(D_0(g))(1) - g(1)| \leq c(\beta+2),$$

contradicting (46). □

References

[1] G.A. Anastassiou and X.M. Yu, Monotone and Probabilistic Wavelet Approximation, *Stochastic Analysis and Applications*, Vol. 10, No. 3, pp. 251–264, 1992.

[2] G.A. Anastassiou and X.M. Yu, Convex and Coconvex—Probabilistic Wavelet Approximation, *Stochastic Analysis and Applications*, Vol. 10, No. 5, pp. 507–521, 1992.

Modified Weighted (0,2) Interpolation

J. Balázs[1]
Department of Numerical Analysis, Eötvös Loránd University,
H-1088 Budapest, Múzeum krt. 6–8, Hungary

Dedicated to the memory of Professor A.K. Varma

Abstract

In this paper we define the modified weighted $(0, 2)$-interpolation. We prove existence, uniqueness theorems and give the explicit formulae for this interpolation.

1 Introduction

By weighted $(0, 2)$ interpolation we mean the solution of the following problem:

Let the system of knots

$$(1.1) \qquad -\infty \le a < x_{n,n} < x_{n-1,n} < \ldots < x_{1,n} < b \le +\infty, \quad (n \in N)$$

be given in the finite or infinite open interval (a, b) and let $w(x) \in C^{(2)}(a, b)$ be a weight function. Find the polynomial $R_n(x)$ of minimal degree satisfying the equalities

$$(1.2) \quad R_n(x_{k,n}) = y_{k,n}; \quad (wR_n)''(x_{k,n}) = y''_{k,n}, \quad (k = 1, 2, \ldots, n; n \in N)$$

where $y_{k,n}, y''_{k,n}$ are arbitrary given real numbers.

The study of this problem was initiated by P.Turán. The first result has been published in [2]. In this paper the author considered the case where the knots are the zeros of the ultraspherical polynomials $P_n^{(\alpha,\alpha)}$ $(\alpha > -1)$ (see Szegő [11]) and the weight function is

$$(1.3) \qquad w(x) = (1 - x^2)^{\frac{1+\alpha}{2}}, \quad (x \in [-1, 1]).$$

[1]Supported by National Scientific Research Funds (OTKA), Grant no. T014244

In [2] it is proved that if $n = 2k$ is even and in addition to the equations (1.2) the equation

$$R_n(0) = \sum_{k=1}^{n} y_{k,n} l_{k,n}^2(0) \tag{1.4}$$

is also satisfied, then the polynomials $R_n(x)$ of minimal degree $2n$ exist and can be determined uniquely. Further, if $n = 2k + 1$ is odd, then the polynomials $R_n(x)$ cannot be determined uniquely. In (1.4) $l_{k,n}$ denotes the basic polynomial of Lagrange interpolation, if the knots (1.1) are the zeros of the polynomials $P_n^{(\alpha,\alpha)}$. The paper [2] presents the explicit expression of the polynomials R_n, further some convergence theorems. If the equation (1.4) does not hold, then the polynomials R_n of minimal degree $2n - 1$ fail to exist.

In the paper [7] J.Prasad proved the results of [2] for the case, if the knots (1.1) are the zeros of the Jacobi-polynomials $P_n^{(\alpha,-\alpha)}$ $(0 < |\alpha| \leq \frac{1}{2})$ (see Szegő [11]). In the case if $(a, b) = (-\infty, +\infty)$ and the knots are the zeros of the Hermite polynomials J. Prasad [6] first proved the results of [2]. L.Szili [12] first proved a convergence theorem. In [4] I. Joó sharpened this convergence theorem. The weight function in this case is $w(x) = \exp(-\frac{x^2}{2})$ $(x \in \mathbf{R})$.

In [5] I.Joó and L.Szili solved the problem in the case, if the knots (1.1) are the zeros of the Jacobi-polynomials $P_n^{(\alpha,\beta)}$ $(\alpha, \beta > -1)$ and $P_n^{(\alpha,\beta)}(0) \neq 0$. (For the Jacobi-polynomials see the book of Szegő [11].) Here the weight function is

$$w(x) = (1-x)^{\frac{\alpha+1}{2}}(1+x)^{\frac{\beta+1}{2}}, \quad (x \in [-1,1]; \alpha, \beta > -1).$$

They proved that in this case there exists a unique polynomial R_n of degree $2n$ satisfying the equations (1.2) and (1.4). Here in equation (1.4) $l_{k,n}$ denotes the basic polynomial of Lagrange-interpolation on the zeros of the polynomials $P_n^{(\alpha,\beta)}$. Further they proved that under the given conditions the polynomial R_n of degree at most $2n - 1$ satisfying the equations (1.2) only (**not** satisfying the equations (1.4)) fails to exist. Moreover the authors determined the explicit form of the polynomials R_n. They also proved the following convergence theorem.

Theorem. *(I.Joó - L.Szili) If $f : [-1, 1] \to \mathbf{R}$ is continuously differentiable and*

$$y_{k,n} = f(x_{k,n}); \;\; y''_{k,n} = \mathcal{O}(1)(1-x_{k,n})^{\frac{\alpha-1}{2}}(1+x_{k,n})^{\frac{\beta-1}{2}} n\omega\Big(f'; \frac{1}{2n}\Big),$$

then the weighted $(0,2)$ *interpolational polynomials* $R_{n_{\alpha,\beta}}(f;x)$ *satisfy the following relation*

$$(1.5)\qquad |f(x)-R_{n_{\alpha,\beta}}(f;x)| = \mathcal{O}(1)\Big(\omega\Big(f';\frac{1}{2n_{\alpha,\beta}}\Big)+\frac{\log n_{\alpha,\beta}}{n_{\alpha,\beta}^{2(\gamma+1)}}\Big),$$

$$(x\in[-1+\epsilon,1-\epsilon],\epsilon\in(0,1)\ \text{is fixed})$$

where $\omega(f';\delta)$ *is the modulus of continuity of* f' *and* $O(1)$ *is independent of* $n_{\alpha,\beta}$ *and* x.

In (1.5) $n_{\alpha,\beta}$ is an odd number if $\alpha-\beta=4l+2$ ($l\in\mathbf{Z}$) and it is an even number if $\alpha-\beta=4l$ ($l\in\mathbf{Z}$); otherwise it is an arbitrary natural number and

$$\gamma=\begin{cases}\min(\alpha,\beta) & \text{if}\ \ \min(\alpha,\beta)<-\frac{1}{2}\\ -\frac{1}{2} & \text{if}\ \ \min(\alpha,\beta)\geq-\frac{1}{2}\end{cases}$$

We note that if $\alpha=\beta$, then in the case $n=2k+1$ it follows $P_n^{(\alpha,\alpha)}(0)=0$ and the polynomials R_n cannot be determined uniquely. Further, if $\alpha=\beta$ and $n=2k$, then there always exists a unique polynomial R_n, as in this case $P_n^{(\alpha,\alpha)}(0)\neq 0$.

2 The definition of the modified weighted $(0,2)$ interpolation, results

By modified weighted $(0,2)$ interpolation we mean the solution of the following problem:

Let the system of knots

$$(2.1)\qquad -\infty\leq a<x_{n,n}<x_{n-1,n}<\ldots<x_{1,n}<b\leq+\infty,\ \ (n\in\mathbf{N})$$

be given in the finite or infinite open interval (a,b) and let $w(x)\in C^{(2)}(a,b)$ be a weight function. Find the polynomial $R_n(x)$ of minimal degree at most $2n-1$ satisfying the equalities

$$(2.2)\qquad R_n(x_{k,n})=y_{k,n};\ \ (k=1,2,\ldots,n;n\in\mathbf{N}),$$

$$(2.3)\qquad R_n'(x_{n,n})=y_{n,n}';\ \ (n\in\mathbf{N}),$$

$$(2.4)\qquad (wR_n)''(x_{k,n})=y_{k,n}'',\ \ (k=1,2,\ldots,n-1;\ n-1\in\mathbf{N})$$

where $y_{k,n},y_{n,n}',y_{k,n}''$ are arbitrary given real numbers.

Let $\omega_{n-1}(x)$ denote the polynomial of degree $n-1$ having the zeros $x_{k,n},\ k=1,2,\ldots,n-1$ in (2.1); that is,

$$\omega_{n-1}(x) = c\prod_{k=1}^{n-1}(x - x_{k,n}).$$

If there exists a weight function $w(x) \in C^{(2)}(a,b)$ satisfying

$$\{w\omega_{n-1}\}''_{x_{k,n}} = 0 \ \ (k = 1,2,\ldots,n-1; n-1 \in \mathbf{N}), \tag{2.5}$$

and

$$w(x_{k,n}) \neq 0, \ \ (k = 1,2,\ldots,n-1; n-1 \in \mathbf{N}), \tag{2.6}$$

then the following theorem holds.

Theorem 1 *If there exists a weight function* $w(x) \in C^{(2)}(a,b)$ *satisfying the conditions* (2.5) *and* (2.6) *then there exist the modified weighted* $(0,2)$ *interpolational polynomials* R_n *of degree at most* $2n-1$ *corresponding to the system of knots* (2.1) *and satisfying the equations* (2.2), (2.3) *and* (2.4). *Further they can be determined uniquely (existence, uniqueness).*

We note, that if the zeros of the polynomial $\omega_{n-1}(x)$ are the zeros of the classical orthogonal polynomials, then the weight functions $w(x) \in C^{(2)}(a,b)$ satisfying the conditions (2.5) and (2.6) do always exist. We will show this in the sequel.

Theorem 1 implies that also in the case $P_n^{(\alpha,\beta)}(0) = 0$ the modified weighted $(0,2)$ interpolational polynomials R_n of degree at most $2n-1$ do exist and the condition (1.4) is not necessary. This will be clear from the proof of Theorem 1.

3 Proof of Theorem 1

In what follows we assume that for the basic system of knots (2.1) there exists a twice continuously differentiable weight function $w(x) \in C^{(2)}(a,b)$ satisfying the conditions (2.5) and (2.6). Further the polynomial $\omega_{n-1}(x)$ of degree $n-1$ has the form

$$\omega_{n-1}(x) = c\prod_{k=1}^{n-1}(x - x_{k,n}), \ \ (c \neq 0 \ \text{constant}) \tag{3.1}$$

Let $A_{k,n}(x)$ $(k = 1,2,\ldots,n)$, $\overline{A}_{n,n}(x)$ and $B_{k,n}(x)$ $(k = 1,2,\ldots,n-1)$ denote the polynomials of degree $2n-1$ satisfying the equations

$$A_{k,n}(x_{j,n}) = \delta_{j,k}; A'_{k,n}(x_{n,n}) = 0, (k,j = 1,2,\ldots,n; n \in \mathbf{N}) \tag{3.2}$$

(3.3) $$(wA_{k,n})''(x_{j,n}) = 0, (k = 1, 2, \ldots, n; j = 1, 2, \ldots, n-1; n \in \mathbf{N})$$

(3.4) $$\overline{A}_{n,n}(x_{j,n}) = 0; \overline{A}'_{n,n}(x_{n,n}) = 1, (j = 1, 2, \ldots, n; n \in \mathbf{N}),$$

(3.5) $$(w\overline{A}_{n,n})''(x_{j,n}) = 0, (j = 1, 2, \ldots, n-1; n \in \mathbf{N})$$

(3.6) $$B_{k,n}(x_{j,n}) = 0, (k = 1, 2, \ldots, n-1; j = 1, 2, \ldots, n; n \in \mathbf{N})$$

(3.7) $$B'_{k,n}(x_{n,n}) = 0, (k = 1, 2, \ldots, n-1),$$

(3.8) $$(wB_{k,n})''(x_{j,n}) = \delta_{j,k}, (j, k = 1, 2, \ldots, n-1; n \in \mathbf{N}).$$

Here
$$\delta_{j,k} = \begin{cases} 0 & j \neq k \\ 1 & j = k. \end{cases}$$

In the sequel $l_{k,n}$ $(k = 1, 2, \ldots, n-1; n \in \mathbf{N})$ denotes the basic Lagrange-interpolational polynomial of degree $n-2$, which satisfies $l_{k,n}(x_{j,n}) = \delta_{j,k}$, $(j, k = 1, 2, \ldots, n-1; n \in \mathbf{N})$, that is

(3.9) $$l_{k,n}(x) = \frac{\omega_{n-1}(x)}{\omega'_{n-1}(x_{k,n})(x - x_{k,n})}, (k = 1, 2, \ldots, n-1; n \in \mathbf{N})$$

We prove in the following lemmata the existence of the polynomials $A_{k,n}$, $\overline{A}_{n,n}$ and $B_{k,n}$ of degree $2n-1$ and we give their explicit forms.

Lemma 3.1. *If for the basic system of knots* (2.1) *there exists a weight function* $w(x) \in C^{(2)}(a,b)$ *satisfying* (2.5) *and* (2.6) *then the polynomials* $A_{k,n}$ *satisfying* (3.2) *and* (3.3) *have the form: for* $k = 1, 2, \ldots, n-1$ *we have*

$$A_{k,n}(x) = \frac{x - x_{n,n}}{x_{k,n} - x_{n,n}} l^2_{k,n}(x) +$$

(3.10) $$+\omega_{n-1}(x)\Big[a_k \int_{x_{n,n}}^{x} (t - x_{n,n}) \frac{l'_{k,n}(t) - l'_{k,n}(x_{k,n}) l_{k,n}(t)}{t - x_{k,n}} dt +$$
$$+b_k \int_{x_{n,n}}^{x} l_{k,n}(t)dt + c_k \int_{x_{n,n}}^{x} \omega_{n-1}(t)dt\Big],$$

where

(3.11.1) $$a_k = -\frac{1}{(x_{k,n} - x_{n,n})\omega'_{n-1}(x_{k,n})},$$

$$b_k = -\frac{1}{2w(x_{k,n})\omega'_{n-1}(x_{k,n})}\Big\{w''(x_{k,n})+2w'(x_{k,n})\Big[\frac{1}{x_{k,n}-x_{n,n}}+2l'_{k,n}(x_{k,n})\Big]+$$

$$(3.11.2) \qquad +\frac{4w(x_{k,n})}{x_{k,n}-x_{n,n}}l'_{k,n}(x_{k,n}) + 4w(x_{k,n})l'_{k,n}(x_{k,n})^2\Big\},$$

(3.11.3)
$$c_k = -\frac{1}{\omega_{n-1}(x_{n,n})^2}\Big[\frac{1}{x_{k,n}-x_{n,n}}l^2_{k,n}(x_{n,n}) + b_k\omega_{n-1}(x_{n,n})l_{k,n}(x_{n,n})\Big];$$

and if $k = n$ then

$$(3.12) \qquad A_{n,n}(x) = \frac{\omega_{n-1}(x)}{\omega_{n-1}(x_{n,n})} + a_n\omega_{n-1}(x)\int_{x_{n,n}}^{x} \omega_{n-1}(t)dt,$$

where

$$(3.13) \qquad a_n = -\frac{\omega'_{n-1}(x_{n,n})}{\omega^3_{n-1}(x_{n,n})}.$$

Proof. As by (3.9) we have $l_{k,n}(x_{k,n}) = 1$ and by (2.1) we have $x_{k,n} > x_{n,n}$, using (3.9) and (3.1) it follows that indeed, in (3.10) $A_{k,n}$ is a polynomial of degree $2n-1$. If $x = x_{n,n}$ then obviously $A_{k,n}(x_{n,n}) = 0$ and by $l_{k,n}(x_{j,n}) = \delta_{j,k}$ and $\omega_{n-1}(x_{k,n}) = 0$ $(k = 1,2,\dots,n-1)$ we have $A_{k,n}(x_{j,n}) = \delta_{j,k}$ $(k = 1,2,\dots,n-1; j = 1,2,\dots,n)$.

From (3.10) by differentiation we have for $x = x_{n,n}$ and by (3.11.3)

$$(3.14) \qquad A'_{k,n}(x_{n,n}) = \frac{1}{x_{k,n}-x_{n,n}}l^2_{k,n}(x_{n,n})+$$

$$+\omega_{n-1}(x_{n,n})\Big[b_k l_{k,n}(x_{n,n}) + c_k\omega_{n-1}(x_{n,n})\Big] = 0.$$

If the weight function $w(x) \in C^{(2)}(a,b)$ satisfies (2.5) and (2.6), further as (3.1) implies $\omega_{n-1}(x_{j,n}) = 0$ and (3.9) implies $l_{k,n}(x_{j,n}) = 0$, $j \neq k$, moreover if $j \neq k$ we have

$$(3.14.1) \qquad l'_{k,n}(x_{j,n}) = \frac{\omega'_{n-1}(x_{j,n})}{\omega'_{n-1}(x_{k,n})(x_{j,n}-x_{k,n})}.$$

Then for $j \neq k$ by (3.11.1), (3.14.1) we have

$$\Big(wA_{k,n}\Big)''(x_{j,n}) = w(x_{j,n})\frac{x_{j,n}-x_{n,n}}{x_{k,n}-x_{n,n}}\cdot 2[l'_{k,n}(x_{j,n})]^2 + 2w(x_{j,n})\omega'_n(x_{j,n})\times$$

(3.19) $$\times\left[a_k(x_{j,n}-x_{n,n})\frac{l'_{k,n}(x_{j,n})}{(x_{j,n}-x_{k,n})}\right]=$$

$$=2w(x_{j,n})\frac{x_{j,n}-x_{n,n}}{x_{k,n}-x_{n,n}}\left[l'_{k,n}(x_{j,n})^2-l'_{k,n}(x_{j,n})\frac{\omega'_{n-1}(x_{j,n})}{\omega'_{n-1}(x_{k,n})(x_{j,n}-x_{k,n})}\right]=0.$$

In the case $j=k$ we have

$$l_{k,n}(x_{k,n})=1,\quad \omega_{n-1}(x_{k,n})=0 \text{ and } (w\omega_{n-1})''(x_{k,n})=0.$$

Moreover

$$\lim_{x\to x_{k,n}}\frac{l'_{k,n}(x)-l'_{k,n}(x_{k,n})l_{k,n}(x)}{x-x_{k,n}}=l''_{k,n}(x_{k,n})-l'_{k,n}(x_{k,n})^2.$$

Then we have

$$\Big(wA_{k,n}\Big)''(x_{k,n})=w''(x_{k,n})+2w'(x_{k,n})\Big[\frac{1}{x_{k,n}-x_{n,n}}+2l'_{k,n}(x_{k,n})\Big]+$$

(3.20) $$+w(x_{k,n})\Big[4\frac{1}{x_{k,n}-x_{n,n}}l'_{k,n}(x_{k,n})+2l'_{k,n}(x_{k,n})^2+2l''_{k,n}(x_{k,n})\Big]+$$

$$+2w(x_{k,n})\omega'_{n-1}(x_{k,n})\Big\{a_k(x_{k,n}-x_{n,n})\Big[l''_{k,n}(x_{k,n})-l'_{k,n}(x_{k,n})^2\Big]+b_k\Big\}=0.$$

This equality holds if we replace a_k with the expression (3.11.1) and b_k with the expression (3.11.2).

Hence for the polynomials $A_{k,n}$ $(k=1,2,\ldots,n-1)$ in (3.10) we have $A_{k,n}(x_{j,n})=\delta_{j,k}$, and by (3.14) it follows $A'_{k,n}(x_{n,n})=0$, that is, (3.2) holds. Further, by (3.19) and (3.20) also the equations (3.3) are valid.

If $k=n$, then from (3.12) we have $A_{n,n}(x_{n,n})=1$ and $\omega_{n-1}(x_{j,n})=0$ implies $A_{n,n}(x_{j,n})=0$ for $j=1,2,\ldots,n-1$.

Differentiating the polynomial (3.12) and putting $x=x_{n,n}$ one gets

(3.21) $$A'_{n,n}(x_{n,n})=\frac{\omega'_{n-1}(x_{n,n})}{\omega_{n-1}(x_{n,n})}+a_n\omega^2_{n-1}(x_{n,n})=0,$$

where a_n has the form (3.13).

By the conditions $(w\omega_{n-1})''(x_{j,n})=0$ and $\omega_{n-1}(x_{j,n})=0$ for $j=1,2,\ldots,n-1$, hence

(3.22) $$(wA_{n,n})''(x_{j,n})=2w(x_{j,n})\omega'_{n-1}(x_{j,n})\omega_{n-1}(x_{j,n})=0.$$

This means that the polynomial $A_{n,n}$ in (3.12) satisfies the equations (3.2), (3.3).

Hence Lemma 3.1 is proved.

Lemma 3.2. *Under the conditions of Lemma 3.1 the polynomial of degree* $2n-1$ *satisfying the conditions* (3.4) *and* (3.5) *has the form*

$$\overline{A}_{n,n}(x)=\frac{\omega_{n-1}(x)}{\omega_{n-1}(x_{n,n})^2}\int_{x_{n,n}}^{x}\omega_{n-1}(t)dt. \tag{3.23}$$

Proof. As $\omega_{n-1}(x_{j,n})=0$ $(j=1,2,\ldots,n-1)$, hence $\overline{A}_{n,n}(x_{k,n})=0$ $(k=1,2,\ldots,n)$.

Differentiating the polynomial $A_{n,n}$ in (3.23) and putting $x=x_{n,n}$ we have indeed $\overline{A}'_{n,n}(x_{n,n})=1$.

As $\omega_{n-1}(x_{j,n})=0$, by the condition $(w\omega_{n-1})''(x_{j,n})=0$ $(j=1,2,\ldots,n-1)$, thus we have $(w\overline{A}_{n,n})''(x_{j,n})=0$ for $j=1,2,\ldots,n-1$.

This shows that the polynomial $\overline{A}_{n,n}$ satisfies the equations (3.4), (3.5).

Lemma 3.3. *Under the conditions of Lemma 3.1 the polynomials of degree* $2n-1$ *satisfying the equations* (3.6), (3.7) *and* (3.8) *have the form*

$$B_{k,n}(x)=\omega_{n-1}(x)\Big[\alpha_k\int_{x_{n,n}}^{x}l_{k,n}(t)dt+\beta_k\int_{x_{n,n}}^{x}\omega_{n-1}(t)dt\Big], \tag{3.24}$$

where

$$\alpha_k=\frac{1}{2w(x_{k,n})\omega'_{n-1}(x_{k,n})},$$

$$\beta_k=-\frac{1}{2w(x_{k,n})\omega'_{n-1}(x_{k,n})}\cdot\frac{l_{k,n}(x_{n,n})}{\omega_{n-1}(x_{n,n})}. \tag{3.25}$$

Proof. As $\omega_{n-1}(x_{k,n})=0$ for $k=1,2,\ldots,n-1$, hence we have obviously $B_{k,n}(x_{j,n})=0$, $(k=1,2,\ldots,n-1; j=1,2,\ldots,n)$.

Differentiating (3.24) and putting $x=x_{n,n}$ we have by (3.25)

$$B'_{k,n}(x_{n,n})=\omega_{n-1}(x_{n,n})\Big[\alpha_k l_{k,n}(x_{n,n})+\beta_k\omega_{n-1}(x_{n,n})\Big]=0.$$

By (3.9) we have

$$l_{k,n}(x_{j,n})=\delta_{j,k}\qquad (j,k=1,2,\ldots,n-1)$$

and $(w\omega_{n-1})''(x_{j,n})=0$, further $\omega_{n-1}(x_{j,n})=0$ $(j,k=1,2,\ldots,n-1)$. Hence it follows

$$(wB_{k,n})''(x_{j,n})=\delta_{j,k}.$$

This means that the polynomials $B_{k,n}$ satisfy the equations (3.6), (3.7) and (3.8).

As the polynomials $A_{k,n}$, $\overline{A}_{n,n}$ and $B_{k,n}$ of degree $2n-1$ are basic interpolational polynomials, hence the modified weighted $(0,2)$ interpolational polynomial $R_n(x)$ of degree at most $2n-1$ in Theorem I. can be written in the form

$$(3.26)\quad R_n(x) = \sum_{k=1}^{n} y_{k,n}A_{k,n}(x) + y'_{n,n}\overline{A}_{n,n}(x) + \sum_{k=1}^{n-1} y''_k B_{k,n}(x), \quad (n \in \mathbf{N}).$$

Indeed, by (3.10), (3.12), (3.23) and (3.24) the polynomial $R_n(x)$ satisfies the conditions (2.2), (2.3) and (2.4), where $y_{k,n}$, $y'_{n,n}$ and y''_k are arbitrary real numbers. Hence the existence has been proved. Now we prove the uniqueness.

Suppose that there exists another polynomial $R^*_n(x)$ of degree at most $2n-1$ satisfying the equations (2.2), (2.3) and (2.4), then the polynomial $Q_n(x) = R_n(x) - R^*_n(x)$ is of degree at most $2n-1$ too and then by $Q_n(x_{k,n}) = 0$, $k = 1, 2, \ldots, n$ it follows

$$(3.27)\qquad Q_n(x) = (x - x_{n,n})\omega_{n-1}(x)g_{n-1}(x),$$

where g_{n-1} is a polynomial of degree at most $n-1$. On the other hand, by $Q'_n(x_{n,n}) = 0$ and $\omega_{n-1}(x_{n,n}) \neq 0$ we have necessarily $g_{n-1}(x_{n,n}) = 0$. From the condition

$$(wQ_n)''(x_{k,n}) = 0, \quad (k = 1, 2, \ldots, n-1).$$

Hence, by (3.27) and by the equations

$$(w\omega_{n-1}(x))''(x_{k,n}) = 0, \qquad \omega_{n-1}(x_{k,n}) = 0$$

we obtain

$$(wQ_n)''(x_{k,n}) =$$

$$= 2w(x_{k,n})\omega'_{n-1}(x_{k,n})\Big[(x - x_{n,n})g_{n-1}(x)\Big]'_{x=x_{k,n}} = 0, \quad (k = 1, 2, \ldots, n-1).$$

As $w(x_{k,n}) \neq 0$, $\omega'_{n-1}(x_{k,n}) \neq 0$, hence

$$\Big[(x - x_{n,n})g_{n-1}(x)\Big]'_{x=x_{k,n}} = 0, \quad (k = 1, 2, \ldots, n-1)$$

is satisfied. That is, we have

$$\Big[(x - x_{n,n})g_{n-1}(x)\Big]' = (x - x_{n,n})g'_{n-1}(x) + g_{n-1}(x) = c\omega_{n-1}(x).$$

This implies by the conditions $\omega_{n-1}(x_{n,n}) \neq 0$, $g_{n-1}(x_{n,n}) = 0$ that $c = 0$, that is $\Big[(x - x_{n,n})g_{n-1}(x)\Big] = k$ is a constant. Then by (3.27) $Q_n(x) =$

$k\omega_{n-1}(x)$. On the other hand, as $Q'_n(x_{n,n}) = 0$, $\omega_{n-1}(x_{n,n}) \neq 0$, we necessarily have $k = 0$, that is $Q_n(x) = 0$. This means that the existence and uniqueness of the polynomial R_n in (3.26) has been verified and the proof of Theorem 1 is complete.

Corollary 3.1. *If $V_m(x)$ is an arbitrary polynomial of degree m with $m \leq 2n-1$, then the following identity holds*

$$V_m(x) = \sum_{k=1}^{n} V_m(x_{k,n})A_{k,n}(x) +$$

$$+V'_m(x_{n,n})\overline{A}_{n,n}(x) + \sum_{k=1}^{n-1}(wV_m)''(x_{k,n})B_{k,n}(x). \tag{3.28}$$

Proof. If we let

$$Q_n(x) = V_m(x) - \Big\{\sum_{k=1}^{n} V_m(x_{k,n})A_{k,n}(x) + V'_m(x_{n,n})\overline{A}_{n,n}(x) +$$

$$+\sum_{k=1}^{n-1}(wV_m)''(x_{k,n})B_{k,n}(x)\Big\},$$

then $Q_n(x)$ is a polynomial of degree at most $2n-1$. By Lemma 3.1, Lemma 3.2, Lemma 3.3 and Theorem 1 we obviously have that $Q_n(x) \equiv 0$.

Now we show that if the zeros of the polynomial $\omega_{n-1}(x)$ are the zeros of the classical orthogonal polynomials of degree $n-1$, then the weight functions $w(x) \in C^{(2)}(a,b)$ satisfying the conditions (2.5) and (2.6) always exist. It is known that the zeros of the classical orthogonal polynomials are always real and simple.

Case a.) If $\omega_{n-1}(x) = P_{n-1}^{(\alpha,\beta)}(x)$, $(\alpha, \beta > -1, n = 2,3,\ldots)$, the Jacobi - polynomials, then the basic system of knots in (2.1) is

$$-1 = x_{n,n} < x_{n-1,n} < \ldots < x_{1,n} < 1, \quad (n \in \mathbf{N}) \tag{3.34}$$

and $P_{n-1}^{(\alpha,\beta)}(x_{k,n}) = 0$, $(k = 1,2,\ldots,n-1)$. The weight function satisfying (2.5) and (2.6) in this case has the form

$$w(x) = (1-x)^{\frac{\alpha+1}{2}}(1+x)^{\frac{\beta+1}{2}}. \tag{3.35}$$

Indeed, in this case the function $u = w\omega_{n-1}$ satisfies the differential equation

$$(w\omega_{n-1})'' + \Big\{\frac{1}{4}\frac{1-\alpha^2}{(1-x)^2} + \frac{1}{4}\frac{1-\beta^2}{(1+x)^2} + \\ + \frac{(n-1)(n+\alpha+\beta)+(\alpha+1)(\beta+1)/2}{1-x^2}\Big\} w\omega_{n-1} = 0.$$

(Szegő [11], 4.24.1). This implies that

$$[w\omega_{n-1}]''(x_{k,n}) = 0 \qquad (k = 1, 2, \ldots, n-1)$$

and by (3.34) and (3.35) we have $w(x_{k,n}) \neq 0$ $(k = 1, 2, \ldots, n-1)$.

Case b.) If $\omega_{n-1}(x) = L_{n-1}^{(\alpha)}(x)$, $(\alpha, > -1, n = 2, 3, \ldots)$, the Laguerre - polynomial, then the basic system of knots in (2.1) is

$$0 = x_{n,n} < x_{n-1,n} < \ldots < x_{1,n} < +\infty, \quad (n \in \mathbf{N}) \tag{3.36}$$

and $L_{n-1}^{(\alpha)}(x_{k,n}) = 0$, $(k = 1, 2, \ldots, n-1)$. The weight function satisfying (2.5) and (2.6) in this case has the form

$$w(x) = e^{-\frac{x}{2}} x^{\frac{\alpha+1}{2}}. \tag{3.37}$$

Indeed, in this case the function $u = w\omega_{n-1}$ satisfies the differential equation

$$(w\omega_{n-1})'' + \Big(\frac{n + \frac{\alpha-1}{2}}{x} + \frac{(1-\alpha^2)}{4x^2} - \frac{1}{4}\Big) w\omega_{n-1} = 0.$$

(Szegő [11], 5.1.2). This implies that $[w\omega_{n-1}]''(x_{k,n}) = 0$ $(k = 1, 2, \ldots, n-1)$ and by (3.36) and (3.37) we have $w(x_{k,n}) \neq 0$ $(k = 1, 2, \ldots, n-1)$.

Case c.) If $\omega_{n-1}(x) = H_{n-1}(x)$, $(n = 2, 3, \ldots)$, the Hermite-polynomial, and if $x_{n,n}$ is the smallest zero of the polynomial $H_n(x)$, then the basic system of knots in (2.1) is

$$-\infty < x_{n,n} < x_{n-1,n} < \ldots < x_{1,n} < +\infty, \quad (n \in \mathbf{N}) \tag{3.38}$$

and $H_{n-1}(x_{k,n}) = 0$, $(k = 1, 2, \ldots, n-1)$. The weight function satisfying (2.5) and (2.6) in this case has the form

$$w(x) = e^{-\frac{x^2}{2}} H_{n-1}(x). \tag{3.39}$$

Indeed, in this case the function $u = w\omega_{n-1}$ satisfies the differential equation

$$(w\omega_{n-1})'' + (2n - 1 - x^2) w\omega_{n-1} = 0.$$

(Szegő [11], 5.5.2). This implies that $[w\omega_{n-1}]''(x_{k,n}) = 0$ $(k = 1, 2, \ldots, n-1)$ and by (3.38) and (3.39) we have $w(x_{k,n}) \neq 0$ $(k = 1, 2, \ldots, n-1)$.

Hence in the cases a.), b.) and c.) by Theorem 1. the modified weighted $(0,2)$ interpolational polynomials $R_n(x)$ of degree at most $2n-1$ satisfying (2.2), (2.3) and (2.4) do exist and they are uniquely determined.

We note that Szili [15] has proved, that in the case of the classical orthogonal polynomials the weighted $(0,2)$ interpolational polynomials $R_n(x)$ of degree $2n$ satisfying (1.2) and (1.4) always exist and they are uniquely determined, if $P_n^{(\alpha,\beta)}(0) \neq 0$, $H_n(0) \neq 0$.

However, the problem on the characterization of the basic system of knots (2.1) for which the weight function $w(x) \in C^{(2)}(a,b)$ satisfying (2.5) and (2.6) can be determined, remains open.

The convergence problems and applications will be investigated in a forthcoming paper.

References

[1] P. Bajpai, Weighted $(0,2)$ interpolation on the extended Tchebycheff nodes of second kind, Acta Math. Hungar. 63(2) (1994), 167-181.

[2] J.Balázs, Weighted $(0,2)$ interpolation on the roots of ultraspherical polynomials, Magyar Tud. Akad. Mat. Fiz. Oszt. Közl. 11(1961), 305-338.

[3] S. A. N. Eneduanya, The weighted $(0,2)$ lacunary interpolation, Demonstratio Math. XVIII(1985), 9-21.

[4] I. Joó, On weighted $(0,2)$-interpolation, Annales Univ. Sci. Budapest., 38(1995), 185-222.

[5] I. Joó and L. Szili, Weighted $(0,2)$-interpolation on the roots of Jacobi polynomials, Acta Math. Hungar. 66(1-2) (1995), 25-50.

[6] J. Prasad, Some interpolatory polynomials on Hermite abscissas, Math. Japonicae 12(1) (1967), 73-80.

[7] J. Prasad, On the weighted $(0,2)$ interpolation, SIAM J. Num. Anal. 7(1970), 428-446.

[8] J. Prasad, On the uniform convergence of interpolatory polynomials, J. Austral. Math. Soc. (Series A) 27(1979), 7-16.

[9] J. Prasad and E. J. Eckert, On the representation of functions by interpolatory polynomials, Mathematica (Cluj) 15(1973), 289-305.

[10] A. Sharma, Some poised and nonpoised problems of interpolation, SIAM Review 19(1972), 129-151.

[11] G. Szegö, "Orthogonal polynomials", Amer. Math. Soc. Coll. Publ., New York, 1959.

[12] L. Szili, Weighted $(0,2)$-interpolation on the roots of Hermite polynomials, Annales Univ. Sci. Budapestinensis XXVII(1985), 153-166.

[13] L. Szili, An interpolation process on the roots of the integrated Legendre polynomials, Analysis Mathematica, 9(1983), 235-245.

[14] L. Szili, A convergence theorem for the Pál method of interpolation on the roots of Hermite polynomials, Analysis Mathematica, 11(1985), 75-84.

[15] L. Szili, Weighted $(0,2)$-interpolation on the roots of the classical orthogonal polynomials, Bull. of the Allahabad Math. Soc. 8-9(1993-94), 1-10.

[16] A. K. Varma and S. K. Gupta, An analogue problem of J. Balázs, Studia Sci. Math. Hungar., 5(1970), 215-220.

New Approach to Markov Inequality in L^p Norms

Mirosław Baran*

Jagiellonian University of Cracow, Institute of Mathematics

Reymonta 4, 30-059 Kraków, Poland

Dedicated to Arun Kumar Varma in memory

Abstract

This paper gives a new proof of the Markov inequality for polynomials in the L^p-metric, where $p \geq 2$:

$$\left(\int_{-1}^{1} |Q'(t)|^p dt\right)^{1/p} \leq C_p (\deg Q)^2 \left(\int_{-1}^{1} |Q(t)|^p dt\right)^{1/p}.$$

A simple method that is applied to obtain the constants C_p yields essentially better results (for $p \geq p_0$) than those of a classical theorem of Hille, Szegö and Tamarkin [8] or a result of Gotgheluck [7]. In particular, we get $\lim_{p\to\infty} C_p = 1$.

1 Introduction

Let $\mathbf{R}[t]$ (resp. $\mathbf{C}[t]$) denote the ring of polynomials of one variable over $\mathbf{R}$ (resp. $\mathbf{C}$) and let $\mathcal{T}$ be the ring of trigonometric polynomials. Define, as usualy, the following norms in $\mathbf{C}[t]$ and in $\mathcal{T}$, respectively:

$$\|Q\|_p = \begin{cases} \left(\int_{-1}^{1} |Q(t)|^p dt\right)^{1/p}, & 1 \leq p < \infty; \\ \sup |Q|([-1,1]), & p = \infty, \end{cases}$$

$$\|Q\|_p = \begin{cases} \left(\frac{1}{2\pi}\int_{0}^{2\pi} |Q(\theta)|^p d\theta\right)^{1/p}, & 1 \leq p < \infty; \\ \sup |Q|([0,2\pi]), & p = \infty. \end{cases}$$

*Research suported by the grant 3 PO3A 057 08 from KBN (Committee for Scientific Research) and by the European Programme PECO of the French Government.

The following two inequalities are well-known.

$$\text{(Markov's inequality, 1889)} \quad \|Q'\|_\infty \le (\deg Q)^2 \|Q\|_\infty, \ Q \in \mathbf{C}[t];$$

$$\text{(Bernstein's inequality, 1912)} \quad \|Q'\|_\infty \le (\deg Q) \|Q\|_\infty, \ Q \in \mathcal{T};$$

It seems to be interesting that Markov's inequality can be derived by applying (to appropriate polynomials) only Bernstein's inequality. Namely, as it is well known, if $P \in \mathbf{R}[t]$ and we put $Q(\theta) = P(\cos\theta)$, then we get the inequality (which is also called Bernstein's inequality)

$$|P'(t)| \le (\deg P)(1-t^2)^{-1/2} \|P\|_\infty, \ t \in (-1,1).$$

Applying the Bernstein inequality to the trigonometric polynomial $Q(\theta) = P(\cos\theta)\sin\theta$ we get

$$|P(1)| \le (\deg P + 1) \sup_{|t|\le 1} |Q(t)|\sqrt{1-t^2}.$$

Now replacing the polynomial P by the polynomial $P_x(t) = P(xt)$, for a fixed $x \in [-1,1]$, we obtain

$$|P(x)| = |P_x(1)| \le (\deg P + 1) \sup_{|t|\le 1} |P(xt)|\sqrt{1-t^2}$$

$$\le (\deg P + 1) \sup_{|t|\le 1} |P(xt)|\sqrt{1-(xt)^2} = (\deg P + 1) \sup_{|t|\le |x|} |P(t)|\sqrt{1-t^2}$$

$$\le (\deg P + 1) \sup_{|t|\le 1} |P(t)|\sqrt{1-t^2},$$

i.e.,

$$\text{(Schur's inequality, 1918)} \quad \|P\|_\infty \le (\deg P + 1) \|P(t)\sqrt{1-t^2}\|_{[-1,1]}.$$

It is obvious that Bernstein's inequality together with Schur's inequality (that is applied to the derivative of a polynomial P) give the Markov inequality. The Schur inequality is usually proved separately with Bernstein's inequality. A remark that Bernstein's inequality easily implies Schur's inequality seems to be less known.

Remark 1.1 Applying an interpolation argument (see [3]) *one can show that the following generalization of Schur's inequality holds:*

$$\|P\|_\infty \le (\deg P + 1)^{2\alpha} \|P(t)(1-t^2)^\alpha\|_{[-1,1]}, \ \alpha \ge \frac{1}{2}. \tag{1}$$

Now observe that an application of Markov's inequality to the polynomial $Q(t) = P(t)(1-t^2)$ gives

$$|P(1)| \leq \frac{1}{2}(\deg P + 2)^2 ||P(t)(1-t^2)||_{[-1,1]},$$

which implies

$$||P||_\infty \leq \frac{1}{2}(\deg P + 2)^2 ||P(t)(1-t^2)||_{[-1,1]}$$

whence we obtain

$$||P||_\infty \leq 2^{-\alpha}(\deg P + 2)^{2\alpha} ||P(t)(1-t^2)^\alpha||_{[-1,1]}, \ \alpha \geq 1. \tag{2}$$

In 1932 Bernstein's inequality (for trigonometric polynomials) was extended to the case of p-norms, $1 \leq p < \infty$, by A. Zygmund [13]:

$$(\text{Zygmund's inequality, 1932}) \quad ||Q'||_p \leq (\deg Q)||Q||_p, \ Q \in \mathcal{T}.$$

The case of Markov's inequality in L^p-norms was more complicated. First, E. Schmidt [11] proved Markov's type inequality in the case $p = 2$ with exponent 2. Five years later Hille, Szegö and Tamarkin [8] proved the Markov inequality for all $1 \leq p < \infty$ with the sharp exponent 2:

$$||Q'||_p \leq C_p(\deg Q)^2 ||Q||_p \leq C(\deg Q)^2 ||Q||_p, \ 1 \leq p < \infty,$$

where $C_1 = 8$ and $C_p = 2ep(p-1)^{1/p-1}$, $p > 1$. Here $\lim_{p\to\infty} C_p = 2e$, while the conjecture was that $\lim_{p\to\infty} C_p = 1$. In 1990 P. Goetgheluck [7] found better values for constant C_p, namely he obtained $\lim_{p\to\infty} C_p = 4$. His method was also based on the Zygmund result. In [8, p.730] there we find the following supposition:

Neither customary methods used for the proof of A. Markoff's original theorem ($p = \infty$) nor E. Schmidt's elegant method ($p = 2$) seems to be applicable in the general case.

We shall show that this is only partially true, since one can easily prove Markov's inequality in L^p-norms for $p > 2$ applying only Markov and Bernstein's inequalities (for algebraic polynomials). Moreover we get better constants C_p, so that we obtain $\lim_{p\to\infty} C_p = 1$. From the point of view of functional analysis our methods are based on a simple operator factorization and an application of interpolation techniques.

2 Markov inequality in L^p-norms

Lemma 2.1 *Let $P \in \mathbf{R}[t]$, $\deg P = k \geq 1$. Then for arbitrary $p \geq 1$ we have*

$$||P||_\infty \leq \left[\frac{1}{2}(p+3)^2\right]^{1/p} k^{2/p}||P||_p.$$

Proof. Let $P \in \mathbf{R}[t]$, $\deg P = k \geq 1$. Fix an even $n \in \mathbf{N}$. Define

$$Q(t) := \int_{-1}^{t} P^n(x)dx - \int_{t}^{1} P^n(x)dx.$$

Then $\deg Q = nk+1$, $||Q||_\infty = ||P||_n^n$ and $Q'(t) = 2P^n(t)$. Applying the classical Markov inequality to the polynomial Q we get the inequality

$$||P||_\infty \leq \left[\frac{1}{2}(nk+1)^2\right]^{1/n} ||P||_n. \tag{3}$$

This proves the lemma in the case where p is an even number. In the general case we can choose a natural even number n such that $n-2 < p < n$ and put $\alpha = p/n$. Then we have an interpolation inequality

$$||P||_n \leq ||P||_\infty^{1-\alpha}||P||_p^\alpha.$$

This inequality together with (3) give the required inequality. The proof is completed.

Lemma 2.2 *For each $p > 2$ and for all $P \in \mathbf{C}[t]$ we have the inequlity*

$$||P'||_p \leq \left[4\nu(\frac{2}{p})\right]^{1/p} (\deg P)^{2/q}||P||_\infty, \ \frac{1}{p}+\frac{1}{q} = 1,$$

where $\nu(t) = \pi t/\sin(\pi t)$, $t \in (0,1)$.

Proof. Let us recall that

$$\min(a,b) \leq 2\left[\frac{1}{a}+\frac{1}{b}\right]^{-1}, \ a,b > 0. \tag{4}$$

By the classical Markov and Bernstein inequlities, the following estimate holds :

$$|P'(t)|^p \leq k^p \min\left[(1-t^2)^{-p/2}, k^p\right] ||P||_\infty^p, t \in [-1,1],$$

where $k = \deg P$. Hence, by (4), we obtain

$$|P'(t)|^p \leq 2k^p \left[(1-|t|)^{p/2} + k^{-p}\right]^{-1} ||P||_\infty^p, \ t \in [-1,1],$$

which implies

$$||P'||_p^p \leq 4k^p \int_0^1 (t^{p/2} + k^{-p})^{-1} dt ||P||_\infty^p.$$

Now we calculate

$$\int_0^1 (t^{p/2} + k^{-p})^{-1} dt = k^{p-2} \int_0^{k^2} (1 + t^{p/2})^{-1} dt$$

$$\leq k^{p-2} \int_0^\infty (1 + t^{p/2})^{-1} dt = \frac{2}{p} k^{p-2} \int_0^\infty \frac{y^{2/p-1}}{1+y} dy = \nu(2/p) k^{p-2},$$

whence the inequality from the assertion of the lemma follows.

Remark 2.3 There is known a sharp version of Lemma 2.2 obtained by G. Kristiansen [9] and B. Bojanov [4]:

$$||Q'||_p \leq ||T_k'||_p ||Q||_\infty, \ k = \deg Q, \ p \geq 1,$$

where T_k is the k-th Tchebyshev polynomial $T_k(t) = \cos(k \arccos t)$, $t \in [-1,1]$. However, it is not so easy to deduce Lemma 2.2 from this Kristiansen-Bojanov nice result. Applying the method of the proof of Lemma 2.2 we can derive inequalities which are, up to constants, equivalent to the Kristiansen-Bojanov inequalities.

Theorem 2.4 *If $p > 2$ and $Q \in \mathbf{C}[t]$, then*

$$||Q'||_p \leq \left[2\nu(\frac{2}{p})(p+3)^2\right]^{1/p} (\deg Q)^2 ||Q||_p. \tag{5}$$

Proof. If $Q \in \mathbf{R}[t]$, then (5) immediatelly follows from lemmas above. Now let $Q = R + iS \in \mathbf{C}[t]$, where $R, S \in \mathbf{R}[t]$. It is easy to check that

$$||Q||_\infty = \sup_{\theta \in [0,2\pi]} ||\cos\theta R + \sin\theta S||_\infty \tag{6}$$

and

$$\sup_{\theta\in[0,2\pi]} \|\cos\theta R + \sin\theta S\|_p \leq \|Q\|_p. \tag{7}$$

We have, by Lemma 2.1 and (6),(7),

$$\|Q\|_\infty \leq \sup_{\theta\in[0,2\pi]} \left[\frac{1}{2}(p+3)^2\right]^{1/p} k^{2/p}\|\cos\theta R + \sin\theta S\|_p$$

$$\leq \left[\frac{1}{2}(p+3)^2\right]^{1/p} k^{2/p}\|Q\|_p. \tag{8}$$

Combining (8) with Lemma 2.2 we get (5) in the general case.

Since the function $\nu(t)$ is increasing on the interval $(0, \frac{1}{2}]$, we easily obtain the following.

Corollary 2.5 *If $p \geq 4$, then*

$$\|Q'\|_p \leq \pi^{1/p}(p+3)^{2/p}(\deg Q)^2\|Q\|_p.$$

Remark 2.6 The Kristiansen-Bojanov result [9, 4] shows that the argument of the proof of Theorem 2.4 fails for $1 \leq p \leq 2$: by Z. Ciesielski [5] we have $\|T_k'\|_2 \sim k\sqrt{\log(k+1)}$ and $\|T_k'\|_p \sim k$ for $1 \leq p < 2$, while the inequality in the Lemma 2.1 is valid for $1 \leq p \leq 2$. We shall show below how to omit this difficulty for $p = 2$. The method we present here can also be extended to the case where $p \geq 1$ but we will omit it.

Proposition 2.7 *If $p \geq 2$ is fixed, then for each $Q \in \mathbf{C}[t]$, $\deg Q \leq k$, we have the following Kolmogorov type inequality*

$$\|Q'\|_p \leq 2^{1/p}k^{4/q-2}\|Q\|_\infty^{1-\frac{2}{p}}\left(2k^2\|Q\|_\infty^2 + \|Q''\|_q\|Q\|_p\right)^{1/p}, \quad \frac{1}{p}+\frac{1}{q}=1.$$

Proof. We start from an obvious identity for a real polynomial Q

$$\int_{-1}^{1}(Q'(t))^2dt = (Q(1)Q'(1) - Q(-1)Q'(-1)) - \int_{-1}^{1} Q''(t)Q(t)dt.$$

From this, by Markov and Hölder's inequalities we get

$$\|Q'\|_2 \leq \left(2\|Q\|_\infty\|Q'\|_\infty + \|Q''\|_q\|Q\|_p\right)^{1/2}, \tag{9}$$

$$||Q'||_2 \leq \left(2k^2||Q||_\infty^2 + ||Q''||_q||Q||_p\right)^{1/2}.$$

Since

$$||Q'||_p \leq ||Q'||_\infty^{1-\frac{2}{p}}||Q'||_2^{\frac{2}{p}},$$

we easily complete the proof in the real case (without the factor $2^{1/p}$). Let $Q = R + iS \in \mathbf{C}[\mathbf{t}]$, where $R, S \in \mathbf{R}[t]$. Now the complex case is a consequence of the following inequalities (cf. the proof of theorem 2.4):

$$\sup_{|\theta|\leq\pi} ||\cos\theta R + \sin\theta S||_2 \geq \frac{1}{\sqrt{2}}||Q||_2,$$

$$\sup_{|\theta|\leq\pi} ||\cos\theta R + \sin\theta S||_p \leq ||Q||_p,\ 1 \leq p \leq \infty.$$

The next proposition can be interpretated as a Kristiansen-Bojanov type theorem for the second derivative. Let us note that a sharp version of the result below is unknown. However, in case $p = 2$, there are known very close sharp results obtained by Varma [12] and Dimitrov [6]. Actually, we have found the method presented here when reading Varma's paper.

Proposition 2.8 *If $q > 1$ is fixed, then for $Q \in \mathbf{C}[t]$, $\deg Q \leq k$, we have*

$$||Q''||_q \leq C_q k^{2+2/p}||Q||_\infty,\ \frac{1}{p} + \frac{1}{q} = 1,$$

where

$$C_q := \left(8\nu(\frac{1}{q})\right)^{1/q}.$$

Proof. Applying Bernstein's inequality to a polynomial of the form $Q(\cos\theta)$ (and next using Markov's inequality) we get the inequality

$$||Q''(t)(1-t^2)||_{[-1,1]} \leq 2k^2||Q||_\infty,$$

whence we easily obtain (for $t \in [-1, 1]$)

$$\begin{aligned}|Q''(t)|^q &\leq 2^q k^{2q} \min\left((1-t^2)^{-q}, 2^{-q}k^{2q}\right) ||Q||_\infty^q \\ &\leq 2^{q+1}k^{2q}\left((1-t^2)^q + 2^q k^{-2q}\right)^{-1} ||Q||_\infty^q.\end{aligned}$$

Integrating both sides of the last inequality gives the required inequality (cf. proof of Lemma 2.2).

Put $A_p := 2^{-1/p}(p+3)^{2/p}$, $p \geq 1$. Combining Propositions 2.7, 2.8 with Lemma 2.1 we easily obtain the following theorem.

Theorem 2.9 *If $p \geq 2$, then for arbitrary $Q \in \mathbf{C}[t]$ we have a Markov type inequality*

$$\|Q'\|_p \leq 2^{1/p} A_p^{1-\frac{2}{p}} \left(2A_p^2 + C_q A_p\right)^{1/p} (\deg Q)^2 \|Q\|_p.$$

Corollary 2.10 *If $p \geq 2$, then*

$$\|Q'\|_p \leq (2 + 4\pi p)^{1/p} (p+3)^{2/p} (\deg Q)^2 \|Q\|_p.$$

Remark 2.11 If $p = 2$, (9) yields

$$\|Q'\|_2 \leq \left(2\|Q\|_\infty \|Q'\|_\infty + \|Q''\|_2 \|Q\|_2\right)^{1/2}.$$

Since

$$\|Q''\|_2 \leq \|Q''(t)(1-t^2)^{-1/4}\|_2^{1/2} \|Q''(t)(1-t^2)^{1/4}\|_2^{1/2},$$

we can apply the following (sharp) results:

(Labelle's inequalities [10])

$$\|Q\|_\infty \leq \frac{k+1}{\sqrt{2}} \|Q\|_2,$$

$$\|Q'\|_\infty \leq \frac{1}{2\sqrt{6}} k(k+1)(k+2) \|Q\|_2.$$

(Varma's inequality [12])

$$\|Q''(t)(1-t^2)^{-1/4}\|_2 \leq \sqrt{\frac{2}{15}\pi} \left(k^3(k^4-1)\right)^{1/2} \|Q\|_\infty.$$

(Dimitrov's inequality [6])

$$\|Q''(t)(1-t^2)^{1/4}\|_2 \leq \sqrt{\pi} \left(k^3(k^2-1)\right)^{1/2} \|Q\|_\infty.$$

By the above inequalities we obtain better bounds than those given by the last corollary. However, for $p = 2$ there are known much better (and much more complicated) results than those obtained by our elementary methods (cf. [7] for further information).

3 Markov inequality with respect to the complex equilibrium measure

Let us set $I = [-1, 1]$ and fix a $\beta > -1$. Define $dm_\beta(t) = (1-t^2)^\beta dt,\ t \in I$ and put

$$\|Q\|_{\beta,p} = \left(\int_I |Q|^p dm_\beta\right)^{1/p},\ p \geq 1.$$

By a similar argument to that of the proof of Lemma 2.2 we can easily derive the following.

Proposition 3.1 *If $p > 2(1+\beta)$, then for all $Q \in \mathbf{C}[t]$ we have*

$$\|Q'\|_{\beta,p} \leq (\deg Q)^{2-2(1+\beta)/p} \left[\frac{4}{1+\beta}\nu(2\frac{1+\beta}{p})\right]^{1/p} \|Q\|_\infty.$$

One of the most important measures supported on the interval I is the *complex equilibrium measure* λ_I which is equal to $\lambda_I = 2m_{-\frac{1}{2}}$. Define $\|Q\|_p^* = \left(\int_I |Q|^p d\lambda_I\right)^{1/p}$. Then, as a special case of Proposition 3.1, we have inequality

$$\|Q'\|_p^* \leq (\deg Q)^{2-1/p} \left[16\nu(\frac{1}{p})\right]^{1/p} \|Q\|_\infty,\ p > 1,\ Q \in \mathbf{C}[t]. \tag{10}$$

Now let $P \in \mathbf{R}[t]$. Applying the Bernstein inequality to the polynomial

$$Q(\theta) = \int_0^\theta Q^{2m}(\cos\mu)d\mu - \int_\theta^{2\pi} Q^{2m}(\cos\mu)d\mu$$

we obtain, on a similar way as in the proof of Lemma 2.2, the following result.

Proposition 3.2 *If $Q \in \mathbf{C}[t]$, then for arbitrary $p \geq 1$ we have*

$$\|Q\|_\infty \leq \left(\frac{p}{2}+1\right)^{1/p} (\deg Q)^{1/p} \|Q\|_p^*.$$

Combining Proposition 3.2 with inequality (10) we get the following.

Theorem 3.3 *If $Q \in \mathbf{C}[\mathbf{t}]$, then for an arbitrary $p > 1$ we have*

$$||Q'||_p^* \leq \left[8\nu(\frac{1}{p})\right]^{1/p} (p+2)^{1/p}(\deg Q)^2||Q||_p^*.$$

In particular, if $p \geq 2$, then

$$||Q'||_p^* \leq (4\pi)^{1/p}(p+2)^{1/p}(\deg Q)^2||Q||_p^*.$$

Theorem 3.4 *Let $1 \leq p \leq 2$. Then for all $Q \in \mathbf{C}[t]$, $\deg Q \leq k$, we have Markov's inequality*

$$||Q'||_p^* \leq C_p k^2 ||Q||_p^*,$$

where

$$C_p = 2^{2/p}\left(\frac{p}{2}+1\right)^{\frac{3}{4p}}\left(\frac{32}{5-2p}\right)^{\frac{1}{2p}} \nu(\frac{10}{8p}-\frac{1}{2})^{\frac{1}{2p}}\nu(\frac{1}{2p})^{\frac{1}{4p}}.$$

Proof. Fix $p \in [1,2]$ and define $\beta := \frac{3}{8} + \frac{p}{4}$. Then one can easily check that the following interpolation inequality holds:

$$||Q'||_p \leq 2^{1/p}||Q'||_{-\beta,p}^{\frac{1}{2}}||Q'||_{\frac{p-1}{2},p}^{\frac{1}{4}}||Q'||_{-\frac{3}{4},p}^{\frac{1}{4}}.$$

By Zygmund's inequality we have

$$||Q'||_{\frac{p-1}{2},p} \leq k||Q||_p^*,$$

while the first and the third factor are esimated by Propositions 3.1 and 3.2. Now, by a simple calculation we obtain the required inequality.

References

[1] M. Baran, Bernstein type theorems for compact sets in $\mathbf{R}^n$, J. Approx. Theory 69 (2) (1992), 156-166.

[2] M. Baran, Bernstein type theorems for compact sets in $\mathbf{R}^n$ revisited, J. Approx. Theory 79 (2) (1994), 190-198.

[3] M. Baran, Markov inequality on sets with polynomial parametrization, Ann. Polon. Math. 60 (1) (1994), 69-79.

[4] B. Bojanov, An extension of the Markov inequality, J. Approx. Theory 35 (2) (1982), 181-190.

[5] Z. Ciesielski, On the A. A. Markov inequality for polynomials in the L^p case, in: "Approximation theory", Ed.: G. Anastassiou, pp., 257-262, Marcel Dekker, inc., New York, 1992.

[6] D. K. Dimitrov, Markov Inequalities for Weight Functions of Chebyshev Type, J. Approx. Theory 83 (2) (1995), 175-181.

[7] P. Goetgheluck, On the Markov Inequality in L^p-Spaces, J. Approx. Theory 62 (2) (1990), 197-205.

[8] E. Hille, G. Szegö, J. Tamarkin, On some generalisation of a theorem of A. Markoff, Duke Math. J. 3 (1937), 729-739.

[9] G. K. Kristiansen, Some inequalities for algebraic and trigonometric polynomials, J. London Math. Soc. 20 (2) (1979), 300-314.

[10] G. Labelle, Concerning polynomials on the unit interval, Proc. Amer. Math. Soc. 20 (1969), 321-326.

[11] E. Schmidt, Die asymptotische Bestimmung des Maximums des Integrals über das Quadrat der Ableitung eines normierten Polynoms Sitzungsberichte der Preussischen Akademie, (1932), 287.

[12] A. K. Varma, On Some Extremal Properties of Algebraic Polynomials, J. Approx. Theory 69 (1) (1992), 48-54.

[13] A. Zygmund, A remark on conjugate functions, Proceedings of the London Math. Soc. 34 (1932), 392-400.

A Question in Reliability Theory

Franck Beaucoup and Laurent Carraro
Ecole des Mines de Saint-Etienne, Département de Mathématiques Appliquées
158 cours Fauriel, 42023 Saint-Etienne, France

Abstract

We present a question arising in reliability theory that involves approximation theory. The mathematical question is the following. Given a continuous, increasing function φ on $[0, 1]$, with $\varphi(0) = 0$ and $\varphi(1) = 1$, is it possible to approximate φ uniformly on $[0, 1]$ with functions generated by the "exponential monomials" x^λ, λ real positive, with the following two composition rules : $f \times g = fg$ and $f * g = f + g - fg$?

In this paper we show how this question arises in reliability theory and we consider its polynomial version; that is, the analogous question using only the monomials x^k, k integral ($k \geq 1$), as generating functions.

1 Some basic notions in reliability theory

Let S be a system with a random lifetime T. Suppose that the random variable T has a density function f and a cumulative distribution function F. Then we can define the following functions:

- the *reliability function* of the system S, given by $G = 1 - F$
- the *hazard rate function*, or *failure rate function* of the system S, given by

$$\lambda(t) = \lim_{h \to 0, h > 0} \frac{P(T \in]t, t+h] \mid T > t)}{h} = \frac{f(t)}{G(t)}.$$

The simplest systems to study are those for which the hazard rate function is constant, equal to λ, $\lambda > 0$. In this case, the lifetime distribution T is exponential with parameter λ, so the reliability function is $G(t) = \exp(-\lambda t)$. These systems are referred to as *exponential systems*.

An interesting feature of exponential systems is that the number of failures in the time interval $[0, t]$, when t varies, is a Markov process (actually a Poisson process, see [2]).

Unfortunately, the standard hazard rate function is not constant, but is rather a *bathtub curve*, for which the *useful life period* (in which the hazard rate function is constant) is between a *burn in period* and and a *wear out phase* (see [4]).

2 Formulation of the problem

Erlang represented certain systems as combinations of exponential systems. More precisely, when n systems $S_1, \ldots, S_n$ are put *on standby* (that is, S_{i+1} starts operating when S_i fails), then the global lifetime is $T = T_1 + \ldots + T_n$. If the random variables T_i are exponential with the same parameter λ and are independent, then the density function of the lifetime T of the global system is given by

$$f(t) = \frac{(n\lambda)^n}{(n-1)!} t^{n-1} \exp(-n\lambda t) \tag{1}$$

(this is the *Erlang distribution*). Conversely, it is possible to replace a system whose lifetime is modelled by an Erlang distribution, by n exponential systems on standby. This is the *method of stages* (see [3]).

One might then wonder about the "approximation" of any system by systems built as combinations of exponential systems. It seems more natural to consider only combinations in which the components are linked in series or in parallel (or both). The corresponding operations on the lifetimes and reliability functions are the following.

- Components T_1 and T_2 in series:

  $$T = \min(T_1, T_2),$$

  so

  $$G = G_1 G_2.$$

- Components T_1 and T_2 in parallel:

  $$T = \max(T_1, T_2),$$

  so

  $$1 - G = (1 - G_1)(1 - G_2);$$

  that is,

  $$G = G_1 + G_2 - G_1 G_2.$$

So the question is the following:

Question 1 *Let S be any system, with lifetime T. Does there exist a sequence $(R_n)_{n>0}$ of series-parallel combinations of exponential components, such that if T_n denotes the lifetime of the system R_n, then T_n converges to T in distribution ?*

Let us recall that if G_n denotes the cumulative distribution function of T_n and G that of T, then the convergence of T_n to T in distribution is characterized by the pointwise convergence of G_n to G at every point of continuity of G (see [2]). Hence, if we suppose that G is continuous, then the previous convergence in distribution is equivalent to the pointwise convergence of G_n to G on the real positive axis. That G is continuous will always be assumed hereafter.

3 Reformulation of the question

Transferring the problem onto $[0,1]$ with the change of variable $x = -\log t$, the reliability function of a system is an increasing function φ of x on $[0,1]$, such that $\varphi(0) = 0$ and $\varphi(1) = 1$, and the generating functions are the "exponential monomials" x^λ, where λ is real and positive. The composition rules corresponding to the above combinations are the following:

$$f \times g = fg$$

and

$$f * g = f + g - fg.$$

It is easy to see that every function generated by the exponential monomials with these composition rules is continuous and increasing on $[0,1]$. Hence, if φ is assumed to be continuous, then it is a straightforward exercise to check that the pointwise approximation of φ by functions of our generated set is equivalent to uniform approximation on $[0,1]$ (see [5] for details).

So we get the following question, to which we do not have the answer.

Question 2 *Let φ be a continuous, increasing function on $[0,1]$, with $\varphi(0) = 0$ and $\varphi(1) = 1$. Is it possible to approximate φ uniformly on $[0,1]$ (that is, pointwise) with functions generated by the "exponential monomials" x^λ, λ real positive, with the following two composition rules : $f \times g = fg$ and $f * g = f + g - fg$?*

4 The polynomial version of the problem

Although it is not of particular interest for the applied problem, one might wonder whether the above approximation is possible with only the monomials x^k, k integral ($k \geq 1$), as generating functions. This is the polynomial version of the problem, stated here as Question 3:

Question 3 *Let φ be a continuous, increasing function on $[0,1]$, with $\varphi(0) = 0$ and $\varphi(1) = 1$. Is it possible to approximate φ uniformly on $[0,1]$ (that is, pointwise) with polynomials generated by the monomials x^k, k integral ($k \geq 1$), with the following two composition rules : $P \times Q = PQ$ and $P * Q = P + Q - PQ$?*

In this section we make a few comments about Question 3. Although these do not allow us to answer the question, we hope they will be useful for further investigation of this problem.

The first thing to notice about the polynomials generated by the monomials x^k, $k \geq 1$, using our composition rules, is that they all are increasing, with integer coefficients, and have values 0 at 0 and 1 at 1. So one might wonder whether the reliability function φ on $[0,1]$ can be approximated by polynomials having these three properties. That the answer is positive is shown in the following proposition.

Proposition 4 *Let φ be a continuous, increasing function on $[0,1]$ with integer values at 0 and 1. Then one can approximate φ uniformly on $[0,1]$ by increasing polynomials with integer coefficients.*

Proof. Of course, we can suppose that $\varphi(0) = 0$ and $\varphi(1) = 1$. Then it is well-known (see [1]) that the following polynomials with integer coefficients converge uniformly to φ on $[0,1]$:

$$P_{\varphi,n}(x) = \sum_{k=0}^{n} \lfloor \varphi(\frac{k}{n}) \binom{n}{k} \rfloor x^k (1-x)^{n-k}.$$

However, it is possible that φ is increasing on $[0,1]$ while P_n is not, as shown in the following example. If we take

$$\varphi(t) = \frac{\tanh(10(t - \frac{1}{2})) + \tanh(5)}{2\tanh(5)}$$

and $n = 20$, then the polynomial $P_{\varphi,20}$ is not increasing on $[0,1]$ (one can check that $P'_{\varphi,20}(0.9) < 0$).

However, it is easy to check that the Bernstein polynomials associated with any increasing function ψ, namely

$$B_{\psi,n}(x) = \sum_{k=0}^{n} \psi(\frac{k}{n}) \binom{n}{k} x^k (1-x)^{n-k},$$

are necessarily increasing on $[0,1]$ (and of course they also converge uniformly to ψ on $[0,1]$). Indeed, a simple computation gives

$$\begin{aligned}
B'_{\psi,n}(x) &= \sum_{k=0}^{n} \psi(\frac{k}{n}) \binom{n}{k} k x^{k-1} (1-x)^{n-k} \\
&\quad - \sum_{k=0}^{n} \psi(\frac{k}{n}) \binom{n}{k} (n-k) x^k (1-x)^{n-k-1} \\
&= n \sum_{k=1}^{n} \psi(\frac{k}{n}) \binom{n-1}{k-1} x^{k-1} (1-x)^{n-k} \\
&\quad - n \sum_{k=0}^{n-1} \psi(\frac{k}{n}) \binom{n-1}{k} x^k (1-x)^{n-k-1} \\
&= n \sum_{k=0}^{n-1} \left(\psi(\frac{k+1}{n}) - \psi(\frac{k}{n}) \right) \binom{n-1}{k} x^k (1-x)^{n-k-1}.
\end{aligned}$$

So $B'_{\psi,n}(x) \geq 0$ for every x in $[0,1]$ and $B_{\psi,n}$ is increasing on $[0,1]$.

To prove the proposition, the idea is then the following. For every integer n, we will slightly perturb φ to yield a function ψ_n, which is also continuous and increasing, and whose nth Bernstein polynomial

$$B_{\psi_n,n}(x) = \sum_{k=0}^{n} \psi_n(\frac{k}{n}) \binom{n}{k} x^k (1-x)^{n-k}$$

has integer coefficients; that is, such that $\psi_n(\frac{k}{n})\binom{n}{k}$ is an integer for every k. If the perturbation is small enough (that is, if the functions ψ_n converge uniformly to φ on $[0,1]$), then this will do the trick. Indeed,

$$\|\varphi - B_{\psi_n,n}\| \leq \|\varphi - B_{\varphi,n}\| + \|B_{\varphi,n} - B_{\psi_n,n}\| \leq \|\varphi - B_{\varphi,n}\| + \|\varphi - \psi_n\|$$

where $\|f\| = \sup_{x\in[0,1]} |f(x)|$, so $B_{\psi_n,n}$ converges to φ uniformly on $[0,1]$.

For this purpose, we will choose the values $\psi_n(\frac{k}{n})$ and then extend them to a continuous, piecewise linear function ψ_n on $[0,1]$.

Let us first note that choosing

$$\psi_n(\frac{k}{n}) = \frac{\lfloor \varphi(\frac{k}{n})\binom{n}{k} \rfloor}{\binom{n}{k}},$$

so that the corresponding polynomial $P_{\psi_n,n}$ with integer coefficients is exactly the Bernstein polynomial $B_{\psi_n,n}$, is not suitable because it does not necessarily yield an increasing function ψ_n (take, for example, $n = 4$, $\varphi(\frac{1}{4}) = 0.25$ and $\varphi(\frac{1}{2}) = 0.3$, then $\psi_4(\frac{1}{4}) = \frac{1}{4}$ and $\psi_4(\frac{1}{2}) = \frac{1}{6}$). So we will have to use a more careful process to choose the values $\psi_n(\frac{k}{n})$.

Starting with $\psi_n(0) = 0$ and $\psi_n(1) = 1$, we then set

$$\psi_n(\frac{1}{n}) = \frac{1}{n}\lfloor n\varphi(\frac{1}{n})\rfloor.$$

Then $n\psi_n(\frac{1}{n})$ is an integer,

$$\psi_n(0) \le \psi_n(\frac{1}{n}) \le 1,$$

and

$$|\psi_n(\frac{1}{n}) - \varphi(\frac{1}{n})| \le \frac{1}{n}.$$

Now suppose that $2 \le k \le n-1$ and we have found $\psi_n(\frac{j}{n})$, $1 \le j \le k-1$, such that for every j, $\binom{n}{j}\psi_n(\frac{j}{n})$ is an integer,

$$\psi_n(\frac{j-1}{n}) \le \psi_n(\frac{j}{n}) \le 1,$$

and

$$|\psi_n(\frac{j}{n}) - \varphi(\frac{j}{n})| \le \varepsilon_j$$

where $\varepsilon_j = \sum_{i=1}^{j} \frac{1}{\binom{n}{i}}$. Let m be the smallest integer such that both $m\binom{n}{k}^{-1} \ge \psi_n(\frac{k-1}{n})$ and $(m+1)\binom{n}{k}^{-1} \ge \varphi(\frac{k}{n})$. If $m\binom{n}{k}^{-1} > 1$, set $\psi_n(\frac{k}{n}) = 1$. If $\varphi(\frac{k}{n}) < m\binom{n}{k}^{-1} \le 1$, set $\psi_n(\frac{k}{n}) = m\binom{n}{k}^{-1}$. If $m\binom{n}{k}^{-1} \le \varphi(\frac{k}{n})$, set $\psi_n(\frac{k}{n}) = \binom{n}{k}^{-1}\lfloor\varphi(\frac{k}{n})\binom{n}{k}\rfloor$.

In any case, $\binom{n}{k}\psi_n(\frac{k}{n})$ is an integer and it is straightforward to verify that

$$\psi_n(\frac{k-1}{n}) \le \psi_n(\frac{k}{n}) \le 1$$

and

$$|\psi_n(\frac{k}{n}) - \varphi(\frac{k}{n})| \le \sum_{i=1}^{k} \frac{1}{\binom{n}{i}} = \varepsilon_k.$$

By induction, we have determined $\varphi(\frac{k}{n})$, $1 \le k \le n-1$, satisfying the requirements above.

Now

$$\varepsilon_{n-1} = \sum_{j=1}^{n-1} \frac{1}{\binom{n}{j}} \le \frac{2}{n} + (n-3)\frac{1}{\binom{n}{2}},$$

so

$$\lim_{n\to\infty} \max_{0\le k\le n} \left|\psi_n(\frac{k}{n}) - \varphi(\frac{k}{n})\right| = 0.$$

We now extend the values $\psi_n(\frac{k}{n})$ to a continuous, piecewise linear function ψ_n on $[0,1]$. Since φ is uniformly continuous on $[0,1]$ and ψ_n approximates φ well at the points k/n, $0 \le k \le n$, standard arguments show that the piecewise linear extensions of the ψ_n converge to φ uniformly on $[0,1]$. Therefore the Bernstein polynomials $B_{\psi_n,n}$ associated with ψ_n are increasing, have integer coefficients, and converge uniformly to φ on $[0,1]$. This completes the proof of the proposition.

Consequently, the answer to Question 3 would be positive if our composition rules generated, from the monomials, the whole set E of polynomials that have integer coefficients, are increasing on $[0,1]$ and have values 0 at 0 and 1 at 1. Unfortunately, this is not the case. For example, the polynomial

$$P(x) = x^3 - x^2 + x$$

is in E, but cannot be generated by the monomials x^k, $k \ge 1$, with our composition rules. The only polynomials of degree 3 that can be generated are the following: x^3, $-x^3 + x^2 + x$, $-x^3 + 2x^2$ and $x^3 - 3x^2 + 3x^3$.

Of course, Proposition 1 also ensures that to give a positive answer to Question 3, it would suffice to show that one can approximate every polynomial in E with our generated polynomials. However, we do not know whether this is true.

The following computations were made in order to compare, for small degrees, the number of generated polynomials with the number of polynomials in E.

For small values of n, we computed all generated polynomials of degree at most n, and also all the polynomials of the form

$$P(x) = \sum_{k=0}^{n} \alpha_k x^k (1-x)^{n-k} \tag{2}$$

with non-negative integer α_k's such that $\alpha_0 = 0$, $\alpha_n = 1$ and the following monotonicity condition holds:

$$\frac{\alpha_k}{\binom{n}{k}} \le \frac{\alpha_{k+1}}{\binom{n}{k+1}}, \quad 0 \le k \le n-1.$$

Note that these polynomials are exactly those used in the proof of the proposition above. In particular, it is shown in this proof that they are all in the set E (in fact, we will see below that every polynomial in E has such a representation). It is worth mentioning that n might be greater than the degree of P.

The results of these computations are given in the following table.

n	generated polynomials of degree at most n	polynomials of the form (2)
1	1	1
2	3	3
3	7	10
4	17	50
5	41	374
6	105	4631

This table shows that there are very few polynomials in E that we can generate by the monomials with our composition rules, but of course this does not give an answer to Question 3.

Finally, let us state the following representation proposition, that shows that formula (2) actually gives a characterization of the polynomials in E.

Proposition 5 *Let P be a polynomial with integer coefficients, that is increasing on $[0,1]$ and has values 0 at 0 and 1 at 1. Then P has a representation of the form*

$$P(x) = \sum_{k=0}^{n} \alpha_k x^k (1-x)^{n-k}$$

with non-negative integer α_k's such that $\alpha_0 = 0$, $\alpha_n = 1$ and

$$\frac{\alpha_k}{\binom{n}{k}} \leq \frac{\alpha_{k+1}}{\binom{n}{k+1}}, \qquad 0 \leq k \leq n-1.$$

Proof. Since P is increasing, its derivative P' is non-negative on $[0,1]$. From a representation theorem due to Bernstein (see [1]), it follows that P' has a representation of the form

$$P'(x) = \sum_{k=0}^{m} \beta_k x^k (1-x)^{m-k},$$

with non-negative β_k's (the smallest integer m for which such a representation exists is called the *Lorentz degree* of P', see [1]).

Now, if we define the sequence of α_k's by

$$\alpha_0 = 0$$

and

$$(k+1)\alpha_{k+1} - (m+1-k)\alpha_k = \beta_k, \quad 0 \le k \le m,$$

then the polynomial

$$Q(x) = \sum_{k=0}^{m+1} \alpha_k x^k (1-x)^{m+1-k}$$

satisfies $Q' = P'$. Since $P(0) = 0 = Q(0)$, it follows that $P = Q$.

Now, since P has integer coefficients, so has Q, and it is easy to check that this ensures that the α_k's are all integers. Moreover, we have $\alpha_{m+1} = Q(1) = P(1) = 1$.

Finally, the condition

$$\frac{\alpha_k}{\binom{m+1}{k}} \le \frac{\alpha_{k+1}}{\binom{m+1}{k+1}}, \quad 0 \le k \le m$$

holds since

$$\frac{\alpha_k}{\binom{m+1}{k}} - \frac{\alpha_{k+1}}{\binom{m+1}{k+1}} = -\frac{k!(m-k)!}{(m+1)!}\beta_k.$$

This completes the proof of the proposition.

In conclusion, let us recall that from a practical point of view, Question 3 is not especially interesting. The question that one would really like to answer is Question 2. In fact, even a positive answer to any of these questions would be of no practical interest without an efficient algorithm to build an approximation.

References

[1] P. Borwein and T. Erdélyi, "Polynomials and Polynomial Inequalities", Graduate Texts in Mathematics, Springer Verlag, New York, 1995.

[2] L. Breiman, "Probability", Classics in Applied Mathematics, SIAM, 1992.

[3] L. Klienrock, "Queuing systems", Vol. 1, J. Wiley and Sons, 1975.

[4] P. E. Pfeiffer, "Probability for applications", Springer Texts in Statistics, Springer Verlag, New York, 1990.

[5] G. Pólya and G. Szegö, "Problems and Theorems in Analysis", Vol. 1, Springer Verlag, Berlin, 1972.

Notes on Miscellaneous Approximation Problems

Borislav Bojanov*
Department of Mathematics, University of Sofia
Blvd. James Boucher 5, 1126 Sofia, BULGARIA

Abstract

We consider several disconnected problems in approximation theory concerning topics as parametric approximation of $|x|$, the Bernstein Polynomials, the Walsh estimate of the zeros of a polynomial, the coefficients in the Gauss quadrature, mean interpolation by splines.

1 Parametric approximation of $|x|$

In this paper we shall denote by π_n the set of all algebraic polynomials of degree n with real coefficients.

Given a function f defined on $[-1,1]$, consider the problem of minimizing the quantity

$$||f(P) - Q|| := \max_{x\in[-1,1]} |f(P(x)) - Q(x)|$$

over all pairs of polynomials $P, Q \in \pi_n$ such that

$$P(-1) = -1, \quad P(1) = 1, \quad P'(x) \geq 0 \quad \text{on} \quad [-1,1].$$

The minimal value of $||f(P)-Q||$ is called *best parametric approximation* of f. For each continuous function $f \in C[-1,1]$ there exists a pair (P,Q) for which the deviation $||f(P)-Q||$ attains its minimal value. This fact follows easily from the compactness of the set of uniformly bounded polynomials of fixed degree n.

*The research was supported by the Bulgarian Ministry of Science under Contract No. MM–414.

The problem of parametric approximation was posed by Bl. Sendov in [10] (also in [9]) and studied latter by many authors (see, for example, [2], [3], [4], [7], [8], [5]), [11]). This note is devoted to the parametric approximation of $|x|$. It was quite natural to check the new way of approximation on this particular function. The results on Bernstein [1] on the uniform approximation of $|x|$ by algebraic polynomials, as well as the result of Newman [6] and others on the rational approximation of $|x|$ play a crucial role in Approximation Theory. However, despite the effort of many mathematicians the exact value of the approximation and the explicit form of the elements of best approximations are not known yet in both cases. So, it was rather striking that the study of the parametric approximation of $|x|$ goes so smoothly. Sendov found in [10] an explicit expression for the pair (P, Q) of best parametric approximation to $|x|$ and the exact value of this approximation for each odd n. Namely this elegant result inspired immediately an interest in the subject. The case of even n stayed as an open problem for some time and was discussed in the research seminar on Approximation theory, run by Bl. Sendov. In [2] we found the extremal pair and the value of the best approximation in the even case. Both the papers of Sendov and [2] have been published in Bulgarian, and in addition, the exposition in my paper is not very detailed which makes the text difficult to understand. Meanwhile we found a simpler proof and the purpose of this note is to present it here.

As usual, denote by $T_n(x)$ the Tchebycheff polynomial of degree n,

$$T_n(x) := \cos(n \arccos x) \quad \text{for} \quad -1 \le x \le 1.$$

Introduce the polynomials

$$P_T(x) := \delta \frac{T_n(2x+1) - T_n(-2x+1)}{2},$$

$$Q_T(x) := \delta \frac{T_n(2x+1) + T_n(-2x+1)}{2},$$

where δ is chosen to normalize P_T by the condition $P_T(1) = 1$. Clearly

$$\delta = \frac{2}{T_n(3) - T_n(-1)} = \frac{2}{T_n(3) - (-1)^n}$$

and therefore

$$\|\,|P_T| - Q_T\| = \delta = \frac{2}{T_n(3) - (-1)^n} =: E_T$$

Sendov [10] proved that P_T, Q_T is the unique pair of best parametric approximation to $|x|$ in case of odd n and conjectured that this should be true for all n.

We shall consider a slightly more general problem allowing P to vary in the wider class H_n of polynomials $P \in \pi_n$ satisfying the requirements:

$$P(-1) = -1, \quad P(1) = 1, \quad |P(x)| \le 1 \quad \text{on } [-1,1],$$

$$P \ \text{ changes sign only at one point in } (-1,1).$$

The point of sign change of P will be denoted in this note by $\alpha = \alpha(P)$. We are going to give a simple proof to the following assertion.

Theorem 1 *The equality*

$$E_{nn}(|x|) := \inf_{P \in H_n, Q \in \pi_n} \| \, |P| - Q \| = \| \, |P_T| - Q_T \| = \frac{2}{T_n(3) - (-1)^n}$$

holds for every natural n. *The element of best approximation* (P_T, Q_T) *is unique in the set* $H_n \times \pi_n$.

For convenience, let us introduce first some notations and prove an auxiliary lemma.

With any pair (P, Q) of admissible approximating polynomials we associate the polynomials

$$W_1(x) := -P(x) + Q(x)$$

$$W_2(x) := P(x) + Q(x).$$

Our new proof of Theorem 1 is based on the following lemma.

Lemma 2 *Assume that* W_1 *and* W_2 *are two polynomials from* π_n *satisfying*

$$\begin{aligned} &|W_1(x)| \le E \ \ on \ \ [\alpha, 1], \qquad |W_1(x)| > W_2(x) \ on \ [1, \alpha), \\ &|W_2(x)| \le E \ \ on \ \ [-1, \alpha], \ \ |W_2(x)| > W_1(x) \ on \ (\alpha, 1], \end{aligned} \tag{1.1}$$

with some E *and* $\alpha \in (-1,1)$. *Let*

$$\begin{aligned} W_2(1) - W_1(1) &> 2, \\ W_1(-1) - W_2(-1) &> 2. \end{aligned} \tag{1.2}$$

Then there exists a pair of polynomials $p \in H_n$ *and* $q \in \pi_n$ *such that*

$$\| \, |p| - q \| < E \quad on \quad [-1,1].$$

Proof. Evidently there is a constant $\lambda < 1$ such that the polynomials

$$V_1(x) := \lambda W_1(x), \qquad V_2(x) := \lambda W_2(x)$$

satisfy still (1.2) and (1.1) with $E_0 := \lambda E < E$. Introduce the polynomials

$$P_0(x) := V_2(x) - V_1(x), \qquad Q_0(x) := V_1(x) + V_2(x).$$

Clearly $P_0(x)$ changes sign only once on $[-1, 1]$ (at α). In view of the assumptions on W_1 and W_2, and the definition of V_1, V_2, we have

$$\begin{aligned} P_0(1) &= \frac{V_2(1) - V_1(1)}{2} > 1, \\ P_0(-1) &= \frac{V_2(-1) - V_1(-1)}{2} < -1. \end{aligned}$$

Thus, there exist points a, b, $-1 < a < \alpha < b < 1$, such that $P_0(a) = -1$, $P_0(b) = 1$. Then the polynomials

$$p(x) := P_0(\frac{b-a}{2}x + \frac{a+b}{2}), \quad q(x) := Q_0(\frac{b-a}{2}x + \frac{a+b}{2})$$

provide the required approximation. This completes the proof.

Proof of Theorem 1. Let P, Q be an arbitrary admissible pair of approximating polynomials. Assume that

$$\| |P| - Q\| = E.$$

Without loss of generality we may assume that $\alpha(P) \geq 0$. It was noted by Sendov [10] that

$$E_1 \geq \frac{2}{T_n(3) + 1}. \tag{1.3}$$

The proof is very simple. We clearly have

$$|P(x) + Q(x)| \leq E \quad \text{for} \quad \alpha \in [-1, \alpha]$$

and

$$P(1) + Q(1) = 1 + Q(1) \geq 1 + 1 - E = 2 - E.$$

Since

$$\|E\, T_n(2x + 1)\| = E \quad \text{on} \quad [-1, 0],$$

we conclude from a basic property of the Tchebycheff polynomials that

$$|E\, T_n(2x + 1)| \geq |P(x) + Q(x)| \quad \text{for all} \quad x > 0.$$

In particular for $x = 1$ we get

$$E\,T_n(3) \geq |P(1) + Q(1)| \geq 2 - E$$

which implies (1.3).

The theorem follows from (1.3) if n is an odd number. Assume further that n is even.

Let P, Q be an extremal pair of polynomials, that is,

$$E := \|\,|P| - Q\| = \inf_{R \in H_n, S \in \pi_n} \|\,|R| - S\|.$$

Assume without loss of generality that $\alpha(P) \geq 0$. We shall show that (P, Q) coincides with the "Tchebycheff" pair (P_T, Q_T). In order to do this consider the corresponding polynomials $W_1(x), W_2(x)$ associated with (P, Q). Define new polynomials $\tau_1(x), \tau_2(x)$ by the following conditions: τ_1 is the Tchebycheff polynomial for the interval $[\alpha, 1+\varepsilon_1]$ normalized by $\|\tau_1\| = E$ on $[\alpha, 1+\varepsilon_1]$. Here $\varepsilon_1 \geq 0$ is chosen so that $\tau_1(1) = W_1(1)$ and $\tau_1'(x) \neq 0$ on $(1, \infty)$. Similarly, $\tau_2(x)$ is the Tchebycheff polynomial for the interval $[-1-\varepsilon_2, \alpha]$ normalized by $\|\tau_2\| = E$ on $[-1-\varepsilon_2, \alpha]$, where $\varepsilon_2 \geq 0$ is chosen so that $\tau_2(-1) = W_2(-1)$ and $\tau_2(x) \neq 0$ on $(-\infty, -1)$.

It is easily seen that

$$\tau_2(x) \geq W_2(x) \quad \text{for } x \geq \alpha, \tag{1.4}$$
$$\tau_1(x) \geq W_1(x) \quad \text{for } x \leq \alpha.$$

Let us prove the first relation. The proof of the second one is similar.

It follows from the oscillating property of the Tchebycheff polynomials that $\tau_2(x) - W_2(x)$ vanishes at n distinct points in $[-1, \alpha]$. Since $\tau_2(\alpha) = E \geq W_2(\alpha)$, the assumption $W_2(1) > \tau_2(1)$ will lead to existence of one additional zero of $\tau_2 - W_2$ and consequently, to contradiction. Similarly the assumption $W_2(1) = \tau_2(1)$ leads to $W_2 \equiv \tau_2$. The relation (1.4) is proved.

If $\tau_2(1) > W_2(1)$ and $\tau_1(-1) > W_1(-1)$, then

$$\begin{aligned} \tau_1(-1) - \tau_2(-1) &> W_1(-1) - W_2(-1) = 2 \\ \tau_2(1) - \tau_1(1) &> W_2(1) - W_1(1) = 2 \end{aligned}$$

and, by Lemma 2, the approximation E provided by P, Q can be improved. Then P, Q is not a extremal pair, a contradiction. Moreover, even if only one of the above inequalities is strict, say $\tau_2(1) \geq W_2(1)$, then $\lambda\tau_2(x)$ with some λ very close to 1, $0 < \lambda < 1$, will satisfy the conditions

$$\lambda\tau_2(1) - \tau_1(1) > 2, \quad \tau_1(-1) - \lambda\tau_2(-1) > 2$$

and we get a contradiction too, by Lemma 2. Therefore, the only possibility for the polynomials W_1 and W_2 is

$$W_1 \equiv \tau_1, \qquad W_2 \equiv \tau_2.$$

Next we shall show that $\varepsilon_1 = \varepsilon_2 = 0$. Indeed, assume for definiteness that $|\tau_2'(\alpha)| \geq |\tau_1'(\alpha)|$ (the opposite case is studied similarly). Then, in view of the known property of the Tchebycheff polynomial $|T_n'(x)| \leq T_n'(1)$ for all $x \in [-1, 1]$, we have

$$\tau_1'(-1) > \tau_2'(\alpha) \geq \tau_2'(x) \quad \text{for all} \quad x \in [-1-\varepsilon_2, \alpha]$$

and thus

$$\tau_2(-1-\varepsilon_2) - \tau_1(-1-\varepsilon_2) > \tau_2(-1) - \tau_1(-1) = 2.$$

Now the linear transformation of the polynomials τ_1, τ_2 from $[-1-\varepsilon_2, 1+\varepsilon_1]$ to $[-1, 1]$ would induce by Lemma 2 a better approximation than E. Therefore $\varepsilon_1 = 0$.

So, we arrived in the case when one of the polynomials W coincides with one of τ. We shall show that the other couple coincides too. Indeed, assume for definiteness that $W_2(x) \equiv \tau_2(x)$. Precisely, according to the definition of τ_2,

$$W_2(x) = ET_n\left(\frac{2}{1+\alpha}x + \frac{1-\alpha}{1+\alpha}\right).$$

Since n is even, we have $W_2(1) \geq 2 - E$. Thus

$$T_n\left(\frac{3-\alpha}{1+\alpha}\right) \geq \frac{2}{E} - 1 \geq \frac{2}{E_T} - 1 = T_n(3) - 2$$

and therefore

$$\begin{aligned} 2 &\geq T_n(3) - T_n\left(\frac{3-\alpha}{1+\alpha}\right) \\ &\geq \left(3 - \frac{3-\alpha}{1+\alpha}\right) \cdot T_n'(1) = \frac{4\alpha}{1+\alpha}n^2. \end{aligned}$$

Hence $1 + \alpha \geq 2\alpha n^2$ and this yields $\alpha \leq \frac{1}{2n^2-1}$. Since n is even, $n \geq 2$ and thus

$$\alpha \leq \frac{1}{7}. \tag{1.5}$$

Having this estimation of α it is not difficult to show now that $W_2'(1) > W_1'(1)$. We shall show even more, namely

$$W_2'(1) > ET_n'\left(\frac{2}{1-\alpha}x - \frac{1+\alpha}{1-\alpha}\right)\Big|_{x=1} \cdot \frac{2}{1-\alpha}.$$

Indeed, assume the contrary, that is,

$$W_2'(1) = T_n'\left(\frac{3-\alpha}{1+\alpha}\right)\frac{2}{1+\alpha} \leq \frac{2}{1-\alpha}T_n'(1) = \frac{2}{1-\alpha}n^2.$$

This is equivalent to

$$T_n'\left(\frac{3-\alpha}{1+\alpha}\right) \leq \frac{1+\alpha}{1-\alpha}n^2.$$

But, by Taylor's expression,

$$T_n'\left(\frac{3-\alpha}{1+\alpha}\right) > T_n'(1) + T_n''(1)\left(\frac{3-\alpha}{1+\alpha} - 1\right).$$

Combining both last inequalities and taking into account (1.5), we get

$$\frac{2n^2+1}{3} \leq \left(\frac{1+\alpha}{1-\alpha}\right)^2 \leq \left(\frac{1+1/7}{1-1/7}\right)^2$$

which leads to $n < 2$. Thus $W_2'(1) > W_1'(1)$ for all even natural numbers n. Now we derive from this that $W_2(1+h) - W_1(1+h) > 2$ for a very small $h > 0$ and then the polynomials W_2 and W_1, considered on $[-1, 1+h]$, will induce by Lemma 2 a better approximation to $|x|$ than P, Q, provided $W_1(1) < E$. Thus $W_1(1) = E$ and hence $\varepsilon_2 = 0$. So we proved that W_1 and W_2 coincide with the Tchebycheff polynomials for the intervals $[-1, \alpha]$ and $[\alpha, 1]$, respectively. Since $W_2(1) = W_1(-1) = 2 - E$ we conclude that $\alpha = 0$ and therefore $(P, Q) = (P_T, Q_T)$. The proof is complete.

References

[1] S. N. Bernstein, Sur la valeur asymptotique de la meilleure approximation de $|x|$, Acta Mathematica 37,(1913), 1-57.

[2] B. Bojanov, Uniform parametric approximation of $|x|$ by algebraic polynomials, Ann. Univ. Sofia, Faculté de Math. 64(1971), 331-337 (in Bulgarian).

[3] G. Iliev, An improvement of Sendov's estimation for parametric approximation of partially analytic functions, Pliska, Studia math. bulg. 1(1977), 93-99.

[4] G. Iliev, Estimation of the parametric approximation of analytic functions in an open interval, Compt. rend. Acad. bulg. Sci. 30, 5(1977), 653-655.

[5] S. Konyagin, Parametric approximation of piecewise analytic functions, Mat. Zametki 48, 4(1990), 58-68 (in Russian).

[6] D. J. Newman, Rational approximation to $|x|$, Michigan Math. J. 11(1964), 11-14.

[7] M. Nikolcheva, Compt. rend. Acad. bulg. Sci. 29, 4(1976), 469.

[8] V. Popov and G. Iliev, Parametric approximation of piecewise analytic functions, Pliska, Studia math. bulg. 1(1977), 72-78.

[9] Bl. Sendov, Certain questions in the theory of approximation of functions and sets in Hausdorff metric, Uspehi Mat. Nauk v. XXIV, 5(149) (1969), 141-178 (in Russian).

[10] Bl. Sendov, Parametric approximation, Ann. Univ. Sofia, Faculté de Math. 64(1971), 237-247. (in Bulgarian).

[11] J. Szabados, On parametric approximation, Acta Math. Acad. Sci. Hung. 23(1972), 275-287.

2 On the Walsh estimation for the zeros of polynomials

We proved in [1] the following: *Let x, y, z be the positive roots of the equations*

$$\begin{aligned} x^n &= a_1 x^{n-1} + a_2^2 x^{n-2} + \cdots + a_n^n \quad (a_i \geq 0 \textit{ for all } i), \\ y^n &= b_1 y^{n-1} + b_2^2 y^{n-2} + \cdots + b_n^n \quad (b_i \geq 0 \textit{ for all } i), \\ z^n &= (a_1 + b_1) z^{n-1} + (a_2 + b_2)^2 z^{n-2} + \cdots + (a_n + b_n)^n, \end{aligned}$$

respectively. Then

$$z \leq x + y. \tag{2.1}$$

The proof is short and we shall reproduce it here for completeness. The first two equations can be rewritten in the form

$$\begin{aligned} 1 &= \sum_{k=1}^{n} \left(\frac{a_k}{x}\right)^k, \\ 1 &= \sum_{k=1}^{n} \left(\frac{a_k}{x}\right)^k. \end{aligned}$$

Since the function $\varphi(t) := t^k$ is convex on $(0, \infty)$, that is,

$$\alpha\varphi(t_1) + (1-\alpha)\varphi(t_2) \geq \varphi(\alpha t_1 + (1-\alpha)t_2)) \quad (0 \leq \alpha \leq 1),$$

we add the above equations multiplied by $\alpha := x/(x+y)$ and $1-\alpha$, respectively, and get

$$1 = \sum_{k=1}^{n} \left[\frac{x}{x+y}\left(\frac{a_k}{x}\right)^k + \frac{y}{x+y}\left(\frac{b_k}{y}\right)^k\right] \geq \sum_{k=1}^{n} \left(\frac{a_k + b_k}{x+y}\right)^k$$

which is equivalent to

$$(x+y)^n \geq \sum_{k=1}^{n} (a_k + b_k)^k \, (x+y)^{n-k}.$$

The latter shows that $z \leq x + y$ and (2.1) is proved.

Let us state (2.1) in its more general form.

Theorem 3 *Let $\{x_j\}_{j=1}^m$ be the positive roots of the equations*

$$x^n = \sum_{k=1}^{n} a_{jk}^k x^{n-k}, \quad j = 1, \ldots, m,$$

where $\{a_{jk}\}$ are non-negative numbers. Then the positive root z of the equation

$$x^n = \sum_{k=1}^{n} \left(\sum_{j=1}^{m} a_{jk}^k\right)^k x^{n-k}$$

satisfies the inequality

$$z \leq x_1 + \cdots + x_m.$$

The assertion clearly follows by repeated application of (2.1).

Now we shall show that the following known estimation due to Walsh [3] is a simple consequence of Theorem 3.

Walsh estimation. *Let z be any root of the equation*

$$x^n + a_1 x^{n-1} + \cdots + a_n = 0$$

where $a_1, \ldots, a_n$ are arbitrary complex numbers. Then

$$|z| \leq |a_1| + |a_2|^{\frac{1}{2}} + |a_3|^{\frac{1}{3}} + \cdots + |a_n|^{\frac{1}{n}}.$$

Proof. By the classical Cauchy theorem (see for example [2]), $|z| \leq R$, where R is the unique positive root of the equation

$$x^n = |a_1|x^{n-1} + \cdots + |a_{n-1}|x + |a_n|.$$

We need only to apply Theorem 2.1 to the equations

$$x^n = |a_k|x^{n-k} \quad \text{(with a root } x_k = |a_k|^{\frac{1}{k}}\text{)}$$

for $k = 1, \ldots, n$, in order to get (2.2).

References

[1] B. Bojanov, On an estimation of the roots of algebraic equations, Zastos. Mat. XI, 2(1970), 195-205.

[2] M. Marden, "Geometry of Polynomials", Mathematical Surveys, Number 3, Amer. Math. Soc., Providence, 1966.

[3] J. L. Walsh, An inequality for the roots of an algebraic equation, Ann. of Math. 25(1924), 285-286.

3 On the coefficients of the Gauss quadrature formula

Let $\mu(x)$ be a given weight function on $[a, b]$. Denote by $P_0(x), P_1(x), \ldots$ the sequence of orthogonal polynomials associated with $\mu(x)$ and $[a, b]$. Assume that the polynomials $\{P_n\}$ are orthonormalized, that is,

$$\int_a^b \mu(x) P_n^2(x)\, dx = 1 \quad \text{for all} \quad n.$$

For convenience, we shall further assume that P_n is of the form

$$P_n(x) = \alpha_n (x - x_1) \cdots (x - x_n)$$

with $\alpha_n > 0$. Let

$$\int_a^b \mu(x) f(x)\, dx \approx \sum_{k=1}^n A_k f(x_k) \tag{3.1}$$

be the Gauss quadrature formula with a weight $\mu(x)$ in $[a,b]$. It is known that the nodes $\{x_k\}_1^n$ are located at the zeros of the polynomial P_n. The coefficients $\{A_k\}$ of (3.1) can be computed by the formula

$$A_k = \frac{\alpha_n}{\alpha_{n-1}} \frac{1}{P_n'(x_k) P_{n-1}(x_k)}, \quad k = 1, \ldots, n. \tag{3.2}$$

This is a well-known fact and the proof can be found in any text on numerical integration. All proofs we have seen rely on the Cristoffel-Darboux theorem for orthogonal polynomials. The goal of this note is to present a simple direct proof of (3.2) which uses only properties of the divided difference.

Proof of (3.2). Consider the summation functional

$$D[f] := \alpha_{n-1} \sum_{k=1}^n A_k\, P_{n-1}(x_k) f(x_k). \tag{3.3}$$

If $f \in \pi_{n-1}$ then $fP_{n-1} \in \pi_{2n-2}$, and thus (1) integrates fP_{n-1} exactly. Therefore

$$D[f] = \alpha_{n-1} \int_a^b \mu(x) f(x) P_{n-1}(x)\, dx$$

and thus, in view of the orthogonality of P_{n-1}, we get $D[f] = 0$ for $f(x) = 1, x, \ldots, x^{n-2}$. In addition, for $f(x) = x^{n-1}$, we have

$$D[x^{n-1}] = \alpha_{n-1} \int_a^b \mu(x) x^{n-1} P_{n-1}(x)\, dx = \int_a^b \mu(x) P_{n-1}^2(x)\, dx = 1.$$

These two properties characterize completely the divided difference $f[x_1, \ldots, x_n]$. Therefore

$$D[f] = f[x_1, \ldots, x_n] = \alpha_n \sum_{k=1}^n \frac{f(x_k)}{P_n'(x_k)}.$$

Now comparing the coefficients of $f(x_k)$ in this expression and those in (3.3), we get (3.2). The proof is complete.

Remark: This observation was mentioned in [1].

Reference ,

[1] Borislav Bojanov, "Lectures on Numerical Analysis", Darba, Sofia, 1995.

4 Smoothing of T-systems by Bernstein polynomials

A set of functions $u_0(t), \ldots, u_N(t)$ is called a *Tchebycheff system (or briefly, T-system)* on the interval $\mathcal{A}$ if any non-zero linear combination

$$u(t) = a_0 u_0(t) + \cdots + a_N u_N(t)$$

with real coefficients $\{a_k\}$ has no more than N distinct zeros in $\mathcal{A}$.

Tchebycheff systems are widely used in Analysis and Applications. They are natural extensions of the algebraic system $1, x, x^2, \ldots, x^n$. A stronger requirement concerning the zero counting of the smooth functions leads to the following notion.

The set $\{u_j\}_{j=0}^N$ of N times continuously differentiable functions on the interval $\mathcal{A}$ is called an *extended Tchebycheff system (ET-system)* if every non-zero $u(t)$ from span $\{u_j\}_{j=0}^N$ has no more than N zeros in $\mathcal{A}$ counting the multiplicities.

The main trick in working with non-smooth T-systems is to approximate the original functions $u_0, \ldots, u_N$ by some smooth functions $\hat{u}_0, \ldots, \hat{u}_N$ which preserve the characteristic property of the T-systems and then work with the latter. The commonly used smoothing operator is defined by the Gaussian transform

$$f \to G_\varepsilon(f; x) := \frac{1}{\sqrt{2\pi}\,\varepsilon} \int_0^1 f(t) e^{-(x-t)^2/2\varepsilon^2}\, dt\,.$$

It is known that it transforms an arbitrary T-system into ET-system. The purpose of this note is to show that the Bernstein polynomials possess also this useful property.

The Bernstein polynomial $B_n(f; t)$ of degree n for the function f, defined on $[0, 1]$, is given by the formula

$$B_n(f; t) = \sum_{k=0}^{n} \binom{n}{k} f\left(\frac{k}{n}\right) t^k (1-t)^{n-k}.$$

$B_n(f; t)$ can be presented as an expansion along the powers of t in the following way

$$B_n(f;t) = \sum_{k=0}^{n} \binom{n}{k} \Delta^k f_0 \, t^k. \tag{4.1}$$

Here $\Delta^m f_0$ is the finite difference of the function f at the points $0, \frac{1}{n}, \ldots, \frac{m}{n}$. The proof of this known formula can be found for example in the book of Natanson [1].

Theorem 4 *Let $\{u_i\}_0^N$ be a Tchebycheff system on $[0,1]$. Then for each natural n the Bernstein transforms*

$$\{B_n(u_0), B_n(u_1), \ldots, B_n(u_N)\}$$

of $\{u_i\}_0^N$ form an extended Tchebycheff system on $[0,1]$.

Proof. We shall denote by $Z(f;(a,b))$ the total number of zeros of f in (a,b) (counting the multiplicity). According to the classical *Descartes' rule* the polynomial $c_0x^n + \ldots + c_{n-1}x + c_n$ has no more positive zeros (counting multiplicities) than the number $S^-(c_0, \ldots, c_n)$ of strong sign changes in the sequence $c_0, \ldots, c_n$ of its coefficients. Using this result one can show that

$$Z(q;(0,1)) \leq S^-(a_0, \ldots, a_n) \tag{4.2}$$

for each polynomial q of the form

$$q(x) = \sum_{k=0}^{n} a_k x^k (1-x)^{n-k}.$$

The proof is very simple: Set $x = 1/(1+t)$. Then

$$g(t) := q\left(\frac{1}{1+t}\right) = (1+t)^{-n} \sum_{k=0}^{n} a_k t^{n-k}$$

and by the Descartes rule,

$$Z(q;(0,1)) = Z(g;(0,\infty)) \leq S^-(a_0, \ldots, a_n).$$

Now we are ready to prove the theorem. Assume that

$$f(t) = \sum_{k=0}^{N} \alpha_k B_n(u_k;t)$$

is an arbitrary linear combination of $\{B_n(u_k;t)\}_{k=0}^N$. We have to show that f has no more than N zeros in $[0,1]$ counting the multiplicities . In order to do this observe first that f may be written as a transform of the polynomial $p(t) = \sum_{k=0}^{N} \alpha_k u_k(t)$. Precisely, $f(t) = B_n(p;t)$ and hence

$$f(t) = \sum_{k=0}^{n} \binom{n}{k} p\left(\frac{k}{n}\right) t^k (1-t)^{n-k}.$$

Then, according to the estimation (4.2) just proved, we have

$$Z(f;(0,1)) \leq S^-(p(0), p\left(\frac{1}{n}\right), \ldots, p(1)). \tag{4.3}$$

Now suppose that f vanishes at 0 with multiplicity j. Since

$$f^{(m)}(0) = B_n^{(m)}(p;0) \quad \text{for} \quad m = 0, \ldots, n,$$

and, in view of the presentation (4.1),

$$B_n^{(m)}(p;0) = m! \binom{n}{m} \Delta^m p_0,$$

the assumption $f^{(m)}(0) = 0$ leads to $\Delta^m p_0 = 0$. But clearly the equations $\Delta^m p_0 = 0$ for $m = 0, \ldots, j-1$ are equivalent to

$$p(0) = p\left(\frac{1}{n}\right) = \ldots = p\left(\frac{j-1}{n}\right) = 0. \tag{4.4}$$

Similarly, if f has a zero at 1 of multiplicity i, then

$$p(1) = p\left(\frac{n-1}{n}\right) = \ldots = p\left(\frac{n-i-1}{n}\right) = 0. \tag{4.5}$$

Since $p \in$ span $\{u_k\}_0^N$, and $\{u_k\}_0^N$ is a T-system, p has no more than N distinct zeros in $[0,1]$. But because of (4.4) and (4.5), p has no more than $N-i-j$ distinct zeros in $\left(\frac{j-1}{n}, \frac{n-i+1}{n}\right)$ and consequently, no more than $N-i-j$ sign changes there. Thus

$$\begin{aligned} S^-\left(p(0), p\left(\frac{1}{n}\right), \ldots, p(1)\right) &= S^-\left(p\left(\frac{j-1}{n}\right), \ldots, p\left(\frac{n-i+1}{n}\right)\right)) \\ &\leq N-i-j. \end{aligned}$$

This and (4.3) yield $Z(f;(0,1)) \leq N-i-j$ and therefore

$$Z(f;[0,1]) = i + j + Z(f;(0,1)) \leq N.$$

The theorem is proved.

Reference

[1] I. P. Natanson, "Constructive Function Theory", Vols. I, II, III, Ungar, New York, 1964, 1965.

5 Mean interpolation by splines

Denote by $S_{r-1}(x_1, \ldots, x_n)$ the linear space of all polynomial splines of degree $r-1$ with knots at $x_1 \leq \cdots \leq x_n$ ($x_i < x_{i+r}$ for all i). Let $\{I_k\}_{k=1}^{n}$ be a given sequence of ordered closed subintervals, that is,

$$I_k = [\alpha_k, \beta_k], \quad k = 1, \ldots, n+r, \quad \beta_k \leq \alpha_{k+1}.$$

Consider the interpolation problem

$$\text{(5.1)} \qquad \int_{I_k} s(t)\,dt = F_k, \quad k = 1, \ldots, n+r, \quad s \in S_{r-1}(x_1, \ldots, x_n).$$

Kobza and Zapalka [2] studied this problem in case of splines of low degree. We shall show here that the characterization of the mean interpolation problem (5.1) is easily reduced to a treatment of the ordinary spline interpolation.

Theorem 5 *The interpolation problem (5.1) has a unique solution for each given set of values $\{F_k\}$ if and only if the knots $\{x_k\}$ and the subintervals $\{I_k\}_{k=1}^{n+r}$ satisfy the interlacing conditions:*

$$I_k \cap (x_{k-r}, x_k) \neq \emptyset, \quad k = 1, \ldots, n+r,$$

where $x_i := -\infty$ for $i = 0, \ldots, r-1$ and $x_j := \infty$ for $j = n+1, \ldots, n+r$.

Proof. Any spline $s(t)$ from $S_{r-1}(x_1, \ldots, x_n)$ can be presented (see, for example, [1]) in the form

$$s(t) = \sum_{i=1}^{r} a_i t^{i-1} + \sum_{k=1}^{n} c_k (x_k - t)_+^{r-1}$$

with some real coefficients $\{a_i\}$ and $\{c_k\}$. Thus the interpolation conditions (5.1) can be rewritten as a system of $n+r$ linear equations with respect to the unknowns $\{a_i\}$ and $\{c_k\}$:

$$\sum_{i=1}^{r} a_i \int_{I_k} t^{i-1}\,dt_k + \sum_{k=1}^{n} c_k \int_{I_k} (x_k - t)_+^{r-1}\,dt_k = F_k, \quad k = 1, \ldots, n+r.$$

Here t_k are independent variables running over I_k, respectively. Since $\{I_k\}$ are ordered and not overlapping, we clearly have

$$t_1 \le t_2 \le \cdots \le t_{n+r}. \tag{5.2}$$

Denote by D the determinant of the last linear system. For the sake of convenience, set

$$\begin{aligned} u_j(t) &:= t^{j-1} \text{ for } j = 1, \dots, r, \\ u_k(t) &:= (x_k - t)_+^{r-1} \text{ for } k = r+1, \dots, n+r. \end{aligned}$$

Observe that

$$D = \begin{bmatrix} \int_{I_1} u_1(t_1)\,dt_1 & \dots & \int_{I_1} u_{n+r}(t_1)\,dt_1 \\ \int_{I_2} u_1(t_2)\,dt_2 & \dots & \int_{I_2} u_{n+r}(t_2)\,dt_2 \\ \vdots & \ddots & \vdots \\ \int_{I_{n+r}} u_1(t_{n+r})\,dt_{n+r} & \dots & \int_{I_{n+r}} u_{n+r}(t_{n+r})\,dt_{n+r} \end{bmatrix}$$

$$= \int_{I_1}\int_{I_2}\cdots\int_{I_{n+r}} \begin{bmatrix} u_1(t_1) & \dots & u_{n+r}(t_1) \\ \vdots & \ddots & \vdots \\ u_1(t_{n+r}) & \dots & u_{n+r}(t_{n+r}) \end{bmatrix} dt_1\,dt_2 \dots dt_{n+r}.$$

By the total positivity of the truncated power function, it follows from (5.2) that the determinant $U(\mathbf{x}, \mathbf{t})$ under the integrals in the last expression is non-negative. Moreover, by virtue of the Schoenberg-Whitney theorem (see, for example [1]), $U(\mathbf{x}, \mathbf{t}) > 0$ if and only if $t_j \in (x_{j-r}, x_j)$ for all $j = 1, \dots, n+r$, or equivalently, if the condition in the statement of the theorem is fulfilled. Note here that $U(\mathbf{x}, \mathbf{t})$ is continuous from the right as a function of $\mathbf{t} := (t_1, \dots, t_{n+r})$ and thus, if $U(\mathbf{x}, \mathbf{t})$ is positive at one point $\mathbf{t}$ it will be positive on a subinterval, and then the integral (that is, D) will be positive. Therefore $D > 0$ if and only if there exists at least one set of points $t_k \in I_k,\ k = 1, \dots, n+r$ such that $U(\mathbf{x}, \mathbf{t}) > 0$ and hence, if and only if the condition in the statement of the theorem is fulfilled . The proof is complete.

The same method can be applied to the interpolation problem

$$\int_{I_k} \sum_{j=1}^{m} a_m u_m(t)\,dt = F_k, \quad k = 1, \dots, m, \tag{5.3}$$

with respect to any functions $\{u_j\}$ that constitute a weak Tchebycheff system. The following theorem is obviously true.

Theorem 6 *Let* $\{u_k\}_{k=1}^m$ *be a weak Tchebycheff system of continuous functions on* $[a,b]$. *Assume that*

$$\det \{u_k(t_j)\}_{k=1}^{m}{}_{j=1}^{m} > 0$$

at least for one set of points $\{t_1, \ldots, t_m\}$ *such that* $t_k \in I_k$, $k = 1, \ldots, m$. *Then the interpolation problem (5.3) has a unique solution.*

References

[1] B. Bojanov, H. Hakopian, A. Sahakian, "Spline Functions and Multivariate Approximations", Kluwer Academic Publishers, Dordrecht, 1993.

[2] Jiri Kobza and Dusan Zapalka, Natural and smoothing quadratic splines, Appl. Math. 3(1991), 187-204.

Müntz's Theorem on Compact Subsets of Positive Measure

Peter Borwein
Department of Mathematics and Statistics
Simon Fraser University, Burnaby, B. C., Canada V5A 1S6

Tamás Erdélyi
Department of Mathematics, Texas A&M University
College Station, Texas 77843

Abstract

The principal result of this paper is a Remez-type inequality for Müntz polynomials:

$$p(x) := \sum_{i=-n}^{n} a_i x^{\lambda_i},$$

or equivalently for Dirichlet sums:

$$P(t) := \sum_{i=-n}^{n} a_i e^{-\lambda_i t},$$

where $(\lambda_i)_{i=-\infty}^{\infty}$ is a sequence of distinct real numbers. The most useful form of this inequality states that for every sequence $(\lambda_i)_{i=-\infty}^{\infty}$ satisfying

$$\sum_{\substack{i=-\infty \\ \lambda_i \neq 0}}^{\infty} \frac{1}{|\lambda_i|} < \infty$$

there is a constant c depending only on $(\lambda_i)_{i=-\infty}^{\infty}$, A, α, and β (and not on n or A) so that the inequality

$$\|p\|_{[\alpha,\beta]} \leq c\,\|p\|_A$$

holds for every Müntz polynomial p, as above, associated with $(\lambda_i)_{i=-\infty}^{\infty}$, for every set $A \subset [0,\infty)$ of positive Lebessgue measure, and for every

$$[\alpha, \beta] \subset (\text{ess inf } A, \text{ess sup } A).$$

Research of P. Borwein is supported, in part, by NSERC of Canada. Research of T. Erdélyi is supported, in part, by the National Science Foundation of the USA under Grant No. DMS-9623156 and conducted while an International Postdoctoral Fellow of the Danish Research Council at University of Copenhagen.

Here $\|\cdot\|_A$ denotes the supremum norm on A.

This Remez-type inequality allows us to resolve several problems. Most notably we show that the Müntz-type theorems of Clarkson, Erdős, and Schwartz on the denseness of

$$\text{span}\{x^{\lambda_i} : i \in \mathbb{Z}\}, \qquad \lambda_i \in \mathbb{R} \text{ distinct}$$

on $[a, b]$, $a > 0$, remain valid with $[a, b]$ replaced by an arbitrary compact set $A \subset (0, \infty)$ of positive Lebesgue measure. This extends earlier results of the authors under the assumption that the numbers λ_i are nonnegative.

1 Introduction

Müntz's classical theorem characterizes sequences $\Lambda := (\lambda_i)_{i=0}^{\infty}$ with

$$0 = \lambda_0 < \lambda_1 < \lambda_2 < \cdots$$

for which the Müntz space

$$\text{span}\{x^{\lambda_0}, x^{\lambda_1}, \dots\}$$

is dense in $C[0, 1]$. Here, and in what follows, $\text{span}\{x^{\lambda_0}, x^{\lambda_1}, \dots\}$ denotes the collection of finite linear combinations of the functions $x^{\lambda_0}, x^{\lambda_1}, \dots$ with real coefficients, and $C(A)$ is the space of all real-valued continuous functions on $A \subset [0, \infty)$ equipped with the uniform norm. Müntz's Theorem [11, 18, 27, 30] states the following.

Theorem 1.1 *Suppose $(\lambda_i)_{i=0}^{\infty}$ is an increasing sequence of nonnegative real numbers with $\lambda_0 = 0$. The Müntz space* $\text{span}\{x^{\lambda_0}, x^{\lambda_1}, \dots\}$ *is dense in $C[0, 1]$ if and only if $\sum_{i=1}^{\infty} 1/\lambda_i = \infty$.*

The original Müntz Theorem proved by Müntz [18] in 1914, by Szász [27] in 1916, and anticipated by Bernstein [3] was only for sequences of exponents tending to infinity. The point 0 is special in the study of Müntz spaces. Even replacing $[0, 1]$ by an interval $[a, b] \subset (0, \infty)$ in Müntz's Theorem is a non-trivial issue. This is, in large measure, due to Clarkson and Erdős [12] and Schwartz [24] whose works include the result that if $\sum_{i=1}^{\infty} 1/\lambda_i < \infty$, then every function belonging to the uniform closure of $\text{span}\{x^{\lambda_0}, x^{\lambda_1}, \dots\}$ on $[a, b]$ can be extended analytically throughout the region

$$\{z \in \mathbb{C} \setminus (-\infty, 0] : |z| < b\}.$$

There are many generalizations and variations of Müntz's Theorem [1, 4, 5, 6, 7, 8, 9, 16, 17, 19, 24, 26, 28, 29]. There are also still many open problems. For example, the proper generalizations to many variables are still open.

Schwartz [24] extended the results of Clarkson and Erdős to sequences $(\lambda_i)_{i=-\infty}^{\infty}$ of arbitrary real numbers. His main results in this direction are formulated by the next two theorems.

Theorem 1.2 *Suppose $(\lambda_i)_{i=-\infty}^{\infty}$ is a sequence of distinct real numbers. Suppose $0 < a < b$, and $q \in (0, \infty)$. Then*

$$\operatorname{span}\{x^{\lambda_i} : i \in \mathbb{Z}\}$$

is dense in $L^q[a, b]$ if and only if

$$\sum_{\substack{i=-\infty \\ \lambda_i \neq 0}}^{\infty} \frac{1}{|\lambda_i|} = \infty.$$

The same conclusion is valid with $L^q[a, b]$ replaced by $C[a, b]$.

Theorem 1.3 *Suppose $(\lambda_i)_{i=-\infty}^{\infty}$ is a sequence of distinct real numbers satisfying*

$$\sum_{\substack{i=-\infty \\ \lambda_i \neq 0}}^{\infty} \frac{1}{|\lambda_i|} < \infty$$

with $\lambda_i < 0$ for $i < 0$ and $\lambda_i \geq 0$ for $i \geq 0$. Suppose $0 < a < b$, and $q \in (0, \infty)$. Then $\operatorname{span}\{x^{\lambda_i} : i \in \mathbb{Z}\}$ is not dense in $L^q[a, b]$.

Suppose the gap condition

$$\inf\{\lambda_i - \lambda_{i-1} : i \in \mathbb{Z}\} > 0$$

holds. Then every function $f \in L^q[a, b]$ belonging to the $L^q[a, b]$ closure of

$$\operatorname{span}\{x^{\lambda_i} : i \in \mathbb{Z}\}$$

can be represented as

$$f(x) = \sum_{i=-\infty}^{\infty} a_i x^{\lambda_i}, \qquad x \in (a, b).$$

If the above gap condition does not hold, then every function $f \in L^q[a, b]$ belonging to the $L^q[a, b]$ closure of $\operatorname{span}\{x^{\lambda_i} : i \in \mathbb{Z}\}$ can still be represented as an analytic function on

$$\{z \in \mathbb{C} \setminus (-\infty, 0] : a < |z| < b\}.$$

The same conclusion is valid with $L^q[a, b]$ replaced by $C[a, b]$.

In [8] the authors extended Theorem 1.1 and other related results by replacing $[0, 1]$ by an arbitrary compact set $A \subset [0, \infty)$ of positive Lebesgue measure. The main results of this paper, Theorems 3.6 and 3.7, extend Theorems 1.2 and 1.3 to arbitrary compact sets $A \subset (0, \infty)$ of positive Lebesgue measure. Moreover, Theorems 3.6 and 3.7 extend to weighted $L_w^q(A)$ spaces, where w is a nonnegative integrable weight function on A with $\int_A w > 0$.

Theorems 3.6 and 3.7 can be proved fairly simply, once one has established the bounded Remez-type inequality of Theorem 3.1 for non-dense Müntz spaces

$$\operatorname{span}\{x^{\lambda_i} : i \in \mathbb{Z}\}.$$

This is the central result of the paper, and is a result we believe should be a basic tool for dealing with problems about Müntz spaces.

Let $\mathcal{P}_n$ denote the set of all algebraic polynomials of degree at most n with real coefficients. For a fixed $s \in (0,1)$ let

$$\mathcal{P}_n(s) := \{p \in \mathcal{P}_n : m(\{x \in [0,1] : |p(x)| \leq 1\}) \geq s\}$$

where $m(\cdot)$ denotes linear Lebesgue measure. The classical Remez inequality concerns the problem of bounding the uniform norm of a polynomial $p \in \mathcal{P}_n$ on $[0,1]$ given that its modulus is bounded by 1 on a subset of $[0,1]$ of Lebesgue measure at least s. That is, how large can $\|p\|_{[0,1]}$ (the uniform norm of p on $[0,1]$) be if $p \in \mathcal{P}_n(s)$? The answer is given in terms of the Chebyshev polynomials. The extremal polynomials for the above problem are the Chebyshev polynomials $\pm T_n(x) := \pm\cos(n \arccos h(x))$, where h is a linear function that scales $[0,s]$ or $[1-s,1]$ onto $[-1,1]$.

For various proofs, extensions, and applications, see [13, 14, 15, 22, 23].

Our bounded Remez-type inequality of Theorem 3.1 states the following. If $(\lambda_i)_{i=-\infty}^{\infty}$ is a sequence of distinct real numbers satisfying

$$\sum_{\substack{i=-\infty \\ \lambda_i \neq 0}}^{\infty} \frac{1}{|\lambda_i|} < \infty,$$

then there is a constant c depending only on $(\lambda_i)_{i=-\infty}^{\infty}$, A, α, and β (and not on the number of terms in p) so that

$$\|p\|_{[\alpha,\beta]} \leq c\,\|p\|_A$$

for every Müntz polynomial $p \in \operatorname{span}\{x^{\lambda_i} : i \in \mathbb{Z}\}$, for every set $A \subset [0,\infty)$ of positive Lebesgue measure, and for every $[\alpha,\beta] \subset (\operatorname{ess\,inf} A, \operatorname{ess\,sup} A)$.

This extends the Remez-type inequality of the authors [8], where the exponents λ_i are nonnegative. One might note that the existence of such a bounded Remez-type inequality for a Müntz space $\operatorname{span}\{x^{\lambda_i} : i \in \mathbb{Z}\}$ is equivalent to the non-denseness of $\operatorname{span}\{x^{\lambda_i} : i \in \mathbb{Z}\}$ in $C[a,b]$, $0 < a < b$.

The key to the proof of Theorem 3.1 is Theorem 3.2. This theorem states that for the "positive and negative parts" p^+ and p^- of a $p \in \operatorname{span}\{x^{\lambda_i} : i \in \mathbb{Z}\}$, the inequalities

$$\|p^+\|_A \leq c\,\|p\|_A$$

and

$$\|p^-\|_A \leq c\,\|p\|_A$$

hold with a constant c depending only on $(\lambda_i)_{i=-\infty}^{\infty}$ and A (but not on the number of terms in p).

Yet another remarkable consequence of the bounded Remez-type inequality of Theorem 3.1 is that the pointwise and locally uniform convergence of a sequence $(p_n)_{n=1}^{\infty} \subset \text{span}\{x^{\lambda_i} : i \in \mathbb{Z}\}$ on $(0,1)$ are equivalent whenever

$$\sum_{\substack{i=-\infty \\ \lambda_i \neq 0}}^{\infty} \frac{1}{|\lambda_i|} < \infty.$$

See Theorem 3.5. In fact, one can characterize the non-dense Müntz spaces within the Müntz spaces $\text{span}\{x^{\lambda_i} : i \in \mathbb{Z}\}$ as exactly those in which locally uniform and pointwise convergence on $(0,1)$ are equivalent.

2 Notation

The notations

$$\|p\|_A := \sup_{x \in A} |p(x)|,$$
$$\|p\|_{L_w^q(A)} := \left(\int_A |p(x)|^q w(x)\, dx\right)^{1/q},$$

and

$$\|p\|_{L^q(A)} := \left(\int_A |p(x)|^q\, dx\right)^{1/q}$$

are used throughout this paper for measurable functions p defined on a measurable set $A \subset [0,\infty)$, for nonnegative measurable weight functions w defined on A, and for $q \in (0,\infty)$. The space of all real-valued continuous functions on a set $A \subset [0,\infty)$ equipped with the uniform norm is denoted by $C(A)$. If $A := [a,b]$ is a finite closed inerval, then the notation $C[a,b] := C([a,b])$ will be used.

The space $L_w^q(A)$ is defined as the collection of equivalence classes of real-valued measurable functions for which $\|f\|_{L_w^q(A))} < \infty$. The equivalence classes are defined by the equivalence relation $f \sim g$ if $fw = gw$ almost everywhere on A. When $A := [a,b]$ is a finite closed interval, we use the notation $L_w^q[a,b] := L_w^q(A)$. When $w := 1$, we use the notation $L^q[a,b] := L_w^q[a,b]$. Again, it is always our understanding that the space $L_w^q(A)$ is equipped with the $L_w^q(A)$ norm.

The nonnegative-valued functions x^{λ_i} are well-defined on $[0,\infty)$. For a fixed sequence $(\lambda_i)_{i=0}^{\infty}$, the collection of Müntz polynomials

$$p(x) = \sum_{i=0}^{n} a_i x^{\lambda_i}, \qquad a_i \in \mathbb{R},\ \ n \in \mathbb{N}$$

is denoted by

$$\text{span}\{x^{\lambda_0}, x^{\lambda_1}, \dots\}.$$

Similarly, for a fixed sequence $(\lambda_i)_{i=-\infty}^{\infty}$, the collection of Müntz polynomials

$$p(x) = \sum_{i=-n}^{n} a_i x^{\lambda_i}, \qquad a_i \in \mathbb{R},\ \ n \in \mathbb{N}$$

is denoted by

$$\text{span}\{x^{\lambda_i} : i \in \mathbb{Z}\}.$$

The above spaces are called Müntz spaces.

For a measurable set $A \subset \mathbb{R}$, we use the notation

$$\text{ess inf } A := \sup\{x \in \mathbb{R} : m((-\infty, x] \cap A) = 0\}$$

and

$$\text{ess sup } A := \sup\{x \in \mathbb{R} : m([x, \infty) \cap A) = 0\}$$

where $m(\cdot)$ denotes the one-dimensional Lebesgue measure.

3 New Results

The central result of this paper is the following theorem.

Theorem 3.1 *Suppose* $(\lambda_i)_{i=-\infty}^{\infty}$ *is a sequence of distinct real numbers satisfying*

$$\sum_{\substack{i=-\infty \\ \lambda_i \neq 0}}^{\infty} \frac{1}{|\lambda_i|} < \infty\,.$$

Then there is a constant c *depending only on* $(\lambda_i)_{i=-\infty}^{\infty}$, A, α, *and* β *(and not on the number of terms in* p*) so that*

$$\|p\|_{[\alpha,\beta]} \le c\,\|p\|_A$$

for every Müntz polynomial $p \in \text{span}\{x^{\lambda_i} : i \in \mathbb{Z}\}$, *for every set* $A \subset (0, \infty)$ *of positive Lebesgue measure, and for every* $[\alpha, \beta] \subset (\text{ess inf } A, \text{ess sup } A)$.

Theorem 3.2 *Suppose* $(\lambda_i)_{i=-\infty}^{\infty}$ *is a sequence of distinct real numbers satisfying*

$$\sum_{\substack{i=-\infty \\ \lambda_i \neq 0}}^{\infty} \frac{1}{|\lambda_i|} < \infty$$

with $\lambda_i < 0$ *for* $i < 0$ *and* $\lambda_i \ge 0$ *for* $i \ge 0$. *Associated with*

$$p(x) := \sum_{i=-n}^{n} a_i x^{\lambda_i}, \qquad n = 0, 1, \ldots$$

let

$$p^-(x) := \sum_{i=-n}^{-1} a_i x^{\lambda_i} \quad \text{and} \quad p^+(x) := \sum_{i=0}^{n} a_i x^{\lambda_i}.$$

Let $A \subset (0, \infty)$ *be a set of positive Lebesgue measure. Then there exists a constant* c *depending only on* $(\lambda_i)_{i=-\infty}^{\infty}$ *and* A *(and not on the number of terms in* p*) so that*

$$\|p^+\|_A \le c\,\|p\|_A$$

and

$$\|p^-\|_A \le c\,\|p\|_A$$

for every $p \in \operatorname{span}\{x^{\lambda_i} : i \in \mathbb{Z}\}$.

Theorem 3.3 *Suppose* $(\lambda_i)_{i=-\infty}^{\infty}$ *is a sequence of distinct real numbers satisfying*

$$\sum_{\substack{i=-\infty \\ \lambda_i \ne 0}}^{\infty} \frac{1}{|\lambda_i|} < \infty$$

with $\lambda_i < 0$ *for* $i < 0$ *and* $\lambda_i \ge 0$ *for* $i \ge 0$. *Suppose* $A \subset (0,\infty)$ *is a compact set of positive Lebesgue measure. Let* $a := \operatorname{ess\,inf} A$ *and* $b := \operatorname{ess\,sup} A$. *Let* $f \in C(A)$, *and suppose there exist* $p_n \in \operatorname{span}\{x^{\lambda_i} : i \in \mathbb{Z}\}$ *of the form*

$$p_n(x) = \sum_{i=-k_n}^{k_n} a_{i,n} x^{\lambda_i}, \qquad n = 1, 2, \ldots$$

so that $\lim_{n\to\infty} \|p_n - f\|_A = 0$.

Suppose the gap condition

$$\inf\{\lambda_i - \lambda_{i-1} : i \in \mathbb{Z}\} > 0$$

holds. Then f *is of the form*

$$f(x) = \sum_{i=-\infty}^{\infty} a_i x^{\lambda_i}, \qquad x \in (a,b),$$

where

$$\begin{aligned} f^+(x) &:= \textstyle\sum_{i=0}^{\infty} a_i x^{\lambda_i}, \qquad x \in [0,b), \\ f^-(x) &:= \textstyle\sum_{i=-\infty}^{-1} a_i x^{\lambda_i}, \qquad x \in (a,\infty), \qquad \lim_{x\to\infty} f^-(x) = 0. \end{aligned}$$

Furthermore, f *can be extended analytically throughout the region*

$$\{z \in \mathbb{C} \setminus (-\infty, 0] : a < |z| < b\},$$

and

$$\lim_{n\to\infty} a_{i,n} = a_i, \qquad i \in \mathbb{Z}.$$

If the above gap condition does not hold then f *can still be extended analytically throughout the region*

$$\{z \in \mathbb{C} \setminus (-\infty, 0] : a < |z| < b\}.$$

Theorem 3.4 *Suppose* $(\lambda_i)_{i=-\infty}^{\infty}$ *is a sequence of distinct real numbers. Suppose* $A \subset (0,\infty)$ *is a compact set of positive Lebesgue measure. Then*

$$\operatorname{span}\{x^{\lambda_i} : i \in \mathbb{Z}\}$$

is dense in $C(A)$ *if and only if*

$$\sum_{\substack{i=-\infty \\ \lambda_i \neq 0}}^{\infty} \frac{1}{|\lambda_i|} < \infty.$$

Theorem 3.5 *Suppose* $(\lambda_i)_{i=-\infty}^{\infty}$ *is a sequence of distinct real numbers satisfying*

$$\sum_{\substack{i=-\infty \\ \lambda_i \neq 0}}^{\infty} \frac{1}{|\lambda_i|} < \infty.$$

Let $A \subset [0,\infty)$ *be a set of positive Lebesgue measure, and let* $a := \text{ess inf } A$ *and* $b := \text{ess sup } A$. *Assume* $(p_n)_{n=1}^{\infty} \subset \text{span}\{x^{\lambda_i} : i \in \mathbb{Z}\}$ *and*

$$p_n(x) \to f(x), \qquad x \in A.$$

Then $(p_n)_{n=1}^{\infty}$ *converges uniformly on every closed subinterval of* (a,b).

Theorem 3.6 *Suppose* $(\lambda_i)_{i=-\infty}^{\infty}$ *is a sequence of distinct real numbers satisfying*

$$\sum_{\substack{i=-\infty \\ \lambda_i \neq 0}}^{\infty} \frac{1}{|\lambda_i|} < \infty$$

with $\lambda_i < 0$ *for* $i < 0$ *and* $\lambda_i \geq 0$ *for* $i \geq 0$. *Suppose* $A \subset [0,\infty)$ *is a set of positive Lebesgue measure with* $\inf A > 0$, w *is a nonnegative-valued, integrable weight function on* A *with* $\int_A w > 0$, *and* $q \in (0,\infty)$. *Then*

$$\text{span}\{x^{\lambda_i} : i \in \mathbb{Z}\}$$

is not dense in $L_w^q(A)$.

Suppose the gap condition

$$\inf\{\lambda_i - \lambda_{i-1} : i \in \mathbb{Z}\} > 0$$

holds. Then every function $f \in L_w^q(A)$ *belonging to the* $L_w^q(A)$ *closure of*

$$\text{span}\{x^{\lambda_i} : i \in \mathbb{Z}\}$$

can be represented as

$$f(x) = \sum_{i=-\infty}^{\infty} a_i x^{\lambda_i}, \qquad x \in A \cap (a_w, b_w),$$

where

$$a_w := \inf\left\{y \in [0,\infty) : \int_{A\cap(0,y)} w(x)\,dx > 0\right\}$$

and

$$b_w := \sup\left\{y \in [0,\infty) : \int_{A\cap(y,\infty)} w(x)\,dx > 0\right\}.$$

If the above gap condition does not hold, then every function $f \in L_w^q(A)$ belonging to the $L_w^q(A)$ closure of

$$\operatorname{span}\{x^{\lambda_i} : i \in \mathbb{Z}\}$$

can still be represented as an analytic function on

$$\{z \in \mathbb{C} \setminus (-\infty, 0] : a_w < |z| < b_w\}$$

restricted to A.

Theorem 3.7 *Suppose $(\lambda_i)_{i=-\infty}^{\infty}$ is a sequence of distinct real numbers. Suppose $A \subset (0,\infty)$ is a bounded set of positive Lebesgue measure, $\inf A > 0$, w is a nonnegative-valued integrable weight function on A with $\int_A w > 0$, and $q \in (0,\infty)$. Then*

$$\operatorname{span}\{x^{\lambda_i} : i \in \mathbb{Z}\}$$

is dense in $L_w^q(A)$ if and only if

$$\sum_{\substack{i=-\infty \\ \lambda_i \neq 0}}^{\infty} \frac{1}{|\lambda_i|} < \infty.$$

4 Tools

In this section we collect various previously known results concerning Müntz spaces with exponents of the same sign. In Section 5 the proof of the new results from Section 3, which deal with Müntz spaces with arbitrary exponents, will be reduced to the results of this section. Our most important tool is the following Remez-type inequality established in [8].

Theorem 4.1 *Let $(\lambda_i)_{i=0}^{\infty}$ be a sequence of distinct nonnegative exponents satisfying*

$$\sum_{\substack{i=0 \\ \lambda_i \neq 0}}^{\infty} \frac{1}{\lambda_i} < \infty.$$

Then there exists a constant c depending only on $(\lambda_i)_{i=-\infty}^{\infty}$, s, and $\sup A$ (and not on A or the number of terms in p) so that

$$\|p\|_{[0,\inf A]} \leq c\,\|p\|_A$$

for every $p \in \operatorname{span}\{x^{\lambda_0}, x^{\lambda_1}, \ldots\}$ and for every compact set $A \subset (0,\infty)$ of Lebesgue measure at least $s > 0$.

By the substitution $y = x^{-1}$ Theorem 4.1 implies the following.

Theorem 4.2 *Let $(\lambda_i)_{i=0}^\infty$ be a sequence of distinct nonpositive exponents satisfying*

$$\sum_{i=0\,\lambda_i\neq 0}^{\infty} \frac{1}{|\lambda_i|} < \infty.$$

Then there exists a constant c depending only on $(\lambda_i)_{i=-\infty}^\infty$, s, and $\inf A$ (and not on A or the number of terms in p) so that

$$\|p\|_{[\sup A,\infty)} \leq c\,\|p\|_A$$

for every $p \in \operatorname{span}\{x^{\lambda_0}, x^{\lambda_1}, \dots\}$ and for every compact set $A \subset (0,\infty)$ of Lebesgue measure at least $s > 0$.

The following Bernstein-type inequality for non-dense Müntz spaces is also established in [8].

Theorem 4.3 *Let $(\lambda_i)_{i=0}^\infty$ be a sequence of distinct nonnegative exponents satisfying $\sum_{i=1}^\infty 1/\lambda_i < \infty$. Suppose $\lambda_0 = 0$ and $\lambda_1 \geq 1$. Then for every $\varepsilon \in (0,1)$, there is a constant c_ε depending only on ε and $(\lambda_i)_{i=-\infty}^\infty$ (but not on the number of terms in p) so that*

$$\|p'\|_{[0,1-\varepsilon]} \leq c_\varepsilon \|p\|_{[0,1]}$$

for every $p \in \operatorname{span}\{x^{\lambda_0}, x^{\lambda_1}, \dots\}$.

Theorem 4.4 *Let $\Lambda := (\lambda_i)_{i=0}^\infty$ be a sequence of distinct nonnegative exponents satisfying*

$$\sum_{\substack{i=0\\ \lambda_i\neq 0}}^{\infty} \frac{1}{\lambda_i} < \infty.$$

Let $0 \leq a < b$. Suppose

$$(p_n)_{n=1}^\infty \subset \operatorname{span}\{x^{\lambda_0}, x^{\lambda_1}, \dots\}$$

and $\|p_n\|_{[a,b]} \leq 1$ for each n. Then there is a subsequence of $(p_n)_{n=1}^\infty$ that converges uniformly on every closed subinterval of $[0,b)$.

Proof: Note that the assumptions of the Arzela-Ascoli Theorem are satisfied by Theorems 4.1 and 4.3.

By the substitution $y = x^{-1}$ Theorem 4.4 implies the following.

Theorem 4.5 *Let $(\lambda_i)_{i=0}^\infty$ be a sequence of distinct nonpositive exponents satisfying*

$$\sum_{\substack{i=0\\ \lambda_i\neq 0}}^{\infty} \frac{1}{|\lambda_i|} < \infty.$$

Let $0 \leq a < b$. Suppose

$$(p_n)_{n=1}^\infty \subset \operatorname{span}\{x^{\lambda_0}, x^{\lambda_1}, \dots\}$$

and $\|p_n\|_{[a,b]} \le 1$ *for each* n. *Then there is a subsequence of* $(p_n)_{n=1}^\infty$ *that converges uniformly on every closed subinterval of* (a,∞).

The following theorem is from Schwartz [24].

Theorem 4.6 *Let* $(\lambda_i)_{i=0}^\infty$ *be a sequence of distinct nonnegative exponents satisfying*

$$\sum_{\substack{i=0\\ \lambda_i\neq 0}}^{\infty} \frac{1}{\lambda_i} < \infty.$$

Let $0 \le a < b$. *Suppose the sequence*

$$(p_n)_{n=1}^\infty \subset \operatorname{span}\{x^{\lambda_0}, x^{\lambda_1}, \dots\}$$

converges to a function f *uniformly on* $[a,b]$. *Then* f *can be extended analytically throughout the region*

$$\{z \in \mathbb{C} \setminus (-\infty, 0] : |z| < b\}.$$

By the substitution $y = x^{-1}$ Theorem 4.6 implies the following.

Theorem 4.7 *Let* $(\lambda_i)_{i=0}^\infty$ *be a sequence of distinct nonpositive exponents satisfying*

$$\sum_{\substack{i=0\\ \lambda_i\neq 0}}^{\infty} \frac{1}{|\lambda_i|} < \infty.$$

Let $0 \le a < b$. *Suppose the sequence*

$$(p_n)_{n=1}^\infty \subset \operatorname{span}\{x^{\lambda_0}, x^{\lambda_1}, \dots\}$$

converges to a function f *uniformly on* $[a,b]$. *Then* f *can be extended analytically throughout the region*

$$\{z \in \mathbb{C} \setminus (-\infty, 0] : a < |z|\}.$$

The following two results are also from Schwartz [24].

Theorem 4.8 *Let* $(\lambda_i)_{i=0}^\infty$ *be a sequence of distinct nonnegative exponents satisfying*

$$\sum_{\substack{i=0\\ \lambda_i\neq 0}}^{\infty} \frac{1}{\lambda_i} < \infty.$$

Let $(\gamma_i)_{i=0}^\infty$ *be a sequence of distinct negative exponents satisfying*

$$\sum_{\substack{i=0\\ \gamma_i\neq 0}}^{\infty} \frac{1}{|\gamma_i|} < \infty.$$

Let $0 \lneq a < b$. Suppose $f \in C[a,b]$ is a function so that both of the sequences

$$(p_n)_{n=1}^{\infty} \subset \operatorname{span}\{x^{\lambda_0}, x^{\lambda_1}, \dots\}$$

and

$$(q_n)_{n=1}^{\infty} \subset \operatorname{span}\{x^{\gamma_0}, x^{\gamma_1}, \dots\}$$

converge to f uniformly on $[a,b]$. Then $f = 0$ on $[a,b]$.

Theorem 4.9 *Suppose $(\lambda_i)_{i=-\infty}^{\infty}$ is a set of distinct real numbers satisfying*

$$\sum_{\substack{i=-\infty \\ \lambda_i \neq 0}}^{\infty} \frac{1}{|\lambda_i|} < \infty$$

with $\lambda_i < 0$ for $i < 0$ and $\lambda_i \geq 0$ for $i \geq 0$. Suppose $0 < a < b$. Let $f \in C[0,1]$, and suppose there exist $p_n \in \operatorname{span}\{x^{\lambda_i} : i \in \mathbb{Z}\}$ of the form

$$p_n(x) = \sum_{i=-k_n}^{k_n} a_{i,n} x^{\lambda_i}, \qquad n = 1, 2, \dots$$

so that $\lim_{n\to\infty} \|p_n - f\|_{[a,b]} = 0$.

Suppose the gap condition

$$\inf\{\lambda_i - \lambda_{i-1} : i \in \mathbb{Z}\} > 0$$

holds. Then f is of the form

$$f(x) = \sum_{i=-\infty}^{\infty} a_i x^{\lambda_i}, \qquad x \in (a,b),$$

where

$$\begin{aligned} f^+(x) &:= \textstyle\sum_{i=1}^{\infty} a_i x^{\lambda_i}, \qquad x \in [0,b), \\ f^-(x) &:= \textstyle\sum_{i=-\infty}^{-1} a_i x^{\lambda_i}, \qquad x \in (a,\infty), \qquad \lim_{x\to\infty} f^-(x) = 0, \end{aligned}$$

f can be extended analytically throughout the region

$$\{z \in \mathbb{C} \setminus (-\infty, 0] : a < |z| < b\},$$

and

$$\lim_{n\to\infty} a_{i,n} = a_i, \qquad i \in \mathbb{Z}.$$

If the above gap condition does not hold then f can still be extended analytically throughout the region

$$\{z \in \mathbb{C} \setminus (-\infty, 0] : a < |z| < b\}.$$

5 Proofs

Proof of Theorem 3.2 It is sufficient to prove only the first inequality, the second inequality follows from the first one by the substitution $y = x^{-1}$. If the first inequality fails to hold then there exists a sequence $(p_n)_{n=1}^{\infty} \subset \text{span}\{x^{\lambda_i} : i \in \mathbb{Z}\}$ so that

$$\|p_n^+\|_A = 1, \quad n = 1, 2, \dots, \quad \text{and} \quad \lim_{n\to\infty} \|p_n\|_A = 0.$$

Since $p = p^+ + p^-$, the above relations imply that

$$\|p_n^-\|_A \le K < \infty, \qquad n = 1, 2, \dots .$$

For the sake of brevity, let $a :=$ ess inf A and $b :=$ ess sup A. By Theorems 4.1, 4.2, 4.4, and 4.5, there exists a subsequence $(p_{n_i}^+)_{i=1}^{\infty}$ that converges uniformly to a function f on every closed subinterval of $[0, b)$, while $(p_{n_i}^-)_{i=1}^{\infty}$ converges uniformly to a function g on every closed subinterval of (a, ∞). Now $\lim_{i\to\infty} \|p_{n_i}\|_A = 0$ and $p_{n_i} = p_{n_i}^+ + p_{n_i}^-$ imply that $f + g = 0$ on $A \cap (a, b)$. By Theorem 4.6, f is analytic on $(0, b)$. By Theorem 4.7, g is analytic on (a, ∞). So $f+g$ is analytic on (a, b). Since $f+g = 0$ on $A\cap(a, b)$, and since $m(A\cap(a, b)) > 0$, we conclude by the Unicity Theorem that $f + g = 0$ on (a, b). Now Theorem 4.8 implies that $f = g = 0$ on (a, b).

Hence, for every $y \in (a, b)$,

$$\lim_{i\to\infty} \|p_{n_i}^+\|_{[\inf A,\, y]} = 0$$

and

$$\lim_{i\to\infty} \|p_{n_i}^+\|_{[y,\, \sup A]} = \lim_{i\to\infty} \|p_{n_i} - p_{n_i}^-\|_{A\cap[y,\, \sup A]} = 0.$$

Therefore

$$\lim_{i\to\infty} \|p_{n_i}^+\|_A = 0$$

which contradicts the fact that $\|p_n^+\|_A = 1, \quad n = 1, 2, \dots .$

Proof of Theorem 3.1 The result is a straightforward consequence of Theorems 4.1, 4.2, and 3.2.

Proof of Theorem 3.3 The result is a straightforward consequence of Theorems 3.1 and 4.9.

Proof of Theorem 3.4 Suppose

$$\sum_{\substack{i=-\infty \\ \lambda_i \ne 0}}^{\infty} \frac{1}{|\lambda_i|} = \infty.$$

Let $f \in C(A)$. By Tietze's Theorem there exists an $\tilde{f} \in C[\inf A, \sup A]$ so that $\tilde{f}(x) = f(x)$ for every $x \in A$. By Müntz's Theorem there is a sequence

$$(p_n)_{n=1}^{\infty} \subset \text{span}\{x^{\lambda_i} : i \in \mathbb{Z}\}$$

so that

$$\lim_{n\to\infty} \|\tilde{f} - p_n\|_{[0,1]} = 0.$$

Therefore

$$\lim_{n\to\infty} \|f - p_n\|_A = 0,$$

which finishes the trivial part of the theorem.

Suppose now that

$$\sum_{\substack{i=-\infty \\ \lambda_i \neq 0}}^{\infty} \frac{1}{|\lambda_i|} < \infty.$$

Then Theorem 3.3 yields that

$$\operatorname{span}\{x^{\lambda_i} : i \in \mathbb{Z}\}$$

is not dense in $C(A)$.

Proof of Theorem 3.5 Let $[\alpha, \beta] \subset (a, b)$. Egoroff's Theorem and the definition of a and b imply the existence of sets $B_1 \subset A \cap (0, \alpha)$ and $B_2 \subset A \cap (\beta, \infty)$ of positive Lebesgue measure so that $(p_i)_{i=1}^{\infty}$ converges uniformly on $B := B_1 \cap B_2$, hence it is uniformly Cauchy on B. Now Theorem 3.1 yields that $(p_i)_{i=1}^{\infty}$ is uniformly Cauchy on $[\alpha, \beta]$, which proves the theorem.

Proof of Theorem 3.6 Suppose $f \in L_w^q(A)$ and suppose there is a sequence

$$(p_n)_{n=1}^{\infty} \subset \operatorname{span}\{x^{\lambda_i} : i \in \mathbb{Z}\}$$

so that

$$\lim_{n\to\infty} \|f - p_n\|_{L_w^q(A)} = 0.$$

Minkowski's Inequality (if $q \in (0,1)$, then a multiplicative factor $2^{1/q-1}$ is needed) yields that $(p_i)_{i=1}^{\infty}$ is a Cauchy sequence in $L_w^q(A)$. The assumptions on w imply that for every $(\alpha, \beta) \subset [a, b]$ there exists a $\delta > 0$ so that the sets

$$B_1 := \{x \in A \cap (\beta, \infty) : w(x) > \delta\}$$

and

$$B_2 := \{x \in A \cap (0, \alpha) : w(x) > \delta\}$$

are of positive Lebesgue measure. Note that

$$\|p\|_{L^q(B_i)} \le \delta^{-1} \|p\|_{L_w^q(B_i)} \le \delta^{-1} \|p\|_{L_w^q(A)}, \qquad i = 1, 2,$$

for every $p \in L_w^q(A)$. Therefore, $(p_n)_{n=1}^{\infty}$ is a Cauchy sequence in $L^q(B)$, where $B := B_1 \cap B_2$. So, by Theorem 3.1, $(p_n)_{n=1}^{\infty}$ is uniformly Cauchy on $[\alpha, \beta]$. The theorem now follows from Theorem 3.3.

Proof of Theorem 3.7 Suppose

$$\sum_{\substack{i=-\infty \\ \lambda_i \neq 0}}^{\infty} \frac{1}{|\lambda_i|} = \infty.$$

Let $f \in L^q_w(A)$. It is standard measure theory to show that for every $\varepsilon > 0$ there exists a $g \in C[\inf A, \sup A]$ so that

$$\|f - g\|_{L^q_w(A)} < \frac{\varepsilon}{2}.$$

Now Müntz's Theorem implies that there exists a $p \in \operatorname{span}\{x^{\lambda_i} : i \in \mathbb{Z}\}$ so that

$$\|g - p\|_{L^q_w(A)} \le \|g - p\|_A \left(\int_A w\right)^{1/q} < \frac{\varepsilon}{2}.$$

Therefore $\operatorname{span}\{x^{\lambda_i} : i \in \mathbb{Z}\}$ is dense in $L^q_w(A)$.

Suppose now that

$$\sum_{\substack{i=-\infty \\ \lambda_i \neq 0}}^{\infty} \frac{1}{|\lambda_i|} < \infty.$$

Then Theorem 3.6 yields that $\operatorname{span}\{x^{\lambda_i} : i \in \mathbb{Z}\}$ is not dense in $L^q_w(A)$.

References

[1] J. M. Anderson, *Müntz-Szász type approximation and the angular growth of lacunary integral functions*, Trans. Amer. Math. Soc. **169** (1972), 237–248.

[2] J. Bak and D. J. Newman, *Rational combinations of $x^{\lambda_k}, \lambda_k \ge 0$ are always dense in $C[0,1]$*, J. Approx. Theory **23** (1978), 155–157.

[3] S. N. Bernstein, *Collected Works: Vol 1. Constructive Theory of Functions (1905-1930), English Translation*, Atomic Energy Commission, Springfield, Va, 1958.

[4] R. P. Boas, *Entire Functions*, Academic Press, New York, 1954.

[5] P. B. Borwein, *Zeros of Chebyshev polynomials in Markov Systems*, J. Approx. Theory **63** (1990), 56–64.

[6] P. B. Borwein, *Variations on Müntz's theme*, Can. Math. Bull. **34** (1991), 305–310.

[7] P. B. Borwein and T. Erdélyi, *Notes on lacunary Müntz polynomials*, Israel J. Math. **76** (1991), 183–192.

[8] P. B. Borwein and T. Erdélyi, *Generalizations of Müntz's Theorem via a Remez-type inequality for Müntz spaces*, J. Amer. Math. Soc. **10** (1997), 327-349.

[9] P. B. Borwein and T. Erdélyi, *Polynomials and Polynomial Inequalities*, Springer-Verlag, New York, N.Y., 1995.

[10] P. B. Borwein, T. Erdélyi, and J. Zhang, *Müntz systems and orthogonal Müntz polynomials*, Trans. Amer. Math. Soc. **342** (1994), 523-542.

[11] E. W. Cheney, *Introduction to Approximation Theory*, McGraw-Hill, New York, 1966.

[12] J. A. Clarkson and P. Erdős, *Approximation by polynomials*, Duke Math. J. **10** (1943), 5–11.

[13] T. Erdélyi, *Remez-type inequalities on the size of generalized polynomials*, J. London Math. Soc. **45** (1992), 255–264.

[14] T. Erdélyi, *Remez-type inequalities and their applications*, J. Comp. and Applied Math. **47** (1993), 167–210.

[15] G. Freud, *Orthogonal Polynomials*, Pergamon Press, Oxford, 1971.

[16] A. Kroó and F. Peherstorfer, *On the distribution of extremal points of general Chebyshev polynomials*, Trans. Amer. Math. Soc. **329** (1992), 117-130.

[17] W. A. J. Luxemburg and J. Korevaar, *Entire functions and Müntz-Szász type approximation*, Trans. Amer. Math. Soc. **157** (1971), 23–37.

[18] C. Müntz, *Über den Approximationsatz von Weierstrass*, H. A. Schwartz Festschrift, Berlin, 1914.

[19] D. J Newman, *Derivative bounds for Müntz polynomials*, J. Approx. Theory **18** (1976), 360–362.

[20] D. J. Newman, *Approximation with rational functions*, vol. 41, Regional Conference Series in Mathematics, Providence, Rhode Island, 1978.

[21] G. Nürnberger, *Approximation by Spline Functions*, Springer-Verlag, Berlin, 1989.

[22] E. J. Remez, *Sur une propriété des polynômes de Tchebyscheff*, Comm. Inst. Sci. Kharkow **13** (1936), 93–95.

[23] T. J. Rivlin, *Chebyshev Polynomials, 2nd ed.*, Wiley, New York, 1990.

[24] L. Schwartz, *Etude des Sommes d'Exponentielles*, Hermann, Paris, 1959.

[25] P. W. Smith, *An improvement theorem for Descartes systems*, Proc. Amer. Math. Soc. **70** (1978), 26–30.

[26] G. Somorjai, *A Müntz-type problem for rational approximation*, Acta. Math. Hung. **27** (1976), 197–199.

[27] O. Szász, *Über die Approximation steliger Funktionen durch lineare Aggregate von Potenzen*, **77** (1916), 482–496.

[28] G. Szegő, *On the density of quotients of lacunary polynomials*, Acta Math. Hung. **30** (1922), 149–154.

[29] A. K. Taslakyan, *Some properties of Legendre quasi-polynomials with respect to a Müntz system*, Mathematics **2** (1984), 179-189 Erevan University, Erevan. (Russian, Armenian Summary).

[30] M. von Golitschek, *A short proof of Müntz Theorem*, J. Approx. Theory **39** (1983), 394–395.

Frames and Schauder Bases

Peter G. Casazza
Department of Mathematics, University of Missouri
Columbia, Mo 65211

Ole Christensen
Department of Mathematics, Technical University of Denmark
2800 Lyngby, Denmark

Abstract

We present a rather surprising example of a frame which does not contain a Schauder basis. We also mention a special class of frames which always contains a Schauder basis.

1 Introduction

In all what follows $\mathcal{H}$ denotes a separable Hilbert space with the inner product $< \cdot, \cdot >$ linear in the first entry. For convenience we index all sequences by the natural numbers N. Recall that a family of elements $\{f_i\}_{i=1}^{\infty}$ in $\mathcal{H}$ is a *Schauder basis*, if for every $f \in \mathcal{H}$ there exists unique coefficients $\{c_i\}$ such that $f = \sum_{i=1}^{\infty} c_i f_i$. The definition can also be used in a Banach space setting, but we do not need it here.

A *Riesz basis* is a set of the form $\{Te_i\}_{i=1}^{\infty}$, where $\{e_i\}_{i=1}^{\infty}$ is an orthonormal basis for $\mathcal{H}$ and T is a bounded invertible operator on $\mathcal{H}$. Remember that $\{f_i\}_{i=1}^{\infty}$ is *unconditional* if a series $\sum_{i=1}^{\infty} c_i f_i$ converges unconditionally whenever it converges. As shown in [7, 3], a Schauder basis $\{f_i\}_{i=1}^{\infty}$ is a Riesz basis if and only if $\{f_i\}_{i=1}^{\infty}$ is unconditional and

$$\exists A, B > 0: \quad A \leq ||f_i|| \leq B, \ \forall i \in N.$$

A *frame* is a family of elements $\{f_i\}_{i=1}^{\infty} \subseteq \mathcal{H}$ such that

$$\exists A, B > 0: \quad A||f||^2 \leq \sum_{i=1}^{\infty} | < f, f_i > |^2 \leq B||f||^2, \ \forall f \in \mathcal{H}.$$

A and B are called the *frame bounds*. If we can choose $A = B$, the frame is called *tight*.

The invertibility and positivity of the frame operator

$$S : \mathcal{H} \to \mathcal{H},\ Sf = \sum_{i=1}^{\infty} < f, f_i > f_i$$

lead to a representation of any $f \in \mathcal{H}$ as an infinite linear combination of the frame elements ([7]):

$$f = SS^{-1}f = \sum_{i=1}^{\infty} < f, S^{-1}f_i > f_i,\quad \forall f \in \mathcal{H}.$$

A frame can be *overcomplete*, i.e., given $f \in \mathcal{H}$, there might exist coefficients $\{c_i(f)\} \neq \{< f, S^{-1}f_i >\}$ such that $f = \sum_{i=1}^{\infty} c_i(f) f_i$. Thus it is natural to look at a frame as a kind of "generalized basis".

Riesz bases can be characterized as those frames which are ω-independent, i.e., for which

$$\sum_{i=1}^{\infty} c_i f_i = 0,\ \{c_i\}_{i=1}^{\infty} \in \ell^2(N) \Rightarrow c_i = 0,\ \forall i.$$

The argument is not hard. That a Riesz basis $\{Te_i\}_{i=1}^{\infty}$ is a frame follows directly from the Parseval equality and the fact that T is invertible. On the other hand, if a frame $\{f_i\}_{i=1}^{\infty}$ is ω-independent, we clearly get an incomplete set if we delete any of the elements, and by [7, p. 188] this implies that $\{f_i\}_{i=1}^{\infty}$ is a Riesz basis.

The question of whether a frame contains a Riesz basis has attracted some attention recently. Seip [6] found conditions on a frame of exponentials $\{e^{i\lambda x}\}_{\lambda \in \Lambda}$ for $L^2(-\pi, \pi)$ that guarantee that the frame contains a Riesz basis. He also gave an example of a frame of exponentials without this property. Casazza and Christensen [1] gave a sufficient condition for a frame to contain a Riesz basis, and constructed a frame which does not contain a Riesz basis [2], both in the setting of a general Hilbert space.

The present paper investigates the question of whether there exists a subfamily of a frame which is a Schauder basis. It turns out that the frame constructed in [2] does not contain a Schauder basis. For a special class of frames (which always contains a Riesz basis) we show that a subfamily is a Schauder basis if and only if it is a Riesz basis.

Our work is inspired by work of Olson and Zalik. In [5] they proved that a sequence $\{f(t-\lambda_i\}_{i=1}^{\infty}$ of translates $\{\lambda_i\}_{i=1}^{\infty} \subseteq R$ of the single function $f \in L^p(R)$ can only be a Schauder basis for $L^p(R)$ if $\{\lambda_i\}_{i=1}^{\infty}$ is separated. It turns out that this leads to a proof that there exists no Riesz basis for $L^2(R)$ consisting of translates of a single function. The more general question of the existence of a Schauder basis of this form is still open.

2 The results

We say that a frame $\{f_i\}_{i=1}^{\infty}$ has the *subframe property* if any subfamily $\{f_i\}_{i\in J}, J \subseteq N$ is a frame for $\overline{span}\{f_i\}_{i\in J}$. It was proved in [2] that every frame with the subframe property contains a Riesz basis. As conjectured by Lennard, this actually leads to an equivalent characterization of frames with the subframe property:

Theorem 1 *Let $\{f_i\}_{i=1}^{\infty}$ be a frame. The following are equivalent:*

(1) $\{f_i\}_{i=1}^{\infty}$ has the subframe property.

(2) Every subfamily $\{f_i\}_{i\in J}$ contains a subset which is a Riesz basis for $\overline{span}\{f_i\}_{i\in J}$.

Proof.
(1) $\Rightarrow$ (2) : Let $J \subseteq N$. If (1) is satisfied, $\{f_i\}_{i\in J}$ is a frame for the Hilbert space $\overline{span}\{f_i\}_{i\in J}$ having the subframe property; therefore $\{f_i\}_{i\in J}$ contains a Riesz basis for $\overline{span}\{f_i\}_{i\in J}$.

(2) $\Rightarrow$ (1) : Let again $J \subseteq N$. If (2) is satisfied, there exists a set $K \subseteq J$ such that $\{f_i\}_{i\in K}$ is a Riesz basis for $\overline{span}\{f_i\}_{i\in J}$. In particular, $\{f_i\}_{i\in K}$ is a frame for $\overline{span}\{f_i\}_{i\in J}$. Since $\{f_i\}_{i\in J}$ satisfies the upper frame bound, it is now clear that $\{f_i\}_{i\in J}$ is a frame for $\overline{span}\{f_i\}_{i\in J}$.

Corollary 2 *Assume that a frame $\{f_i\}_{i=1}^{\infty}$ has the subframe property. Then a subfamily $\{f_i\}_{i\in J}$ is a Schauder basis if and only $\{f_i\}_{i\in J}$ is a Riesz basis.*

Proof. Let $\{f_i\}_{i\in J}$ be a Schauder basis; then $\{f_i\}_{i\in J}$ is ω-independent. By the subframe property, $\{f_i\}_{i\in J}$ is also a frame, so we conclude that $\{f_i\}_{i\in J}$ is a Riesz basis.

We observe that the conclusion in Corollary 2 is false for general frames. For example, let $\{e_i\}_{i=1}^{\infty}$ be an orthonormal basis for $\mathcal{H}$ and let

$$\{f_i\}_{i=1}^{\infty} := \{e_1, \frac{1}{\sqrt{2}}e_2, \frac{1}{\sqrt{2}}e_2, \frac{1}{\sqrt{3}}e_3, \frac{1}{\sqrt{3}}e_3, \frac{1}{\sqrt{3}}e_3, ...\},$$

(that is, for $n \in N$ the element $\frac{1}{\sqrt{n}}e_n$ appears n times). Then $\{f_i\}_{i=1}^{\infty}$ is a frame with bounds equal to one. Since a Schauder basis is ω-independent, $\{f_i\}_{i=1}^{\infty}$ only contains the Schauder basis $\{\frac{1}{\sqrt{i}}e_i\}_{i=1}^{\infty}$. This family is not a Riesz basis, since $||\frac{1}{\sqrt{n}}e_n|| \to 0$ for $n \to \infty$.

The main purpose of this paper is to present an example of a frame not containing a Schauder basis. We will use the following characterization of a Schauder basis, which can be found in [4, p. 2]: a sequence $\{f_i\}_{i=1}^{\infty}$ of non-zero vectors spanning $\mathcal{H}$ is a Schauder basis for $\mathcal{H}$ if and only if the basis constant is finite. Recall that the basis constant of $\{f_i\}_{i=1}^{\infty}$ is the smallest constant K satisfying for all natural numbers n, m with $n \geq m$ and all sequences $\{a_i\}$, the following inequality:

$$||\sum_{i=1}^{m} a_i f_i|| \leq K||\sum_{i=1}^{n} a_i f_i||.$$

The main part of the proof of our result is contained in the following lemma:

Lemma 3 *Let $\{e_i\}_{i=1}^{n}$ be an orthonormal basis for a finite dimensional space $\mathcal{H}_n$. Define*

$$f_i = e_i - \frac{1}{n}\sum\nolimits_{j=1}^{n} e_j \text{ for } i = 1, 2, ..., n,$$

and

$$f_{n+1} = \frac{1}{\sqrt{n}}\sum\nolimits_{j=1}^{n} e_j.$$

Then $\{f_i\}_{i=1}^{n+1}$ is a frame for $\mathcal{H}_n$ with bounds $A = B = 1$. Furthermore, any subset of the frame which spans $\mathcal{H}_n$ has basis constant greater than or equal to $\frac{\sqrt{n-2}}{4}$.

Proof. The proof that $\{f_i\}_{i=1}^{n+1}$ is a frame for $\mathcal{H}_n$ with bounds $A = B = 1$ can be found in [2]. The vectors $\{f_i\}_{i=1}^{n}$ are dependent, and any subset of the frame $\{f_i\}_{i=1}^{n+1}$ which spans $\mathcal{H}_n$ must contain f_{n+1} and at least $n-1$ of the other terms. Clearly it is enough to prove the result for subsets $\{g_i\}_{i=1}^{n}$ containing exactly $n-1$ of those terms. Let Δ be a subset of $\{1, 2, \ldots, n\}$ with a number of elements $|\Delta| = n-1$, and let $\Delta^c = \{k\}$ denote its complement. We have

$$||\sum_{i \in \Delta}\left(e_i - \frac{\sum_{j=1}^{n} e_j}{n}\right)|| = ||\frac{1}{n}\sum_{i \in \Delta} e_i - \frac{n-1}{n}e_k||$$

$$= \left[\frac{n-1}{n^2} + \left(\frac{n-1}{n}\right)^2\right]^{1/2} = \sqrt{\frac{n-1}{n}}.$$

Let $[\frac{n-1}{2}]$ denote the integral part of $\frac{n-1}{2}$. For any subset $A \subset \{1, 2, \ldots, n\}$ with $|A| = [\frac{n-1}{2}]$, we have, $|A| \le \frac{n}{2}$ and $|A| \ge \frac{n-1}{2} - \frac{1}{2} = \frac{n}{2} - 1$; therefore

$$\|\sum_{i\in A}\left(e_i - \frac{\sum_{j=1}^n e_j}{n}\right)\| = \|\sum_{i\in A}\left(1 - \frac{|A|}{n}\right)e_i + \sum_{i\notin A}\frac{-|A|}{n}e_i\|$$

$$\ge \left(\sum_{i\in A}(1 - \frac{|A|}{n})^2\right)^{1/2} = |A|^{1/2}(1 - \frac{|A|}{n}) \ge \frac{|A|^{1/2}}{2} \ge \frac{\sqrt{\frac{n}{2}-1}}{2}.$$

When $\{g_i\}_{i=1}^n$ is a spanning subset of the frame $\{f_i\}_{i=1}^{n+1}$, we can choose $m \le n$ so that $\{g_i\}_{i=1}^m$ contains exactly $[\frac{n-1}{2}]$ of the elements of the set $\{f_i\}_{i=1}^n$. Now, there are 2 possibilities.

Case 1 f_{n+1} does not appear in the list $\{g_i\}_{i=1}^m$.

Then as we saw above,

$$\|\sum_{i=1}^m g_i\| \ge \frac{1}{2}\sqrt{\frac{n}{2} - 1},$$

while

$$\|\sum_{i=1}^n g_i\| = \left(\|\frac{\sum_{j=1}^n e_j}{\sqrt{n}}\|^2 + \frac{n-1}{n}\right)^{1/2} = \left(1 + \frac{n-1}{n}\right)^{1/2} \le \sqrt{2}.$$

Therefore

$$\|\sum_{i=1}^m g_i\| \ge \frac{1}{2}\sqrt{\frac{n}{2} - 1}\cdot\frac{\|\sum_{i=1}^n g_i\|}{\sqrt{2}} = \frac{\sqrt{n-2}}{4}\cdot\|\sum_{i=1}^n g_i\|.$$

Hence, the basis constant for $\{g_i\}_{i=1}^n$ is greater than or equal to $\frac{\sqrt{n-2}}{4}$.

Case 2 f_{n+1} appears in the list $\{g_i\}_{i=1}^m$.

Now, since $f_{n+1} \perp f_i, i = 1, 2, ..n$, we have

$$\|\sum_{i=1}^m g_i\| \ge \left(\|\frac{\sum_{j=1}^n e_j}{\sqrt{n}}\|^2 + \left[\frac{1}{2}\sqrt{\frac{n}{2} - 1}\right]^2\right)^{1/2} \ge \frac{\sqrt{\frac{n}{2}-1}}{2},$$

while we still have $\|\sum_{i=1}^n g_i\| \le \sqrt{2}$. So again, the basis constant of $\{g_i\}_{i=1}^n$ is greater than or equal to $\frac{\sqrt{n-2}}{4}$.

We are now ready to prove

Proposition 4 *There is a tight frame $\{f_i\}_{i=1}^\infty$ for which no subset is a Schauder basis for $\mathcal{H}$.*

Proof. Let $\{e_i\}_{i=1}^\infty$ be an orthonormal basis for $\mathcal{H}$ and define

$$\mathcal{H}_n := span\{e_{\frac{(n-1)n}{2}+1}, e_{\frac{(n-1)n}{2}+2}, \ldots, e_{\frac{(n-1)n}{2}+n}\}.$$

So $\mathcal{H}_1 = span\{e_1\}$, $\mathcal{H}_2 = span\{e_2, e_3\}$, etc. By construction,

$$\mathcal{H} = (\sum_{n=1}^\infty \oplus \mathcal{H}_n)_{\mathcal{H}};$$

that is, $g \in \mathcal{H} \Leftrightarrow g = \sum_{n=1}^\infty g_n$, $g_n \in \mathcal{H}_n$, and $||g||^2 = \sum_{n=1}^\infty ||g_n||^2$. We refer to [4] for details about such constructions.
For each space $\mathcal{H}_n$ we construct the sequence $\{f_i^n\}_{i=1}^{n+1}$ as in Lemma 3, starting with the orthonormal basis $\{e_{\frac{(n-1)n}{2}+1}, e_{\frac{(n-1)n}{2}+2}, \ldots, e_{\frac{(n-1)n}{2}+n}\}$. As shown in [2], $\{f_i^n\}_{i=1,n=1}^{n+1,\infty}$ is a frame for $\mathcal{H}$ with bounds $A = B = 1$. Let $\{g_i\}_{i=1}^\infty$ be any spanning subset of this frame. Then the basis constant of $\{g_i\}_{i=1}^\infty$ is greater than or equal to the basis constant of any subset of $\{g_i\}_{i=1}^\infty$. By choosing subfamilies as in Lemma 3, we get basis constants greater than or equal to $\frac{\sqrt{n-2}}{4}$ for all n, implying that the basis constant of $\{g_i\}_{i=1}^\infty$ is infinite. Thus $\{g_i\}_{i=1}^\infty$ is not a Schauder basis for $\mathcal{H}$.

Acknowledgments The authors thank Richard Zalik for many valuable suggestions and Christopher Lennard for a stimulating conversation. The first author was supported by NSF grant DMS 9201357. The second author was supported by the Danish Science Foundation. This paper was written while the second author was visiting the Georgia Institute of Technology and he thanks the Institute, and especially Christopher Heil, for their support.

References

[1] P. G. Casazza and O. Christensen, Hilbert space frames containing a Riesz basis and Banach spaces which have no subspace isomorphic to c_0. J. Math. Anal. Appl. 202 (1996), 940-950.

[2] P. G. Casazza and O. Christensen, Frames containing a Riesz basis and preservation of this property under perturbation. SIAM J. Math. Anal., to appear in 1997.

[3] C. Heil and D. Walnut, Continuous and discrete wavelet transforms, SIAM Review 31, vol. 4 (1989), 628-666.

[4] J. Lindenstrauss and L. Tzafriri, "Classical Banach Spaces" Vol. 1, Springer–Verlag, New York, 1977.

[5] T. E. Olson and R. A. Zalik, Nonexistence of a Riesz basis of translates, in "Approximation Theory" (G. A. Anastassiou, Ed.), 401–408, Lecture Notes in Pure and Applied Mathematics Vol. 138, Marcel Dekker, New York, 1992.

[6] K. Seip, On the connection between exponential bases and certain related sequences in $L^2(-\pi, \pi)$, J. Funct. Anal. 130 (1995), 131-160.

[7] R. M. Young, "An Introduction to Nonharmonic Fourier Series", Academic Press, New York, 1980.

Interpolation on Spheres by Positive Definite Functions

E.W. Cheney
Mathematics Department
University of Texas
Austin, Texas 78712

Xingping Sun
Mathematics Department
Southwestern Missouri State University
Springfield, Missouri, 65804

Abstract

This paper describes recent research on the problem of interpolating data that has been collected on a spherical surface of arbitrary dimension. The interpolation is to be carried out with the aid of a single basic function f that is continuous and real valued on the interval $[-1, 1]$. The interpolating function for nodes $p_1, p_2, \ldots, p_n$ on a sphere is taken to be of the form $\sum_{i=1}^{n} c_i f(\langle x, p_i \rangle)$. We are interested in identifying large classes of functions f that can serve in this capacity.

1 Introduction

Let H be a real Hilbert space, and let S denote the unit sphere of H, i.e.,

$$S := \{x : x \in H \quad \text{and } \langle x, x \rangle = 1\}$$

A function $f : [-1, 1] \to \mathbb{R}$ is called "strictly positive definite" on S if for every natural number n and any n distinct points $x_1, x_2, \ldots, x_n$ in S, the $n \times n$ matrix A with entries

$$A_{ij} = f(\langle x_i, x_j \rangle)$$

is positive definite. From now on, the inner-product of two points x and y in a Hilbert space will be written simply as xy. We shall use S^m for the unit sphere in $\mathbb{R}^{m+1}$ and S^∞ for the unit sphere in the separable Hilbert space ℓ^2.

The matrix A arises naturally in the study of interpolating data on spheres. Let $\mathcal{N} := \{x_1, x_2, \dots, x_n\}$ be a node set of n distinct points on S, and let g be a function defined on $\mathcal{N}$. To interpolate g, one can choose a univariate function $f : [-1,1] \to \mathbb{R}$, and look for an interpolant h in the linear space

$$\mathcal{L} := \text{ span } \left\{ x \mapsto f(xx_i) \; : \; i = 1, 2, \dots, n \right\}$$

The existence of a unique interpolant h in $\mathcal{L}$ is equivalent to the invertibility of the matrix A. If f is strictly positive definite, then the matrix A will be positive definite; it can then be dealt with by many efficient and stable numerical methods.

A function $f : [-1,1] \to \mathbb{R}$ is called "positive definite" on S if it satisfies the weaker condition that for each n and for any n points $x_1, x_2, \dots, x_n$ in S, the $n \times n$ matrix $(f(x_i x_j))$ is nonnegative definite. The continuous positive definite functions on S^m and S^∞ have been characterized by Schoenberg [34]. Let $P_k^{(\lambda)}$ denote the kth-degree Gegenbauer (ultraspherical) polynomial associated with λ [39, page 81]. Schoenberg proved the following seminal result.

Theorem 1.1 *A continuous function $f : [-1,1] \to \mathbb{R}$ is positive definite on S^m if and only if it has the form*

$$f(t) = \sum_{k=0}^{\infty} a_k P_k^{(\lambda)}(t) \tag{1}$$

where $\lambda = (m-1)/2$, $a_k \geq 0$, and $\sum_{k=0}^{\infty} a_k P_k^{(\lambda)}(1) < \infty$.

In the same paper, Schoenberg also characterized the continuous positive definite functions on S^∞ as follows.

Theorem 1.2 *A continuous function $f : [-1,1] \to \mathbb{R}$ is positive definite on S^∞ if and only if it has the form*

$$f(t) = \sum_{k=0}^{\infty} a_k t^k \tag{2}$$

in which $a_k \geq 0$ and $\sum_{k=0}^{\infty} a_k < \infty$.

Schoenberg's results have been generalized in many directions by Bochner [1], Christensen and Ressel [7] and others; see the references in [7].

It is clear from the definitions that "strictly positive definite" functions are also "positive definite." But the other implication is not true. In other words, there exist positive definite functions that are not strictly positive definite. It is important to know what functions in (1) and (2) as characterized by Schoenberg are also strictly positive definite, and therefore can be employed to perform interpolation on spheres.

It turns out that this problem is critically linked to multivariate polynomial interpolation. We will reveal this link in Section 2. Section 3 is devoted to work on the circle S^1, where the group structure allows us to obtain results that are not available on S^m if $m > 1$. In Sections 4 and 5, we discuss results on S^m for $m > 1$, and S^∞, respectively. Section 6 discusses conditionally positive definite functions on spheres.

There are other notions of strict positive definiteness in the literature. For example, Narcowich [27] uses a stronger version of strict positive-definiteness, and is able to characterize the families of such functions in terms of verifiable conditions. We will address this issue briefly in Section 7. Section 8 gives pointers to literature on topics that were peripheral to our main themes.

2 The Connection to Polynomial Interpolation

It is readily established that whether or not a function in (1) or (2) is strictly positive definite does not depend on the actual values of the coefficients a_k, but only on the set

$$K_{m,f} := \{k \in \mathbb{Z}_+ : a_k > 0\}$$

We refer the readers to [31] and [17] for an explanation of this fact. Thus, we find the following definition convenient:

Definition 2.1 *Let K be a subset of $\mathbb{Z}_+$ and let n be a natural number. If the $n \times n$ matrix having ij-entry $f(x_i x_j)$, where*

$$f(t) = \sum_{k \in K} a_k P_k^{(\lambda)}(t) \ , \quad a_k > 0 \ , \quad \sum_{k \in K} a_k P_k^{(\lambda)}(1) < \infty \tag{3}$$

is positive definite for any n distinct points $x_1, \ldots, x_n$ on S^m, then we say that K induces strict positive-definiteness of order n on S^m.

In a similar way, we can define what it means to say that a subset K of $\mathbb{Z}_+$ induces strict positive-definiteness of order n on S^∞.

The connection to multivariate polynomial interpolation is made possible by using spherical harmonics. A spherical harmonic of degree k is, by definition, the restriction to S^m of a homogeneous harmonic polynomial of that degree. Let $\mathcal{H}_k^0$ denote the space of all harmonic polynomials of degree k, and let $\{Y_1^{(k)}, \ldots Y_{h_k}^{(k)}\}$ be an orthonormal basis of $\mathcal{H}_k^0$ (here $h_k := \dim \mathcal{H}_k^0$). The following formula can be found in, for example, Stein and Weiss [37], (page 149), or Müller [26].

$$P_k^{(\lambda)}(xy) = C_{k,\lambda} \sum_{j=1}^{h_k} Y_j^{(k)}(x) Y_j^{(k)}(y) \ , \qquad x, y \in S^m \tag{4}$$

Here $\lambda = \frac{m-1}{2}$, and $C_{k,\lambda}$ is a positive constant depending only on λ and k.

We also need the following result from Stein and Weiss [37].

Lemma 2.2 *Let P_k be a homogeneous polynomial of degree k. Then for each $i = 0, 1, \ldots, \left[\frac{k}{2}\right]$, there is $q_{k-2i} \in \mathcal{H}_{k-2i}^0$, such that for all $x \in S^m$, we have*

$$P_k(x) = \sum_{i=1}^{[k/2]} q_{k-2i}(x)$$

Let Π_k denote the space of all polynomials on $\mathbb{R}^{m+1}$ of degree at most k, and let $\mathcal{H}_k$ denote the space of all harmonic polynomials on $\mathbb{R}^{m+1}$ of degree at most k. Then, Lemma 2.2 shows that on S^m, we have $\Pi_k = \mathcal{H}_k$.

To see whether the set $\{0, 1, \ldots, \ell\}$ induces strict positive-definiteness of order n on S^m, we calculate the quadratic form $\xi^T A \xi$, where $\xi := (\xi_1, \xi_2, \ldots, \xi_n)$ is a nonzero vector in $\mathbb{R}^n$ and A is the $n \times n$ matrix having ij-entry

$$A_{ij} = \sum_{k=0}^{\ell} P_k^{(\lambda)}(x_i x_j)$$

We have

$$\begin{aligned}
\xi^T A\xi &= \sum_{i,j=1}^{n} \xi_i\xi_j A_{ij} \\
&= \sum_{i,j=0}^{n} \xi_i\xi_j \sum_{k=0}^{\ell} P_k^{(\lambda)}(x_i x_j) \\
&= \sum_{k=0}^{\ell} C_{k,\lambda} \sum_{i,j=1}^{n} \xi_i\xi_j \sum_{s=1}^{h_k} Y_s^{(k)}(x_i)Y_s^{(k)}(x_j) \\
&= \sum_{k=0}^{\ell} C_{k,\lambda} \sum_{s=1}^{h_k}\Big(\sum_{j=1}^{n} \xi_j Y_s^{(k)}(x_j)\Big)^2
\end{aligned}$$

Thus, $\xi^T A\xi \geq 0$, and $\xi^T A\xi = 0$ if and only if

$$\sum_{j=1}^{N} \xi_j Y_s^{(k)}(x_j) = 0 \text{ for all } Y_s^{(k)} \in \mathcal{H}_\ell \tag{5}$$

Since $\mathcal{H}_\ell = \Pi_\ell$, Equation (5) is equivalent to the equation $\sum_{j=1}^{n} \xi_j P(x_j) = 0$ for all $P \in \Pi_\ell$. Therefore, we have the following theorem.

Theorem 2.3 *The set $\{0, 1, \ldots, \ell\}$ induces strict positive-definiteness of order n on S^m if and only if for any n distinct points $x_1, \ldots, x_n \in S^m$, the polynomial space Π_ℓ interpolates arbitrary data on $\{x_1, \ldots, x_n\}$.*

The case of $m = 1$ in Theorem 2.3 is due to Xu and Cheney [41], and the general version is due to Ron and Sun [31].

To state an analogous theorem for S^∞ we need the notation Π^K, defined by

$$\Pi^K := \sum_{k\in K} \Pi_k^0$$

Theorem 2.4 *Let K be a subset of $\mathbb{Z}_+$. In order that K induce strict positive-definiteness of order n on S^∞ it is necessary and sufficient that for any n distinct points $x_1, \ldots, x_n$ in S^∞, the polynomial space Π^K interpolate arbitrary data on $\{x_1, \ldots, x_n\}$.*

These two theorems will be exploited later in this article.

3 Strict Positive-Definiteness on the Circle

When we work on the circle S^1, we have

$$\dim \mathcal{H}_k^0 = 2 \text{ for all } k = 1, 2, \ldots$$

and $\mathcal{H}_k^0$ is spanned by the two functions $\cos k\theta$ and $\sin k\theta$, where (r, θ) are the polar coordinates in $\mathbb{R}^2$. The simplicity of spherical harmonics and the group structure on S^1 allow us to obtain sharper results. The following result was proved in [41].

Theorem 3.1 *The set* $\{0, 1, \ldots, [\frac{n}{2}]\}$ *induces strict positive-definiteness of order* n *on* S^1.

A corollary of Theorem 3.1 (also in [41]) is the following.

Corollary 3.2 *The set* $\mathbb{Z}_+$ *induces strict positive-definiteness of all orders on* S^1.

To generalize Corollary 3.2, we need to find other subsets of $\mathbb{Z}_+$ that induce strict positive-definiteness on S^1. The following theorem describes two such sets.

Theorem 3.3 *Let* K *be a subset of* $\mathbb{Z}_+$*, and let* n *be a positive integer. Suppose that one of the following two conditions holds:*

(a) K *contains* n *consecutive integers*

(b) K *contains* n *arithmetic progressions, each of length at least* n*:*

$$d_j, \alpha_j + d_j, \ldots, \alpha_j + (n-1)d_j \ , \qquad j = 1, 2, \ldots, n$$

and the numbers $d_1, d_2, \ldots, d_n$ *are pairwise relatively prime.*

Then K *induces strict positive-definiteness of order* n *on* S^1.

Theorem 3.3 provides many choices of subsets of $\mathbb{Z}_+$ that induce strict positive-definiteness on S^1. Theorem 3.3 was proved in [31]. The results in some special forms were also obtained previously by Menegatto [17] [18].

As a consequence of Theorem 3.3, we can state the following result.

Corollary 3.4 *Let* $K \subset \mathbb{Z}_+$*. If for each positive integer* n*,* K *satisfies one of the two conditions in Theorem 3.3, then* K *induces strict positive-definiteness on* S^1.

Necessary conditions for a subset $K \subset \mathbb{Z}_+$ to induce strict positive-definiteness have also been discovered. The following result was independently proved by Ron and Sun [31] and Menegatto [17] [18].

Proposition 3.5 *In order that a set $K \subset \mathbb{Z}_+$ induce strict positive-definiteness on S^1, it is necessary that K have infinite intersection with every set of the form $k\mathbb{Z}_+$, $k \in \mathbb{N}$ and with every set of the form $k + 2k\mathbb{Z}_+$, $k \in \mathbb{N}$.*

The above result is invalid with respect to any other arithmetic progression. One can show that the set $\mathbb{Z}_+ \setminus (\alpha + k\mathbb{Z}_+)$ induces strict positive-definiteness of all orders on S^1 whenever $\alpha \in \frac{k}{2}\mathbb{Z}_+$; see [32]. Take $k = 4$ for example. In order for a subset $K \subset \mathbb{Z}_+$ to induce strict positive-definiteness on S^1, K must have infinite intersection with $4\mathbb{Z}_+$, as well as with $2+4\mathbb{Z}_+$. It also must have infinite intersection with one of the sets $1+4\mathbb{Z}_+$, $3+4\mathbb{Z}_+$. On the other hand, K may have an empty intersection with one of the two sets $1 + 4\mathbb{Z}_+$ and $3 + 4\mathbb{Z}_+$.

We have sufficient conditions and we also have necessary conditions; but we do not have conditions that are necessary and sufficient. For an arbitrary set $K \subset \mathbb{Z}^+$, there are no existing explicitly-verifiable conditions that reveal whether K induces strict positive-definiteness on S^1. Thus, we can consider that the problem is still open.

4 Results on Multi-dimensional Spheres

The spherical harmonics on S^m, when $m > 1$, no longer have the simple form that they have on the circle. The analysis also turns out to be more subtle and the approach is less straightforward. In general, the results are weaker than those obtained on the circle. The first result in this direction was given in [41].

Theorem 4.1 *The set $\{0, 1, 2, \ldots, n-1\}$ induces strict positive-definiteness of order n on S^m.*

A consequence of Theorem 4.1 is:

Corollary 4.2 *The set $\mathbb{Z}_+$ itself induces strict positive-definiteness on S^m.*

In [41] it was conjectured that the set $\{0, 1, \ldots, [\frac{n}{2}]\}$ induces strict positive-definiteness of order n on S^m when $m > 1$ just as it does on S^1. By using the "least solution for multivariate polynomial interpolation" developed by Ron and de Boor [2], [3], [4], Ron and Sun [31] established that this conjecture is true. We formally state the result in the following theorem.

Theorem 4.3 *The set* $\{0, 1, \ldots, [\frac{n}{2}]\}$ *induces strict positive-definiteness of order* n *on* S^m, *for all* m.

The minimality of the set $\{0, 1, \ldots, [\frac{n}{2}]\}$ was also addressed in [31], but we choose not to include it here due to cumbersome technical details.

In the same paper, Ron and Sun proved the following result:

Theorem 4.4 *Let* $\mathcal{N}$ *be a finite subset of* S^m. *Define*

$$\sigma(\mathcal{N}) := \min\{\#J \ : \ J \subset \mathcal{N}, \quad span\ (\mathcal{N} \setminus J) \neq \mathbb{R}^{m+1}\}$$

Let j *be the smallest integer that satisfies* $\binom{j+m-1}{m} > \sigma(\mathcal{N})$. *Let* f *be a positive definite function on* S^m, *and assume that the set* $\{k \in K_{m,f} : k \geq (\#\mathcal{N}/2)\}$ *contains* j *consecutive even integers as well as* j *consecutive odd integers. Then the matrix having entries* $f(x_i x_j)$, $x_i, x_j \in \mathcal{N}$ *is positive definite.*

Note that $\sigma(\mathcal{N}) \leq n - m$ for any $\mathcal{N} \subset \mathbb{R}^{m+1}$ of cardinality n. Thus the following corollary is obviously true.

Corollary 4.5 *Let* $K \subset \mathbb{Z}_+$, *and let* n *be a positive integer. Let* j *be the least integer that satisfies* $\binom{j+m-1}{m} > n - m$. *If the set* $K \cap \{k \geq n/2\}$ *contains* j *consecutive odd integers and* j *consecutive even integers, then* K *induces strict positive-definiteness on* S^m.

Another corollary of Theorem 4.4 is the following:

Corollary 4.6 *Let* $K \subset \mathbb{Z}_+$. *If* K *contains arbitrarily long sequences of consecutive even and of consecutive odd integers, then* K *induces strict positive-definiteness on* S^m.

Independently, Schreiner [35] proved that a subset $K \subset \mathbb{Z}^+$ induces strict positive-definiteness on S^m if for some positive integer n, the set K contains the set $\mathbb{Z}_+ \setminus \{0, 1, \ldots, n\}$.

We note that Schreiner's result also generalizes Corollary 4.2 obtained by Xu and Cheney [41]. We end this section by giving a remark to Theorem 4.4.

For $m = 1$, the actual requirement is $j \geq n$. Here j is the length of the string of consecutive even integers and of consecutive odd integers required in Theorem 4.4. For $m > 2$, j grows very slowly with n, asymptotically, $j = \mathcal{O}(n^{1/m})$.

5 Results for S^∞

Since each finite-dimensional sphere S^m is contained in S^∞, any function that is strictly positive definite (SPD) on S^∞ is automatically SPD on S^m. Hence great interest attaches to the complete characterization of the SPD functions on S^∞. Menegatto has succeeded in doing so. His theorem is as follows.

Theorem 5.1 *In order that a function f in $C[-1,1]$ be strictly positive definite on the Hilbert sphere S^∞, it is necessary and sufficient that f have the form*

$$f(t) = \sum_{k=0}^{\infty} a_k t^k \qquad (-1 \le t \le 1)$$

in which all coefficients a_k are nonnegative, $\sum a_k < \infty$, and the set $\{k : a_k > 0\}$ contains infinitely many odd and infinitely many even integers.

Several papers have addressed problems connected with the complex Hilbert sphere (i.e., the unit sphere in the complex space ℓ^2.) The papers [7] and [23] are in this category. Menegatto and Sun have recently proved the following theorem.

Theorem 5.2 *For $z \in \mathbb{C}$, let $f(z) = \sum\sum a(m,n) z^m \bar{z}^n$ in which $a(m,n) \ge 0$ for all m and n and $\sum\sum a(m,n) < \infty$. If the set $\{m-n : a(m,n) > 0\}$ contains arbitrarily long sequences of consecutive integers, then f is strictly positive definite on the complex Hilbert sphere.*

6 Conditionally Positive Definite Functions

A function f in $C[-1,1]$ is said to be conditionally positive definite (CPD) on S^m if

$$\sum_{i=1}^{n}\sum_{j=1}^{n} c_i c_j f(x_i x_j) \ge 0 \qquad \text{whenever} \qquad \left(\sum_{i=1}^{n} c_i = 0\right) \tag{6}$$

Here, the points x_i are arbitrary in S^m. In [17] Menegatto proved that $f \in \mathrm{CPD}(S^m)$ if and only if $f + \lambda \in \mathrm{PD}(S^m)$ for some $\lambda \in \mathbb{R}$. Thus, we can say that

$$\mathrm{PD}(S^m) + \mathbb{R} = \mathrm{CPD}(S^m)\ . \tag{7}$$

This result is true for $m = \infty$ as well.

Functions f in the class CPD(S^m) lead to interpolation matrices $A_{ij} = f(x_i x_j)$ whose eigenvalues satisfy

$$\lambda_1 \geq \lambda_2 \geq \cdots \geq \lambda_{n-1} \geq 0 \tag{8}$$

In order that A be nonsingular, we require further that $\lambda_{n-1} > 0 > \lambda_n$ or that $\lambda_n > 0$. The first of these two inequalities will be true if $f(t) \leq 0$ for all t and if f is strictly conditionally positive definite (SCPD), meaning that (6) holds with a strict inequality whenever the points x_i are distinct and $c \neq 0$. For example, one can start with $f \in \text{SPD}(S^m)$ and subtract a constant λ such that $\lambda \geq f(t)$ for all t.

Menegatto also proved that

$$\text{SPD}(S^\infty) + \mathbb{R} = \text{SCPD}(S^\infty) \tag{9}$$

Still open is the question of whether (9) is valid for S^m. The proof of (9) depends upon the following characterization of the space SCPD(S^∞) from [17] and [21].

Theorem 6.1 *In order that a function f in $C[-1,1]$ be strictly conditionally positive definite on S^∞, it is necessary and sufficient that $f(t) = \sum_{k=0}^{\infty} a_k t^k$, where $a \in \ell^1$, $a_k \geq 0$ for $k \geq 1$, and the set $\{k : a_k > 0\}$ contains infinitely many even and infinitely many odd integers.*

A similar characterization for SCPD(S^m) is not known.

7 Generalizations of Positive Definiteness

A number of ways can be employed to generalize the classical concept of a positive definite function. Let us recall the basic definition. If X is a topological space and if $K \in C(X \times X)$, then K is *positive definite* if

$$\sum_{i=1}^{n} \sum_{j=1}^{n} c_i \bar{c}_j K(x_i, x_j) \geq 0 \tag{10}$$

for any finite set of points $x_i \in X$ and any set of complex scalars c_i.

This definition can be restated in terms of functionals if we define x^* to be the point-evaluation functional corresponding to the point x. Thus,

$x^*(\varphi) = \varphi(x)$ for any continuous function φ. Then, let L be a finite sum of the form $L = \sum c_i x_i^*$. Equation (10) now can be written

$$(L \otimes L)(K) \geq 0 \tag{11}$$

for any functional of the form $L = \sum c_i x_i^*$. Certainly this concept can be generalized by allowing more general functionals. Thus, one says that K is *positive definite in the measure sense* if

$$\int K(x,y)\, d\mu(x)\, d\mu(y) \geq 0 \tag{12}$$

for all signed Borel measures μ on X. Equation (12) is a special case of Equation (11) in which $L(\varphi) = \int \varphi(x)\, d\mu(x)$.

A distributional form of positive definiteness restricts the function K to be a "test function." Let X be $\mathbb{R}^n$, and let $\mathcal{D}(\mathbb{R}^n)$ be the family of compactly supported C^∞-functions on $\mathbb{R}^n$. We say that K in $\mathcal{D}(\mathbb{R}^n \times \mathbb{R}^n)$ is *positive definite in the distributional sense* if

$$(T \otimes T)(K) \geq 0$$

for all $T \in \mathcal{D}'(\mathbb{R}^n)$; i.e., all distributions T.

Tensor products of functionals are defined in a two-step process, beginning with the definitions

$$\begin{aligned} (u \otimes v)(x,y) &= u(x)v(y) && (13)\\ (S \otimes T)(u \otimes v) &= S(u)T(v) && (14) \end{aligned}$$

Then extensions are made using linearity and limits. In the case of distributions, the reader can consult [12], [15], for details.

In his important paper [27], Narcowich extends these ideas to closed, compact, connected, orientable, m-dimensional, C^∞-Riemannian manifolds furnished with a C^∞-metric. Let M be such a manifold, and let K be a C^∞-kernel. That is, $K \in C^\infty(M \times M)$. If T is a distribution on M (i.e., $T \in \mathcal{D}'(M)$) then we can write $T^{(1)}(K)$ or $T^{(2)}(K)$ depending on whether T acts on K as a function of its first or second argument. With the notation $K^y(x) = K_x(y) = K(x,y)$ we have

$$(T^{(2)}K)(x) = T(K_x)$$

An interpolation problem is described as follows. Given the kernel K and a linearly independent set of distributions $\{T_1, T_2, \ldots, T_n\}$, we try to solve the interpolation problem

$$T_i(f) = \lambda_i \qquad (1 \le i \le n) \tag{15}$$

with a function f of the form

$$f = \sum_{j=1}^{n} c_j T_j^{(2)}(K) \tag{16}$$

The interpolation matrix that arises will have entries

$$A_{ij} = T_i T_j^{(2)}(K) = (T_i \otimes T_j)(K) \qquad (1 \le i, j \le n) \tag{17}$$

For example, if each T_i is a regular distribution, defined by an integrable function u_i, then Equation (17) takes this form:

$$A_{ij} = \int u_i(x) \int u_j(y) K(x, y)\, dy\, dx \tag{18}$$

Narcowich proves that K is positive definite, as in Equation (10), if and only if it is positive definite in the sense that

$$(T \otimes T)(K) \ge 0$$

for every distribution T. He then goes on to prove the following theorem.

Theorem 7.1 *Let $\{F_j : j \in \mathbb{N}\}$ be the orthonormal base for $L^2(M)$ consisting of eigenfunctions of the Laplace-Beltrami equation. A kernel $K \in C^\infty(M \times M)$ is positive definite if and only if its Fourier coefficients,*

$$a_j = \iint K(x, y) F_j(x) F_j(y)\, dx\, dy \qquad (j = 1, 2, \ldots)$$

are nonnegative. The kernel K is strictly positive definite if and only if $a_j > 0$ for all j.

8 Other Topics

The use of spline functions for approximation on spheres has been vigorously pursued by the Geomathematics Group at the University of Kaiserlautern. The reader wishing to acquaint herself with this work should begin by consulting the lengthy survey of Freeden, Schreiner, and Franke [10].

Wavelets on spheres are the subject of a number of papers, of which a few are [28], [33],[13], [36], [8], and [11].

The subject of numerical integration on spheres has received some recent attention. The following papers are on this topic or are related to it: [30], [14], [9], [16].

The important topic of density or fundamentality of sets of functions is studied on spheres. Recall that a set Y in a normed linear space X is said to be fundamental if the linear span of Y is dense in X. An easy theorem is available (and useful) for the space $C(S^m)$ consisting of all continuous real-valued functions on the sphere S^m, normed with the usual supremum norm.

Theorem 8.1 *Let f be a continuous real-valued function on the interval $[-1,1]$, and suppose that for all non-trivial signed Borel measures μ on S^m we have*

$$\int_{S^m}\int_{S^m} f(xy)\,d\mu(x)\,d\mu(y) > 0$$

Then the set $\{x \mapsto f(xy) \ : \ y \in S^m\}$ is fundamental in $C(S^m)$.

In [38] a characterization is given for the continuous strictly positive definite functions of the type appearing in Theorem 8.1.

References

[1] Bochner, S., *Hilbert distances and positive definite functions*, Ann. of Math., Vol. 42 (1941), 647-656.

[2] de Boor, C. and A. Ron, *On multivariate polynomial interpolation*, Constructive Approximation, Vol. 6 (1990), 287-302.

[3] de Boor, C. and A. Ron, *Computational aspects of polynomial interpolation in several variables*, Mathematics of Computation, Vol. 58, 198 (1992), 705-727.

[4] de Boor, C. and A. Ron, *The least solution of the multivariate polynomial interpolation*, Math. Zeit., Vol. 210 (1992), 347-378.

[5] Cheney, E.W., *Approximation and interpolation on spheres*, Approximation Theory, Wavelets and Applications, S.P. Singh, A. Carbone, and B. Watson, Kluwer Publ. Co., Dordrecht (1995), 47-53.

[6] Cheney, E.W., *Approximation using positive definite functions*, Approximation Theory VIII, C.K. Chui and L.L. Schumaker, World Scientific, Singapore (1996), 145-168.

[7] Christensen, C.R. and P. Ressel, *Positive definite kernels on the complex Hilbert sphere*, Math. Zeit., Vol. 180 (1982), 193-201.

[8] Dahlke, S., W. Dahmen, I. Weinreich, E. Schmitt, *Multiresolution analysis and wavelets on S^2 and S^3*, Numer. Funct. Analy. Optimiz., Vol. 16 (1995), 19-41, MR 96a:42044.

[9] Freeden, W. and J. Fleck, *Numerical integration by means of adapted Euler summation formulas*, Numer. Math., Vol. 51 (1987), 37-64.

[10] Freeden, W., M. Schreiner, and R. Franke, *A survey on spherical spline approximation*, to appear, Surveys on Mathematics in Industry.

[11] Freeden, W. and M. Schreiner, *New wavelet methods for approximating harmonic functions*, preprint.

[12] Friedlander, F.G., Introduction to the Theory of Distributions, Cambridge University Press (1982).

[13] Göttelmann, J., *Locally supported wavelets on the sphere*, preprint, 1996.

[14] Grabner, P.J. and R.F. Tichy, *Spherical designs, discrepancy and numerical integration*, undated preprint.

[15] Hörmander, L., The Analysis of Linear Partial Differential Equations, Springer-Verlag, Berlin (1983).

[16] Madych, W.R. and S.A. Nelson, *Spherical quadrature and inversion of radon transforms*, Proc. Amer. Math. Soc., Vol. 95 (1985), 453-457.

[17] Menegatto, V.A., Interpolation on Spherical Spaces, Ph.D. Dissertation, University of Texas at Austin (August 1992).

[18] Menegatto, V.A., *Interpolation on spherical domains*, Analysis Vol. 14 (1994), 415-424.

[19] Menegatto, V.A., *Interpolation on complex Hilbert spheres using positive definite and conditionally positive definite kernels*, Acta Math. Hung. (to appear).

[20] Menegatto, V.A., *Condition numbers associated with radial function interpolation on spheres*, preprint, 1996.

[21] Menegatto, V.A., *Strictly positive definite kernels on the Hilbert sphere*, Applicable Analysis, Vol. 55 (1994), 91-101.

[22] Menegatto, V.A., *Strictly positive definite kernels on circles*, Rocky Mtn. J. Math. Vol. 25 (1995), 1149-1163.

[23] Menegatto, V.A., *Interpolation on the complex Hilbert sphere*, Approx. Theory and its Appl., Vol. 12 (1996) 2, 31-40.

[24] Menegatto, V.A., *Dense sets of functions on spheres*, preprint, March 1997.

[25] Menegatto, V., and A.P. Peron, *Generalized interpolation on spheres using positive definite and related functions*, preprint, Dec. 1996, to appear, Numer. Funct. Analy. Optimiz.

[26] Müller, C., *Spherical Harmonics*, Lecture Notes in Mathematics, Vol. 17, Springer-Verlag, Berlin, Heidelberg, New York (1966).

[27] Narcowich, F.J., *Generalized Hermite interpolation and positive definite kernels on a Riemannian manifold*, J. Math. Anal. Appl. Vol. 190, (1995), 165-193.

[28] Narcowich, F.J. and J.D. Ward, *Non-stationary wavelets on the m-sphere for scattered data*, preprint.

[29] Narcowich, F.J., N. Sivakumar, J.D. Ward, *Stability results for scattered-data interpolation on Euclidean spheres*,preprint, 1997.

[30] Rakhmanov, E.A., E.B. Saff, Y.M. Zhou, *Minimal discrete energy on the sphere*, Math. Research Letters, Vol. 1 (1994), 647-662.

[31] Ron, A. and X. Sun, *Strictly positive definite functions on spheres in Euclidean spaces*, Math. Comp., Vol. 65 (1996), 1513-1530.

[32] Ron, A. and X. Sun, *Strictly positive definite functions on spheres*, CMS TR 94-6, University of Wisconsin - Madison, February 1994.

[33] Schmitt, E., *Wavelets and multiresolution analysis on sphere-like surfaces*, Wavelet Applications in Signal and Image Processing, 92-101, SPIE, Bellingham, Washington (1995), MR 97b:94008.

[34] Schoenberg, I.J., *Positive definite functions on spheres*, Duke Math. J. , Vol. 9 (1942), 96-108.

[35] Schreiner, M., *On a new condition for strictly positive definite functions on spheres*, to appear, Proc. Amer. Math. Soc.

[36] Schröder, P. and W. Sweldens, *Spherical wavelets: efficiently representing functions on the sphere*, preprint, 1995.

[37] E.M. Stein and G. Weiss, Introduction to Fourier Analysis on Euclidean Spaces, Princeton University Press, Princeton, NJ (1971).

[38] Sun, Xingping and E.W. Cheney, *Fundamental sets of continuous functions on spheres*, preprint Sept. 16, 1994, to appear, Constr. Approx.

[39] Szegö, Gabor, Orthogonal Polynomials (Amer. Math. Colloq. Publ., Vol. 23), Amer. Math. Soc., Providence, RI (1959)

[40] Wahba, G. , *Spline interpolation and smoothing on the sphere*, SIAM J. Sci Stat. Comput., Vol. 2 (1981), 5-16, Erratum: ibid Vol. 3 (1982), 385-6.

[41] Yuan Xu and E.W. Cheney, *"Strictly positive definite functions on spheres*, Proc. Amer. Math. Soc., Vol. 116 (1992), 977-981.

Nonparametric Density Estimation by Polynomials and by Splines

Z. Ciesielski
Instytut Matematyczny Pan
Ul. Abrahama 18
81-825 Sopot, Poland

Abstract

The aim of this paper is to present an approximation argument for an algorithm for determining the degree of a polynomial density estimator on [0,1] constructed in terms of the Bernstein polynomials. The argument applies as well to the local spline density estimators based on the Tchebyshev knots. The algorithm can be implemented on a computer.

1 Introduction

Finite dimensional positive operators from $L^1[0,1]$ onto the space of polynomials over $[0,1]$, of degree not exceeding a given positive integer, or onto the space of splines with Tchebyshev extreme points as knots, can be used to construct nonparametric estimator for the a priori density. Asymptotic properties of estimators of such type were discussed by the author in [1, 2, 3]. For practical reasons it is important to have an automatic way of choosing the 'optimal' degree in the polynomial case and the 'optimal' number of knots in the spline case. Thanks to the new results of Anna Kamont [7] on the approximation by splines with Tchebyshev knots we can apply a similar argument as in [4] to propose a procedure for automatic choice of the 'optimal' degree of the polynomial estimator and the 'optimal' number of the Tchebyshev extreme points of the spline estimator. In both cases, the 'optimal' degree and the 'optimal' number of knots are constructed by means of the given sample. In the generic case of the sample they always exist. The algorithm can be easily implemented on a computer.

2 Preliminaries

The space of all real polynomials of degree not exceeding m is denoted by Π_m. In Π_m we have the Bernstein basis i.e.

$$\Pi_m = \text{span}[B_{i,m}, i = 0, \ldots, m],$$

where

$$B_{i,m}(x) = \binom{m}{i} x^i (1-x)^{m-i}, \quad i = 0, \ldots, m.$$

We need to recall some properties of the Bernstein polynomials. Our attention will be restricted to the interval $I = [0, 1]$ and the following notation will be used

$$(f, g) = \int_I f(x)g(x)\, dx, \quad \|f\|_p = \Big(\int_I |f(x)|^p dx\Big)^{1/p}.$$

We have the following elementary properties of $B_{i,m}$: $B_{i,m}(x) \geq 0$ for $x \in I$; $\sum_{i=0}^m B_{i,m} = 1$ and $(m+1)(B_{i,m}, 1) = 1$ for $i = 0, \ldots, m$. In our construction of the polynomial estimator the following positive kernel is used

$$R_m(x, y) = (m+1) \sum_{i=0}^{m} B_{i,m}(x) B_{i,m}(y).$$

It follows by the definition of $B_{i,m}$ that

$$R_m(x, y) = R_m(y, x), \quad 0 \leq R_m(x, y) \leq m + 1 \quad \text{for} \quad x, y \in I.$$

Now, the Bernstein-Durrmeyer operator is defined as follows

$$(R_m f)(x) = \int_I R_m(x, y) f(y)\, dy = (m+1) \sum_{i=0}^{m} (B_{i,m}, f) B_{i,m}(x).$$

The L^p norm of $R_m : L^p \to \Pi_m$ for $1 \leq p \leq \infty$ is equal to one and R_m takes densities into densities.

The de Casteljau algorithm for given $w \in \Pi_m$, for calculating for fixed x the value

$$w(x) = \sum_{i=0}^{m} w_i B_{i,m}(x),$$

is based on the identity

$$B_{i,m}(x) = (1-x) B_{i,m-1}(x) + x B_{i-1,m-1}(x).$$

Using it we find that for $0 \le k \le m$

$$w(x) = \sum_{i=0}^{m-k} w_i^{(k)}(x) B_{i,m-k}(x),$$

where $w_i^{(k)} \in \Pi_k$, and for $0 \le k < m$ we have

$$w_i^{(k+1)}(x) = (1-x) w_i^{(k)}(x) + x w_{i+1}^{(k)}(x), \qquad i = 0, \ldots, m-k-1.$$

In particular, $w(x) = w_0^{(m)}(x)$.

In the spline case we restrict our attention to the piece-wise linear continuous functions with knots at the extreme points of the Tchebyshev polynomial of the first kind. More precisely, let us define for a positive integer m the set of Tchebyshev knots in I=[0,1] by the formula

$$\pi_m : \quad x_{i,m} = \frac{1}{2}\left(1 - \cos(\pi \frac{i}{m})\right) = \sin^2(\pi \frac{i}{2m}) \quad \text{for} \quad i = 0, \ldots, m,$$

and the auxiliary knots π_{2m}^*: $x_{2i,2m}^* = x_{i,m}$ for $i = 0, \ldots, m$;

$$x_{1,2m}^* = \frac{\sin^2(\frac{\pi}{m})}{3 + 2\cos(\frac{\pi}{m})}, \quad x_{2m-1,2m}^* = 1 - x_{1,2m}^*,$$

and

$$x_{2i-1,2m}^* = \frac{1}{2}\left(1 - \gamma_m \cos(\pi \frac{2i-1}{2m})\right) \quad \text{for} \quad i = 2, \ldots, m-1,$$

where $\gamma_m = \cos(\frac{\pi}{m}) / \cos(\frac{\pi}{2m})$.

Clearly $x_{k-1,2m}^* < x_{k,2m}^*$ for $k = 1, \ldots, 2m$; $x_{0,2m}^* = 0$ and $x_{2m,2m}^* = 1$. Now, denote by $\mathcal{S}_m$ and $\mathcal{S}_{2m}^*$ the linear spaces of piece-wise linear continuous functions with knots π_m and π_{2m}^*, respectively. We also introduce $d_{i,m} = x_{i,m} - x_{i-1,m}$, $d_{k,2m}^* = x_{k,2m}^* - x_{k-1,2m}^*$, $d_{0,m} = 0$ and $d_{m+1,m} = 0$.

In the space $\mathcal{S}_m$ we have the B-spline basis i.e.

$$\mathcal{S}_m = \text{span}[N_{i,m}, i = 0, \ldots, m],$$

where the functions $N_{i,m}(x)$ are defined on I as follows:

$$N_{0,m}(x) = \max\left(0, \frac{x_{1,m} - x}{d_{1,m}}\right) \quad \text{and} \quad N_{m,m}(x) = \max\left(0, \frac{x - x_{m-1,m}}{d_{m,m}}\right)$$

and for $i = 1, \ldots, m-1$

$$N_{i,m}(x) = \min\left[\max\left(0, \frac{x - x_{i-1,m}}{d_{i,m}}\right), \max\left(0, \frac{x_{i+1,m} - x}{d_{i+1,m}}\right)\right].$$

The functions $\{N_{i,m} : i = 0, \ldots, m\}$ form a partition of unity over I. It is convenient to introduce also the functions $N_{i,m}$ normalized in L^1 i.e.

$$M_{i,m} = \frac{2}{d_{i,m} + d_{i+1,m}} N_{i,m} \quad \text{for} \quad i = 0, \ldots, m.$$

Like in the polynomial case the following symmetric and positive kernel will be used

$$Q_m(y, x) = \sum_{i=0}^{m} M_{i,m}(y) N_{i,m}(x).$$

The corresponding operator $Q_m : L^p \to \mathcal{S}_m, \;\; 1 \leq p \leq \infty$, i.e.

$$(Q_m f)(x) = \int_I Q_m(x, y) f(y)\, dy = \sum_{i=0}^{m} (M_{i,m}, f) N_{i,m}(x),$$

has its L^p norm equal to one and it takes densities into densities.

To state the next result we need the definition of the Ditzian-Totik modulus or else the weighted modulus of smothness with the step-weight function $\phi(x) = \sqrt{x(1-x)}$:

$$\omega_{\phi,p}(f, \delta) = \sup_{0<h<\delta/2} \left(\int_{E(h)} |f(x + h\phi(x)) - f(x - h\phi(x))|^p \, dx \right)^{1/p},$$

where $E(h) = \{x : x - h\phi(x), x + h\phi(x) \in I\}$;

The operators R_{m^2} and Q_m have similiar approximation properties. We obtain from [7] and [6] the following:

Proposition 1 *Let $0 < \alpha < 1$ and $1 \leq p \leq \infty$ be given. Then, for $f \in L^p(I)$, the following conditions are equivalent:*

$$\|f - R_{m^2} f\|_p = O(\frac{1}{m^\alpha}) \quad \textit{as} \quad m \to \infty, \tag{1}$$

$$\|f - Q_m f\|_p = O(\frac{1}{m^\alpha}) \quad \textit{as} \quad m \to \infty, \tag{2}$$

Moreover, there are finite constants C and $m_0 > 0$ such that

$$\|f - R_{m^2} f\|_p \leq C\, \omega_{\phi,p}(f, \frac{1}{m}) \quad \textit{for} \quad 1 \leq p \leq 2,\ m \geq m_0, \tag{3}$$

and

$$\|f - Q_m f\|_p \leq C\, \omega_{\phi,p}(f, \frac{1}{m}) \quad \textit{for} \quad 1 \leq p \leq \infty,\ m \geq m_0. \tag{4}$$

Thus, in any L^p norm with $1 \leq p \leq 2$, approximation of f by the spline $Q_m f$ is equally good as approximation by the polynomial $R_{m^2} f$ of degree m^2.

It can be proved by elmentary computation, as it has been communicated to the author by A. Kamont, that:

Proposition 2 *If $1 \leq p \leq \infty$, $m \geq 1$, $g \in S_m$ and*

$$g = \sum_{i=0}^{m} a_i N_{i,m},$$

then for $m \geq m_0$ with some m_0

$$\omega_{\phi,p}(g, \frac{1}{m}) \sim \left(\sum_{i=1}^{m} d_{i,m} |a_i - a_{i-1}|^p \right)^{1/p} \tag{5}$$

with the constants in this equivalence independent of g, p and m.

3 The Estimators

The a priori density of a simple sample $(X_1, \ldots, X_d)$, where $d \in N$, is denoted by f and it is assumed that its support is contained in $I = [0, 1]$. We recall the definition of the empirical distribution

$$F_d(x) = F(X_1, \ldots, X_d; x) = \frac{\#\{j : X_j < x\}}{d}.$$

Now, the polynomial density estimator of degree m for the density f is defined by formula

$$p_{m,d}(x) = p_m(X_1, \ldots, X_d; x) = \int_I R_m(y, x)\, dF_d(y) = \sum_{i=0}^{m} c_i B_{i,m}(x), \tag{6}$$

where

$$c_i = \frac{m+1}{d} \sum_{j=1}^{d} B_{i,m}(X_j). \tag{7}$$

Similarly, the spline estimator corresponding to the knots π_m is given by

$$s_{m,d}(x) = s_m(X_1, \ldots, X_d; x) = \int_I Q_m(y, x)\, dF_d(y) \tag{8}$$

or more explicitly by

$$s_{m,d}(x) = \sum_{i=0}^{d} b_i N_{i,m}(x), \tag{9}$$

where

$$b_i = \int_I M_{i,m}\, dF_d = \frac{1}{d} \sum_{j=1}^{d} M_{i,m}(X_j). \tag{10}$$

The properties of the operators R_m and Q_m ensure that $p_{m,d}$ and $s_{m,d}$ are densities. The asymptotic theory is concerned about the behavior of these density estimators as $d \to \infty$ (see [2, 3, 1]) . In this note we are more interested, given the sample $(X_1, \ldots, X_d)$, in finding a recipe for an 'optimal' choice of the parameter m.

The estimators $p_{m,d}$ and $s_{m,d}$ are symmetric functions in the variables $X_1, \ldots, X_d$ and therefore it is sufficient to treat them as functions of $d+1$ variables over $\Delta_d \times I$, where

$$\Delta_d = \{\mathbf{x} = (x_1, \ldots, x_d) : 0 \le x_1 \le \ldots \le x_d \le 1\}$$

or as vector valued functions e.g. in the spline case

$$s_{m,d} : \mathbf{x} \in \Delta_d \to s_{m,d}(\mathbf{x}; \cdot) \in S_m, \tag{11}$$

where

$$s_{m,d}(\mathbf{x}; x) = \sum_{i=0}^{m} b_i(\mathbf{x}) N_{i,m}(x) \tag{12}$$

with

$$b_i(\mathbf{x}) = \frac{1}{d} \sum_{j=1}^{d} M_{i,m}(x_j). \tag{13}$$

In a similar way we define for $\mathbf{x} \in \Delta_d$ and $x \in I$ the polynomial estimator $p_{m,d}(\mathbf{x}; x)$.

Since we are interested in densities on I it is natural to stay within the $L^1(I)$ space. Thus, for the a priori density f and for a sample $\mathbf{x} \in \Delta_d$ we have

$$\|f - s_{m,d}(\mathbf{x}; \cdot)\|_1 \le \|f - Q_m f\|_1 +$$

$$\|Q_m(f - s_{m,d}(\mathbf{x}; \cdot))\|_1 + \|s_{m,d}(\mathbf{x}; \cdot) - Q_m(s_{m,d}(\mathbf{x}; \cdot))\|_1 .$$

Since we really do not know the a priori density f, there is no hope for finding m for which the right hand side of the last inequality is minimal. Being helpless we invoke the bootstrapping procedure. Namely, we replace at the right hand side of the last inequality the unknown density f by the density we know i.e. by $s_{m,d}(\mathbf{x};\cdot)$. This leads, up to the multiplicative constant 2, to the quantity

$$\|s_{m,d}(\mathbf{x};\cdot) - Q_m(s_{m,d}(\mathbf{x};\cdot))\|_1,$$

which can be estimated, up to a multiplicative constant, with the help of (4) by

$$\omega_{\phi,1}(s_{m,d}(\mathbf{x};\cdot), 1/m).$$

Now, since $s_{m,d}(\mathbf{x};\cdot) \in \mathcal{S}_m$, it follows by (5) and (12) that this in turn is equivalent to

$$C_m(\mathbf{x}) := \sum_{i=1}^{m} d_{i,m}|b_i(\mathbf{x}) - b_{i-1}(\mathbf{x})|. \tag{14}$$

Our aim is to prove, for a given $\mathbf{x} \in \Delta_d$, existence of the smallest $m_0 \in N$ such that

$$C_{m_o}(\mathbf{x}) = \inf_{m \in N} C_m(\mathbf{x}). \tag{15}$$

Definition 3 *An open subset Δ° of the compact $\Delta_d \subset R^d$ is called generic if its complement $\Delta_d \setminus \Delta^\circ$ is nowhere dense in Δ_d.*

Definition 4 *The spline density estimator $s_{m_0,d}(\mathbf{x};\cdot)$ with $\mathbf{x} \in \Delta_d$ defined as in* (12) *and* (13) *is said to be optimal if m_0 is the smallest solution to* (15)*. The polynomial density estimator $p_{m_1,d}(\mathbf{x};\cdot)$ is said to be optimal if $m_1 = m_0^2$, where m_0 is as defined in the first part of this definition.*

We are going to show the existence of a generic set Δ° such that for each $\mathbf{x} \in \Delta^\circ$ there is a finite solution m_0 to (15).

Lemma 5 *Let*

$$M_m(x) = \sum_{i=1}^{m} d_{i,m}\,|M_{i,m}(x) - M_{i-1,m}(x)| \quad \textit{for} \quad x \in I. \tag{16}$$

Then, $M_m \in S_m^*$ and

$$M_m(x_{2i,2m}^*) = M_m(x_{i,m}) = 2 \quad \textit{for} \quad i = 0, \ldots, m, \tag{17}$$

$$M_m(x_{2i-1,2m}^*) = \delta_m \quad \textit{for} \quad i = 2, \ldots, m-1, \tag{18}$$

$$M_m(x_{1,2m}^*) = M_m(x_{2m-1,2m}^*) = \kappa_m, \tag{19}$$

where

$$\delta_m = 1 - \left(\tan(\frac{\pi}{2m})\right)^2 \quad \textit{and} \quad \kappa_m = \frac{2 + 4\cos(\frac{\pi}{m})}{3 + 2\cos(\frac{\pi}{m})} \geq 1 \quad \textit{for} \quad m \geq 3.$$

Consequently,

$$\delta_m \le M_m(x) \le 2 \quad \textit{for} \quad m \ge 3 \quad \textit{and} \quad x \in I. \tag{20}$$

Proof. One checks that

$$M_{i,m}(x^*_{2i-1,2m}) - M_{i-1,m}(x^*_{2i-1,2m}) = 0 \quad \text{for} \quad i = 1, \dots, m.$$

This implies that $M_m \in S^*_m$. Elementary calculations give the desired values $M_m(x^*_{k,2m})$ for $k = 0, \dots, 2m$. Since M_m is continuous and piecewise linear the proof is complete.

Lemma 6 *Let* $\mathbf{x} \in \mathrm{int}\Delta_d$. *Then there is finite* $m^* = m^*(\mathbf{x}) \in N$ *such that*

$$C_m(\mathbf{x}) \ge \delta_m \quad \textit{for} \quad m \ge m^*. \tag{21}$$

Moreover, for all $m \ge 1$ *and* $\mathbf{x} \in \Delta_d$ *we have*

$$C_m(\mathbf{x}) \le \frac{1}{d} \sum_{j=1}^{d} M_m(x_j). \tag{22}$$

Proof. For the given $\mathbf{x} = (x_1, \dots, x_d)$ define

$$\delta = \min\{x_j - x_{j-1} : j = 2, \dots, d\}.$$

Since $\mathbf{x} \in \mathrm{int}\Delta_d$ it follows that $\delta > 0$. Now, we find that the support of $M_{i,m} - M_{i-1,m}$ is equal to $[x_{i-2,m}, x_{i+1,m}]$ with the understanding that $x_{-1,m} = 0$ and $x_{m+1,m} = 1$. For m^* we can now take the smallest m such that $|x_{i+1,m} - x_{i-2,m}| < \delta$ for $i = 1, \dots, m$. This implies that for $m \ge m^*$ and $1 \le i \le m$ we have

$$|b_i(\mathbf{x}) - b_{i-1}(\mathbf{x})| = \frac{1}{d} |\sum_{j=1}^{d} (M_{i,m}(x_j) - M_{i-1,m}(x_j))|$$

$$= \frac{1}{d} \sum_{j=1}^{d} |M_{i,m}(x_j) - M_{i-1,m}(x_j)|,$$

whence by (16), (14) and by(20)

$$C_m(\mathbf{x}) = \frac{1}{d} \sum_{j=1}^{d} M_m(x_j) \ge \delta_m.$$

Inequality (22) follows from the definitions of C_m and M_m.

Definition 7 *Define*

$$\Delta_d^* = \bigcup_{m \geq 3} \{\mathbf{x} \in \Delta_d : C_m(\mathbf{x}) < 1\}.$$

Theorem 8 *The set Δ_d^* defined as in Definition 7 is generic in the sense of Definition 3. Moreover, for each $\mathbf{x} \in \Delta_d^*$ there are optimal spline and polynomial estimators $s_{m_0,d}(\mathbf{x}; \cdot)$ and $p_{m_1,d}(\mathbf{x}; \cdot)$ in the sense of Definition 4.*

Proof. The set Δ_d^* is open. Now, define

$$D = \{\mathbf{x} \in \Delta_d : x_j = x^*_{2i_j-1,2m},\ 1 < i_j < m,\ 1 \leq j \leq d,\ m = 3, 4, \ldots\}.$$

This set is dense in Δ_d. Moreover, Lemma 5 (18) and Lemma 6 (22) imply for given $\mathbf{x} \in D$

$$C_m(\mathbf{x}) \leq \frac{1}{d} \sum_{j=1}^{d} M_m(x^*_{2i_j-1,2m}) = \delta_m < 1,$$

and therefore $D \subset \Delta_d^*$. Thus Δ_d^* is generic in the sense of Definition 3. It remains to check the existence of the optimal estimator in the sense of Definition 4 in the generic case. According to Lemma 6 (21) we obtain for each $\mathbf{x} \in \Delta_d^*$

$$\liminf_{m \to \infty} C_m(\mathbf{x}) \geq \lim_{m \to \infty} \delta_m = 1,$$

and this completes the proof.

4 Comments

We have presented in this note an approximation theory approach to the nonparametric density estimation on the finite interval by polynomials and by splines. In the spline case our attention was restricted to piece-wise linear continuous splines (i.e. splines of the second order). One can use the optimal m_0 found in Theorem 8, for the same reasons as in the polynomial case, in the case of splines of higher orders with the same Tchebyshev knots, as well. It is sufficient to replace the piece-wise linear B-splines $N_{i,m}$ in the definition of the $s_{m,d}$ estimator by B-splines of higher order. Similar technology was implemeted on the computer on the real line with splines with uniformly distributed knots [4]. In implementing the polynomial case the de Casteljau algorithm as presented in Section 2 should be helpful. The spline implementation can be done with the help of known algorithms for the B-splines (see e.g. [5]).

Acknowledgements: The author is indebted to Dr Anna Kamont for stimulating discussions.

References

[1] Z. Ciesielski, Nonparametric polynomial density estimation in the L^p-norm, in "LNM 1354. Proceedings, Havana 1987", pp. 1-10.

[2] Z. Ciesielski, Local spline approximation and nonparametric density estimation, in " Constructive theory of functions'87. Proceedings, Sofia 1988", pp. 79-84.

[3] Z. Ciesielski, Nonparametric polynomial density estimation, Prob. and Math. Stat. 9.1(1988), 1-10.

[4] Z. Ciesielski, Asymptotic nonparametric spline density estimation, Prob. and Math. Stat. 12.1(1991), 1-24.

[5] C. De Boor, "A Practical Guide to Splines", Springer-Verlag, New York, 1978.

[6] Z. Ditzian and K. Ivanov, Bernstein-Type Operators and Their Derivatives, Journal of Approximation Theory 56(1989), 72-90.

[7] A. Kamont, Spline spaces and weighted moduli of smoothness, submitted to a journal.

Birkhoff Type Interpolation on Perturbed Roots of Unity

M.G. de Bruin
Department of Pure Mathematics, Delft University of Technology
P.O. Box 5031, 2600 GA Delft, The Netherlands

A. Sharma
Department of Mathematical Sciences, University of Alberta
Edmonton, Alberta, Canada T6G 2G1

J. Szabados
Mathematical Institute, Hungarian Academy of Sciences
P.O. Box 127, H-1364 Budapest, Hungary

Abstract

Here we prove the regularity of $(0, 1, \ldots, r-2, r)$, $r \geq 2$ interpolation on the $n+1$ nodes $z_k = \frac{\omega_k - \alpha}{1-\alpha\omega_k}$, $0 < \alpha < 1$ $(k = 0, 1, \ldots, n)$ where ω_k is a primitive $(n+1)^{st}$ root of unity. We outline a method for finding the fundamental polynomials and show that $(0, 3)$ interpolation on these nodes is regular except for at most one α. A corresponding Pál type problem is also considered.

1 Introduction

The problem of $(0, 2)$ interpolation on roots of unity was first studied by O. Kis [5] who proved its regularity, gave its fundamental polynomials and a convergence theorem. Later over the years the regularity of $(0, m_1, \ldots, m_q)$ interpolation on roots of unity and a convergence theorem were proved ([6]). Only recently a few papers have appeared dealing with $(0, 1, \ldots, r-2, r)$ interpolation on some non-uniformly distributed nodes on the unit circle. Thus in [9], Xie Sie-Quing following a remark in [8] considered nodes which are obtained by projecting the zeros of $\pi_n(x)$ onto the unit circle. We recall that it was P. Turán who first initiated the problem of lacunary interpolation by first studying $(0, 2)$ interpolation on zeros of $\pi_n(x)$ $(\pi_n(x) = (1-x^2)P'_{n-1}(x)$ where $P_{n-1}(x)$ is Legendre polynomial of degree $n-1)$ [6]. Chen and Sharma [3] have recently proved the regularity of $(0,m)$ interpolation on the zeros of $(z^{2n}+1)(z^2-1)$ and of $(z^{2n}+1)(z^n-1)$ which are non-uniformly distributed on the unit circle. One may also refer to [4]. A general theory of lacunary interpolation on non-uniformly distributed nodes on the unit circle does not exist.

In a recent paper, R. Brück [1] considered the problem of Walsh equiconvergence on nodes which are obtained as the Möbius transform of the roots of unity. He considered the cases of Lagrange and Hermite interpolation on zeros of $w_{n,\alpha}(z)$ (defined in Section 2) and extended the result of Cavaretta et al. [9] on Walsh equiconvergence. Since equiconvergence holds also for lacunary interpolation on roots of unity ([2], [7]), it is natural to ask if $(0,2)$ (or $(0,1,\ldots,r-2,r)$, $r \geq 2$) interpolation on zeros of $w_{n,\alpha}(z)$ is regular.

In Section 2, we state and prove the theorem and give two methods for proving the regularity of the interpolation problem. Section 3 deals with the fundamental polynomials. In this case the fundamental polynomials do not have a neat form and so it is not clear how a convergence result can be derived. In Section 4, we discuss the regularity of $(0,3)$ interpolation on the zeros of $w_{n,\alpha}(z)$. In this case the method of solving a linear differential equation as in Section 2 does not work. However the second method used in Section 2 is applicable and we show that the problem is regular for all α except at most one. In Section 5, we discuss Pál type interpolation and show its close connection with the results of Sections 2 and 4.

2 Main Result

Let ω_k be the zeros of $z^{n+1}-1$ and let $z_k = \frac{\omega_k - \alpha}{1-\alpha\omega_k}$ for some α, $0<\alpha<1$. Then $\omega_k = (z_k+\alpha)/(1+\alpha z_k)$ so that z_k is a zero of the polynomial $\omega_{n,\alpha}(z)$ where

$$\omega_{n,\alpha}(z) = (z+\alpha)^{n+1} - (1+\alpha z)^{n+1}, \quad 0<\alpha<1. \tag{2.1}$$

We shall prove that for any positive integer $r \geq 2$, we have

Theorem 2.1 *For any* $0<\alpha<1$, *the problem of* $(0,1,\ldots,r-2,r)$ *interpolation on zeros of* $\omega_{n,\alpha}(z)$ *is regular.*

Proof. We shall show that if $P(z)$ is a polynomial of degree $\leq r(n+1)-1$ and satisfies

$$P^{(\nu)}(z_k) = 0, \quad k=0,1,\ldots,n, \quad \nu = 0,1,\ldots,r-2,r, \tag{2.2}$$

then $P(z)$ is identically zero. To prove this we set

$$P(z) := (\omega_{n,\alpha}(z))^{r-1}Q(z), \quad Q(z) \in \pi_n.$$

If we determine $Q(z)$ by the requirement that $P^{(r)}(z_k) = 0, \quad k = 0, 1, \ldots, n,$ then we obtain

$$\binom{r}{r-1} \left(\omega_{n,\alpha}^{r-1}(z)\right)_{z_k}^{(r-1)} Q'(z_k) + \left(\omega_{n,\alpha}^{r-1}(z)\right)_{z_k}^{(r)} Q(z_k) = 0. \tag{2.3}$$

It is easy to check that

$$\begin{aligned} \left(\omega_{n,\alpha}^{r-1}(z)\right)_{z_k}^{(r-1)} &= (r-1)!\left(\omega'_{n,\alpha}(z_k)\right)^{r-1} \\ \left(\omega_{n,\alpha}^{r-1}(z)\right)_{z_k}^{(r)} &= \frac{r(r-1)}{2}(r-1)!\left(\omega'_{n,\alpha}(z_k)\right)^{r-2}\omega''_{n,\alpha}(z_k). \end{aligned}$$

Using these and simplifying, we get from (2.3) the following $n+1$ conditions

$$\frac{r-1}{2}\,\omega''_{n,\alpha}(z_k)Q(z_k) + \omega'_{n,\alpha}(z_k)Q'(z_k) = 0, \quad k = 0, 1, \ldots, n. \tag{2.4}$$

¿From (2.1) and (2.4) on simplifying, we now have

$$2(1+\alpha z_k)(z_k+\alpha)Q'(z_k) + n(r-1)(1+2\alpha z_k+\alpha^2)Q(z_k) = 0 \\ (k = 0, 1, \ldots, n). \tag{2.4a}$$

It follows that $Q(z)$ satisfies the differential equation:

$$2(1+\alpha z)(z+\alpha)Q'(z) + (r-1)n(1+2\alpha z+\alpha^2)Q(z) = c\omega_{n,\alpha}(z). \tag{2.5}$$

Multiplying both sides by the integrating factor $[(1+\alpha z)(z+\alpha)]^{n(r-1)/2}$ and integrating we obtain

$$\begin{aligned} &[(1+\alpha z)(z+\alpha)]^{n(r-1)/2}Q(z) \\ &\quad = \tfrac{c}{2}\int_{-1/\alpha}^{z} \omega_{n,\alpha}(t)[(1+\alpha t)(t+\alpha)]^{\frac{n(r-1)}{2}-1}dt + D. \end{aligned}$$

Since the left side vanishes for $-1/\alpha$ and $-\alpha$, it follows that $D = 0$ and that

$$\frac{c}{2}\int_{-1/\alpha}^{-\alpha} \omega_{n,\alpha}(t)[(1+\alpha t)(t+\alpha)]^{\frac{n(r-1)}{2}-1}dt = 0. \tag{2.6}$$

We shall now show that the integral on the left in (2.6) is not zero for $0 < \alpha < 1$. This will complete the proof of uniqueness.

For this and for later purposes we need the integral

$$\begin{aligned} I(a,b): &= \int_{-1/\alpha}^{-\alpha} (t+\alpha)^a(1+\alpha t)^b dt \\ &= \frac{(1-\alpha^2)^{a+b+1}}{\alpha^{a+1}} \cdot \frac{\Gamma(a+1)\Gamma(b+1)}{\Gamma(a+b+2)}(-1)^a, \\ &\quad (a,b>0) \end{aligned} \tag{2.7}$$

which is easily checked after the substitution $t \to \frac{1-\alpha^2}{\alpha}t - \frac{1}{\alpha}$ and using the well-known formula for the Euler integral of the first kind ([10], formula (1.7.5)). Hence the integral in (2.6) is equal to

$$I\Big(\frac{n(r+1)}{2}, \frac{n(r-1)}{2} - 1\Big) - I\Big(\frac{n(r-1)}{2} - 1, \frac{n(r+1)}{2}\Big) =$$
$$= \frac{(1-\alpha^2)^{nr}}{\alpha^{n(r+1)/2}+1} \frac{\Gamma(n(r-1)/2)\Gamma(n(r+1)/2)}{\Gamma(nr+1)}\{1+(-1)^n\alpha^{n+1}\} \neq 0,$$

since $0 < \alpha < 1$. □

Another and an alternative approach is to set

$$Q(z) = \sum_{j=0}^{n} \alpha^j c_j (z+\alpha)^j (1+\alpha z)^{n-j}.$$

Then since

$$z_k + \alpha = \frac{(1-\alpha^2)\omega_k}{1-\alpha\omega_k}, \quad 1+\alpha z_k = \frac{1-\alpha^2}{1-\alpha\omega_k},$$

we have

$$Q(z_k) \quad = \quad (1-\alpha^2)^n \sum_{j=0}^{n} \alpha^j c_j \omega_k^j \Big/ (1-\alpha\omega_k)^n$$

$$Q'(z_k) = (1-\alpha^2)^{n-1}\sum_{j=0}^{n-1}\alpha^{j+1}[(j+1)c_{j+1}+(n-j)c_j]\omega_k^j \Big/ (1-\alpha\omega_k)^{n-1}.$$

Then (2.4a) yields on simplification after cancelling the common factor

$$2\omega_k \sum_{j=0}^{n-1}\alpha^{j+1}[(j+1)c_{j+1}+(n-j)c_j]\omega_k^j + (r-1)n(1-\alpha\omega_k)\sum_{j=0}^{n}\alpha^j c_j\omega_k^j = 0.$$

Simplifying further we get

$$2\sum_{j=1}^{n}\alpha^j[jc_j+(n+1-j)c_{j-1}]\omega_k^j + (r-1)n\sum_{j=0}^{n}\alpha^j c_j\omega_k^j + (r-1)n\sum_{j=1}^{n}\alpha^j c_{j-1}\omega_k^j + (r-1)n\alpha^{n+1}c_n = 0$$
$$(k=0,1,\ldots,n).$$

Hence we have a polynomial of degree $\leq n$ vanishing at the $n+1$ zeros ω_k $(k=0,1,\ldots,n)$. This leads to the following system of equations:

$$\begin{cases} c_0+\alpha^{n+1}c_n = 0 \\ [(r-1)n+2j]c_j + [(r+1)n+2-2j]c_{j-1} = 0, \\ \quad j=1,\ldots,n. \end{cases}$$

¿From the last n equations, we have

$$c_1 = -\frac{(r+1)n}{(r-1)n+2}c_0, \qquad c_2 = -c_1\frac{(r+1)n-2}{(r-1)n+4}, \ldots$$
$$c_{n-1} = -\frac{(r-1)n+4}{(r+1)n-2}c_{n-2}, \quad c_n = -\frac{(r-1)n+2}{(r+1)n}c_{n-1}.$$

Multiplying these we get $c_n = (-1)^n c_0$ and from the first equation we have

$$c_n = -c_0/\alpha^{n+1}.$$

Hence $c_0(1+(-1)^n\alpha^{n+1}) = 0$, whence $c_0 = 0$ $(0<\alpha<1)$, and so all the c_j's are zero.

3 Fundamental Polynomial

There are r different types of fundamental polynomials $\rho_{\mu,\nu}(z) \in \pi_{r(n+1)-1}$ determined by the conditions:

$$\rho_{\mu,\nu}^{(j)}(z_k) = \delta_{\mu j}\cdot\delta_{\nu k} \ (k,\nu = 0,1,\ldots,n;\ \mu,j = 0,1,\ldots,r-2,r).$$

First we determine $\rho_{r,\nu}(z)$ for a fixed ν $(0 \le \nu \le n)$. Clearly

$$\rho_{r,\nu}(z) = (\omega_{n,\alpha}(z))^{r-1} Q(z), \; Q(z) \in \pi_n \tag{3.1}$$

and satisfies the following conditions (see (2.3)):

$$(r-1)\omega''_{n,\alpha}(z_k)Q(z_k) + 2\omega'_{n,\alpha}(z_k)Q'(z_k) = 2\,\frac{\delta_{\nu,k}}{r!\omega'_{n,\alpha}(z_\nu)^{r-2}}\,. \tag{3.2}$$

If we set

$$\ell_{\nu,\alpha}(z) := \frac{\omega_{n,\alpha}(z)}{(z-z_\nu)\omega'_{n,\alpha}(z_\nu)} = \sum_{k=0}^{n} \frac{\omega_\nu^{n-k}(z+\alpha)^k(1+\alpha z)^{n-k}}{(1-\alpha^2)^n} \tag{3.3}$$

then from (3.2) we see that $Q(z)$ satisfies the following differential equation:

$$\begin{aligned} 2(1+\alpha z)(z+\alpha)Q'(z) + (r-1)n(1+2\alpha z+\alpha^2)Q(z) \\ = g_\nu(z) + c_\nu \omega_{n,\alpha}(z) \end{aligned} \tag{3.4}$$

where

$$g_\nu(z) = \frac{2}{r!\omega'_{n,\alpha}(z_\nu)^{r-2}}\,\ell_{\nu,\alpha}(z).$$

Then as in Section 2, we have

$$\begin{aligned} &Q(z)[(z+\alpha)(1+\alpha z)]^{\frac{n(r-1)}{2}} \\ &= \textstyle\int_{-1/\alpha}^{z} [g_\nu(t) + c_\nu\omega_{n,\alpha}(t)][(t+\alpha)(1+\alpha t)]^{\frac{n(r-1)}{2}-1}dt. \end{aligned} \tag{3.5}$$

The constant c_ν will be determined by the condition

$$\int_{-1/\alpha}^{-\alpha} [g_\nu(t) + c_\nu\omega_{n,\alpha}(t)][(t+\alpha)(1+\alpha t)]^{\frac{n(r-1)}{2}-1}dt = 0. \tag{3.6}$$

Hence using (2.3) - (2.9) as well as (3.3), we get

$$c_\nu = -\frac{a_\nu \cdot \alpha^{nr+2}}{(1-\alpha^2)^2} \cdot \frac{\sum_{k=0}^{n} \frac{\omega_\nu^{n-k}}{\alpha^k}\Gamma(\frac{n(r-1)}{2}+k)\Gamma(\frac{n(r+1)}{2}-k)}{[1+(-1)^n\alpha^{n+1}]\Gamma(\frac{n(r-1)}{2})\Gamma(\frac{n(r+1)}{2})}\,.$$

With this value of c_ν, we get $Q(z)$ from (3.5) and then $\rho_{r,\nu}(z)$ is given by (3.1).

The other fundamental polynomials $\rho_{\mu,\nu}(z)$ $(\mu = 0, 1, \ldots, r-2)$ can be represented as

$$\begin{gathered}\rho_{\mu,\nu}(z) = H_{\mu,\nu,r}(z) - \textstyle\sum_{k=0}^{n} H_{\mu,\nu,r}^{(r)}(z_k)\rho_{r,k}(z) \\ (\mu = 0, 1, \ldots, r-2,\ \nu = 0, 1, \ldots, n)\end{gathered} \tag{3.7}$$

where $H_{\mu,\nu,r}(z) \in \pi_{r(n+1)-1}$ is the Hermite fundamental polynomial of interpolation having the properties:

$$H_{\mu,\nu,r}^{(j)}(z_k) = \delta_{\mu,j}\delta_{\nu,k}\ (\mu, j = 0, 1, \ldots, r-1;\ \nu, k = 0, 1, \ldots, n).$$

The polynomials $H_{\mu,\nu,r}(z)$ can also be explicitly written as:

$$H_{\mu,\nu,r}(z) := \frac{(\ell_\nu(z))^r}{\mu!} \sum_{s=0}^{r-\mu-1} \frac{[\ell_\nu^{-r}(z)]_{z=z_\nu}^{(s)}}{s!} (z - z_\nu)^{s+\mu}.$$

4 $(0,3)$ Interpolation

In view of the result of Theorem 1, it seems natural to ask if $(0,3)$ interpolation on zeros of $\omega_{n,\alpha}(z)$ is regular. In order to answer this question, we set

$$P(z) = \omega_{n,\alpha}(z)Q(z), \quad Q(z) \in \pi_n, \tag{4.1}$$

and require that

$$P'''(z_\nu) = 0, \quad \nu = 0, 1, \ldots, n.$$

Using the method of Section 2, we see easily that $Q(z)$ satisfies a differential equation of order 2 :

$$\begin{aligned}&3(z+\alpha)^2(1+\alpha z)^2 Q''(z)\\ &\quad +3n(z+\alpha)(1+\alpha z)(1+2z\alpha+\alpha^2)Q'(z)\\ &\quad +n(n-1)[(1+\alpha z)^2 + (1+\alpha z)\alpha(z+\alpha) + \alpha^2(z+\alpha)^2]Q(z)\\ &\quad = [A(z+\alpha)+B]\omega_{n,\alpha}(z).\end{aligned} \tag{4.2}$$

If we set $Q(z) := \sum_{\nu=0}^{n} \alpha^\nu c_\nu (z+\alpha)^\nu (1+\alpha z)^{n-\nu}$, then easy computation shows that

$$
\begin{aligned}
&3n(z+\alpha)(1+\alpha z)(1+2\alpha z+\alpha^2)Q'(z) \\
&= \; 3n\Big[\alpha(c_1+nc_0)(z+\alpha)(1+\alpha z)^{n+1} \\
&\quad +\sum_{\nu=2}^{n}\alpha^{\nu}\{\nu c_\nu+nc_{\nu-1}+(n+2-\nu)c_{\nu-2}\} \\
&\quad \times(z+\alpha)^{\nu}(1+\alpha z)^{n+2-\nu} \\
&\quad +\alpha^{n+1}\{nc_n+c_{n-1}\}(z+\alpha)^{n+1}(1+\alpha z)\Big].
\end{aligned}
$$

Similarly,

$$
\begin{aligned}
3(z+\alpha)^2(1+\alpha z)^2Q''(z) \;=\;& 3\sum_{\nu=2}^{n}\alpha^{\nu}[\nu(\nu-1)c_\nu+2(\nu-1)(n+1-\nu)c_{\nu-1} \\
& +(n+2-\nu)(n+1-\nu)c_{\nu-2}](z+\alpha)^{\nu}(1+\alpha z)^{n+2-\nu}
\end{aligned}
$$

Using the above expressions in the differential equation (4.2) on putting z_k for z and recalling that

$$
z_k+\alpha=\frac{(1-\alpha^2)\omega_k}{1-\alpha\omega_k}, \quad 1+\alpha z_k=\frac{1-\alpha^2}{1-\alpha\omega_k}
$$

we get an equation in powers of ω_k. Keeping in mind that $\omega^{n+1}=1$, we get a system of $n+1$ homogeneous equations to determine the c_j's :

$$
\left\{
\begin{aligned}
&(n-1)c_0+(n+2)\alpha^{n+1}c_{n-1}+(4n-1)\alpha^{n+1}c_n=0 \\
&c_{\nu-2}\{3(n+2-\nu)(n+1-\nu)+3n(n+2-\nu)+n(n-1)\} \\
&\quad +c_{\nu-1}\{6(\nu-1)(n+1-\nu)+3n^2+n(n-1)\} \\
&\quad +c_\nu\{3\nu(\nu-1)+3n\nu+n(n-1)\}=0, \quad \nu=2,\ldots,n \\
&(4n-1)c_0+(n+2)c_1+(n-1)\alpha^{n+1}c_n=0
\end{aligned}
\right.
\tag{4.3}
$$

Notice that the $(n+1)^{\text{th}}$ order determinant $\Delta_n(\alpha)$ of the above homogeneous system of equations is a reciprocal polynomial. The $(n+2-\nu)^{\text{th}}$ row of the determinant is the same as the νth row in the reverse order for $2\le\nu\le n$. Expanding the determinant in terms of the elements of the first two rows we see that

$$
\Delta_n(\alpha)=A_n+B_n\alpha^{n+1}+A_n\alpha^{2n+2}
$$

where $A_n = (n-1)(n+2)\prod_{\nu=2}^{2}\{3\nu(\nu-1)+3n\nu+n(n-1)\} > 0.$

In order to say something more about the determinant we write

$$\Delta_n(\alpha) = A_n \cdot \begin{vmatrix} 1 & 0 & 0 & \dots & & a_{0,n-1}\alpha^{n+1} & a_{0,n}\alpha^{n+1} \\ a_{10} & 1 & 0 & \dots & & 0 & a_{1n}\alpha^{n+1} \\ a_{20} & a_{21} & 1 & 0 & & 0 & 0 \\ \dots & \dots & \dots & \dots & \dots & \dots & \dots \\ 0 & 0 & 0 & 0 & a_{n,n-2} & a_{n,n-1} & 1 \end{vmatrix} \tag{4.4}$$

where $a_{0,n-1} = \frac{n+2}{n-1}$, $a_{0n} = \frac{4n-1}{n-1}$, $a_{1n} = \frac{n-1}{n+2}$ and each of the remaining $n-1$ rows has only three non-zero terms $a_{\nu,\nu-2}, a_{\nu,\nu-1}, 1$ which are bounded numbers. From (4.4), we see that the sum of the squares of the non-zero elements in the νth row is

$$\begin{aligned} & a^2_{\nu,\nu-2} + a^2_{\nu,\nu-1} + 1 \\ &= \frac{(7n^2 - 9n\nu + 3\nu^2 + 14n - 9\nu + 6)^2 + (4n^2 + 6n\nu - 6\nu^2 - 7n + 12\nu - 6)^2}{(n^2 + 3n\nu + 3\nu^2 - n - 3\nu)^2} + 1 \\ &< \frac{(7n^2 - 8n\nu + 3\nu^2)^2 + (4n^2 + 7n\nu - 6\nu^2)^2}{n^4} + 1 \\ &= \frac{65n^4 - 56n^3\nu - 6n^2\nu^2 - 132n\nu^3 + 45\nu^4}{n^4} + 1 < 66. \end{aligned}$$

Now expanding $\Delta_n(\alpha)$ in terms of the first two rows we have 6 non-zero terms, two of which are A_n and $A_n\alpha^{2n+2}$. The other four terms are

$$A_n\Big(\sum_{j=1}^{4} A_j D_j\Big)$$

where A_j's are two by two determinants from the first two rows and D_j's are $n-1$ order determinant (minors of A_j's) from the last $n-1$ rows. By Hadamard's theorem

$$D_j^2 \le 66^{n-1}$$

and

$$|A_j| < \Big(\frac{4n-1}{n+2}\cdot\frac{4n-1}{n-1} - \frac{n-1}{n-2}\Big)\alpha^{n+1} \le 15\alpha^{n+1}.$$

Hence

$$\Delta_n(\alpha) > A_n(1 - 4 \cdot 15\alpha^{n+1} \cdot 66^{n+1}).$$

Thus if $\alpha < \frac{1}{\sqrt{66}}$, then $\Delta_n(\alpha) > 0$ and we have regularity.

We offer the conjecture that $\Delta_n(\alpha) > 0$ for all $0 < \alpha < 1$ when n is odd, but has only one zero when n is even. In support of the conjecture we calculate $\Delta_n(\alpha)$ when $n = 3$ and 4. We find that:

$$\Delta_n(\alpha) = \begin{cases} c_1(17 + 62\alpha^4 + 17\alpha^8), & c_1 > 0, \quad \text{when} \quad n = 3 \\ c_2(11 - 47\alpha^5 + 11\alpha^{10}), & c_2 > 0, \quad \text{when} \quad n = 4. \end{cases}$$

5 A Pál Type Interpolation Problem

Let $\{z_k\}_{k=0}^n$ denote the zeros of $\omega_{n,\alpha}(z)$ and let $\{z_k'\}_0^n$ denote the zeros of $v_{n,\alpha}(z) := (z+\alpha)^{n+1} + (1+\alpha z)^{n+1}$. A Pál type interpolation problem is to find a polynomial $P(z)$ of degree $\leq 2n-1$ which interpolates to a given data in the zeros of $\omega_{n,\alpha}(z)$ (say) and whose derivative interpolates a given data in the zeros of $v_{n,\alpha}(z)$. We propose to *show that this problem is regular.*

In order to prove our assumption we set

$$P(z) = \omega_{n,\alpha}(z)Q(z), \quad Q(z) \in \pi_n \tag{5.1}$$

where we require the condition A: $P'(z_k') = 0,\ k = 0, 1, \ldots, n$. Since

$$\begin{aligned} \omega_{n,\alpha}'(z_k') &= -\frac{(n+1)(1+\alpha z_k')^n}{z_k' + \alpha}(1 + 2\alpha z_k' + \alpha^2) \\ \omega_{n,\alpha}(z_k') &= -2(1+\alpha z_k')^{n+1}, \end{aligned}$$

and

$$P'(z_k') = \omega_{n,\alpha}'(z_k')Q(z_k') + \omega_{n,\alpha}(z_k')Q'(z_k') = 0, \quad k = 0, 1, \ldots, n$$

it follows that

$$2(1+\alpha z_k')^{n+1}Q'(z_k') + \frac{(n+1)(1+\alpha z_k')^n}{z_k' + \alpha}(1 + 2\alpha z_k + \alpha^2)Q(z_k') = 0.$$

Thus $Q(z)$ satisfies the differential equation

$$2(z+\alpha)(1+\alpha z)Q'(z) + (n+1)(1 + 2\alpha z + \alpha^2)Q(z) = cv_{n,\alpha}(z).$$

Then as in Section 2, we obtain

$$Q(z)[(z+\alpha)(1+\alpha z)]^{(n+1)/2}$$
$$= c\int_{-1/\alpha}^{z} v_{n,\alpha}(t)[(t+\alpha)(1+\alpha t)]^{(n-1)/2}dt.$$

In order that $Q(z) \in \pi_n$, it is necessary and sufficient that

$$\int_{-1/\alpha}^{-\alpha} v_n^\alpha(t)[(t+\alpha)(1+\alpha t)]^{(n-1)/2}dt = 0.$$

Following the integral (2.7), it is easy to see that the above integral is not zero and this completes the proof of regularity.

The fundamental polynomials can also be found by the method used in Section 3.

An analogous problem is to replace condition A after (5.1) by condition (5.2) below:

$$P''(z_k') = 0, \quad k = 0, 1, \ldots, n. \tag{5.2}$$

Thus we seek to show that $P(z) \in \pi_{2n+1}$ is identically zero if $P(z_k) = 0, \quad k = 0, 1, \ldots, n$ and satisfies (5.2). We set $P(z) = \omega_{n,\alpha}(z)Q(z)$, $Q(z) \in \pi_n$.

Since

$$\omega''_{n,\alpha}(z_k') = -\frac{(n+1)n(1+\alpha z_k')^{n-1}}{(z_k'+\alpha)^2}[(1+\alpha z_k')^2 + \alpha^2(z_k'+\alpha)^2],$$

and since (5.2) is equivalent to

$$\omega''_{n,\alpha}(z_k')Q(z_k') + 2\omega'_{n,\alpha}(z_k')Q'(z_k') + \omega_{n,\alpha}(z_k')Q''(z_k') = 0$$

we see on simplifying that $Q(z)$ satisfies the following $n+1$ conditions

$$\begin{aligned}&2(z_k'+\alpha)^2(1+\alpha z_k')^2Q''(z_k')\\ &\quad+2(n+1)(1+\alpha z_k')(z_k'+\alpha)(1+2\alpha z_k'+\alpha^2)Q'(z_k')\\ &\quad+(n+1)n[(1+\alpha z_k')^2+\alpha^2(z_k'+\alpha)^2]Q(z_k') = 0\\ &\quad(k = 0, 1, \ldots, n),\end{aligned} \tag{5.3}$$

Since $z_k' + \alpha = \frac{1-\alpha^2}{1-\alpha\omega_k'}\omega_k'$, $\quad 1+\alpha z_k' = \frac{1-\alpha^2}{1-\alpha\omega_k'}$, where ${\omega_k'}^{n+1} = -1$ we first set

$$Q(z) = \sum_{\nu=0}^{n} \alpha^\nu c_\nu (z+\alpha)^\nu (1+\alpha z)^{n-\nu}$$

so that

$$Q(z_k') = \frac{(1-\alpha^2)^n}{(1-\alpha\omega_k')^n} \sum_{\nu=0}^{n} \alpha^\nu c_\nu {\omega_k'}^\nu .$$

Similarly, an easy calculation shows that

$$\begin{aligned}
&(z_k'+\alpha)(1+\alpha z_k')(1+2\alpha z_k'+\alpha^2)Q'(z_k') \\
&\quad = \frac{(1-\alpha^2)^{n+2}}{(1-\alpha\omega_k')^{n+2}} \Big[\sum_{\nu=1}^{n} \alpha^\nu [\nu c_\nu + (n+1-\nu)c_{\nu-1}] {\omega_k'}^\nu \\
&\quad + \sum_{\nu=2}^{n+1} \alpha^\nu [(\nu-1)c_{\nu-1} + (n+2-\nu)c_{\nu-2}] {\omega_k'}^\nu \Big],
\end{aligned}$$

and

$$\begin{aligned}
&(z_k'+\alpha)^2(1+\alpha z_k')^2 Q''(z_k') \\
&\quad = \frac{(1-\alpha^2)^{n+2}}{(1-\alpha\omega_k')^{n+2}} \Big[\sum_{\nu=2}^{n} \alpha^\nu \{\nu(\nu-1)c_\nu \\
&\quad +2(\nu-1)(n+1-\nu)c_{\nu-1} \\
&\quad +(n+2-\nu)(n+1-\nu)c_{\nu-2}\Big\} {\omega_k'}^\nu \Big].
\end{aligned}$$

Using these values in (5.3) and recalling that ${\omega_k'}^{n+1} = -1$, we see that a polynomial of degree n satisfies $n+1$ homogeneous conditions and so must be identically zero. This yields a homogeneous system of $n+1$ equations which are listed below:

$$\begin{aligned}
&nc_0 + \cdots && +(-(n+2)\alpha^{n+1})c_{n-1} - 2n\alpha^{n+1}c_n = 0 \\
&2nc_0 + (n+2)c_1 + && \qquad \ldots \quad +0 - n\alpha^{n+1}c_n = 0
\end{aligned}$$

and

$$\begin{aligned}
&\{2(n+2-\nu)(2n+2-\nu) + n(n+1)\}c_{\nu-2} \\
&\quad +\{4(\nu-1)(n+1-\nu) + 2n(n+1)\}c_{\nu-1} \\
&\quad +\{2\nu(n+\nu) + n(n+1)\}c_\nu = 0.
\end{aligned}$$

The structure of the determinant of this system of equations is similar but different from the determinant $\Delta_n(\alpha)$ in (4.4). It can be seen easily that when written as a polynomial in α, it is of the form $A_n + B_n\alpha^{n+1} +$

$A_n\alpha^{2n+2}$, $A_n > 0$. The reasoning in Section 4, shows that this can have at most one zero. So the problem is regular except perhaps for one value of α between 0 and 1

Remark. The similarity between the Pál type problems and the cases of $(0,2)$ and $(0,3)$ interpolation on the zeros is interesting. The method used in Sections 2 and 4 can be used to settle the problem of the regularity of $(0,2,3)$ interpolation in this case, but this becomes very tedious.

References

[1] R. Brück, Lagrange interpolation in non-uniformly distributed nodes on the unit circle, Analysis 16(1996), 273-282.

[2] A.S. Cavaretta, Jr., A. Sharma and R.S. Varga, Interpolation in the roots of unity: An extension of a theorem of J.L. Walsh. Resultate du Mathematik 3(1981), 151-191.

[3] W. Chen and A. Sharma, Lacunary interpolation on some non uniformly distributed nodes on the unit circle, Annales Univ. Sci. Budapest. 16(1996), 69-82.

[4] C.K. Chui, Xie-Chang Shen and Lefan Zhong, On Lagrange interpolation at disturbed roots of unity, C.A.T. Report 209.

[5] O. Kis, Notes on interpolation (in Russian), Acta Math. Acad. Sci. Hungar. 11 1960, 49-64.

[6] G.G. Lorentz, S.D. Riemenschneider and K. Jetter, Birkhoff Interpolation, Addison Wesley Pub. Co., Mass., U.S.A., 1983.

[7] R.B. Saxena, A. Sharma and Z. Ziegler, Hermite-Birkhoff interpolation on the roots of unity and Walsh equiconvergence, Linear Algebra and Applications 52/53 (1985), 603-615.

[8] Xie-Chang Shen, Introduction of a new class of interpolants, Birkhoff interpolants in the complex plane (Chinese, English Summary), Advances in Math, Beijing 18 (4) (1989), 412-432.

[9] Xie Siquing, Regularity of $(0,1,\ldots,r-2,r)$ and $(0,1,\ldots,r-2,r)^*$ interpolations on some sets of the unit circle, to appear in Jour. Approx. Theory.

[10] G. Szegö, Orthogonal Polynomials, AMS Coll. Publ., Vol. 23, Providence, RI, 1959.

Approximation by Entire Functions with Only Real Zeros

L.T. Dechevsky
Dépt. de Math. et de Stat., Université de Montréal
Montréal H3C 3J7, Canada

D.P. Dryanov
Dépt. de Math. et de Stat., Université de Montréal
Montréal H3C 3J7, Canada

Q.I. Rahman
Dépt. de Math. et de Stat., Université de Montréal
Montréal H3C 3J7, Canada

Abstract

It was shown by J.G. Clunie that if f is a polynomial with only real zeros, and $x = -1$, $x = 1$ are consecutive zeros, then

$$\int_{-1}^{1} |1 - f(x)|\, \mathrm{d}x > \frac{1}{10}\,.$$

We refine and also extend this inequality in various ways.

1 Introduction

A few years ago, in a lecture on convergence properties of polynomials with only real zeros, given at the Université de Montréal, Prof. J.G. Clunie mentioned the following proposition as an auxiliary result, essential for the proof of the main theorem.

Proposition 1. *If f is a polynomial with only real zeros, having $x = -1$ and $x = 1$ as consecutive zeros, then*

$$\int_{-1}^{1} |1 - f(x)|\, \mathrm{d}x > \frac{1}{10}\,. \tag{1.1}$$

The lecture was based on a manuscript which has never been submitted for publication for reasons explained in [2, p. 110]. Inequality (1.1) was good enough to establish the main result for which it was used as a lemma, but it is **not best possible**. In this paper we extend it in various ways. It is clearly desirable to obtain the sharp version of (1.1). Although we are unable to prove it at this stage, we believe that in the extremal case, the polynomial f should be of the form $c\,(1-x^2)$ where c is an appropriate constant. Indeed, we conjecture that if f is a polynomial with only real zeros, and $x=-1$, $x=1$ are consecutive zeros, then for each $p \geq 1$ the integral $\int_{-1}^{1} |1-f(x)|^p \, \mathrm{d}x$ is minimized by some constant multiple of $(1-x^2)$. We shall show that this is true at least for $p=2$.

2 Statement of results

Sometimes, the best way to solve a problem is to think of an appropriate generalization. With this in mind we remark that one of the most important properties of polynomials with only real zeros is that they are logarithmically concave between two consecutive zeros. Indeed, if f is such a polynomial with zeros $x_1, \ldots, x_n$, then

$$\left\{\frac{f'(x)}{f(x)}\right\}' = -\sum_{\nu=1}^{n} \frac{1}{(x-x_\nu)^2}$$

is negative at each point of the real line where it is defined. So we generalize our problem by considering the class $\mathfrak{A}$ of all functions of the form $f(x) := (1-x^2)\psi(x)$ where ψ is positive and logarithmically concave on $(-1,1)$. We wish to mention an important subclass of $\mathfrak{A}$. An entire function f is said to belong to the Laguerre-Pólya class, $\mathfrak{L}-\mathfrak{P}$ for short, if it is the local uniform limit in $\mathbb{C}$ of a sequence of polynomials with only real zeros (see [1] and the literature cited therein for additional facts about $\mathfrak{L}-\mathfrak{P}$). Let us denote by $(\mathfrak{L}-\mathfrak{P})_1$ the set of all functions in $\mathfrak{L}-\mathfrak{P}$ which have $x=-1$, $x=1$ as consecutive zeros and are positive in $(-1,1)$. Functions in $(\mathfrak{L}-\mathfrak{P})_1$ can be written as $f(x) = (1-x^2)\psi(x)$, where $\psi(x) > 0$ on $(-1,\,1)$ and

$$\psi(z) = c\,\mathrm{e}^{-az^2+bz} \prod_{k=1}^{\infty} (1-\zeta_k z)\,\mathrm{e}^{\zeta_k z} \quad (-1 \leq \zeta_k \leq 1,\ c>0,\ a \geq 0,\ b \in \mathbb{R})\,.$$

Note that ψ is logarithmically concave between any two consecutive zeros. Thus the result which we prove here for functions in $\mathfrak{A}$ are *a fortiori* true for those belonging to the important subclass $(\mathfrak{L}-\mathfrak{P})_1$.

Instead of minimizing $\int_{-1}^{1} |1-f(x)| \, \mathrm{d}x$ over polynomials with only real zeros and having $x = -1$, $x = 1$ as consecutive zeros, we minimize it over $\mathfrak{A}$. Let $\mathfrak{B}$ denote the subclass of $\mathfrak{A}$ consisting of functions of the form $(1-x^2)\,\mathrm{e}^{Ax+B}$, where $A \geq 0$, $B \in \mathbb{R}$. We prove

Theorem 2. *Given any f belonging to $\mathfrak{A}\backslash\mathfrak{B}$, there exist constants $A \geq 0$, $B \in \mathbb{R}$ such that*

$$\int_{-1}^{1} F(|1-f(x)|)\,\mathrm{d}x > \int_{-1}^{1} F(|1-(1-x^2)\,\mathrm{e}^{Ax+B}|)\,\mathrm{d}x \tag{2.1}$$

if $F : [0,\infty) \mapsto [0,\infty)$ is strictly increasing.

Theorem 3. *Let $\mathfrak{A}_e$ consist of all functions in $\mathfrak{A}$ which are even. If $f \in \mathfrak{A}_e\backslash\mathfrak{B}$, then we can find a positive constant c such that*

$$\int_{-1}^{1} F(|1-f(x)|)\,\mathrm{d}x > \int_{-1}^{1} F(|1-c\,(1-x^2)|)\,\mathrm{d}x \tag{2.2}$$

if $F : [0,\infty) \mapsto [0,\infty)$ is strictly increasing. If moreover, F is strictly convex then there exists a unique constant $c(F)$ such that

$$\int_{-1}^{1} F(|1-f(x)|)\,\mathrm{d}x > \int_{-1}^{1} F(|1-c(F)(1-x^2)|)\,\mathrm{d}x \tag{2.2$'$}$$

for all $f \in \mathfrak{A}_e\backslash\{c(F)(1-x^2)\}$.

Remark: Inequalities (2.1) and (2.2) may not be strict if F is only non-decreasing. For example, if $F(t) \equiv 1$, then $\int_{-1}^{1} F(|1-f(x)|)\,\mathrm{d}x = 2$ for all $f \in \mathfrak{A}$.

The following result is a special case of Theorem 3.

Corollary 4. *For each $p \in (0,\infty)$, we have*

$$\inf_{f\in\mathfrak{A}_e} \int_{-1}^{1} |1-f(x)|^p\,\mathrm{d}x = \inf_{c>0} \int_{-1}^{1} |1-c\,(1-x^2)|^p\,\mathrm{d}x\,. \tag{2.3}$$

At least for $1 < p < \infty$, the infimum on the right is attained for a unique $c = c_p$.

We believe that the left-hand side of (2.3) can be replaced by

$$\inf_{f\in\mathfrak{A}} \int_{-1}^{1} |1-f(x)|^p\,\mathrm{d}x\,,$$

if $p \geq 1$. At this stage we are only able to prove this conjecture for $p = 2$ (see Theorem 9 below). However, in the following theorem we give a lower bound for $\inf_{f\in\mathfrak{A}} \int_{-1}^{1} F(|1-f(x)|)\,\mathrm{d}x$ under very general assumptions.

Theorem 5. *Let $F : [0, \infty) \mapsto [0, \infty)$ be nondecreasing. If F is convex or satisfies the triangle inequality*

$$F(a+b) \le F(a) + F(b) \quad \textit{for} \quad a \ge 0,\ b \ge 0\,,$$

like $F(t) := t^p$, $0 < p \le 1$ or $F(t) := t/(1+t)$, then

$$\inf_{f \in \mathfrak{A}} \int_{-1}^{1} F(|1 - f(x)|)\, \mathrm{d}x \ge \frac{1}{2} \inf_{c>0} \int_{-1}^{1} F(|1 - c\,(1 - x^2)|)\, \mathrm{d}x\,. \tag{2.4}$$

In particular, for all $p \in (0, \infty)$, we have

$$\inf_{f \in \mathfrak{A}} \int_{-1}^{1} |1 - f(x)|^p\, \mathrm{d}x \ge \frac{1}{2} \inf_{c>0} \int_{-1}^{1} |1 - c\,(1 - x^2)|^p\, \mathrm{d}x\,. \tag{2.5}$$

Remark: From the proofs of Theorems 2, 3 and 5 it will be clear that the Lebesgue measure $\mathrm{d}x$ appearing in (2.1) – (2.5) can be replaced by more general measures, for example by $w(x)\,\mathrm{d}x$ where w is continuous, positive and even on $[-1, 1]$. We wish to add that in Theorem 2 the approximant (the constant function 1) can be replaced by any function g which is positive and logarithmically convex on $(-1, 1)$. The same can be said about Theorems 3 and 5 if, in addition, we require g to be even.

The case $0 < p \le 1$ of the next theorem provides an example of a situation where the existence and uniqueness of $c(F)$ in (2.2′) is assured without F being strictly convex.

Theorem 6. *For each $p \in (0, \infty)$ there exists a constant $c_p > 0$ such that for every $f \in \mathfrak{A}_e$, different from $c_p\,(1 - x^2)$, we have*

$$\int_{-1}^{1} |1 - f(x)|^p\, \mathrm{d}x > \int_{-1}^{1} |1 - c_p\,(1 - x^2)|^p\, \mathrm{d}x\,. \tag{2.6}$$

Here is an immediate consequence of Theorem 6.

Corollary 7. *Let $(\mathfrak{L} - \mathfrak{P})_{1,e}$ consist of functions in $(\mathfrak{L} - \mathfrak{P})_1$ which are even. Then the approximation problem*

$$\min_{f \in (\mathfrak{L} - \mathfrak{P})_{1,e}} \int_{-1}^{1} |1 - f(x)|^p\, \mathrm{d}x \qquad (0 < p < \infty) \tag{2.7}$$

has a unique solution which is of the form $c_p\,(1 - x^2)$.

Let $\mathcal{P}_R$ be the class of all polynomials with only real zeros, vanishing at $x = -1$, $x = 1$ and different from zero on $(-1, 1)$. The next corollary is also an immediate consequence of Theorem 6.

Corollary 8. *Let $\mathcal{P}_{R,e}$ consist of polynomials in $\mathcal{P}_R$ which are even. Then the solution of the approximation problem*

$$\min_{f\in\mathcal{P}_{R,e}} \int_{-1}^{1} |1-f(x)|^p \,\mathrm{d}x \qquad (0<p<\infty) \tag{2.7'}$$

is unique and is of the form $c_p\,(1-x^2)$.

It is likely that (2.7) and the more general problem

$$\min_{f\in\mathcal{P}_R} \int_{-1}^{1} |1-f(x)|^p \,\mathrm{d}x \qquad (0<p<\infty)$$

have the same solution, though we are able to prove this only for $p=2$. Indeed, this is contained in the following result.

Theorem 9. *The approximation problem*

$$\min_{f\in\mathfrak{A}} \int_{-1}^{1} (1-f(x))^2 \,\mathrm{d}x$$

has the unique solution $f^*(x) := (5/4)(1-x^2)$.

The following result improves upon (1.1) and also generalizes it.

Theorem 10. *If $f\in\mathfrak{A}$, then*

$$\int_{-1}^{1} |1-f(x)| \,\mathrm{d}x \geq \frac{1}{3}\,\frac{4\xi_1^3-3\xi_1^2+1}{1-\xi_1^2}, \tag{2.8}$$

where $\xi_1 = 2\sin(\pi/18)$.

It may be added that the right hand side of (2.8) is larger than 3/10.

3 Proofs of the results

3.1 *Proof of Theorem 2.* Let $f(x) := (1-x^2)\psi(x)$. Then $1-f(x)=0$ if and only if

$$\log\psi(x) = -\log(1-x^2)\,. \tag{3.1}$$

Since the left-hand side of (3.1) is concave whereas the right-hand side is strictly convex, equation (3.1) can have at most two solutions in $(-1,1)$. Thus on $(-1,1)$, the curve $y=f(x)$ may cut the line $y=1$ twice, touch it once or remain below it.

Case (i). Let the equation (3.1) have two solutions $x_1 < x_2$ in $(-1, 1)$. Consider the straight line passing through the points $(x_1, -\log(1 - x_1^2))$ and $(x_2, -\log(1 - x_2^2))$, which may be expressed as

$$y = Ax + B = -\log(1 - x_1^2) + \frac{x - x_1}{x_2 - x_1} \log \frac{1 - x_1^2}{1 - x_2^2}.$$

We note that on replacing $\log \psi(x)$ by $Ax + B$ in $|1 - (1 - x^2) e^{\log \psi(x)}|$ we obtain a better pointwise approximation of the constant function 1. To see this note that the function $-\log(1 - x^2)$ is strictly convex whereas $\log \psi$ is concave and so on $[-1, x_1) \cup (x_2, 1]$ we have

$$\log \psi(x) \leq Ax + B < -\log(1 - x^2),$$

i.e.

$$\psi(x) \leq e^{Ax+B} < \frac{1}{1 - x^2},$$

i.e.

$$(1 - x^2)\psi(x) \leq (1 - x^2) e^{Ax+B} < 1.$$

Furthermore, there exists a non-degenerate subinterval $[\alpha, \beta]$ contained in $[-1, x_1) \cup (x_2, 1]$ on which the first inequality is also strict. Besides, on (x_1, x_2) we have

$$\log \psi(x) \geq Ax + B > -\log(1 - x^2),$$

i.e.

$$(1 - x^2)\psi(x) \geq (1 - x^2) e^{Ax+B} > 1.$$

Hence

$$|1 - (1 - x^2)\psi(x)| \geq |1 - (1 - x^2) e^{Ax+B}| \quad \text{for} \quad -1 \leq x \leq 1 \tag{3.2}$$

with strict inequality on (α, β).

Case (ii). Let the equation (3.1) have a unique solution x_0 in $(-1, 1)$. In this case we consider the straight line $y = Ax + B$ which is tangent to the curve $y = -\log(1 - x^2)$ at the point $(x_0, -\log(1 - x_0^2))$. Arguing as in the previous case we see that (3.2) holds in this case also and that the inequality is strict on some non-degenerate subinterval of $(-1, 1)$.

Case (iii). Let the equation (3.1) have no solution in $(-1, 1)$, i.e.

$$(1 - x^2)\psi(x) < 1 \quad \text{for all} \quad x \in [-1, 1].$$

This means that the graphs of $y = -\log(1-x^2)$ and $y = \log\psi(x)$ have no common points. The function $-\log(1-x^2)$ being convex and $\log\psi(x)$ concave we can find a line $y = Ax + B$ such that

$$\log\psi(x) < Ax + B < -\log(1-x^2) \quad \text{for} \quad x \in [-1,1]\,.$$

Thus (3.2) holds again and the inequality is strict for all $x \in [-1,1]$.

Since F is strictly increasing it follows that for $-1 \le x \le 1$, we have

$$F\left(\left|1-(1-x^2)\psi(x)\right|\right) \ge F\left(\left|1-(1-x^2)\,\mathrm{e}^{Ax+B}\right|\right) \tag{3.3}$$

and the inequality is strict on some non-degenerate subinterval. Besides, the increasing nature of F implies that the two sides of (3.3) are integrable. The inequality in (3.3) being strict on an interval of positive length we obtain (2.1) since

$$\int_{-1}^{1} F\left(\left|1-(1-x^2)\,\mathrm{e}^{Ax+B}\right|\right)\,\mathrm{d}x = \int_{-1}^{1} F\left(\left|1-(1-x^2)\,\mathrm{e}^{-Ax+B}\right|\right)\,\mathrm{d}x\,.$$

3.2 *Proof of Theorem 3.* The function $\log\psi(x)$ is even by hypothesis. In view of the fact that $-\log(1-x^2)$ is also even, the line $y = Ax+B$ appearing in *cases* (i) and (ii) of the proof of (2.1) must be horizontal whereas in *case* (iii) it can be chosen to be so. This means that (3.2) holds with $A = 0$, i.e. for some positive c,

$$|1-(1-x^2)\psi(x)| \ge |1-c\,(1-x^2)| \quad \text{for} \quad -1 \le x \le 1 \tag{3.4}$$

and the inequality is strict on some non-degenerate subinterval of $(-1,1)$. Hence (2.2) holds.

The existence and uniqueness of $c(F)$ when F is strictly convex follows from the observation that

$$J(c) := \int_{-1}^{1} F\left(|1-c\,(1-x^2|)\right)\,\mathrm{d}x$$

is a strictly convex function on $[0,\infty)$.

3.3 *Proof of Theorem 5.* For any given $A \ge 0,\ B \in \mathbb{R}$ let

$$\phi_{A,B}(x) := (1-x^2)\,\mathrm{e}^{Ax+B}\,.$$

If $-1 < x < 1$, then with

$$\lambda(x) := \frac{\sqrt{\phi_{A,B}(x)}}{\sqrt{\phi_{A,B}(x)}+\sqrt{\phi_{A,B}(-x)}}\,,$$

where the square roots are all positive, we have

$$\begin{aligned}\left|1-\sqrt{\phi_{A,B}(x)\phi_{A,B}(-x)}\right| &= |1-\{(1-\lambda(x))\phi_{A,B}(x)+\lambda(x)\phi_{A,B}(-x)\}| \\ &= |(1-\lambda(x))(1-\phi_{A,B}(x))+\lambda(x)(1-\phi_{A,B}(-x))| \\ &\leq (1-\lambda(x))|1-\phi_{A,B}(x)|+\lambda(x)|1-\phi_{A,B}(-x)| \qquad (3.5)\end{aligned}$$

because $0 < \lambda(x) < 1$.

Case (i). Let F be convex. Using the fact that F is nondecreasing and convex we obtain

$$\begin{aligned}F(|1-\sqrt{\phi_{A,B}(x)\phi_{A,B}(-x)}|) &\leq F((1-\lambda(x))|1-\phi_{A,B}(x)|+\lambda(x)|1-\phi_{A,B}(-x)|) \\ &\leq (1-\lambda(x))F(|1-\phi_{A,B}(x)|)+\lambda(x)F(|1-\phi_{A,B}(-x)|) \\ &\leq F(|1-\phi_{A,B}(x)|)+F(|1-\phi_{A,B}(-x)|)\,.\end{aligned}$$

Hence

$$\begin{aligned}\int_{-1}^{1} F(|1-\sqrt{\phi_{A,B}(x)\phi_{A,B}(-x)}|)\,\mathrm{d}x &\leq \int_{-1}^{1} F(|1-\phi_{A,B}(x)|)\,\mathrm{d}x+\int_{-1}^{1} F(|1-\phi_{A,B}(-x)|)\,\mathrm{d}x \\ &= 2\int_{-1}^{1} F(|1-\phi_{A,B}(x)|)\,\mathrm{d}x\,,\end{aligned}$$

i.e.

$$\int_{-1}^{1} F(|1-(1-x^2)\,\mathrm{e}^{B}|)\,\mathrm{d}x \leq 2\int_{-1}^{1} F(|1-(1-x^2)\,\mathrm{e}^{Ax+B}|))\,\mathrm{d}x\,.$$

Combining this with (2.1) and taking note of the remark following the statement of Theorem 3 we see that (2.4) holds in the case when F is convex.

Case (ii). Let F satisfy the triangle inequality. Since F is also nondecreasing it follows from (3.5) that

$$\begin{aligned}F(|1-\sqrt{\phi_{A,B}(x)\phi_{A,B}(-x)}|) &\leq F\left((1-\lambda(x))\,|1-\phi_{A,B}(x)|+\lambda(x)\,|1-\phi_{A,B}(-x)|\right) \\ &\leq F\left(|1-\phi_{A,B}(x)|+|1-\phi_{A,B}(-x)|\right) \\ &\leq (|1-\phi_{A,B}(x)|)+F\left(|1-\phi_{A,B}(-x)|\right)\,,\end{aligned}$$

and so

$$\int_{-1}^{1} F(|1-(1-x^2)\,e^{B}|)\,dx$$
$$\leq \int_{-1}^{1} \left\{F\left(\left|1-(1-x^2)\,e^{Ax+B}\right|\right) + F\left(\left|1-(1-x^2)\,e^{-Ax+B}\right|\right)\right\} dx$$
$$= 2\int_{-1}^{1} F\left(\left|1-(1-x^2)\,e^{Ax+B}\right|\right)\, dx\,,$$

which completes the proof of (2.4) in *case* (ii).

3.4 *Proof of Theorem 6.* In view of Theorem 3 it is enough to prove that for each $p \in (0,\infty)$ the infimum of $\int_{-1}^{1} |1-c\,(1-x^2)|^p\,dx$ over all positive c is attained for one value of c which we denote by c_p.

Let $\phi_p(c) := \int_{-1}^{1} |1-c\,(1-x^2)|^p\,dx$. Then

$$|\phi_p(c_1)-\phi_p(c_2)| \leq |c_1-c_2|^p \int_{-1}^{1} (1-x^2)^p\,dx$$

for $0<p<1$, whereas

$$|(\phi_p(c_1))^{1/p}-(\phi_p(c_2))^{1/p}| \leq |c_1-c_2| \left(\int_{-1}^{1} (1-x^2)^p\,dx\right)^{1/p}$$

for $1 \leq p < \infty$. Hence ϕ_p is a continuous function of c for all $p \in (0,\infty)$. It is easily seen that if $p \in (0,\infty)$ is fixed, then $\phi_p(c) \to \infty$ as $c \to \infty$ and so

$$\inf_{c>0} \phi_p(c) = \inf_{0<c<c_p^*} \phi_p(c)\,,$$

where c_p^* is some positive constant which we may assume to be greater than 1. Besides, ϕ_p is a decreasing function of c for $0<c<1$ and so

$$\inf_{c>0} \phi_p(c) = \inf_{1\leq c\leq c_p^*} \phi_p(c)\,.$$

There exists, therefore, a constant $c_p \in [1, c_p^*]$ such that

$$\inf_{c>0} \phi_p(c) = \phi_p(c_p)\,.$$

Since the infimum of $\phi_p(c)$ over the open interval $(0,\infty)$ is attained, the point c_p must be a root of the equation

$$\frac{d\,\phi_p}{dc} = 0\,,$$

which is equivalent to

$$\int_{-1}^{1} |1 - c\,(1-x^2)|^{p-1} \mathrm{sgn}\,(1 - c\,(1-x^2))(1-x^2)\,\mathrm{d}x = 0\,.$$

This can hold only if the integrand changes sign in $(-1,1)$, i.e. c_p must be greater than 1. Therefore, the curve $y = c_p\,(1-x^2)$ must intersect the line $y = 1$ twice, say at $x = -\xi_p$ and $x = \xi_p$ for some $\xi_p \in (0,1)$. Since $c_p = 1/(1-\xi_p^2)$, it would be sufficient to show that for each p the minimum of

$$\omega(\xi) := \int_0^1 \left| \frac{x^2 - \xi^2}{1 - \xi^2} \right|^p \mathrm{d}x$$

over $(0,1)$ is attained for only one value of ξ. Putting $\eta := \xi^2$ and making the substitution $u = x^2$ we see that

$$\omega(\xi) = \frac{1}{2}\,\Omega(\eta) := \int_0^1 u^{-\frac{1}{2}} \left| \frac{u-\eta}{1-\eta} \right|^p \mathrm{d}u\,.$$

So we may show that the minimum of $\Omega(\eta)$ over $(0,1)$ is attained for one and only one value of η. The minimum of $\Omega(\eta)$ can occur only at the roots of the equation

$$\frac{d\,\Omega}{d\eta} = 0\,,$$

i.e. of

$$\int_0^1 u^{-\frac{1}{2}} \left| \frac{u-\eta}{1-\eta} \right|^{p-1} \mathrm{sgn}\left(\frac{u-\eta}{1-\eta} \right) \frac{u-1}{(1-\eta)^2}\,\mathrm{d}u = 0\,.$$

Since $0 < \eta < 1$, this is equivalent to the equation

$$\int_0^1 (1-u)u^{-\frac{1}{2}}\,|u-\eta|^{p-1} \mathrm{sgn}\,(u-\eta)\,\mathrm{d}u = 0\,.$$

Replacing η by $(1+\beta)/2$ and making the substitution $u = (v+1)/2$ we are led to the equation

$$\int_{-1}^{1} (1-v)(1+v)^{-\frac{1}{2}} |v-\beta|^{p-1} \mathrm{sgn}\,(v-\beta)\,\mathrm{d}v = 0$$

and we have to show that it is satisfied for only one value of β in $(-1,1)$. Writing $\beta = \cos\theta_1, (0 < \theta_1 < \pi)$ and putting $v = \cos\theta$, the equation is easily seen to be equivalent to

$$\int_{-\pi}^{\pi} \lambda(\theta)\,\cos\frac{\theta}{2}\,\mathrm{d}\theta = 0\,, \tag{3.6}$$

where

$$\begin{aligned}\lambda(\theta) \quad := \quad & \cos^2\frac{\theta}{2}\left|\sin\frac{\theta-(\pi-\theta_1)}{2}\sin\frac{\theta+(\pi-\theta_1)}{2}\right|^{p-1} \\ & \times \ \operatorname{sgn}\left[\sin\frac{\theta-(\pi-\theta_1)}{2}\sin\frac{\theta+(\pi-\theta_1)}{2}\right].\end{aligned}$$

Since the function $\lambda(\theta)$ is even, we also have

$$\int_{-\pi}^{\pi}\lambda(\theta)\,\sin\frac{\theta}{2}\,\mathrm{d}\theta = 0\,. \tag{3.7}$$

According to a result contained in [3], although not explicitly stated there, if $\alpha\in[0,1]$ and $\alpha_\mu>0$ for $\mu=0,\cdots,m$, then there exists a unique $(t_1^*,\cdots,t_m^*)\in\{\vec{t}:\ -\pi\le t_1\le\ldots\le t_m\le\pi\}$ which satisfies the system of equations

$$\begin{aligned}\int_{-\pi}^{\pi} & \left|\cos\frac{\theta}{2}\right|^{\alpha(\alpha_0-1)}\prod_{\mu=1}^{m}\left|\sin\frac{\theta-t_\mu}{2}\right|^{\alpha(\alpha_\mu-1)} \\ & \times\operatorname{sgn}\left(\prod_{\mu=1}^{m}\sin\frac{\theta-t_\mu}{2}\right)\mathrm{e}^{\frac{\mathrm{i}(m-1-2k)}{2}\theta}\,\mathrm{d}\theta=0 \quad (k=0,\ldots,m-1)\,;\end{aligned}$$

furthermore, $-\pi<t_1^*<\cdots<t_m^*<\pi$.

Taking $m=2$, $\alpha=1$, $\alpha_0=3$, $\alpha_1=\alpha_2=p$ we obtain, in particular, that the system of equations

$$\begin{aligned}\int_{-\pi}^{\pi} & \left|\cos\frac{\theta}{2}\right|^{2}\left|\sin\frac{\theta-t_1}{2}\sin\frac{\theta-t_2}{2}\right|^{p-1} \\ & \times\operatorname{sgn}\left(\sin\frac{\theta-t_1}{2}\sin\frac{\theta-t_2}{2}\right)\mathrm{e}^{\pm\frac{\mathrm{i}}{2}\theta}\,\mathrm{d}\theta=0\,,\end{aligned} \tag{3.8}$$

has a unique solution $-\pi<t_1^*<t_2^*<\pi$. Replacing θ by $-\theta$ in (3.8) we see that $-\pi<-t_2^*<-t_1^*<\pi$ is also a solution of the same system. Hence by the uniqueness of the solution, t_2^* must be equal to $-t_1^*$. We can now claim that the system of equations (3.6) – (3.7) has a unique solution and because (3.7) is satisfied for all values of θ_1, equation (3.6) must have a unique solution.

3.5 *Proof of Theorem 9.* According to Theorem 2, if

$$\phi(c,\rho):=\int_{-1}^{1}\left(1-c\,(1-x^2)\,\mathrm{e}^{\rho x}\right)^2\,\mathrm{d}x\,,$$

then

$$\inf_{f\in\mathfrak{A}}\int_{-1}^{1}(1-f(x))^2\,\mathrm{d}x = \inf_{c>0,\,\rho\in\mathbb{R}}\phi(c,\rho)\,.$$

Let $\rho \in \mathbb{R}$ be fixed. Note that $\phi(c,\rho)$ tends to infinity as $c \to \infty$ and is a decreasing functions of c in a small interval of the form $(0, 2\delta)$. Hence there exists a positive number Δ_ρ such that

$$\inf_{c>0}\phi(c,\rho) = \min_{\delta<c<\Delta_\rho}\phi(c,\rho).$$

The minimum on the right is attained for just one value of c, say $c(\rho)$ which is the root of the equation

$$\frac{\partial\phi}{\partial c} = -2\int_{-1}^{1}\left(1-c\,(1-x^2)\,\mathrm{e}^{\rho x}\right)(1-x^2)\,\mathrm{e}^{\rho x}\mathrm{d}x = 0\,,$$

i.e.

$$c(\rho) = \frac{\left(\int_{-1}^{1}(1-x^2)\,\mathrm{e}^{\rho x}\,\mathrm{d}x\right)^2}{\int_{-1}^{1}(1-x^2)^2\,\mathrm{e}^{2\rho x}\,\mathrm{d}x}\,.$$

It is a matter of simple calculation that

$$\begin{aligned}\phi(c(\rho),\rho) &= \int_{-1}^{1}\left\{1-2c(\rho)\,(1-x^2)\,\mathrm{e}^{\rho x}+c^2(\rho)(1-x^2)^2\,\mathrm{e}^{2\rho x}\right\}\mathrm{d}x\\ &= 2-\frac{\left(\int_{-1}^{1}(1-x^2)\,\mathrm{e}^{\rho x}\,\mathrm{d}x\right)^2}{\int_{-1}^{1}(1-x^2)^2\,\mathrm{e}^{2\rho x}\,\mathrm{d}x}\,.\end{aligned}$$

We will show that if ρ is any real number different from zero then,

$$\phi(c(\rho),\rho) > \phi(c(0),0)\,. \tag{3.9}$$

It is easily seen that for all $\rho \neq 0$,

$$\begin{aligned}&\left(\int_{-1}^{1}(1-x^2)\,\mathrm{e}^{\rho x}\,\mathrm{d}x\right)^2\\ &\qquad= \frac{4}{\rho^4}\left\{\mathrm{e}^{2\rho}\left(1-\frac{2}{\rho}+\frac{1}{\rho^2}\right)+\mathrm{e}^{-2\rho}\left(1+\frac{2}{\rho}+\frac{1}{\rho^2}\right)+2-\frac{2}{\rho^2}\right\},\end{aligned}$$

$$\begin{aligned}&\int_{-1}^{1}(1-x^2)^2\,\mathrm{e}^{2\rho x}\,\mathrm{d}x\\ &\qquad= \frac{1}{\rho^3}\left\{\mathrm{e}^{2\rho}\left(1-\frac{3}{2\rho}+\frac{3}{4\rho^2}\right)-\mathrm{e}^{-2\rho}\left(1+\frac{3}{2\rho}+\frac{3}{4\rho^2}\right)\right\}.\end{aligned}$$

Since $c(0) = 5/4$, $\phi(c(0), 0) = 1/3$ we need to prove that for all $\rho \neq 0$ we have

$$2 - \frac{\frac{4}{\rho^4}\left\{e^{2\rho}(1 - \frac{2}{\rho} + \frac{1}{\rho^2}) + e^{-2\rho}(1 + \frac{2}{\rho} + \frac{1}{\rho^2}) + 2 - \frac{2}{\rho^2}\right\}}{\frac{1}{\rho^3}\left\{e^{2\rho}(1 - \frac{3}{2\rho} + \frac{3}{4\rho^2}) - e^{-2\rho}(1 + \frac{3}{2\rho} + \frac{3}{4\rho^2})\right\}} > \frac{1}{3},$$

i.e. (since $\int_{-1}^{1}(1-x^2)^2\, e^{2\rho x}\, dx > 0$)

$$\begin{aligned} g(\rho) \quad &:= \quad e^{2\rho}(5\rho^3 - \frac{39}{2}\rho^2 + \frac{111}{4}\rho - 12) \\ &\qquad -e^{-2\rho}(5\rho^3 + \frac{39}{2}\rho^2 + \frac{111}{4}\rho + 12) + 24(1-\rho^2) > 0\,. \end{aligned}$$

Set

$$h(\rho) := e^{2\rho}(5\rho^3 - \frac{39}{2}\rho^2 + \frac{111}{4}\rho - 12)\,,$$

so that

$$g(\rho) = h(\rho) + h(-\rho) + 24(1-\rho^2)\,,\ \ g(0) = 0\,.$$

Thus g is an even entire function and so its Maclaurin expansion has the form

$$g(\rho) \;=\; \sum_{k=1}^{\infty} \frac{1}{(2k)!} g^{(2k)}(0)\rho^{2k}\,.$$

Note that $g^{(2k)}(0)$ is equal to $2h^{(2k)}(0) - 48$ if $k = 1$ and is equal to $2h^{(2k)}(0)$ if $k \geq 2$. Calculating the even order derivatives of h we see that

$$\begin{aligned} h^{(2)}(\rho) \;&=\; e^{2\rho}(20\rho^3 - 18\rho^2 - 15\rho + 24), & g^{(2)}(0) = 0\,; \\ h^{(4)}(\rho) \;&=\; e^{2\rho}(20\rho^3 + 42\rho^2 - 21\rho), & g^{(4)}(0) = 0\,; \\ h^{(6)}(\rho) \;&=\; 16\, e^{2\rho}(20\rho^3 + 102\rho^2 + 93\rho), & g^{(6)}(0) = 0\,. \end{aligned}$$

Now we make the important observation that the coefficients of ρ^3, ρ^2, ρ in the expression for $h^{(6)}(\rho)$ are all positive. In fact, this implies (as is clear from *the form* of the expression) that if $k \geq 4$, then $h^{(2k)}(\rho)$ must be of the form

$$e^{2\rho}\left(a_k\rho^3 + b_k\rho^2 + c_k\rho + d_k\right)\,,$$

where the coefficients a_k, b_k, c_k, d_k are all positive and so

$$g^{(2k)}(0) \;=\; d_k > 0\,,\ \ \text{for}\ \ k \geq 4\,.$$

Thus

$$g(\rho) \;=\; \sum_{k=4}^{\infty} \frac{1}{(2k)!}\, d_k\, \rho^{2k}$$

is positive for all $\rho \neq 0$.

3.6 *Proof of Theorem 10.* From (2.5) it follows that

$$\int_{-1}^{1} |1 - f(x)| \mathrm{d}x \geq \frac{1}{2} \inf_{c>0} \int_{-1}^{1} \left|1 - c\,(1 - x^2)\right| \, \mathrm{d}x \,.$$

According to Theorem 6, there exists a unique constant $c_1 > 0$ such that,

$$\inf_{c>0} \int_{-1}^{1} \left|1 - c\,(1 - x^2)\right| \, \mathrm{d}x = \int_{-1}^{1} \left|1 - c_1(1 - x^2)\right| \, \mathrm{d}x \,.$$

Here we will determine the constant c_1 explicitly and evaluate the integral $\int_{-1}^{1} \left|1 - c_1(1 - x^2)\right| \, \mathrm{d}x$. We know that if $L(c) := \int_{-1}^{1} \left|1 - c\,(1 - x^2)\right| \, \mathrm{d}x$, then c_1 must satisfy the equation

$$\frac{\mathrm{d}L}{\mathrm{d}c} = - \int_{-1}^{1} \operatorname{sgn}\bigl(1 - c\,(1 - x^2)\bigr)\,(1 - x^2)\,\mathrm{d}x = 0$$

and so $c_1 > 1$. Now let $-\xi$, $\xi > 0$ be the points in $(-1, 1)$, where the curve $y = c\,(1 - x^2)$ cuts the line $y = 1$, i.e. $c = 1/(1 - \xi^2)$. Then the preceeding equation can be seen to be equivalent to

$$\int_{0}^{1} \operatorname{sgn}(x - \xi)(1 - x^2) = 0 \,,$$

i.e. to

$$\xi^3 - 3\xi + 1 = 0 \,, \quad 0 < \xi < 1 \tag{3.10}$$

which we may write in the form

$$4w^3 - 3w + \frac{1}{2} = 0$$

by setting $\xi = 2w$. Then putting $w = \cos\gamma$ we obtain

$$\cos 3\gamma = -\frac{1}{2}$$

which has the roots $\gamma_1 = 2\pi/9$, $\gamma_2 = 4\pi/9$, $\gamma_3 = 8\pi/9$ in $[0, \pi]$. Hence (3.10) has one root, namely $\xi_1 = 2\cos(4\pi/9) = 2\sin(\pi/18)$ in $(0, 1)$. From what has been said above it follows that

$$c_1 = \frac{1}{1 - \xi_1^2} = \frac{1}{1 - 4\sin^2 \frac{\pi}{18}} \,.$$

A straightforward calculation now gives

$$\int_{-1}^{1} \left|1 - c_1(1 - x^2)\right| \, \mathrm{d}x = \frac{2}{3} \, \frac{4\xi_1^3 - 3\xi_1^2 + 1}{1 - \xi_1^2} \,.$$

4 Concluding remarks

Let $\mathfrak{A}_1$ be the class of all functions of the form $(1-x^2)\psi(x)$ where $\psi(x)$ is positive and concave on $(-1,1)$. The idea to prove Theorems 2 and 5 also shows that if F is nondecreasing and convex, then

$$\inf_{f\in\mathfrak{A}_1}\int_{-1}^{1} F\left(|1-f(x)|\right)\,\mathrm{d}x = \inf_{c>0}\int_{-1}^{1} F\left(\left|1-c\left(1-x^2\right)\right|\right)\,\mathrm{d}x\,. \tag{4.1}$$

Furthermore, the infimum on the right-hand side of (4.1) is attained for a unique constant $c = c(F)$ if F is strictly convex. Compare (4.1) with (2.4) wherein we would have liked to have 1 in place of 1/2. We wish to remark that in both cases the problem is first reduced to a finite dimensional one. However, there is one important difference. In the concave case the reduced problem is linear whereas in the logarithmically concave case it is non-linear. This seems to be the basic difference between the two problems which somehow makes the logarithmically concave case harder.

References

[1] J.G. Clunie, Convergence of polynomials with restricted zeros , in "Functional Analysis and Operator Theory" (B.S. Yadav and D. Singh, Eds.), pp. 100-105, Lecture Notes in Math. No. **1511**, Springer-Verlag, Berlin, 1992.

[2] J.G. Clunie and A.B.J. Kuijlaars, Approximation by polynomials with restricted zeros , J. Approx. Theory **79** (1994), 109-124.

[3] D. Dryanov, Polynomials of minimal L_α-deviation, $\alpha > 0$, Constr. Approx. **10** (1994), 377-409.

Nonlinear Means in Geometric Modeling

Michael S. Floater
SINTEF
P. O. Box 124 Blindern, 0314 Oslo, Norway
E-mail: Michael.Floater@math.sintef.no

Charles A. Micchelli
IBM Corporation
T.J. Watson Research Center
Department of Mathematical Sciences
P. O. Box 218
Yorktown Heights, N. Y. U.S.A. 10598
E-mail: cam@watson.ibm.com

Abstract

In this paper we introduce a notion of nonlinear mean and begin to explore its use in curve design by developing a nonlinear version of the de Casteljau algorithms for the evaluation of the Bernstein-Bézier representation of curves.

1 Introduction

Many recursive algorithms for polynomial representation of curves and surfaces have at their core, the simple process of taking convex combinations of two vectors, [4]. Likewise this basic step also appears in stationary subdivision methods, [4], for example the line average algorithm of [1] for cube spline surfaces. The thought arises whether a notion of *nonlinear average* or *mean* would allow for the recursive generation of practical representations of curves and surfaces.

In this paper we begin to explore this question by introducing a notion of nonlinear means and show how it can be used in curve design by describing a nonlinear version of de Casteljau's algorithm for the recursive generation of the Bernstein-Bézier representation of curves, [4]. In the first section we define a family of nonlinear means and study their properties. In the second section we provide a nonlinear version of de Casteljau's algorithm.

2 Nonlinear Average

In this section we formalize a notion of nonlinear average which we shall work with throughout the paper. We divide our discussion between the univariate case and the multivariate case. The former will serve as a motivation for the latter. Our first goal is to give a nonlinear version to the process of taking a convex combination of two numbers x and y in $\mathbb{R}$ with weight $t \in [0,1]$, a process central to many algorithms in geometric modeling, c.f. [4]. For this process, we use the notation

$$a(t;x,y) := (1-t)x + ty \tag{2.1}$$

where t, x and y are in $\mathbb{R}$. A core idea in the use of this linear averaging process is its *repetition* under function composition. In this regard, the easily verified *functional* identity is central. For any $r, s, t \in I := [0,1]$ and $x, y \in \mathbb{R}$ we have that

$$a(r; a(s;x,y), a(t;x,y)) = a(a(r;s,t);x,y). \tag{2.2}$$

This equation has a geometric interpretation. The left hand side of equation (2.2) computes a convex combination with weight r of two numbers in $\mathbb{R}$ each of which are convex combinations of x and y with weights s and t respectively. Equation (2.2) says that this is the same as a convex combination of x and y with weight $a(r;s,t)$. We use this equation to introduce a notion of nonlinear convex combination in $\mathbb{R}^d$ which we call a *mean*.

Definition 1 *Let J be a subset of $\mathbb{R}^d$. A function $m : I \times J \times J \to J$ which satisfies for every $r, s, t \in I$, and $x, y \in J$ the equations*

$$m(r; m(s;x,y), m(t;x,y)) = m(a(r;s,t);x,y)$$

and

$$m(0;x,y) = x, \quad m(1;x,y) = y$$

is called a mean. Moreover, we say that $m(t;x,y)$ is the mean of x and y with weight t.

The class of all means associated with a set J has an important invariance property which will be useful in *constructing* specific means.

Proposition 2 *Let J be a subset of $\mathbb{R}^d$ and m a mean on J. Let K be a subset of $\mathbb{R}^d$ and suppose $f : K \to J$ is a bijection. Then the function m_f defined for $t \in I$, $x, y \in K$ by the equation*

$$m_f(t; x, y) := f^{-1}(m(t; f(x), f(y))$$

is a mean on K.

Proof: First note for any x and y in K that

$$m_f(0; x, y) = f^{-1}(m(0; f(x), f(y)) = f^{-1}(f(x)) = x$$

and likewise

$$m_f(1; x, y) = f^{-1}(m(1; f(x), f(y)) = f^{-1}(f(y)) = y.$$

Also, for $r, s, t \in I$ we have that

$$\begin{aligned} &m_f(r;\, m_f(s; x, y), m_f(t; x, y)) \\ &\quad \equiv f^{-1}\big(m(r; f(m_f(s; x, y)), f(m_f(t; x, y)))\big) \\ &\quad = f^{-1}\big(m(r; m(s; f(x), f(y)), m(t; f(x), f(y)))\big) \\ &\quad = f^{-1}\big(m(a(r; s, t); f(x), f(y))\big) \\ &\quad = m_f(a(r; s, t), x, y). \end{aligned}$$

■

In particular, for any *convex* J and bijection $f : K \to J$ the function $b := a_f$ is a mean on K. As an example, if $f_1 : \mathbb{R} \to \mathbb{R}$ is any linear function of the form $f_1(x) = \alpha x + \beta$, $\alpha \neq 0$, then $a_{f_1} = a$. A more interesting example is found from taking the function $f_2(x) = 1/x$, $f_2 : \mathbb{R}_+ \to \mathbb{R}_+$, $\mathbb{R}_+ = (0, \infty)$, in which case

$$a_{f_2}(t; x, y) = \frac{xy}{(1-t)y + tx}$$

and a_{f_2} is a mean on $\mathbb{R}_+$. With $f_3(x) = \ln x$, $f_3 : \mathbb{R}_+ \to \mathbb{R}_+$, we find

$$a_{f_3}(t; x, y) = x^{1-t} y^t$$

and a_{f_3} is a mean on $\mathbb{R}_+$. When $t = 1/2$ all these means are familiar: $a(1/2; x, y)$ is the arithmetic mean, $a_{f_2}(1/2; x, y)$ the harmonic mean, and $a_{f_3}(1/2; x, y)$ the geometric mean. As a further example, if $f_4 : \mathbb{R} \to (-\frac{\pi}{2}, \frac{\pi}{2})$ is the function $f_4(x) = \tan^{-1} x$ then

$$a_{f_4}(t; x, y) = \tan((1-t)\tan^{-1} x + t \tan^{-1} y)$$

is a mean on $\mathbb{R}$.

We now show in the univariate case that a large class of means are of the form a_f.

Proposition 3 *Let* $J = [x_0, y_0]$ *and suppose* m *is a mean on* J. *If* $g(t) := m(t; x_0, y_0)$ *is a bijection* $g : I \to J$ *with inverse* $f : J \to I$ *then for all* $x, y \in J$,

$$m(t; x, y) = f^{-1}((1-t)f(x) + tf(y)). \tag{2.3}$$

Proof: Since g is invertible, there exist uniquely s and t in I such that $g(s) = x$ and $g(t) = y$. Then for $r \in I$,

$$\begin{aligned} m(r; x, y) &= m(r; m(s; x_0, y_0), m(t; x_0, y_0)) \\ &= m((1-r)s + rt; x_0, y_0) \\ &= g((1-r)g^{-1}(x) + rg^{-1}(y)) \\ &= f^{-1}((1-r)f(x) + rf(y)). \end{aligned}$$

■

We note that $m(t; x_0, y_0)$ need not be increasing in t when $x_0 < y_0$ even if it is invertible in t. Indeed if $f : \mathbb{R} \to \mathbb{R}$ is the bijection

$$f(x) = \begin{cases} x & x \geq 0; \\ x^{-1} & x < 0, \end{cases}$$

then a_f is a mean on $\mathbb{R}$ and

$$a_f(t; -1, 1) = \begin{cases} (2t-1)^{-1} & 0 \leq t < \frac{1}{2}; \\ 2t - 1 & \frac{1}{2} \leq t \leq 1. \end{cases}$$

However if $m(t; x_0, y_0)$ is (strictly) increasing in t then we can deduce further properties of m as follows.

Corollary 4 *Let* J *and* m *be as in Proposition 3 and suppose further that* $g(t) := m(t; x_0, y_0)$ *is an increasing function for* $t \in I$. *Then for* $x, y \in J$: *if* $x < y$, $m(t; x, y)$ *is increasing in* t; *if* $x > y$, $m(t; x, y)$ *is decreasing in* t; *and if* $x = y$, $m(t; x, y) = m(t; x, x) = x$.

Proof: From Proposition 3,

$$m(r; x, y) = g((1-r)g^{-1}(x) + rg^{-1}(y)),$$

for $x, y \in J$. Let $s = g^{-1}(x)$ and $t = g^{-1}(y)$ and suppose $0 \leq r_1 < r_2 \leq 1$. Then

$$m(r_2; x, y) - m(r_1; x, y) = g((1-r_2)s + r_2 t) - g((1-r_1)s + r_1 t)$$

and so, since g is increasing, if $x < y$ we have that $s < t$ and so $m(r_2; x, y) > m(r_1; x, y)$. Conversely if $x < y$ we have that $m(r_2; x, y) < m(r_1; x, y)$. Therefore $m(t; x, y)$ is monotonic when $x \neq y$. Furthermore, we have that

$$m(r_2; x, x) - m(r_1; x, x) = g(s) - g(s) = 0.$$

■

The class of means of the form a_f is intimitely connected to second order systems of differential equations reminicent of the geodesic equation on a Riemannian manifold. We have in mind the following fact.

Proposition 5 *Let $J \subseteq \mathbb{R}^d$ and $f : J \to J$ be a bijection which has continuous second derivatives and an everywhere nonsingular jacobian. Then there is a continuous tensor $\Gamma = (\Gamma_{ijk})$, $i, j, k = 1, 2, \ldots, d$ on J such that the mean $b = a_f := (b_1, b_2, \ldots, b_d)$ of x and y in J satisfies for $t \in I$ and $i = 1, 2, \ldots, d$ the differential equations*

$$\ddot{b}_i(t) + \sum_{j,k=1}^{d} \Gamma_{ijk}(b(t))\, \dot{b}_j(t)\dot{b}_k(t) = 0$$

and the boundary conditions $b(0) = x$ and $b(1) = y$.

Proof: By definition for $t \in I$,

$$f(b(t)) = (1-t)f(x) + tf(y).$$

We differentiate both sides of this equation with respect to t and conclude that

$$\mathrm{J}(b(t))\, \dot{b}(t) = f(y) - f(x),$$

where $\mathrm{J} = (\mathrm{J}_{ij})_{ij}$ denotes the Jacobian matrix of f. Differentiating once again, we find

$$\mathrm{J}(b(t))\, \ddot{b}(t) + c = 0,$$

where c is the column vector whose r-th element is

$$c_r := \sum_{j,k=1}^{d} \frac{\partial^2 f_r}{\partial b_j \partial b_k}(b(t))\, \dot{b}_j(t)\dot{b}_k(t) = \sum_{j,k=1}^{d} (\partial_j \mathrm{J}_{rk})(b(t))\, \dot{b}_j(t)\dot{b}_k(t).$$

Multiplying by the inverse matrix J^{-1}, we find

$$\ddot{b}(t) + \mathrm{J}^{-1}(b(t))\, c = 0,$$

and so for $i = 1, 2, \ldots, d$,

$$\ddot{b}_i(t) + \sum_{r=1}^{d} (\mathrm{J}^{-1})_{ir}(b(t)) \sum_{j,k=1}^{d} (\partial_j \mathrm{J}_{rk})(b(t))\, \dot{b}_j(t)\dot{b}_k(t) = 0$$

which yields the result where, for $i, j, k = 1, 2, \ldots, d$,

$$\Gamma_{ijk} := \sum_{r=1}^{d} (\mathrm{J}^{-1})_{ir}\, \partial_j \mathrm{J}_{rk}.$$

■

In the univariate case $d = 1$ this equation has the simple form

$$\ddot{u}(t) + \frac{f''(u(t))}{f'(u(t))}\,\dot{u}^2(t) = 0$$

which is satisfied by the function

$$u(t) = f^{-1}((1-t)f(x) + tf(y))$$

where x and y are scalars and f is a twice differentiable function. In this case there is a converse which we describe next.

Proposition 6 *Let $J = [x_0, y_0]$ and $g \in L^1(J)$. Then there is a bijection $f : J \to J$ such that for any x and $y \in J$ the function*

$$u(t) = f^{-1}((1-t)f(x) + tf(y)) \tag{2.4}$$

satisfies the second order semilinear differential equation

$$\ddot{u}(t) + g(u(t))\dot{u}^2(t) = 0 \tag{2.5}$$

subject to the boundary conditions $u(0) = x$, $u(1) = y$.

Proof: For $x \in [x_0, y_0]$ we let

$$f(x) = x_0 + (y_0 - x_0)\,\frac{\int_{x_0}^{x} \exp \int_{x_0}^{r} g(s)ds\,dr}{\int_{x_0}^{y_0} \exp \int_{x_0}^{r} g(s)ds\,dr}.$$

Then f is an increasing function which maps J onto itself and

$$\frac{f''}{f'} = g.$$

■

Recently, Gonsor and Neamtu [2] presented a unified view of several types of spline functions including polynomial splines as well as trigonometric and hyperbolic ones. Their work is based on blossoming where the fundamental function is the solution to a second order *linear* equation. The only case in which equation (2.5) is linear in u is $g = 0$. In this case that solution u is linear as well. So the only solutions of (2.5) in common with those used by Gonsor and Neamtu are linear means.

3 Nonlinear de Casteljau Algorithm

In this section we consider de Casteljau's evaluation algorithm for means. Given a mean m on a subset J of $\mathbb{R}^d$, control points $x^0, x^1, \ldots, x^n \in J$ and $t \in I$ we define for $r = 0, 1, \ldots, n$ and $i = 0, 1, \ldots, n-r$,

$$\begin{aligned} p_{i,0}(t) &= x^i, \quad i = 0, 1, \ldots, n, \\ p_{i,r}(t) &= m(t; p_{i,r-1}(t), p_{i+1,r-1}(t)). \end{aligned} \tag{3.1}$$

When $K = \mathbb{R}^d$ and

$$m(t; x^0, x^1) = (1-t)x^0 + tx^1$$

it is well-known that

$$p_{i,r}(t) := \sum_{j=0}^{r} b_{j,r}(t) x^{i+j}$$

where

$$b_{j,r}(t) = \binom{r}{j} t^j (1-t)^{r-j}$$

are the Bernstein polynomials of degree r; c.f. [4]. In particular, we have that

$$p_{0,n}(t) := \sum_{j=0}^{n} b_{j,n}(t) x^j.$$

In the general case we recursively define the *n-th order mean* $m : I \times J^{n+1} \to J$ from a given mean $m : I \times J^2 \to J$ by the expression

$$m(t; x^0, \ldots, x^n) = m(t; m(t; x^0, \ldots, x^{n-1}), m(t; x^1, \ldots, x^n)) \tag{3.2}$$

for $x^0, \ldots, x^n$ in J, t in I, and $n \geq 2$. Then $m(t; x^0, \ldots, x^n)$ is simply the function $p_{0,n}(t)$ defined in (3.1). We can express the n-th mean as an $(n-r)$-th mean of r-th means for any $r = 1, \ldots, n-1$:

Proposition 7 *Let J be a subset of $\mathbb{R}^d$ and let m be a mean on J. Then for any $n \geq 2$, the associated n-th mean satisfies for any $r = 1, \ldots, n-1$ the identity*

$$m(t; x^0, \ldots, x^n) = m(t; m(t; x^0, \ldots, x^r), \ldots, m(t; x^{n-r}, \ldots, x^n)). \tag{3.3}$$

Proof: Fix $r \geq 1$ and consider $n > r$. When $n = r + 1$, (3.3) is equivalent to (3.2). For $n \geq r + 2$, we have by induction on n,

$$\begin{aligned} & m(t; x^0, \ldots, x^n) \\ &= m(t; m(t; x^0, \ldots, x^{n-1}), m(t; x^1, \ldots, x^n)) \\ &= m(t; m(t; m(t; x^0, \ldots, x^r), \ldots, m(t; x^{n-r-1}, \ldots, x^{n-1})), \\ &\qquad m(t; m(t; x^1, \ldots, x^{r+1}), \ldots, m(t; x^{n-r}, \ldots, x^n))) \\ &= m(t; m(t; x^0, \ldots, x^r), \ldots, m(t; x^{n-r}, \ldots, x^n)). \end{aligned}$$

■

Now let us discuss some properties the n-th mean $m(t; x^0, \ldots, x^n)$ shares with the n-th linear mean

$$a(t; x^0, \ldots, x^n) = \sum_{j=0}^{n} b_{j,n}(t) x^j.$$

Firstly, $m(t; x^0, \ldots, x^n)$ has the property of *end point interpolation*, that is

$$m(0; x^0, \ldots, x^n) = x^0, \qquad m(1; x^0, \ldots, x^n) = x^n.$$

Another property enjoyed by $a(t; x^0, \ldots, x^n)$ is *linear precision*, that is, if

$$x^i = \frac{n-i}{n} x^0 + \frac{i}{n} x^n, \qquad i = 1, \ldots, n-1,$$

then

$$a(t; x^0, \ldots, x^n) = (1-t) x^0 + t x^n.$$

The n-th mean $m(t; x^0, \ldots, x^n)$ has an analogous property stated in the next result.

Proposition 8 *Let J be a subset of $\mathbb{R}^d$ and m a mean on J. For $r_0, r_1, \ldots, r_n$ in I, $t \in I$ and x, $y \in J$*

$$m(t; m(r_0; x, y), m(r_1; x, y), \ldots, m(r_n; x, y)) = m\Big(\sum_{j=0}^{n} b_{j,n}(t) r_j; x, y\Big). \tag{3.4}$$

Proof: When $n = 1$, equation (3.4) holds by Definition 1. Otherwise $n > 1$ and then by induction on n,

$$\begin{aligned} & m(t; m(r_0; x, y), \ldots, m(r_n; x, y)) \\ &\quad = m(t; m(t; m(r_0; x, y), \ldots, m(r_{n-1}; x, y)), \end{aligned}$$

$$
\begin{aligned}
&m(t; m(r_1; x, y), \ldots, m(r_n; x, y))) \\
&= m\Big(t; m\Big(\sum_{j=0}^{n-1} b_{j,n-1}(t) r_j; x, y\Big), m\Big(\sum_{j=0}^{n-1} b_{j,n-1}(t) r_{j+1}; x, y\Big)\Big) \\
&= m\Big((1-t)\sum_{j=0}^{n-1} b_{j,n-1}(t) r_j + t\sum_{j=0}^{n-1} b_{j,n-1}(t) r_{j+1}; x, y\Big),
\end{aligned}
$$

which yields equation (3.4). ■

In particular, letting $r_i = \frac{i}{n}$, $i = 0, 1, \ldots, n$, we obtain a generalization of linear precision specifically, if $x^i = m(\frac{i}{n}; x^0, x^n)$ for $i = 1, \ldots, n-1$, then

$$m(t; x^0, \ldots, x^n) = m(t; x^0, x^n).$$

Given a curve $p : I \to J$, $J \subseteq \mathbb{R}^d$, and a mean $m : I \times J^2 \to J$, let us define the n-th Bernstein approximation of p with respect to m to be the curve

$$m_n(p, t) := m(t; p(0), \ldots, p(\frac{i}{n}), \ldots, p(1)). \tag{3.5}$$

When $m = a$, m_n is the n-th Bernstein operator

$$a_n(p, t) = \sum_{j=0}^{n} b_{j,n}(t) p(\frac{j}{n}).$$

and hence converges uniformly to p as $n \to \infty$, if p is continuous. This is Bernstein's approximation theorem, cf. [3], extended to vector valued functions.

As observed in Section 2 many means $m : I \times J^2 \to J$ are of the form a_f for some bijection $f : J \to K$ and convex K. The n-th mean in this case has a particularly simple form.

Lemma 9 *Let $a_f : I \times J^2 \to J$ be a mean where $f : J \to K$ is bijective and K is convex. For $x^0, \ldots, x^n$ in J and t in I,*

$$a_f(t; x^0, \ldots, x^n) = f^{-1}\Big(\sum_{j=0}^{n} b_{j,n}(t) f(x^j)\Big).$$

Proof: From equation (3.2) we find by induction that for $n \geq 2$,

$$
\begin{aligned}
&a_f(t; x^0, \ldots, x^n) \\
&= \quad f^{-1}\Big((1-t) f(m(t; x^0, \ldots, x^{n-1})) + t f(m(t; x^1, \ldots, x^n))\Big)
\end{aligned}
$$

$$= f^{-1}\Big((1-t)\sum_{j=0}^{n-1} b_{j,n-1}(t)f(x^j) + t\sum_{j=0}^{n-1} b_{j,n-1}(t)f(x^{j+1})\Big)$$

$$= f^{-1}\Big(\sum_{j=0}^{n} b_{j,n}(t)f(x^j)\Big).$$

■

From Lemma 9 we obtain a simple extension of Bernstein's approximation theorem to nonlinear averages of the form (2.3).

Proposition 10 *Suppose the same hypothesis as Lemma 9, and both $f : J \to K$ and its inverse $f^{-1} : K \to J$ are continuous. Then, denoting by $(a_f)_n(p)$ the n-th Bernstein approximation of p with respect to the mean a_f, it follows that*

$$\lim_{n\to\infty} (a_f)_n(p,t) = p(t), \qquad t \in [0,1].$$

Proof: From Lemma 9 we find that

$$(a_f)_n(p,t) = f^{-1}\Big(\sum_{j=0}^{n} b_{j,n}(t)f(p(\frac{j}{n}))\Big) = f^{-1}(a_n(f\circ p,t)) \tag{3.6}$$

and since $f \circ p : [0,1] \to K$ and $f^{-1} : K \to J$ are continuous, we deduce that for t in I,

$$\lim_{n\to\infty} (a_f)_n(p,t) = f^{-1}(\lim_{n\to\infty} a_n(f\circ p,t)) = f^{-1}((f\circ p)(t)) = p(t).$$

■

Furthermore, Voronovskaya [5] showed that the rate of convergence of Bernstein approximation is $1/n$ and provided p is C^2,

$$\lim_{n\to\infty} n(a_n(p,t) - p(t)) = \frac{t(1-t)}{2}p''(t). \tag{3.7}$$

Let us suppose that $d = 1$ and that both p and f are C^2. From equation (3.7), we have

$$a_n(f\circ p,t) - f(p(t)) = O(n^{-1}).$$

Therefore, using equation (3.6) and making a Taylor expansion of f^{-1} about $f(p(t))$ we find that

$$(a_f)_n(p,t) = f^{-1}(f(p(t)) + \big(a_n(f\circ p,t) - f(p(t))\big)\,(f^{-1})'(f(p(t)) + O(n^{-2}).$$

It follows, using again the asymptotic formula (3.7), that

$$\lim_{n\to\infty} n\big((a_f)_n(p,t) - p(t)\big) = \frac{t(1-t)}{2}\,(f\circ p)''(t)\,(f^{-1})'(f(p(t)).$$

Since

$$(f \circ p)''(t) = f'(p(t))p''(t) + f''(p(t)(p'(t))^2$$

and

$$(f^{-1})'(f(p(t)) = \frac{1}{f'(p(t))}$$

we obtain the result that

$$\lim_{n\to\infty} n((a_f)_n(p,t) - p(t)) = \frac{(1-t)t}{2}\left(p''(t) + \frac{f''(p(t))}{f'(p(t))}(p'(t))^2\right). \quad (3.8)$$

In particular, if f is a linear function then $a_f = a$ and $f'' = 0$ and in that case (3.8) reduces to (3.7).

Acknowledgement

Much of this work was done during the Spring of 1996 when the authors were visiting the University of Zaragoza. We wish to thank our host, Mariano Gasca, for providing a friendly environment that led to this stimulating scientific exchange.

We would also like to thank Narendra Govil for his generous and indispensable help in the preparation of this manuscript.

References

[1] W. Dahmen and C. A. Micchelli, Subdivision algorithms for the generation of box spline surfaces, Comput. Aided Geo. Design, 1 (1984), 115-129.

[2] D. Gonsor and M. Neamtu, Null spaces of differential operators, polar forms, and splines, J. Approx. Theory **86** (1996), 81–107.

[3] G. G. Lorentz, Approximation of Functions, Holt, Rinehart and Winston, 1966.

[4] C. A. Micchelli, *Mathematical Aspects of Geometric Modeling*, CBMS-NSF Regional Conference Series in Applied Mathematics, Vol 65, SIAM, Philadelphia, 1995.

[5] E. Voronskaya, Détermination de la forme asymptotique d'approximation des fonctions par le polynômes de M. Bernstein, Doklady Akademii Nauk SSSR (1932), 79-85.

Nonlinear Stationary Subdivision

Michael S. Floater
SINTEF
P. O. Box 124 Blindern, 0314 Oslo, Norway
E-mail: michael.floater@math.sintef.no

Charles A. Micchelli
IBM Corporation
T.J. Watson Research Center
Department of Mathematical Sciences
P. O. Box 218
Yorktown Heights, N. Y. U.S.A. 10598
E-mail: cam@watson.ibm.com

Abstract

In this paper we study a concrete interpolatory subdivision scheme based on rational interpolation and show that it preserves convexity.

1 Introduction

There are two ideas which interest us in this paper. The first is to explore concrete possibilities of *nonlinear* stationary subdivision strategies. The linear case is treated in some detail in [7] where its connection to modeling of curves as well to wavelet construction is highlighted. The second idea that is featured here is that of convexity preserving interpolatory subdivision, as studied, for example, in [6].

One of the most celebrated concrete interpolatory subdivision schemes was introduced in [1]. This scheme is based upon *local polynomial interpolation* and is intimately connected to orthonormal wavelet construction. In [8] local *exponential* interpolation with *real* frequencies was considered and also shown to lead to a multiparameter family of wavelets of minimal support. Further developments of these ideas appear in [5] and [9]. The paper [9] reveals their applicability to wavelet construction in Sobolev spaces while [5] focuses upon the connection to the construction of conjugate filters with prescribed zeros.

In this paper we propose to generate nonlinear interpolatory subdivision by *rational* interpolation. We study one such example and prove that when the data is convex the interpolant to the data generated by the subdivision scheme is likewise convex. A counterexample to higher order convexity preservation is also presented.

2 Nonlinear Subdivision

In this section we formulate a notion of nonlinear stationary subdivision. All the methods we consider are *stationary* and *homogeneous*. This terminology, borrowed from the theory of Markoff chains, means that we always iterate the *same* operator–*homogeniety* and this operator commutes with shift by an integer—*stationarity*. Stationarity of subdivision must be formulated with care. To be precise, we consider the linear space X of all bi-infinite real sequences $x = (x_j : j \in \mathbb{Z})$. On this space acts the forward shift operator $T : X \to X$ defined by the equation

$$(Tx)_j = x_{j+1}, \qquad j \in \mathbb{Z}. \tag{2.1}$$

Let $F : X \to X$ be *any* mapping from X into itself. We say F is *stationary* provided that it commutes with the shift operator T, in other words

$$TF = FT.$$

Every such mapping is determined by *one scalar-valued* function $f : X \to \mathbb{R}$ by the formula

$$(Fx)_i = f(T^i x), \qquad i \in \mathbb{Z}.$$

For instance, if $a = (a_i : i \in \mathbb{Z})$ is a bi-infinite vector of compact support, that is $a_j = 0$, $j < l$ or $j > m$ for some integers l and m, and f is given by

$$f(x) = \sum_{j \in \mathbb{Z}} a_{-j} x_j,$$

then the function $F(x)$ becomes the *convolution* of a with x, that is,

$$(Fx)_i = \sum_{j \in \mathbb{Z}} a_j x_{i-j}, \qquad i \in \mathbb{Z}. \tag{2.2}$$

This is the only linear and stationary mapping.

A subdivision operator $S_{\mathbf{h}} : X \to X$ is determined by *two* scalar-valued mappings $\mathbf{h} := (h_0, h_1) : X \to \mathbb{R}^2$ and is defined by the formula

$$(S_{\mathbf{h}}x)_i := \begin{cases} h_0(T^j x), & i = 2j; \\ h_1(T^j x), & i = 2j+1. \end{cases} \tag{2.3}$$

If we introduce the bi-infinite sequence $(y_k(x) : k \in \mathbb{Z}/2)$ defined by

$$y_k(x) := (S_{\mathbf{h}}x)_{2k}, \qquad k \in \mathbb{Z}/2 \tag{2.4}$$

then it easily follows that $y_{k+1}(x) = y_k(Tx)$, that is, $S_{\mathbf{h}}$ is *stationary* when the vector in the range of $S_{\mathbf{h}}$ is indexed over the fine lattice $\mathbb{Z}/2$. When each of the functions h_0 and h_1 are linear, in particular, $h_\ell(x) = \sum_{k\in\mathbb{Z}} a_{2k+\ell}x_{-k}$, for $\ell \in \{0,1\}$, $x = (x_k : k \in \mathbb{Z}) \in X$ and $a = (a_i : i \in \mathbb{Z})$ is a prescribed bi-infinite vector of finite support then $S_{\mathbf{h}}$ has the familiar form

$$(L_a x)_i = \sum_{k\in\mathbb{Z}} a_{i-2k}x_k. \tag{2.5}$$

Similar to the linear case we say that $S_{\mathbf{h}}$ converges with respect to some subspace $Y \subset X$ if for every $x \in Y$ there is a function f_x which is continuous on $\mathbb{R}$ such that

$$\lim_{r\to\infty} \sup\{|(S_{\mathbf{h}}^r x)_j - f_x(j/2^r)| : j \in \mathbb{Z}\} = 0 \tag{2.6}$$

and for some $x \in Y$ we have that $f_x \neq 0$.

We will now give an example of a nonlinear subdivision scheme generated by *local rational interpolation.* To this end, let us consider the following problem. Given points $x_0, x_1, \ldots, x_{n+1}$ we wish to find a rational function R of the form P/Q where P is a polynomial of degree at most n and Q has degree at most one such that

$$R(t_j) = x_j, \qquad j = 0, 1, \ldots, n+1 \tag{2.7}$$

where $t_0 < t_1 < \cdots < t_{n+1}$ are prescribed. There is a special circumstance which should be considered separately. To this end, for $i = 0$ or 1 we let $w_i = [x_i, x_{i+1}, \ldots, x_{i+n}]$, be the divided difference of $x_i, x_{i+1}, \ldots, x_{i+n}$ at $t_i, t_{i+1}, \ldots, t_{i+n}$. Recall that

$$w := \frac{w_1 - w_0}{t_{n+1} - t_0} = [x_0, x_1, \ldots, x_{n+1}]$$

is the leading coefficient of the polynomial of degree at most $n+1$ (one more than we require for P above) which satisfies the interpolation conditions (2.7). Hence, when $w_1 = w_0$ this polynomial has at most degree n and is the solution R to our interpolation problem. There remains the case that $w_1 \neq w_0$. We only discuss the case that $w_1 w_0 > 0$. In this case our rational interpolant is given by the formula

$$R(t) = \frac{w_1(t_{n+1} - t)p_-(t) + w_0(t - t_0)p_+(t)}{w_1(t_{n+1} - t) + w_0(t - t_0)} \tag{2.8}$$

where in this formula p_- and p_+ are polynomials of degree at most n which interpolate the $n+1$ data $x_0, x_1, \dots, x_n$ at $t_0, t_1, \dots, t_n$ and $x_1, x_2, \dots, x_{n+1}$ at $t_1, t_2, \dots, t_{n+1}$, respectively. It is important to rewrite this function in another form. To this end we let p be the unique polynomial of degree at most $n-1$ which solves the interpolation problem

$$p(t_j) = x_j, \qquad j = 1, 2, \dots, n.$$

Proposition 2.1.

$$R(t) = p(t) + (t - t_1)\cdots(t - t_n)\frac{w_0 w_1}{w_0\lambda_0(t) + w_1\lambda_1(t)}$$

where

$$\lambda_1(t) = \frac{t_{n+1} - t}{t_{n+1} - t_0}, \qquad \lambda_0(t) = \frac{t - t_0}{t_{n+1} - t_0}$$

are the barycentric coordinates of t relative to t_0 and t_{n+1}.

Proof: Using the Newton form of polynomial interpolation we have that

$$p_+(t) = p(t) + w_1(t - t_1)\cdots(t - t_n)$$

and

$$p_-(t) = p(t) + w_0(t - t_1)\cdots(t - t_n).$$

Substituting these two formulas into equation (2.8) and simplifying the resulting expression proves the formula. □

To motivate the use of this result for subdivision we restrict ourselves to the case that $n = 2$. We choose $t_i = i - 1$, $i = 0, 1, 2, 3$, interpolate the data x_{-1}, x_0, x_1, x_2 and evaluate the rational interpolant at $t = 1/2$. This gives us the formula

$$R(1/2) = \frac{x_0 + x_1}{2} - \frac{1}{8}H(x_{-1} - 2x_0 + x_1,\, x_0 - 2x_1 + x_2) \tag{2.9}$$

where $H(a,b)$ is the *harmonic mean* defined for $a,b \in \mathbb{R}_+$ by the formula

$$H(a,b) = \begin{cases} \frac{2ab}{a+b}, & (a,b) \neq (0,0); \\ 0, & (a,b) = (0,0). \end{cases} \tag{2.10}$$

To ensure the validity of formula (2.9) we must demand that both $x_1 - 2x_0 + x_{-1}$ and $x_0 - 2x_1 + x_2$ are positive. Note the important fact that

$$H(a,b) \leq 2 \min(a,b), \qquad a,b \in R_+.$$

In formula (2.9) we think of the nonlinear term as a *perturbation* of the linear term. This suggests to us to multiply the nonlinear term by a relaxation parameter and also use this formula to generate the following nonlinear subdivision scheme.

Let $E : \mathbb{R} \times \mathbb{R} \to \mathbb{R}$ be a given continuous function. For any bi-infinite vector $x = (x_i : i \in \mathbb{Z})$ we set for $l \in \mathbb{Z}$

$$\Delta^2 x_l = x_l - 2x_{l+1} + x_{l+2}.$$

Choose $w \in \mathbb{R}$ and define the bi-infinite vector $y = (y_i : i \in \mathbb{Z})$ by the formulas

$$y_{2i} = x_i, \qquad i \in \mathbb{Z}$$

$$y_{2i+1} = \frac{x_i + x_{i+1}}{2} - \lambda E(\Delta^2 x_{i-1}, \Delta^2 x_i), \qquad i \in \mathbb{Z}. \tag{2.11}$$

This defines a nonlinear subdivision scheme. The special case above would correspond to $\lambda = 1/8$ and

$$E(a,b) := H(|a|,|b|). \tag{2.12}$$

Hence, when the vector x is convex, that is $\Delta^2 x_i \geq 0, i \in \mathbb{Z}$, and $\lambda = 1/8$, (2.11) reduces to (2.9). In this case the scheme reproduces rational polynomials whose numerators are of degree at most two and whose denominators have degree at most one.

We also remark that when

$$E(a,b) = \frac{1}{2}(a+b), \qquad a,b \in \mathbb{R}$$

and $\lambda = 1/8$ the linear subdivision scheme (2.11) is obtained by interpolating the data $x_{i-1}, x_i, x_{i+1}, x_{i+2}$ at the points $i-1$, i, $i+1$, $i+2$, respectively by a *cubic* polynomial and evaluating the resulting polynomial at the point $i + 1/2$. This is the special case of the interpolatory subdivision scheme introduced in [1]

Returning to (2.11) we shall rewrite it in another form. To this end, we introduce the sequence $m = (m_j : j \in \mathbb{Z})$ defined by setting $m_0 = 1, m_1 = m_{-1} = 1/2$ and $m_j = 0,\ j \notin \{-1, 0, 1\}$. Then the subdivision scheme

$$(L_m x)_i = \sum_{j \in \mathbb{Z}} m_{i-2j} x_j \tag{2.13}$$

corresponds to the linear term in (2.11) and L_m converges with limit

$$f_x(t) = \sum_{j \in \mathbb{Z}} x_j M(t - j), \qquad t \in \mathbb{R} \tag{2.14}$$

where $M(t) = \max(1 - |t|, 0),\ t \in \mathbb{R}$. The function f_x is the piecewise linear function with breakpoints at integers such that $f_x(j) = x_j,\ j \in \mathbb{Z}$. Also, we set for $x \in X$

$$\mathbf{e}(x) = (0, E(x_{-1} - 2x_0 + x_1, x_0 - 2x_1 + x_2)). \tag{2.15}$$

This mapping $\mathbf{e} : X \to \mathbb{R}^2$ determines a nonlinear subdivision scheme and thus our scheme (2.11) has the form

$$F_m(x) = L_m(x) - \lambda S_{\mathbf{e}}(x), \qquad x \in X.$$

When E is *nonnegative* and $\lambda \geq 0$ then formula (2.11) shows for $i \in \mathbb{Z}$ that

$$F_m(x) \leq L_m(x).$$

For λ and E further constrained the subdivision scheme (2.11) has the useful property of being *convexity preserving*. This fact is proved next. We let

$$C_2 = \{x : x \in X, \Delta^2 x_i \geq 0, i \in \mathbb{Z}\}.$$

Proposition 2.2. Suppose there is a constant $\mu > 0$ such that for $a, b \in \mathbb{R}_+$

$$0 \leq E(a, b) \leq \mu \min(a, b). \tag{2.16}$$

Then for $0 \leq \lambda \leq \frac{1}{4\mu}$,

$$F_m(C_2) \subset C_2.$$

Proof. First, we observe for all $j \in \mathbb{Z}$ that

$$\begin{aligned}(\Delta^2 F_m(x))_{2j} &= (F_m(x))_{2j} - 2(F_m(x))_{2j+1} + (F_m(x))_{2j+2} \\ &= x_j - 2\{\frac{x_j + x_{j+1}}{2} - \lambda E(\Delta^2 x_{j-1}, \Delta^2 x_j)\} + x_{j+1} \\ &= 2\lambda E(\Delta^2 x_{j-1}, \Delta^2 x_j) \geq 0\end{aligned}$$

and then we verify that

$$\begin{aligned}&(\Delta^2 F_m(x))_{2j-1} \\ &= (F_m(x))_{2j-1} - 2(F_m(x))_{2j} + (F_m(x))_{2j+1} \\ &= \frac{x_{j-1} + x_j}{2} - 2x_j + \frac{x_j + x_{j+1}}{2} \\ &\quad -\lambda\{E(\Delta^2 x_{j-2}, \Delta^2 x_{j-1}) + E(\Delta^2 x_{j-1}, \Delta^2 x_j)\} \\ &= \frac{1}{2}\Delta^2 x_{j-1} - \lambda\{E(\Delta^2 x_{j-2}, \Delta^2 x_{j-1}) + E(\Delta^2 x_{j-1}, \Delta^2 x_j)\} \\ &\geq \frac{1}{2}\Delta^2 x_{j-1} - 2\lambda\mu\Delta^2 x_{j-1} = \frac{1}{2}(1 - 4\lambda\mu)\Delta^2 x_{j-1}.\end{aligned}$$

□

Remark 2.3. When in addition E has the property that $E(a, b) > 0$ whenever a and b are positive then strictly convex data, $\Delta^2 x_i > 0$, $i \in \mathbb{Z}$ is mapped into such by the the nonlinear scheme (2.11).

From Proposition 2.2 follows the next result.

Theorem 2.4. Let $x^0 \in C_2$, define $x^r := F_m^r(x^0), r \in \mathbb{Z}_+$ and suppose the hypothesis of Proposition 2.2 holds. Then the sequence of polygonal lines

$$f^r(t) := \sum_{j \in \mathbb{Z}} x_j^r \, M(2^r t - j), \qquad t \in \mathbb{R}$$

for $r \in \mathbb{Z}_+$ satisfies the following properties.

(i) For all $t \in \mathbb{R}$ and $r \in \mathbb{Z}_+$

$$f^r(t) \geq f^{r+1}(t).$$

(ii) For all $r \in \mathbb{Z}_+$, f^r is convex on $\mathbb{R}$.

(iii) $\lim_{r \to \infty} f^r(t) = f_x(t)$, $t \in \mathbb{R}$, f_x is continuous on $\mathbb{R}$ and $f_x(j) = x_j$, $j \in \mathbb{Z}$.

Proof : Since M is the refinable function associated with the linear subdivision scheme L_m we have for $t \in \mathbb{R}$ and $r \in \mathbb{Z}_+$ that

$$f^{r+1}(t) - f^r(t) = \sum_{j \in \mathbb{Z}} (F_m^{r+1}(x^0) - L_m F_m^r(x^0))_j M(2^{r+1}t - j). \quad (2.17)$$

But

$$F_m^{r+1}(x^0) = F_m(F_m^r(x^0)) \leq L_m(F_m^r(x^0)) \quad (2.18)$$

and so (i) follows.

The function f^r is convex if and only if $x^r \in C_2$ and that is assured by Proposition 2.2 which takes care of (ii). Finally, considering (iii), for $i \in \mathbb{Z}$ let $l_i : \mathbb{R} \to \mathbb{R}$ be the unique linear interpolant satisfying $l_i(i) = x_i$ and $l_i(i+1) = x_{i+1}$ and let $b : \mathbb{R} \to \mathbb{R}$ be the piecewise linear function defined for $t \in [j, j+1]$, $j \in \mathbb{Z}$, as $b(t) = \max(l_{j-1}(t), l_{j+1}(t))$. Due to standard properties of convex functions, all convex interpolants $g : \mathbb{R} \to \mathbb{R}$, $g(j) = x_j$, are bounded below by b and therefore the sequence of functions f^r is bounded below by b. Hence the limit in (iii) exists and must be continuous since it is convex. □

Remark 2.5. For any $p > 0$ we define

$$E_p(a,b) = \phi^{-1}\left(\frac{\phi(|a|) + \phi(|b|)}{2}\right) = \frac{2^{1/p}\,|a|\,|b|}{(|a|^p + |b|^p)^{1/p}} \quad (2.19)$$

where $\phi(t) := t^{-p}$, $t \in \mathbb{R}_+$. Then for any $a, b \in \mathbb{R}$

$$0 \leq E_p(a,b) \leq 2^{1/p} \min(|a|, |b|)$$

so that this family of means satisfies the hypothesis of Proposition 2.2 with $\mu = 2^{1/p}$ and so the corresponding scheme produces a convex interpolant to convex data $\{x_j : j \in \mathbb{Z}\}$ for $0 \leq \lambda \leq 2^{-(2+1/p)}$. When

$$E(a,b) = \frac{1}{2}(a+b), \qquad a, b \in \mathbb{R} \quad (2.20)$$

the hypothesis of Proposition 2.2 is *not* satisfied. This *linear* subdivision scheme is precisely the one studied in [2] in which $w = \lambda/2$ plays the role of tension parameter. This scheme does not preserve convexity even when it converges.

The case of the nonlinear means above for $p = 1$ corresponding to the harmonic mean was independently considered in [3]. We fell upon it through the process of rational interpolation. This fact seems to have not been noticed in [3] as well as the monotonicity of the polygonal lines embodied in (i) of Theorem 2.4. Since then we received [4] where similar issues are studied further.

We remark that the proof of Proposition 2.2 shows that for arbitrary $\lambda \geq 0$, property (i) still holds. Hence the subdivision scheme converges for every $x \in X$ in the sense that

$$\lim_{r \to \infty} f^r(t)$$

exists for each $t \in \mathbb{R}$ but we cannot rule out that the limit function may be $-\infty$ at some point and also not continuous. For a more restrictive range of λ we prove convergence in terms of our definition (2.6). To this end we define the subspace of X

$$Y_2 = \{x : x \in X, ||\Delta^2 x||_\infty < \infty\},$$

where $||x||_\infty := \sup\{|x_i| : i \in \mathbb{Z}\}$ for a bi-infinite sequence $x = \{x_i : i \in \mathbb{Z}\}$.

Theorem 2.6. Suppose there exists a constant $\rho > 0$ such that for all $a, b \in \mathbb{R}$

$$|E(a,b)| \leq \rho \max(|a|, |b|). \tag{2.21}$$

Then for $|\lambda| < (4\rho)^{-1}$, the subdivision scheme (2.11) converges with respect to Y_2 and the limit function f_x interpolates x_j at $t = j$, $j \in \mathbb{Z}$, that is,

$$f_x(j) = x_j, \qquad j \in \mathbb{Z}.$$

Remark 2.7. When ϕ is a monotonic function on $\mathbb{R}_+$ (either increasing or decreasing) then

$$E(a,b) = \phi^{-1}\left(\frac{\phi(|a|) + \phi(|b|)}{2}\right)$$

has the property that

$$E(a,b) \leq \max\{|a|, |b|\}.$$

In fact, if $|a| \leq |b|$ and ϕ increases then ϕ^{-1} also does so that

$$\phi^{-1}\left\{\frac{\phi(|a|) + \phi(|b|)}{2}\right\} \leq \phi^{-1}\left\{\frac{2\phi(|b|)}{2}\right\} = |b|.$$

Likewise, if $|a| \leq |b|$ and ϕ decreases then so does ϕ^{-1} and so since $\phi(|a|) \geq \phi(|b|)$,

$$\frac{\phi(|a|) + \phi(|b|)}{2} \geq \phi(|b|)$$

from which we get again

$$E(a,b) \le |b|.$$

Hence (2.11) converges for $|\lambda| < \frac{1}{4}$.

When $E(a,b) = \frac{1}{2}(a+b)$, (2.11) is the scheme in [2]. In this case the equivalent result that the scheme converges for $|w| < \frac{1}{8}$ is proved there.

Alternatively, this linear scheme can be written in the standard form (2.5) where $a_j = 0$, $|j| > 3$ and $a_{-3} = a_3 = -w$, $a_{-2} = a_2 = 0$, $a_{-1} = a_1 = \frac{1}{2} + w$, $a_0 = 1$. In this case, the symbol of the scheme is given by

$$a(z) = \sum_{j \in \mathbb{Z}} a_j z^j = 1 + (w + \frac{1}{2})(z^1 + z^{-1}) - w(z^3 + z^{-3}).$$

When $z = e^{i\theta}$ and $x = \cos\theta$ we have that

$$a(z) = (1+x)(1+8wx(1-x)).$$

A direct computation confirms that when $-\frac{1}{2} < w \le \frac{1}{16}$, $a(e^{i\theta}) \ge 0$ for $|\theta| \le \pi$ with equality if and only if $x = -1$. Hence it follows that the scheme converges for this range, [8]. Moreover, the refinable function f corresponding to the vector $x_j = 0$, $j \in \mathbb{Z}\backslash\{0\}$, $x_0 = 1$ which is supported on $(-3,3)$ is the autocorrelation of a refinable function ϕ_w which yields an orthonormal wavelet of finite support (for an explanation of wavelet construction see Chapter 2 of [7]) . When $w = \frac{1}{16}$ this wavelet is one of the family constructed by Daubechies, [8]. The special case $w = \frac{1}{16}$ corresponds to local cubic interpolation and is a special case of the schemes appearing in [1].

Proof: According to formula (2.17) we have that

$$\|f^{r+1} - f^r\|_\infty \le |\lambda|\,\rho\,\|\Delta^2 x^r\|_\infty$$

where $x^r = F_m^r(x)$ and $\|f\|_\infty := \sup\{|f(t)| : t \in \mathbb{R}\}$. Also, returning to the proof of Proposition 2.2 we see for $j \in \mathbb{Z}$ and $r \in \mathbb{N}$ that

$$|(\Delta^2 x^r)_{2j}| \le 2|\lambda|\,\rho\,\|\Delta^2 x^{r-1}\|_\infty \tag{2.22}$$

and

$$|(\Delta^2 x^r)_{2j+1}| \le (\frac{1}{2} + 2|\lambda|\rho)\,\|\Delta^2 x^{r-1}\|_\infty. \tag{2.23}$$

Combining the inequalities (2.22) and (2.23) we obtain for $r \in \mathbb{Z}_+$ the inequality

$$\|f^{r+1} - f^r\|_\infty \le |\lambda|\rho\,\gamma^r\,\|\Delta^2 x^0\|_\infty \tag{2.24}$$

where $\gamma := \frac{1}{2} + 2|\lambda|\rho$. Since by hypothesis, $\gamma < 1$, this proves $\lim_{r\to\infty} f^r = f_x$ uniformly on $\mathbb{R}$.

Next, we shall show f_x is Hölder continuous and afterwards that it is the limit of the subdivision scheme in the sense of (2.6). For the first claim we note that since $f^r(\frac{t}{2^r})$ is a piecewise linear function which interpolates x_j^r at $t = j$, $j \in \mathbb{Z}$ it follows for $r \in \mathbb{Z}_+$, $t, s \in \mathbb{R}$ that

$$|f^r(t) - f^r(s)| \leq 2^r \|\Delta x^r\|_\infty |t - s|.$$

From the formula (2.11) and our hypothesis it follows for $r \in \mathbb{N}$ that

$$\|\Delta x^r\|_\infty \leq \frac{1}{2}\|\Delta x^{r-1}\|_\infty + |\lambda|\rho \|\Delta^2 x^{r-1}\|_\infty.$$

Consequently, we have that

$$\begin{aligned}\|\Delta x^r\|_\infty &\leq (\frac{1}{2} + 2|\lambda|\rho)\|\Delta x^{r-1}\|_\infty \\ &\leq (\frac{1}{2} + 2|\lambda|\rho)^r \|\Delta x^0\|_\infty\end{aligned}$$

and so for $r \in \mathbb{Z}_+$, $t, s \in \mathbb{R}$ we have that

$$|f^r(t) - f^r(s)| \leq (2\gamma)^r \|\Delta x^0\|_\infty |t - s|. \tag{2.25}$$

Combining this inequality with (2.24) we conclude with the help of Lemma 2.1 of [7], p.82, that f_x is Hölder continuous with exponent $\mu = -\log_2 \gamma$ which is positive since $\gamma < 1$. Thus there is a constant $C > 0$ such that

$$|f_x(t) - f_x(s)| \leq C|t - s|^\mu,$$

for all $t, s \in \mathbb{R}$. To finish the proof we note that the function

$$g^r(t) = \sum_{j\in\mathbb{Z}} f_x(\frac{j}{2^r})M(2^r t - j), \qquad t \in \mathbb{R}$$

has the property that

$$|g^r(t) - f_x(t)| \leq 2^{-\mu r}C, \qquad t \in \mathbb{R}.$$

Hence, we have the bound

$$\begin{aligned}\sup\{|x_j^r &- f_x(\frac{j}{2^r})| : j \in \mathbb{Z}\} \\ &\leq \sup\{|\sum_{j\in\mathbb{Z}}(x_j^r - f_x(\frac{j}{2^r}))M(2^r t - j)| : t \in \mathbb{R}\} \\ &\leq \|f^r - f_x\|_\infty + 2^{-\mu r}C, \qquad r \in \mathbb{Z},\end{aligned}$$

and sending $r \to \infty$ proves the result. $\square$

One may be optimistic that interpolatory subdivision based on rational interpolation with one pole as appears in Proposition 2.1 with $n \geq 3$ would preserve higher order convexity. Unfortunately, this conjecture fails as the next observation indicates. Specifically, instead of the nonlinear scheme of the form (2.11) we consider, for a given function $F : \mathbb{R} \times \mathbb{R} \to \mathbb{R}$ which is *symmetric*, i.e. $F(s,t) = F(t,s)$, $s,t \in \mathbb{R}$, and a constant $\mu \in \mathbb{R}$, the nonlinear subdivision scheme

$$y_{2i} = x_i, \qquad i \in \mathbb{Z},$$

$$y_{2i+1} = A(x_i, x_{i+1}) - \frac{1}{8}A(\Delta^2 x_{i-1}, \Delta^2 x_i) + \mu F(\Delta^4 x_{i-2}, \Delta^4 x_{i-1}), \quad i \in \mathbb{Z}. \tag{2.26}$$

where

$$A(s,t) = \frac{1}{2}(t+s), \qquad t,s \in \mathbb{R}.$$

For the choice $\mu = 3/128$ and $F(s,t) = A(s,t)$ this becomes the linear interpolatory subdivision scheme in [1] corresponding to quintic interpolation. For the choice $\mu = 3/128$ and $F(s,t) = H(s,t)$, with H given in (2.10), the scheme corresponds to rational interpolation with rational polynomials P/Q where P is quartic and Q linear. When $\mu = 0$, the scheme reduces to that of [1] corresponding to cubic interpolation, or the scheme (2.11) with $\lambda = 1/8$.

For ease of notation we set $\alpha_i = \Delta^4 x_i$ and $F_i = F(\alpha_{i-2}, \alpha_{i-1}), i \in \mathbb{Z}$ in the computations we perform next.

Let us now investigate the higher order convexity preservation of this scheme. To this end, we recall that

$$\Delta^4 y_\ell = y_\ell - 4y_{\ell+1} + 6y_{\ell+2} - 4y_{\ell+3} + y_{\ell+4}, \qquad \ell \in \mathbb{Z}$$

and therefore we obtain

$$\Delta^4 y_{2i} = \frac{1}{4}\alpha_{i-1} - 4\mu(F_i + F_{i+1}), \qquad i \in \mathbb{Z} \tag{2.27}$$

and

$$\Delta^4 y_{2i+1} = -\frac{1}{8}A(\alpha_{i-1}, \alpha_i) + \mu(F_i + 6F_{i+1} + F_{i+2}), \qquad i \in \mathbb{Z} \tag{2.28}$$

We note for the case $\mu = 3/128$ and $F(t,t) = t$, $t \in \mathbb{R}$ that whenever $x_i = p(i), i \in \mathbb{Z}$, $p \in \pi_4$, and $\alpha_i = k$, we have that

$$\Delta^4 y_i = \frac{1}{16}k, \qquad i \in \mathbb{Z}$$

which means the scheme reproduces quartic polynomials.

If the subdivision procedure (2.26) preserves four-convexity, that is, whenever $\Delta^4 x_i \geq 0$, $i \in \mathbb{Z}$ it follows that $\Delta^4 y_i \geq 0$, $i \in \mathbb{Z}$ we conclude first from (2.27) that for any sequence $\alpha_i \geq 0$, $i \in \mathbb{Z}$ we have

$$\alpha_{i-1} \geq 16\mu(F_i + F_{i+1}), \qquad i \in \mathbb{Z} \tag{2.29}$$

and then from (2.28) that

$$\alpha_{i-1} + \alpha_i \leq 16\mu(F_i + 6F_{i+1} + F_{i+2}), \qquad i \in \mathbb{Z}. \tag{2.30}$$

For any $a, b \geq 0$, letting $\alpha_{i-1} = a$, $\alpha_{i-2} = \alpha_i = b$ in (2.29) yields

$$a \geq 32\mu F(a, b) \tag{2.31}$$

and letting $\alpha_{i-2} = \alpha_i = a$, $\alpha_{i-1} = \alpha_{i+1} = b$ in (2.30) yields

$$a + b \leq 128\mu F(a, b). \tag{2.32}$$

Combining (2.31) and (2.32) we conclude for $a, b \geq 0$ that

$$b \leq 3a,$$

an apparent contradiction.

3 Numerical Examples

The subdivision scheme (2.11) was applied to two data sets (i) and (ii), tabulated in Tables 1 and 2 respectively

t_i	-2.0	-1.0	0.0	1.0	2.0	3.0	4.0	5.0	6.0
x_i	0.00	0.00	0.00	0.15	0.50	2.00	5.00	13.0	21.0

Table 1. Data set (i)

t_i	-2.0	-1.0	0.0	1.0	2.0	3.0	4.0	5.0	6.0
x_i	0.00	0.00	0.00	0.15	1.00	2.00	5.00	13.0	21.0

Table 2. Data set (ii)

Figures 1, 2, 3, 4, and 5 show the result of applying the scheme to data set (i) while Figures 6 and 7 depict the ouput when starting with data set (ii). In Figures 1 and 6 we used the linear scheme in which $E(a, b)$ is the arithmetic mean (2.20) and $\lambda = 1/8$. Figures 2 and 7 show the result of the rational convexity preserving scheme defined by setting $E(a, b)$ to be the harmonic mean (2.12) and $\lambda = 1/8$. In Figure 3 the harmonic mean was used again, this time setting $\lambda = 1/16$, thereby increasing the tension. This scheme also preserves convexity. The harmonic mean (2.12) is equivalent to the function E_1 where E_p is defined in (2.19). Figure 4 shows the result of applying instead E_2 and this time $\lambda = 1/(4\sqrt{2})$ which is the upper limit of the range of λ for which Proposition 2.2 guarantees convexity preservation. Figure 5 on the hand shows the output using $E_{1/2}$ and $\lambda = 1/16$ which again is the limit of the range in Proposition 2.2.

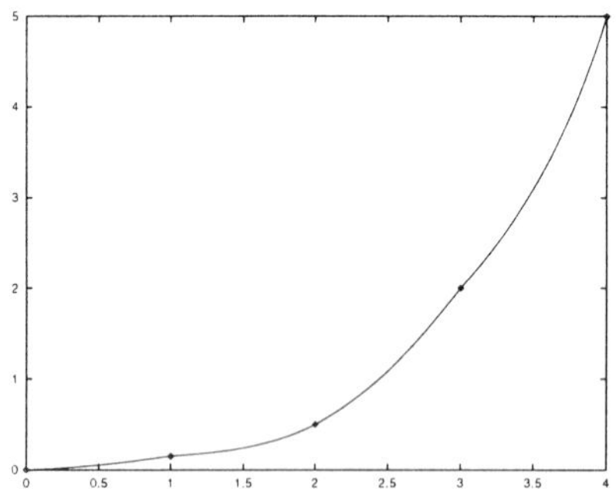

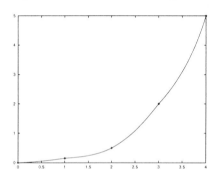

Fig. 1. Linear, $\lambda = 1/8$, data (i).

Fig. 2. Harmonic, $\lambda = 1/8$, data (i).

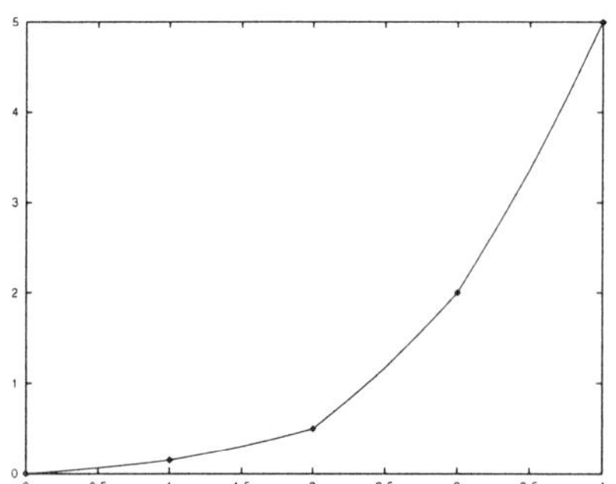

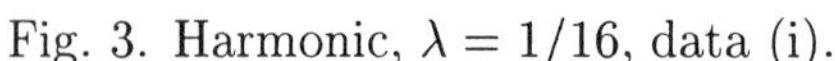

Fig. 3. Harmonic, $\lambda = 1/16$, data (i).

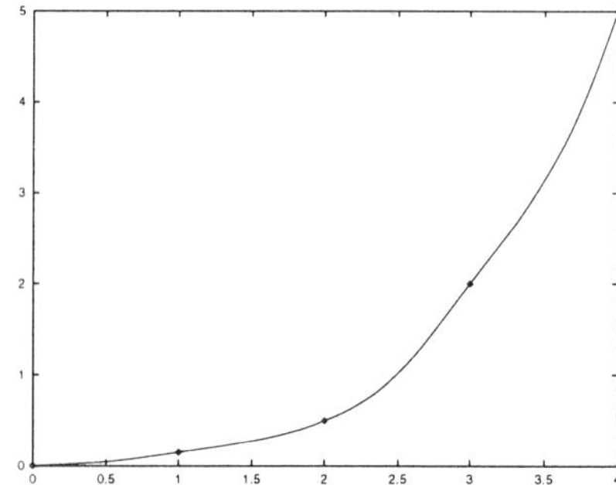

Fig. 4. E_2, $\lambda = 1/(4\sqrt{2})$, data (i).

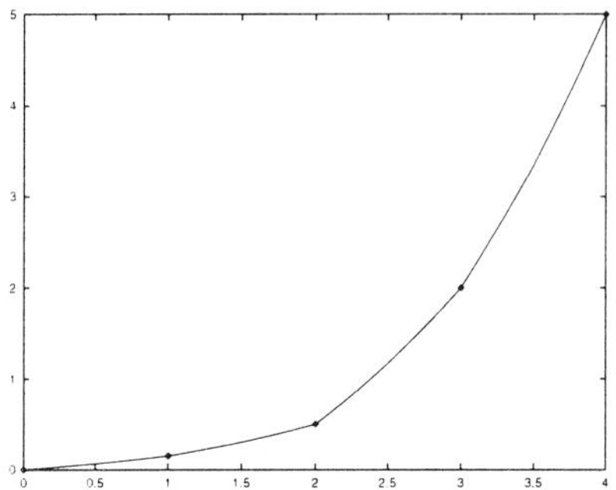

Fig. 5. $E_{1/2}$, $\lambda = 1/16$, data (i).

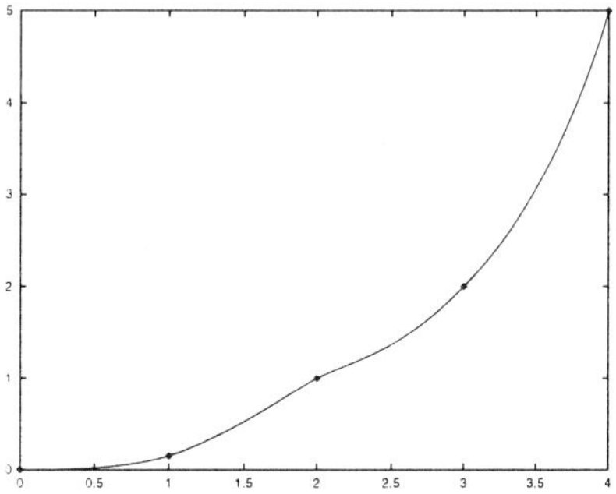

Fig. 6. Linear, $\lambda = 1/8$, data (ii).

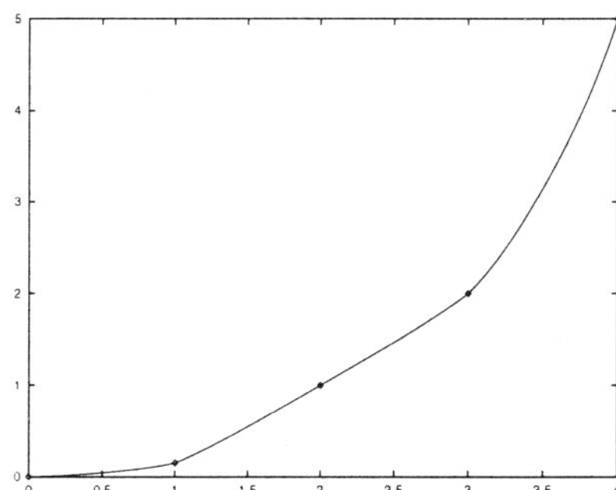

Fig. 7. Harmonic, $\lambda = 1/8$, data (ii).

Acknowledgement

Much of this work was done during the Spring of 1996 when the authors were visiting the University of Zaragoza. We wish to thank our host, Mariano Gasca, for providing a friendly environment that led to this stimulating scientific exchange.

We would also like to thank Narendra Govil for his generous and indispensable help in the preparation of this manuscript.

References

[1] G. Deslauriers and S. Dubuc Symmetric iterative interpolation processes, Constr. Approx. **5** (1989), 49-68.

[2] N. Dyn, D. Levin and J. A. Gregory, A 4-point interpolatory subdivision scheme for curve design, Computer Aided Geometric Design, **4** (1987), 257-268.

[3] F. Kuyt and R. van Damme, Smooth interpolation by a convexity preserving nonlinear subdivision algorithm, preprint 1996.

[4] F. Kuyt and R. van Damme, Convexity preserving interpolatory subdivision schemes, Memorandum No. 1357, University of Twente, The Netherlands, 1996.

[5] W. Lawton and C. A. Micchelli, Construction of Conjugate Quadrature Filters with Specified Zeros, to appear in Numerical Algorithms.

[6] A. Le Méhauté and F. I. Utreras, Convexity-preserving interpolatory subdivision, Computer Aided Geometric Design, **11** (1994), 17-37.

[7] C. A. Micchelli, Mathematical Aspects of Geometric Modeling, CBMS-NSF Regional Conference Series in Applied Mathematics, Vol 65, SIAM, Philadelphia, 1995.

[8] C. A. Micchelli, Interpolatory subdivision schemes and wavelets, J. Approx. Theory **86** (1996), 41–71.

[9] C. A. Micchelli, On a family of filters arising in wavelet construction, Applied and Computational Harmonic Analysis **4** (1997), 38-50.

Convex Univalent Functions and Omitted Values

Richard Fournier *
Centre de Recherches Mathématiques, Université de Montréal
Montréal H3C 3J7, Canada

Jinxi Ma †
Mathematisches Institut, Universität Würzburg
D-97074 Würzburg, Germany

Stephan Ruscheweyh
Mathematisches Institut, Universität Würzburg
D-97074 Würzburg, Germany

Abstract

Recently Chuaqui and Osgood [4] have shown that $-1/a_2 \notin f(\mathbb{D})$ for any function $f = z + \sum_{k=2}^{\infty} a_k z^k$ in the Nehari class $\mathcal{N}$. We show that this can be strongly refined if f is restricted to the class $\mathcal{C}$ of normalized convex univalent functions in $\mathbb{D}$, but this latter refinement does not admit any immediate extension back into $\mathcal{N}$. We also deal with bounded and unbounded convex univalent functions. So we prove that

$$|a_4| \leq \frac{7}{3}|a_2| + \frac{2}{3M}, \quad M := \sup_{z\in\mathbb{D}} |f(z)|,$$

holds in all of $\mathcal{C}$, where the constant $\frac{2}{3}$ is best possible, but that $|a_4| \leq 2|a_2|$ is true and sharp for the subclass $\mathcal{C}_u$ of unbounded normalized convex univalent functions in $\mathbb{D}$. We also discuss some other problems in the vicinity of those mentioned above.

*This author acknowledges support of FCAR (Quebec).
†Supported by German Academic Exchange Service (DAAD)

1 Introduction and statement of the results

The general theme of this paper is to study convex univalent functions in the unit disk $\mathbb{D}$ omitting or assuming certain values. This concerns the omission of specific complex numbers from the range of these functions, but also the property to be bounded or unbounded. In a recent paper, Chuaqui and Osgood [4] showed that

$$f(z) \neq \frac{-1}{a_2(f)}, \quad z \in \mathbb{D}, \tag{1}$$

holds for any function

$$f(z) = z + \sum_{k=2}^{\infty} a_k(f) z^k \tag{2}$$

in the Nehari class $\mathcal{N}$, i.e. analytic in $\mathbb{D}$ and satisfying

$$|S_f[z]| \leq \frac{2}{(1-|z|^2)^2}, \quad z \in \mathbb{D},$$

where

$$S_f[z] := \left(\frac{f''(z)}{f'(z)}\right)' - \frac{1}{2}\left(\frac{f''(z)}{f'(z)}\right)^2$$

is the Schwarzian Derivative. It is well known that $\mathcal{N} \supset \mathcal{C}$, the class of normalized convex univalent functions. We became interested in the question to which extent the above result can be improved in the subclass $\mathcal{C}$, and how sharp (1) is within $\mathcal{N}$. These considerations lead us to the following notation and results.

For any set $\mathcal{W}$ of functions f analytic in $\mathbb{D}$ and normalized as in (2) we define the set

$$\sigma(\mathcal{W}) := \bigcup_{f \in \mathcal{W}} \{a_2(f) f(z) : z \in \mathbb{D}\}.$$

Theorem 1 *We have*

$$\sigma(\mathcal{C}) = \left\{ w : \operatorname{Re} w > -\frac{1}{2} \right\}. \tag{3}$$

If $f \in \mathcal{C}$ maps $\mathbb{D}$ onto an unbounded convex wedge-shaped polygon (sector mapping) then the vertex of $a_2(f) f(\partial\mathbb{D})$ lies in $\partial\sigma(\mathcal{C})$.

An immediate consquence of Theorem 1 is the subordination

$$a_2(f)\, f(z) \prec \frac{z}{1-z}, \quad f \in \mathcal{C},$$

which nicely complements the well known

$$\frac{f(z)}{z} \prec \frac{1}{1-z}, \quad f \in \mathcal{C}.$$

Our next theorem slightly extends the Chuaqui-Osgood result.

Theorem 2 *We have* $\sigma(\mathcal{N}) = \mathbb{C} \setminus \{-1\}$.

It would be interesting to determine $\sigma(\mathcal{W})$ for other sets $\mathcal{W}$. We recall that a function f as in (2) is starlike of order $\alpha < 1$ (i.e. $f \in \mathcal{S}_\alpha$) if and only if

$$\operatorname{Re} \frac{zf'(z)}{f(z)} > \alpha, \quad z \in \mathbb{D}.$$

One can readily show that $\sigma(\mathcal{S}_0) = \mathbb{C}$, so the interesting sets $\mathcal{W}$ with non-trivial $\sigma(\mathcal{W})$ must be small.

Let $\mathcal{C}_b$ and $\mathcal{C}_u$ denote the sets of bounded respectively unbounded functions in $\mathcal{C}$. The following result is implicitly known (compare Paatero [10]). We denote by $\mathcal{H}(\mathbb{D})$ the space of analytic functions in the unit disk, endowed with the topology of locally uniform convergence.

Theorem 3 $f \in \mathcal{C}_u$ *if and only if there exist* ζ *with* $|\zeta| = 1$ *and* $g \in \mathcal{S}_{\frac{1}{2}}$ *such that* $f(0) = 0$ *and*

$$zf'(z) = \frac{g(z)}{1-\zeta z}, \quad z \in \mathbb{D}. \tag{4}$$

In particular, $\mathcal{C}_u$ *is a compact set in* $\mathcal{H}(\mathbb{D})$.

Note that $\mathcal{C}$ is also compact, but $\mathcal{C}_b$ is not. This is a very special feature of convex univalent functions which apparently has not found much attention yet.

It is known (and easily verified from (4)) that the only functions $f \in \mathcal{C}_u$ with $a_2(f) = 0$ are the odd functions

$$z \mapsto \frac{1}{2\zeta} \log\left(\frac{1+\zeta z}{1-\zeta z}\right), \quad |\zeta| = 1. \tag{5}$$

This means, for instance, that for $f \in \mathcal{C}_u$ we have

$$a_2(f) = 0 \Rightarrow a_{2k}(f) = 0, \; k = 2, 3, 4, \ldots,$$

and so there is a chance to obtain estimates of $|a_{2k}(f)|$ in terms of $|a_2(f)|$ in $\mathcal{C}_u$. Indeed, we get more:

Theorem 4 *Let $f \in C_u$ be as in (2). Then*

$$|a_{mk}(f)| \le m\,|a_k(f)|, \quad m, k = 2, 3, \dots. \tag{6}$$

These are all best possible.

In general we can relate the coefficients $a_2(f)$, $a_4(f)$ to

$$M(f) := \sup\{|f(z)| \,:\, z \in \mathbb{D}\} \le \infty.$$

Theorem 5 *Let $f \in C$ be as in (2). Then we have*

$$|a_4(f)| \le \frac{7}{3}|a_2(f)| + \frac{2}{3M(f)}. \tag{7}$$

The constant $\frac{2}{3}$ is best possible.

There is some evidence that the constant $\frac{7}{3}$ in (7) can be replaced by 2, which then would be best possible (compare (6) for $m = k = 2$). An interesting consequence of Theorem 5 is

Theorem 6 *Let $f \in C$ be as in (2), and assume $a_2(f) = 0$. Then*

$$|a_4(f)|\, M(f) \le \frac{2}{3}. \tag{8}$$

The constant $\frac{2}{3}$ is (still!) best possible.

From Theorems 1 and 5 one can also deduce

Theorem 7 *There exists a constant $C \le \frac{11}{3}$ such that for each $f \in C$ as in (2) we have*

$$\left|\frac{a_4(f)f(z)}{1 + a_2(f)f(z)}\right| \le C. \quad z \in \mathbb{D}. \tag{9}$$

In particular,

$$f(z) \ne \frac{-1}{a_2(f) + \zeta a_4(f)}, \quad z \in \mathbb{D},\ |\zeta| \le \frac{1}{C}. \tag{10}$$

The best possible value for C is not known. It cannot be < 2, however. We shall discuss these questions in greater detail after the proof of Theorem 5 in Sect. 5.

The functional considered in (9) can be seen to not be bounded over the Nehari class $\mathcal{N}$. Clearly, also (10) generalizes (1) for $C \subset \mathcal{N}$. For related results on omitted values of Moebius transforms of convex univalent functions we refer f.i. to [1],[2],[3],[5],[6].

Acknowledgement The authors should like to thank Richard Greiner, who posed a problem which triggered the present investigation.

2 Proof of Theorem 1 and some related remarks

Proof. Let $f \in \mathcal{C}$ be as in (2), and $x \in \mathbb{D}$ be fixed. It is known (cf. Sheil-Small [12], Suffridge [13]) that the function

$$F(z) := \frac{xz}{f(x)} \frac{f(z) - f(x)}{z - x} \tag{11}$$

belongs to $\mathcal{S}_{\frac{1}{2}}$. Hence

$$G(z) := \frac{zF'(z)}{F(z)} = 1 + \sum_{k=1}^{\infty} a_k(G) z^k$$

satisfies $\operatorname{Re} G(z) > \frac{1}{2}$ in $\mathbb{D}$. An application of the Schwarz-Pick Lemma to $-1 + 1/G(z)$ yields

$$\left|a_2(G) - a_1^2(G)\right| \leq 1 - |a_1(G)|^2 . \tag{12}$$

Now

$$a_1(G) = \frac{1}{x} - \frac{1}{f(x)}, \quad a_2(G) = \frac{1}{x^2} - \frac{1}{f(x)} \left(2a_2(f) + \frac{1}{f(x)} \right),$$

and inserting this into (12) gives

$$\begin{aligned} |x|^2 \left(2\mathrm{Re}\, \frac{f(x)}{x} - 1 \right) &- (1 - |x|^2)|f(x)|^2 \\ &\geq \left|2xf(x) - 2x^2 \left(1 + a_2(f)f(x)\right)\right| \qquad (13) \\ &\geq 2|x|^2 \mathrm{Re}\, \frac{f(x)}{x} - 2|x|^2 \mathrm{Re}\, \left(1 + a_2(f)f(x)\right), \end{aligned}$$

and therefore

$$\mathrm{Re}\, \left(a_2(f)\, f(x)\right) \geq -\frac{1}{2} + \frac{1}{2}(1 - |x|^2) \left|\frac{f(x)}{x}\right|^2 \geq -\frac{1}{2},$$

for arbitrary $x \in \mathbb{D}$. This proves

$$\sigma(\mathcal{C}) \subset \left\{ w : \mathrm{Re}\, w > -\frac{1}{2} \right\},$$

and looking at $\frac{z}{1-z} \in \mathcal{C}$ we conclude that indeed equality holds in (3). The sector mappings mentioned in the statement of Theorem 1 are given by

$$f_{c,\theta}(z) := \frac{1}{c(1+e^{i\theta})}\left[\left(\frac{1+z}{1-e^{i\theta}z}\right)^c - 1\right], \quad 0 < c \leq 1,\ 0 < \theta < \pi,$$

and elementary rotations of those (which leave the functional $a_2(f)f(z)$ invariant). In this case we have

$$a_2(f_{c,\theta}) = \frac{1}{2}\left(c - 1 + e^{i\theta}(1+c)\right),$$

and calculation yields

$$a_2(f)f_{c,\theta}(-1) = -\frac{1}{2} - i\,\frac{\tan(\theta/2)}{c}.$$

It may be of interest to point out that (3) is, in a sense, also characteristic for $\mathcal{C}$: if $\mathcal{W}$ is a linear invariant family of functions (2) satisfying

$$\sigma(\mathcal{W}) \subset \left\{w : \operatorname{Re} w > -\frac{1}{2}\right\}, \tag{14}$$

then $\mathcal{W} \subset \mathcal{C}$. This follows from the fact that (14) implies that $|a_2(f)| \leq 1$ for $f \in \mathcal{W}$ which implies this conclusion in a linear invariant family.

The following result explains the significance of the functional $a_2(f)$ in the context of Theorem 1, and therefore in general.

Theorem 8 *Let λ be a continuous linear functional on $\mathcal{H}(\mathbb{D})$, and assume that*

$$\sigma_\lambda := \bigcup_{f\in C} \lambda(f)f(\mathbb{D})$$

is contained in a halfplane. Then $\lambda(f) = C_1\,a_1(f) + C_2\,a_2(f)$, where C_1, C_2 are constants.

Proof. We may assume that $\lambda \not\equiv 0$, and therefore that 0 is an interior point of σ_λ. We can assume that σ_λ is contained in the half-plane $\operatorname{Re} w > -1$. By a theorem of Toeplitz [14] we have the existence of a unique function $g \in \mathcal{H}(\overline{\mathbb{D}})$ with

$$\lambda(f) = (g * f)(1), \quad f \in \mathcal{H}(\mathbb{D}),$$

where $*$ denotes the Hadamard product. If we assume, as we may, that $\lambda(c) = 0$ for all constant functions in $\mathcal{H}(\mathbb{D})$, then $g(0) = 0$. Using our assumptions for the functions $z/(1-e^{i\theta}z) \in \mathcal{C}$, $\theta \in \mathbb{R}$, then we obtain

$$\operatorname{Re}\left(1 + e^{-i\theta}g(e^{i\theta})\frac{z}{1-e^{i\theta}z}\right) \geq 0, \quad \theta \in \mathbb{R},\ z \in \mathbb{D}.$$

With $|z| \to 1$ we deduce that

$$0 \le e^{-2i\theta} g(e^{i\theta}) \le 2, \quad \theta \in \mathbb{R}.$$

The function

$$G(z) := \frac{g(z)}{z^2} - \frac{a_1(g)}{z} - \overline{a_1(g)} z$$

is in $\mathcal{H}(\overline{\mathbb{D}})$ and takes real values on $\partial\mathbb{D}$, and must be therefore equal to a real constant t. Hence

$$g(z) = a_1(g)z + tz^2 + \overline{a_1(g)} z^3.$$

If $a_1(g) = 0$ we are done, so assume the contrary. Then for the odd strip-mappings

$$f_\phi(z) := \frac{e^{-i\phi}}{2} \log\left(\frac{1 + ze^{i\phi}}{1 - ze^{i\phi}}\right)$$

our assumption yields

$$\mathrm{Re}\,(1 + a_1(g) f_\phi(z)) > 0, \quad z \in \mathbb{D}, \ \phi \in \mathbb{R},$$

which is impossible.

Note that the linearity of λ was essential in the proof just given. Without this assumption, the conclusion is not valid, as simple examples show.

3 Proof of Theorem 2

Proof. We only need to show that for every $\zeta \ne -1$ we find $f \in \mathcal{N}$ and $z \in \mathbb{D}$ with $\zeta = a_2(f)f(z)$. This is clear for $\zeta = 0$, which we exclude from now on. Assume the contrary: $\zeta \notin a_2(f)f(\mathbb{D})$ for all $f \in \mathcal{N}$. Fix $g \in \mathcal{N}$ and let $w \in g(\mathbb{D})^c$. Then

$$\frac{wg(z)}{w - g(z)} = z + (a_2(g) + \frac{1}{w})z^2 + \dots$$

is in $\mathcal{N}$ too, and therefore our assumption holds for this function as well. It eventually yields

$$\frac{(1+\zeta)g(z)}{-\zeta + a_2(g)g(z)} \ne w, \quad z \in \mathbb{D},$$

and therefore

$$\frac{(1+\zeta)g(z)}{-\zeta + a_2(g)g(z)} \prec g(z). \tag{15}$$

Now let $f \in \mathcal{N}$ and set

$$g(z) := \frac{f(z)}{1 + a_2(f)f(z)}.$$

Then $a_2(g) = 0$, and (15) becomes

$$\frac{1+\zeta}{-\zeta} \frac{f(z)}{1 + a_2(f)f(z)} \prec \frac{f(z)}{1 + a_2(f)f(z)}.$$

We now apply this to a function $F \in \mathcal{C}_u \subset \mathcal{N}$ with $a_2(F) \neq 0$. Let $z_n \in \mathbb{D}$ be a sequence with $F(z_n) \to \infty$. Then

$$\frac{1+\zeta}{-\zeta} \frac{F(z_n)}{1 + a_2(F)F(z_n)} = \frac{F(v(z_n))}{1 + a_2(F)F(v(z_n))},$$

where $v(z)$ satisfies $|v(z)| \leq |z|$ in $\mathbb{D}$. The assumption $\limsup |F(v(z_n))| = \infty$ leads to an immediate contradiction, therefore we can find $z \in \overline{\mathbb{D}}$ with

$$\frac{1+\zeta}{-\zeta} \frac{1}{a_2(F)} = \frac{F(z)}{1 + a_2(F)F(z)},$$

or, equivalently,

$$-\zeta - 1 \in a_2(F)F(\overline{\mathbb{D}}), \quad F \in \mathcal{C}_u, \; a_2(F) \neq 0.$$

Looking at strip-mappings with $a_2(F) \to 0$ it is clear that this can only hold for $-\zeta - 1$ purely imaginary. This, however can be ruled out by a sequence of wedge-mappings $a_2(F)F(z)$, symmetric to the real axis and with their opening going to zero (recall from Theorem 1 that the vertex is in the point $-\frac{1}{2}$).

4 Unbounded convex univalent functions

Proof. (Theorem 3) Let $f \in \mathcal{C}_u$. Without loss of generality we can assume that the positive real axis is contained in $f(\mathbb{D})$. By the covering theorem valid in $\mathcal{C}$ we know that the disk $\{z : |z| < \frac{1}{2}\}$ is also contained in this range. Using the convexity of $f(\mathbb{D})$ we then see that $f(\mathbb{D})$ contains the interior of the set

$$W := \mathbb{D}_{\frac{1}{2}} \bigcup \left\{ z : \operatorname{Re} z \geq 0, \; |\operatorname{Im}(z)| \leq \frac{1}{2} \right\}.$$

Hence we can approximate f locally uniformly in $\mathbb{D}$ by a sequence of polygonal mappings $f_n \in \mathcal{C}$, with $f_n(0) = 0, f_n'(0) > 0, f_n'(0) \to 1$, which also satisfy $W \subset f_n(\mathbb{D})$, and which are therefore unbounded. Each f_n has a Schwarz-Christoffel representation

$$f_n(z) = f_n'(0) \int_0^z \frac{1}{(1-\zeta_{n,1}t)^{\lambda_{n,1}}} \prod_{j=2}^{j_n} \frac{1}{(1-\zeta_{n,j}t)^{\lambda_{n,j}}} dt,$$

where $|\zeta_{n,j}| = 1$, $\lambda_{n,j} \in (0,2)$, for $j = 1, \ldots, j_n$, and $\sum_{j=1}^{j_n} \lambda_{n,j} = 2$. Furthermore we have chosen $\lambda_{n,1} = \max_{1 \le j \le j_n} \lambda_{n,j}$, so that the unboundedness of f_n forces $\lambda_{n,1} \ge 1$. Then we can write

$$zf_n'(z) = \frac{f_n'(0)}{1-\zeta_{n,1}z} g_n(z),$$

where

$$g_n(z) := \frac{z}{(1-\zeta_{n,1}z)^{\lambda_{n,1}-1}} \prod_{j=2}^{j_n} \frac{1}{(1-\zeta_{n,j}z)^{\lambda_{n,j}}},$$

and a simple verification shows that $g_n \in \mathcal{S}_{\frac{1}{2}}$. Without loss of generality we may assume that

$$\lim_{n\to\infty} \zeta_{n,1} =: \zeta, \qquad \lim_{n\to\infty} g_n =: g \in \mathcal{S}_{\frac{1}{2}},$$

exist, so that

$$zf_n'(z) \to \frac{g(z)}{1-\zeta z} = zf'(z).$$

If, on the other hand, zf' has the representation (4) one readily verifies that $zf' \in \mathcal{S}_0$ and therefore $f \in \mathcal{C}$. Without loss of generality we assume $\zeta = 1$ and obtain

$$\begin{aligned} M(r,f) := \max_{\phi\in\mathbb{R}} |f(re^{i\phi})| &= \max_{\phi\in\mathbb{R}} \left| \int_0^r f'(te^{i\phi})dt \right| \\ &\ge \operatorname{Re} \int_0^r f'(t)dt \\ &\ge \int_0^r \frac{1}{1-t} \operatorname{Re} \frac{g(t)}{t} dt \\ &\ge \int_0^r \frac{1}{1-t^2} dt \\ &= \frac{1}{2} \log\left(\frac{1+r}{1-r}\right), \end{aligned}$$

where we made use of the well-known relation

$$\operatorname{Re} \frac{g(z)}{z} > \frac{1}{2}, \quad g \in \mathcal{S}_{\frac{1}{2}},\ z \in \mathbb{D}.$$

Hence $f \in \mathcal{C}_u$.

For the proof of Theorem 4 we make use of an estimate for funtions with positive real part in $\mathbb{D}$. In spite of its simplicity we do not know of an explicit reference for it (cf. Ruscheweyh [11] or Hallenbeck-MacGregor [7] for a related result), so we include a proof.

Lemma 9 *Let* $H(z) = 1 + \sum_{k=1}^{\infty} a_k(H) z^k$ *be analytic in* $\mathbb{D}$, *satisfying* $\operatorname{Re} H(z) > 0,\ z \in \mathbb{D}$. *Then*

$$|a_{mn}(H) - 2| \leq m^2 \, |a_n(H) - 2| \, , \quad m, n \in \mathbb{N}. \tag{16}$$

These inequalities are best possible.

Proof. Let $g \in \mathcal{S}_0$. It has been shown by Hummel [8] that then the function $G(z) := (1-z)^2 \frac{g(z)}{z}$ is close-to-convex in $\mathbb{D}$. Since this function does not vanish in $\mathbb{D}$, it is clear that also $F(z) = \log G(z)$ is univalent in $\mathbb{D}$. Therefore, for the function

$$zF'(z) = \frac{zg'(z)}{g(z)} - \frac{1+z}{1-z} = \sum_{k=1}^{\infty} a_k(zF') z^k,$$

de Branges' Theorem (Bieberbach's conjecture) yields

$$|a_m(zF')| \leq m^2 |a_1(zF')|, \, m \in \mathbb{N}. \tag{17}$$

Now let $H(z)$ and assume $n > 1$. Then

$$H_n(z) := \frac{1}{n} \sum_{k=1}^{n} H(e^{2\pi i k/n} z^{1/n}) = 1 + \sum_{k=1}^{\infty} a_{kn}(H) z^k$$

is also analytic in $\mathbb{D}$, and has positive real part there. Hence we need to prove (16) only for $n = 1$. Now let $n = 1$ and H be given. We define g as the analytic solution of $H(z) = \frac{zg'(z)}{g(z)}$ with $g(0) = 0$, $g'(0) = 1$. Then $g \in \mathcal{S}_0$, and in the notation of above we have $a_k(zF') = a_k(H) - 2,\ k \in \mathbb{N}$, and (17) gives the asserted inequality. That this is sharp can be seen from the functions

$$H(z) := \frac{1 - z^{2n}}{1 - 2\cos(\phi) z^n + z^{2n}}, \quad \phi \to 0.$$

Proof. (Theorem 4) For $f(z) = \sum_{k=1}^{\infty} a_k(f) z^k$ we have, by Theorem 3,

$$\sum_{k=1}^{\infty} k a_k(f) z^k = z f'(z) = \frac{g(z)}{1 - \zeta z} = \sum_{k=1}^{\infty} \sum_{j=1}^{k} a_j(g) \overline{\zeta}^j \, (\zeta z)^k.$$

Thus, writing $\pi_j(z) := \sum_{k=1}^{j} a_h(g)z^k$, we need to prove

$$|\pi_{nm}(\overline{\zeta})| \leq m^2|\pi_n(\overline{\zeta})|, \quad |\zeta| = 1. \tag{18}$$

Now, because $g \in \mathcal{S}_{\frac{1}{2}}$, we know that for $\zeta \in \mathbb{D}$ the function

$$G_\zeta(z) := 2\frac{1}{g(\zeta)}\frac{g(z\zeta) - g(\zeta)}{z-1} - 1 = 1 + \sum_{k=1}^{\infty} a_k(G_\zeta)z^k$$

is analytic in $\mathbb{D}$ and satisfies $\operatorname{Re} G_\zeta > 0$ there. A simple calculation yields

$$a_j(G_\zeta) = 2 - 2\frac{\pi_j(\zeta)}{g(\zeta)},$$

and therefore, using Lemma 9, we obtain

$$\left|2\frac{\pi_{mn}(\zeta)}{g(\zeta)}\right| = |a_{mn}(G_\zeta) - 2| \leq m^2\,|a_n(G_\zeta) - 2| = m^2\left|2\frac{\pi_n(\zeta)}{g(\zeta)}\right|,$$

which, taking limits properly, gives (18).

To prove sharpness fix $n \geq 2$ and set

$$f_x(z) := \int_0^z \frac{dt}{(1-t)\sqrt{1 - 2xe^{i\phi(x)}t + e^{2i\phi(x)}t^2}} = \sum_{k=0}^{\infty} a_k(f_x)z^k, \quad 0 < x \leq 1,$$

with

$$\phi(x) := \frac{2\pi}{n} + \frac{1}{2}(x-1)\cot\frac{\pi}{n}.$$

We then have

$$ka_k(f_x) = \sum_{j=0}^{k-1} P_j(x)e^{ij\phi(x)},$$

where P_j denotes the Legendre polynomial of degree j. Let $m \in \mathbb{N}$. Since

$$\lim_{x\to 1} a_{mn}(f_x) = 0$$

we use l'Hospital's rule and

$$\left.\frac{d}{dx}a_{mn}(f_x)\right|_{x=1} = \frac{1}{mn}\sum_{j=0}^{mn-1}\left(P_j'(1) + \frac{1}{2}ij\cot\frac{\pi}{n}\right)e^{\frac{2\pi ij}{n}} = \frac{mn}{2 - 2e^{2\pi i/n}}.$$

to obtain

$$\lim_{x\to 1}\left|\frac{a_{mn}(f_x)}{a_n(f_x)}\right| = m,$$

so that f_x, for n fixed and $x \to 1$, serves as example to prove that the estimates given are best possible.

5 Proofs of Theorems 5-7

Proof. (Theorem 5) Let $f(z) = z + \sum_{k=2}^{\infty} a_k(f) z^k \in \mathcal{C}$, and

$$1 + \frac{zf''(z)}{f'(z)} = \frac{1+v(z)}{1-v(z)}, \tag{19}$$

where $v(z) = \sum_{k=1}^{\infty} a_k(v) z^k \in \mathcal{H}(\mathbb{D})$ with $|v(z)| \leq |z|$ in $\mathbb{D}$. Define numbers $\rho_j \in \mathbb{R}$ such that

$$\begin{aligned} |a_1(v)| &= \rho_1, \\ |a_2(v)| &= \rho_2(1-\rho_1^2), \\ \left| a_3(v) + \frac{\overline{a_1(v)} a_2^2(v)}{1-\rho_1^2} \right| &= \rho_3(1-\rho_1^2)(1-\rho_2^2). \end{aligned}$$

Then, by the Schwarz-Pick Lemma, we obtain $\rho_j \in [0,1]$, $j = 1,2,3$, and

$$|v(z)| \leq v_0(|z|), \quad z \in \mathbb{D}, \tag{20}$$

where

$$v_0(z) := z \frac{z\dfrac{z+\rho_2}{1+\rho_2 z} + \rho_1}{1 + \rho_1 z \dfrac{z+\rho_2}{1+\rho_2 z}}.$$

Using suitable values $\theta_j \in \mathbb{R}$ we find

$$\begin{aligned} a_1(v) &= \rho_1 e^{i\theta_1}, \\ a_2(v) &= \rho_2(1-\rho_1^2) e^{i\theta_2}, \\ a_3(v) &= -\rho_2^2 \rho_1 (1-\rho_1^2) e^{i(2\theta_2-\theta_1)} + \rho_3(1-\rho_1^2)(1-\rho_2^2) e^{i\theta_3}, \end{aligned}$$

and via (19),

$$\begin{aligned} a_2(f) &= a_1(v) = \rho_1 e^{i\theta_1}, \\ a_4(f) &= \frac{1}{6} a_3(v) + \frac{5}{6} a_2(v) a_1(v) + a_1(v)^3 \\ &= \frac{5}{6} \rho_2 \rho_1 (1-\rho_1^2) e^{i(\theta_1+\theta_2)} + \rho_1^3 e^{3i\theta_1} - \frac{1}{6} \rho_2^2 \rho_1 (1-\rho_1^2) e^{i(2\theta_2-\theta_1)} \\ &\quad + \frac{1}{6} \rho_3 (1-\rho_1^2)(1-\rho_2^2) e^{i\theta_3}. \end{aligned}$$

Note that the last relation implies, in particular,

$$|a_4(f)| \leq \rho_1 + \frac{1}{6}(1-\rho_2^2). \tag{21}$$

If we construct f_0 from v_0 in a similar fashion as f and v are related, then we obtain

$$f_0(z) = \int_0^z \frac{dt}{(1-t)^{2\lambda}(1+2ct+t^2)^{1-\lambda}} = z + \sum_{k=2}^{\infty} a_k(f_0)z^k \in \mathcal{C}, \qquad (22)$$

where

$$\begin{aligned} \lambda &= \frac{(1+\rho_1)(1+\rho_2)}{3+\rho_2-\rho_1+\rho_1\rho_2} \in [0,1], \\ c &= \frac{1}{2}(1+\rho_2-\rho_1+\rho_1\rho_2) \in [0,1]. \end{aligned}$$

In particular, we have

$$\begin{aligned} a_2(f_0) &= \rho_1, \\ a_4(f_0) &= \frac{5}{6}\rho_2\rho_1(1-\rho_1^2) + \rho_1^3 - \frac{1}{6}\rho_2^2\rho_1(1-\rho_1^2) + \frac{1}{6}(1-\rho_1^2)(1-\rho_2^2), \end{aligned}$$

and therefore

$$|a_4(f)| \le |a_4(f_0)| + \frac{1}{3}\rho_1. \qquad (23)$$

Furthermore, by (20), we see that

$$M = M(f) := \sup_{z\in\mathbb{D}} |f(z)| \le f_0(1) = M(f_0). \qquad (24)$$

We now turn to the actual proof of (7). Assume first that f_0 is unbounded. Then, by Theorem (4) and (23)

$$|a_4(f)| \le |a_4(f_0)| + \frac{1}{3}|a_2(f)| \le \frac{7}{3}|a_2(f_0)| \le \frac{7}{3}|a_2(f)| + \frac{2}{3M},$$

the assertion.

From now on assume that $f_0(1) < \infty$, which implies that $\lambda < \frac{1}{2}$. Furthermore we note that $\rho_1 = \lambda(1+c) - c$, and therefore

$$0 \le \frac{c}{1+c} \le \lambda < \frac{1}{2}. \qquad (25)$$

Using (21) we find

$$\begin{aligned} \frac{|a_4(f)| - \frac{7}{3}|a_2(f)|}{1-2\lambda} &\le \frac{1-\rho_2^2-8\rho_1}{6(1-2\lambda)} \\ &= \frac{4c - 4\lambda(1+c) + 2\dfrac{\lambda(1-c^2)}{(\lambda(1+c)+1-c)^2}}{3(1-2\lambda)} \\ &=: A(c,\lambda). \end{aligned}$$

We wish to show that

$$A(c,\lambda) \leq \frac{2}{3}c(1+c) \tag{26}$$

under the condition (25). Writing $u := \lambda(1+c)+1-c$ this turns out to be equivalent to

$$-2u^3 + (2-c)u^2 + u - (1-c) \leq 0, \quad 1 \leq u \leq \frac{3-c}{2},$$

which is readily verified.

Now let c, λ satisfy (25) and define

$$B(c,\lambda) := (1-2\lambda)f_0(1) = \int_0^1 \frac{(1-2\lambda)dt}{(1-t)^{2\lambda}(1+2ct+t^2)^{1-\lambda}}.$$

Using

$$\frac{c}{(1+2ct+t^2)^{1-\lambda}} \leq \frac{c}{c+t} \leq \frac{1}{1+t}$$

we obtain:

$$\begin{aligned}
B(c,\lambda) &\leq \frac{1}{c}\int_0^1 \frac{(1-2\lambda)dt}{(1-t)^{2\lambda}(1+t)} \\
&= \frac{1}{c}\left(1 - \int_0^1 (1-t)^{1-2\lambda}(1+t)^{-2}dt\right) \\
&\leq \frac{1}{c}\left(1 - \int_0^1 (1-t)^{\frac{1-c}{1+c}}(1+t)^{-2}dt\right) \\
&= \frac{1}{c}\left(1 - \frac{1+c}{2}F(1,2,2+\frac{1-c}{1+c},-1)\right) \\
&= \frac{1}{c}\left(1 - \frac{1+c}{4}F(1,1+\frac{1-c}{1+c},2+\frac{1-c}{1+c},\frac{1}{2})\right) \\
&\leq \frac{1}{c}(1 - \frac{1+c}{4}) \\
&\leq \frac{1}{c(1+c)},
\end{aligned}$$

where F denotes the hypergeometric function.

Putting things together we get

$$(|a_4(f)| - \frac{7}{3}|a_2(f)|)M(f) \leq A(c,\lambda)\cdot B(c,\lambda) \leq \frac{2}{3},$$

the asserted estimate.

We now turn to the discussion of the sharpness of the constants involved in (7). In fact, the sharpness of the factor $\frac{2}{3}$ will be obtained in the proof of Theorem 6. And this then implies the ***absolute*** sharpness of this constant in the following sense: whenever we have a general estimate

$$|a_4(f)| \leq C_1|a_2(f)| + \frac{C_2}{M(f)}, \quad f \in \mathcal{C}, \tag{27}$$

then $C_2 \geq \frac{2}{3}$.

The constant $\frac{7}{3}$ in (7), however, is not likely to be sharp. Instead, we make the following

Conjecture 1 *For $f \in \mathcal{C}$ we have*

$$|a_4(f)| \leq 2|a_2(f)| + \frac{2}{3M(f)}. \tag{28}$$

The constants 2 *and* $\frac{2}{3}$ *are sharp in the strong sense mentioned above.*

For the truth of this conjecture we have the following evidence. First, it is true and sharp for $f \in \mathcal{C}_u$ (Theorem 4). And secondly it is true and sharp for the functions $f_0 \in \mathcal{C}$ introduced in (22). This can be shown using exactly the same steps as in the proof of Theorem 5, just replacing (21) by

$$|a_4(f_0)| \leq \frac{2}{3}\rho_1 + \rho_1^3 + \frac{1}{6}(1-\rho_2^2).$$

It can then be shown that the (correspondingly adjusted) function

$$A^*(c,\lambda) := \frac{|a_4(f_0)| - 2|a_2(f_0)|}{1-2\lambda}$$

satisfies the same relation as A in (26). The functions f_0 map the unit disk onto elongated triangles symmetric to the real axis, and there is a certain chance that this is the extremal configuration for the problem under discussion.

Proof. (Theorem 6) The estimate (8) follows immediately from Theorem 5. It remains to prove sharpness. Using f_0 as in (22) with $|a_2(f_0)| = \rho_1 = 0$, so that

$$a_4(f_0) = \frac{1-\rho_2^2}{6}, \; \lambda = \frac{1+\rho_2}{3+\rho_2}, \; c = \frac{1+\rho_2}{2},$$

we obtain

$$\lim_{\rho_2 -> 1} a_4(f_0) f_0(1) = \frac{2}{3}.$$

Proof. (Theorem 7) We have

$$\left|\frac{a_4(f)f(z)}{1+a_2(f)f(z)}\right| = \frac{|a_4(f)f(z)|}{\frac{2}{7}+|a_2(f)f(z)|}\cdot\frac{\frac{2}{7}+|a_2(f)f(z)|}{|1+a_2(f)f(z)|} = Q_1\cdot Q_2,$$

say. Then $Q_1 \leq \frac{7}{3}$ by Theorem 5 and using $\mathrm{Re}\,(a_2(f)f(z)) > \frac{-1}{2}$ we find $Q_2 \leq \frac{11}{7}$. A discussion similar to the one at the end of the proof of Theorem 5, using proper sequences of functions f_0, shows that C cannot be smaller than 2.

We note that, as a consequence of a recent result of J. Ma [9], the extremal situation in Theorem 7 can be approximated through bounded convex functions mapping $\mathbb{D}$ onto polygons with at most 4 edges. But even this reduced problem seems to be difficult handling.

References

[1] R. M. Ali, A distortion theorem for a class of Moebius transformations of convex maps, Rocky Mountain J. Math. 19 (1989), 1083–1094.

[2] R. W. Barnard and G. Schober, Moebius transformations for convex mappings, Complex Variables 3 (1984), 55–69.

[3] R. W. Barnard and G. Schober, Moebius transformations for convex mappings II, Complex Variables 7 (1986), 205–214.

[4] M. Chuaqui and B. Osgood, Sharp distortion theorems associated with the Schwarzian derivative, J. London Mat. Soc. (2) 48 (1993), 289–295.

[5] R. R. Hall, On a conjecture of Clunie and Sheil-Small, Bull. London Math. Soc. 12 (1980), 25–28.

[6] R. R. Hall and St. Ruscheweyh, On transformations of functions with bounded boundary rotation, Indian J. Pure Appl. Math. 16 (1985), 1317–1325.

[7] D. J. Hallenbeck and T.H. MacGregor, Linear problems and convexity techniques in geometric function theory, Pitman, Boston, 1984.

[8] J. A. Hummel, Multivalent starlike functions, J. Analyse Math. 18 (1967), 133–160.

[9] J. Ma, Julia variations for extremal problems of convex univalent functions, Thesis, Würzburg 1997.

[10] V. Paatero, Über die konforme Abbildung von Gebieten, deren Ränder von beschränkter Drehung sind, Ann. Acad. Sci. Fenn., Ser. A. 33, 9 (1933), 78 p.

[11] St. Ruscheweyh, Convolutions in geometric function theory, Les Presses de l'Université de Montréal, 1982.

[12] T. B. Sheil-Small, Convolutions of convex functions, J. London Math. Soc. 2 (1969), 483–492.

[13] T. J. Suffridge, Some remarks on convex maps of the unit disc, Duke Math. J. 37 (1970), 775–777.

[14] O. Toeplitz, Die linearen vollkommenen Räume der Funktionentheorie, Comment. Math. Helv. 23 (1949), 222-242.

Total Positivity and Total Variation

T. N. T. Goodman
Department of Mathematics and Computer Science, University of Dundee
Dundee DD1 4HN, Scotland

Abstract

It is shown that applying a totally positive matrix decreases the total angle turned through by a sequence of vectors but, under certain restrictions, it increases the total angle turned through by the vector products of consecutive pairs of the vectors. This implies results for curves constructed either from totally positive bases or by totally positive subdivision matrices. Essentially this says that the total angle turned through by the curve is bounded by the total angle turned through by its control polygon whereas, under certain restrictions, the total angle turned through by its binormal bounds that for the control polygon.

1 Introduction

The variation diminishing property states that if T is a totally positive banded matrix and v is a vector for which Tv is defined, then the number of sign changes in Tv is bounded by the number in v. Some history of this result is given in [4]. The variation diminishing property has useful applications in approximation theory and computer aided geometric design, see [3], and it was these applications which led to a generalisation of the property in [2], where v is replaced by a more general matrix. To illustrate some applications of this generalisation, suppose that $(\phi_1, \ldots, \phi_n)$ form a partition of unity on an interval and have totally positive collocation matrices. It is shown in [2] that, under certain conditions, if $r = \sum_{i=1}^{n} P_i\phi_i$ for $P_1, \ldots, P_n$ in R^2, then the number of inflections in the curve r is bounded by the number of inflections in the polygonal arc $P_1, \ldots, P_n$, while if $P_1, \ldots, P_n$ are in R^3, then the number of changes in sign of the torsion of r is bounded by that for $P_1, \ldots, P_n$.

A property related to the variation diminishing property concerns the total variation $\sum_{j=2}^{n} |a_j - a_{j-1}|$ of a sequence $(a_1, \ldots, a_n)$. If T and v are as before and T is stochastic, then the total variation of Tv is bounded by that of v. It follows that if $(\phi_1, \ldots, \phi_n)$ are as above and $f = \sum_{i=1}^{n} a_i \phi_i$ for $a_1, \ldots, a_n$ in R, then the variation of the function f is bounded by the total variation of the sequence $(a_1, \ldots, a_n)$, a result first shown by Schoenberg in [5].

The above considerations motivated us to attempt to extend results on the diminution of total variation in a similar manner to the extension in [2] of the variation diminishing property. However the situation turned out to be surprisingly different, as we show in this paper.

In Section 2 we make some basic definitions and prove some simple preliminary results. Then in Section 3 we prove two results on diminishing the total angle turned through by a sequence of vectors. The latter result shows that if $(\phi_1, \ldots, \phi_n)$ are as above and $r = \sum_{i=1}^{n} P_i \phi_i$ for $P_1, \ldots, P_n$ in R^d, $d \geq 2$, then the total angle turned through by the curve r is bounded by that turned through by the polygonal arc $P_1 \ldots P_n$. (This was stated but not proved in [3].) We note that for a smooth enough curve, the total angle it turns through equals the integral with respect to arc length of the magnitude of its curvature. In Section 4 we investigate related results when the curvature is replaced by torsion, i.e. we consider the angle turned through by the binormal to the curve, which we could think of as the total "twist" of the curve. Surprisingly we show that, under conditions similar to those in [2], the total twist is *increased* by the application of a totally positive matrix, so giving a *variation enlarging property*.

2 Preliminaries

We recall that a matrix is *totally positive* (TP) if all its minors are non-negative. A matrix is *one-banded* if its only non-zero elements lie in two consecutive diagonals and it is *stochastic* if the elements in each row sum to one. It is known (see [1], [3, Theorem 2.3]) that a finite (stochastic) matrix is TP if and only if it is a product of one-banded, non-negative (stochastic) matrices.

Now take $p \geq 1$. Following [2, Definitions 2.1 and 2.5], we say that a matrix is *p-restricted* if any consecutive rows of rank p vanish outside some p consecutive columns, while it is *strongly p-restricted* if any p consecutive rows are linearly independent and vanish outside some p consecutive columns. For strongly p-restricted TP matrices, [2] gives a stronger factorisation result in terms of the following types of one-banded non-negative matrices.

Type 1 A square non-singular one-banded non-negative matrix with at most one non-zero element not on the main diagonal.

Type 2 A matrix $T = (T_{ij})_{i=1}^{q+1}{}_{j=1}^{q}$ where for some $p \le s \le q$,

$$\begin{aligned} T_{jj} &> 0, \quad j = 1, \ldots, s \\ T_{j+1,j} &> 0, \quad j = s - p + 1, \ldots, q, \end{aligned}$$

and all other elements are zero.

It is shown in Theorem 2.2 of [2] that any $m \times n$ TP matrix of rank n which is strongly p-restricted for some p, $1 \le p \le n$, is a product of matrices of Type 1 or Type 2.

Now let Δ_n denote the $(n-1) \times n$ matrix given by

$$(\Delta_n v)_i = v_{i+1} - v_i, \quad i = 1, \ldots, n-1.$$

For any stochastic $m \times n$ matrix T, $m, n \ge 2$, it is easily seen that there is a unique $(m-1) \times (n-1)$ matrix S such that

$$\Delta_m T = S \Delta_n. \tag{1}$$

Indeed applying (1) to $v = (v_i)_{i=1}^n$, where $v_i = 0$, $i = 1, \ldots, j$, $v_i = 1$, $i = j+1, \ldots, n$, gives

$$\begin{aligned} S_{ij} &= \sum_{k=j+1}^{n} (T_{i+1,k} - T_{ik}) \\ &= \sum_{k=1}^{j} (T_{i,k} - T_{i+1,k}), \quad i = 1, \ldots, m-1, \quad j = 1, \ldots, n-1, \end{aligned} \tag{2}$$

on recalling that $\sum_{k=1}^n T_{ik} = 1$, $i = 1, \ldots, m$.

From (2) we see that if rows $i, \ldots, i+q-1$ of T vanish outside columns $j, \ldots, j+p-1$, then rows $i, \ldots, i+q-2$ of S vanish outside columns $j, \ldots, j+p-2$. Now take $1 \le i < i+p-1 \le m$. From (1) we have

$$\Delta_p(T[i, \ldots, i+p-1|1, \ldots, n]) = (S[i, \ldots, i+p-2|1, \ldots, n-1])\Delta_n, \tag{3}$$

where $T[i, \ldots, i+p-1|1, \ldots, n]$ denotes the submatrix of T comprising rows $i, \ldots, i+p-1$ and columns $1, \ldots, n$. If rows $i, \ldots, i+p-1$ of T have rank r, then the left-hand side of (3) has rank $\ge r-1$. Since the right-hand side of (3) then has rank $\ge r-1$, rows $i, \ldots, i+p-2$ of S must have rank $\ge r-1$.

Lemma 1 *If T is p-restricted (respectively strongly p-restricted), then S is $(p-1)$-restricted (respectively strongly $(p-1)$-restricted).*

Proof. Suppose that T is strongly p-restricted and consider rows $i, \ldots, i+p-2$ of S. Since rows $i, \ldots, i+p-1$ of T have rank p, it follows from our above remark that rows $i, \ldots, i+p-2$ of S have rank $p-1$. Moreover since rows $i, \ldots, i+p-1$ of T vanish outside some columns $j, \ldots, j+p-1$, rows $i, \ldots, i+p-2$ of S vanish outside columns $j, \ldots, j+p-2$. Thus S is strongly $(p-1)$-restricted.

Next suppose that T is p-restricted and suppose that rows $i, \ldots, i+q-1$ of S have rank $(p-1)$. By our above remark, rows $i, \ldots, i+q$ of T have rank $\leq p$ and so vanish outside some columns $j, \ldots, j+p-1$. Thus rows $i, \ldots, i+q-1$ of S vanish outside columns $j, \ldots, j+p-2$, and so S is $(p-1)$-restricted.

Lemma 2 *If T is a TP matrix, then S is also TP.*

Proof. Suppose that $T = T_1 \ldots T_r$, where $T_1, \ldots, T_r$ are one-banded, non-negative, stochastic matrices. Let $S_1, \ldots, S_r$ be the corresponding matrices as in (1). Then successive application of (1) gives $\Delta_m T_1 \ldots T_r = S_1 \ldots S_r \Delta_n$ and so $S = S_1 \ldots S_r$. Thus it is sufficient to prove that S in (1) is TP when T is one-banded, non-negative, stochastic. In this case we may assume, with suitable indexing, that for each i,

$$T_{ii} = \lambda_i, \quad T_{i,i+1} = 1 - \lambda_i,$$

for $0 \leq \lambda_i \leq 1$, and $T_{ij} = 0$ otherwise. Then for each i and suitable vector v,

$$\begin{aligned}
(\Delta_m T v)_i &= (Tv)_{i+1} - (Tv)_i \\
&= \lambda_{i+1} v_{i+1} + (1 - \lambda_{i+1}) v_{i+2} - \lambda_i v_i - (1 - \lambda_i) v_{i+1} \\
&= \lambda_i (v_{i+1} - v_i) + (1 - \lambda_{i+1})(v_{i+2} - v_{i+1}) \\
&= \lambda_i (\Delta_n v)_i + (1 - \lambda_{i+1})(\Delta_n v)_{i+1}.
\end{aligned}$$

Comparing with (1) shows that S is a one-banded, non-negative matrix, and the proof is complete.

3 Variation Diminution

For non-zero vectors u, v in R^d, $d \geq 2$, we let $a(u, v)$ denote the angle between u and v, where $0 \leq a(u, v) \leq \pi$, i.e.

$$a(u, v) = \cos^{-1} \left(\frac{uv}{|u||v|} \right).$$

We shall need the following "triangle inequality," which is straightforward to prove.

Lemma 3 *Any non-zero vectors u, v, w in R^d, $d \geq 2$, satisfy $a(u,w) \leq a(u,v) + a(v,w)$ with equality if and only if $v = \lambda v + \mu w$ for some $\lambda, \mu \geq 0$.*

Now let $v = (v_1, \ldots, v_n)^T$, where $v_1, \ldots, v_n \in R^d$, $d \geq 2$. Letting $\tilde{v}_1, \ldots, \tilde{v}_m$ denote the non-zero elements of v, in order, we define

$$A(v) = \sum_{i=1}^{m-1} a(\tilde{v}_i, \tilde{v}_{i+1}).$$

Theorem 4 *If T is a $m \times n$ TP matrix and $v = (v_1, \ldots, v_n)^T$, $v_1, \ldots, v_n \in R^d$, $d \geq 2$, then*

$$A(Tv) \leq A(v). \tag{4}$$

Proof. Since non-zero entries are omitted in calculating $A(v)$, we remove zero entries of v and the corresponding column of T. Since T is a product of one-banded non-negative matrices, it is sufficient to prove (4) when T is such a matrix. Since zero entries are omitted in calculating $A(Tv)$ we may remove any zero rows from T. The effect of T is to successively insert positive combinations of two consecutive elements of v and to remove elements of v. To be precise we can factor T into strictly positive diagonal matrices and matrices T of the following forms.
(A) T is a $(q+1) \times q$ matrix, where for some r, $2 \leq r \leq q$,

$$\begin{aligned} (Tv)_i &= v_i, \quad i = 1, \ldots, r-1, \\ (Tv)_r &= \lambda v_{r-1} + \mu v_r, \quad \lambda, \mu \geq 0, \\ (Tv)_i &= v_{i-1}, \quad i = r+1, \ldots, q+1. \end{aligned}$$

(B) T is a $q \times (q+1)$ matrix, where for some r, $1 \leq r \leq q+1$,

$$\begin{aligned} (Tv)_i &= v_i, \quad i = 1, \ldots, r-1, \\ (Tv)_i &= v_{i+1}, \quad i = r, \ldots, q. \end{aligned}$$

If T is a strictly positive diagonal matrix, then $A(Tv) = A(v)$. In Case (A), when $(Tv)_r = 0$, then clearly $A(Tv) = A(v)$. In Case (A), when $(Tv)_r \neq 0$, then $A(Tv) = A(v)$ by Lemma 3. In Case 2, $A(Tv) \leq A(v)$ by Lemma 3.

Theorem 5 *If T is a $m \times n$ TP matrix, $m, n \geq 2$, and $v = (v_1, \ldots, v_n)^T$, $v_1, \ldots, v_n \in R^d$, $d \geq 2$, then*

$$A(\Delta_m T v) \leq A(\Delta_n v).$$

Proof. Let S be the $(m-1) \times (n-1)$ matrix as in (1). By Lemma 2, S is TP. So by (1) and Theorem 4,

$$A(\Delta_m Tv) = A(S\Delta_n v) \leq A(\Delta_n v).$$

Theorems 4 and 5 can be applied to give variation diminishing properties of curves by taking T to be either a collocation matrix or a subdivision matrix, in a similar manner to the final sections of [2] and [3]. Here we shall only consider briefly the case of Theorem 5 and collocation matrices.

For any curve $r : [a, b] \to R^d$, $d \geq 2$, we define $\theta(r)$ to be $\sup A(r(t_2) - r(t_1), \ldots, r(t_m) - r(t_{m-1}))$, where the supremum is taken over all increasing sequences $(t_1, \ldots, t_m)$ in $[a, b]$, all $m \geq 3$. Thus $\theta(r)$ can be considered as the angle turned through by r. Now take functions $(\phi_1, \ldots, \phi_n)$ on $[a, b]$ with $\sum_{i=1}^n \phi_i = 1$ and suppose that for any points $t_1 < \ldots < t_m$ in $[a, b]$, the collocation matrix $(\phi_j(t_i))_{i=1}^m{}_{j=1}^n$ is TP. (For examples of such functions, e.g. B-splines, see [3].) Then it follows easily from Theorem 5 that if

$$r = \sum_{i=1}^n P_i \phi_i, \quad P_i \in R^d, \quad d \geq 2, \quad i = 1, \ldots, n,$$

then

$$\theta(r) \leq A(P_2 - P_1, \ldots, P_n - P_{n-1}).$$

Essentially this says that the angle turned through by the curve r is bounded by the angle turned through by the polygonal arc $P_1 \ldots P_n$.

4 Variation Enlargement

We shall consider $n \times 3$ matrices A which satisfy the following property for some $3 \leq p \leq n$.

For any $1 \leq i < j < k \leq n$ with $k - i \leq p - 1$,

$$\det A[i, j, k|1, 2, 3] > 0. \tag{5}$$

Lemma 6 *Take $3 \leq p \leq n$ and suppose that A is a $n \times 3$ matrix satisfying (5) while T is a strongly p-restricted TP $m \times n$ matrix, $m \geq p$. Then TA satisfies (5).*

Proof. Choose $1 \leq i < j < k \leq m$ with $k - i \leq p - 1$. Choose l with $1 \leq l \leq i < k \leq l + p - 1 \leq m$. By the definition of strongly p-restricted, rows $l, \ldots, l + p - 1$ of T are linearly independent and vanish outside some p columns, say $r, \ldots, r + p - 1$. Then

$$\det TA[i, j, k|1, 2, 3] =$$
$$\textstyle\sum_{r \leq \alpha < \beta < \gamma \leq r+p-1} \det T[i, j, k|\alpha, \beta, \gamma] \det A[\alpha, \beta, \gamma|1, 2, 3].$$

By (5), each term $\det A[\alpha, \beta, \gamma|1, 2, 3]$ in this summation is > 0. Since T is TP, each term $\det T[i, j, k|\alpha, \beta, \gamma]$ in the summation is ≥ 0. Moreover since rows i, j, k of T are linearly independent, there is some choice of α, β, γ, $r \leq \alpha < \beta < \gamma \leq r + p - 1$, with $\det T[i, j, k|\alpha, \beta, \gamma] > 0$. Thus the summation is > 0, which shows that TA satisfies (5).

Lemma 7 *Take* u, v, w, z *in* R^3 *and suppose that* $\det(u, v, w)$, $\det(v, w, z)$, $\det(u, v, z) > 0$. *Then for any* $\lambda, \mu > 0$ *there exist* $\alpha, \beta > 0$ *so that*

$$v \times w = \alpha(u \times v) + \beta(v \times (\lambda w + \mu z)). \tag{6}$$

Proof. Choose $k > 0$ with

$$\det(u, v, w) = k \det(v, w, z) = \det(kz, v, w).$$

Then $\det(u - kz, v, w) = 0$ and so we may choose c with

$$v \times w = c((u - kz) \times v) = c(u \times v) + ck(v \times z). \tag{7}$$

Now choose $K > 0$ with $\det(u, v, w) = K \det(u, v, z)$. Then $\det(u, v, w - Kz) = 0$ and so we may choose C with

$$v \times (w - Kz) = C(u \times v)$$

and so

$$v \times w = C(u \times v) + K(v \times z). \tag{8}$$

Since $\det(u, v, z) \neq 0$, the expressions (7) and (8) must coincide. So $ck = K > 0$ and since $k > 0$, we must have $c > 0$, i.e. $C > 0$. Now (8) is equivalent to

$$v \times w = \frac{C\mu}{\mu + K\lambda}(u \times v) + \frac{K}{\mu + K\lambda}(v \times (\lambda w + \mu z)),$$

which is of form (6).

Now for $v_1, \ldots, v_n \in R^3$ and $A = (v_1, \ldots, v_n)^T$, we define

$$V(A) = \sum_{i=2}^{n-1} a(v_{i-1} \times v_i, v_i \times v_{i+1}).$$

Lemma 8 *Suppose that* A *is a* $n \times 3$ *matrix which satisfies (5) for some* $4 \leq p \leq n$. *Suppose that* T *is a* $n \times n$ *matrix of Type 1 or a* $(n+1) \times n$ *matrix of Type 2. Then*

$$V(TA) \geq V(A). \tag{9}$$

Proof. Let $A = (v_1, \ldots, v_n)^T$ and $TA = (v'_1, \ldots, v'_m)^T$ for $m = n$ or $n+1$. We note that by (5),

$$\det(v_i, v_j, v_k) > 0 \tag{10}$$

for $1 \le i < j < k \le n$ with $k - i \le 3$.

First suppose that T is upper triangular of Type 1. Then for some $1 \le l \le n-1$,

$$\begin{aligned} v'_l &= \alpha_l v_l + \beta_l v_{l+1}, \\ v'_i &= \alpha_i v_i, \quad 1 \le i \le n, \quad i \ne l, \end{aligned} \tag{11}$$

for some $\beta_l > 0$, $\alpha_i > 0$, $i = 1, \ldots, n$. Recalling (10) we can apply Lemma 7 to $v_{l-2}, \ldots, v_{l+1}$, to give

$$v_{l-1} \times v_l = \alpha(v_{l-2} \times v_{l-1}) + \beta(v_{l-1} \times v'_l), \tag{12}$$

for some $\alpha, \beta > 0$. Now

$$\begin{aligned} & a(v_{l-2} \times v_{l-1}, v_{l-1} \times v_l) + a(v_{l-1} \times v_l, v_l \times v_{l+1}) \\ \le\; & a(v_{l-2} \times v_{l-1}, v_{l-1} \times v_l) + a(v_{l-1} \times v_l, v_{l-1} \times v'_l) \\ & + a(v_{l-1} \times v'_l, v_l \times v_{l+1}) \\ =\; & a(v_{l-2} \times v_{l-1}, v_{l-1} \times v'_l) + a(v_{l-1} \times v'_l, v_l \times v_{l+1}), \end{aligned}$$

by (12). Recalling (11) this becomes

$$\begin{aligned} & a(v_{l-2} \times v_{l-1}, v_{l-1} \times v_l) + a(v_{l-1} \times v_l, v_l \times v_{l+1}) \le \\ & a(v'_{l-2} \times v'_{l-1}, v'_{l-1} \times v'_l) + a(v'_{l-1} \times v'_l, v'_l \times v'_{l+1}). \end{aligned}$$

For $1 \le i \le l-2$ and $l+1 \le i \le n$ we have from (11),

$$a(v_{i-1} \times v_i, v_i \times v_{i+1}) = a(v'_{i-1} \times v'_i, v'_i \times v'_{i+1})$$

and so (9) holds. Similarly we can derive (9) when T is lower triangular of Type 1.

Now suppose that T is of Type 2. So for some l, $2 \le l \le n-p-1$,

$$\begin{aligned} v'_i &= \beta_i v_i, \quad i = 1 \ldots, l-1, \\ v'_i &= \alpha_i v_{i-1} + \beta_i v_i, \quad i = l, \ldots, l+p-2, \\ v'_i &= \alpha_i v_{i-1}, \quad i = l+p-1, \ldots, n+1, \end{aligned}$$

for $\beta_i > 0$, $i = 1, \ldots, l+p-2$, $\alpha_i > 0$, $i = l, \ldots, n+1$. Now

$$\begin{aligned} & a(v'_{i-1} \times v'_i, v'_i \times v'_{i+1}) \\ & = a(v_{i-1} \times v_i, v_i \times v_{i+1}), \quad i = 1, \ldots, l-1, \\ & = a(v_{i-2} \times v_{i-1}, v_{i-1} \times v_i), \quad i = l+p-1, \ldots, n+1. \end{aligned}$$

Thus it suffices to show

$$\sum_{i=l}^{l+p-3} a(v_{i-1} \times v_i, v_i \times v_{i+1}) \leq \sum_{i=l}^{l+p-2} a(v'_{i-1} \times v'_i, v'_i \times v'_{i+1}). \tag{13}$$

We shall prove by induction that for $r = l, \ldots, l+p-3$,

$$\sum_{i=l}^{r} a(v_{i-1} \times v_i, v_i \times v_{i+1}) \leq \\ \sum_{i=l}^{r} a(v'_{i-1} \times v'_i, v'_i \times v'_{i+1}) + a(v'_r \times v'_{r+1}, v'_{r+1} \times v_{r+1}). \tag{14}$$

Now

$$\begin{aligned} & a(v_{l-1} \times v_l, v_l \times v_{l+1}) \\ & = a(v'_{l-1} \times v'_l, v'_{l+1} \times v_{l+1}) \\ & \leq a(v'_{l-1} \times v'_l, v'_l \times v'_{l+1}) + a(v'_l \times v'_{l+1}, v'_{l+1} \times v_{l+1}), \end{aligned}$$

and so (14) holds for $r = l$. Next suppose that (14) holds for some r, $l \leq r \leq l+p-4$. We first note that by (10),

$$\det(v'_r, v'_{r+1}, v_{r+1}), \det(v'_{r+1}, v_{r+1}, v_{r+2}), \det(v'_r, v'_{r+1}, v_{r+2}) > 0,$$

and so we can apply Lemma 7 to $v'_r, v'_{r+1}, v_{r+1}, v_{r+2}$ to give

$$v'_{r+1} \times v_{r+1} = \alpha(v'_r \times v'_{r+1}) + \beta(v'_{r+1} \times v'_{r+2}), \tag{15}$$

for some $\alpha, \beta > 0$. Now

$$\begin{aligned} & a(v'_r \times v'_{r+1}, v'_{r+1} \times v_{r+1}) + a(v_r \times v_{r+1}, v_{r+1} \times v_{r+2}) \\ = \; & a(v'_r \times v'_{r+1}, v'_{r+1} \times v_{r+1}) + a(v'_{r+1} \times v_{r+1}, v'_{r+2} \times v_{r+2}) \\ \leq \; & a(v'_r \times v'_{r+1}, v'_{r+1} \times v_{r+1}) + a(v'_{r+1} \times v_{r+1}, v'_{r+1} \times v'_{r+2}) \\ & + a(v'_{r+1} \times v'_{r+2}, v'_{r+2} \times v_{r+2}) \\ = \; & a(v'_r \times v'_{r+1}, v'_{r+1} \times v'_{r+2}) + a(v'_{r+1} \times v'_{r+2}, v'_{r+2} \times v_{r+2}), \end{aligned}$$

by (15). So (14) holds with r replaced by $r+1$ which completes the proof by induction.

Thus (14) holds for $r = l+p-3$. Noting that

$$a(v'_{l+p-3} \times v'_{l+p-2}, v'_{l+p-2} \times v_{l+p-2}) = a(v'_{l+p-3} \times v'_{l+p-2}, v'_{l+p-2} \times v'_{l+p-1}),$$

we deduce (13), which completes the proof.

Theorem 9 *Suppose that T is a TP $m \times n$ matrix of rank n which is strongly p-restricted for some p, $4 \leq p \leq n$. Suppose that A is a $n \times 3$ matrix satisfying (5). Then*

$$V(TA) \leq V(A).$$

Proof. Since T can be factored into a product of matrices of Types 1 and 2, the result follows by successive application of Lemmas 6 and 8.

Theorem 10 *Suppose that T and A are as in Theorem 9 except that T is merely p-restricted. If any three consecutive rows of T are linearly independent, then*

$$V(TA) \geq V(A).$$

Proof. Take any three consecutive rows of T. Since $p \geq 4$, they have rank $< p$ and since T is p-restricted, they must vanish outside some p consecutive columns. Following the same proof as that of Lemma 6, we can see that the corresponding rows of TA are linearly independent. Thus $V(TA)$ is well- defined and depends continuously on the entries of T. The result then follows from Theorem 9 and Theorem 2.9 of [2].

Theorem 11 *Suppose that T is a stochastic, TP, $m \times n$ matrix of rank n which is p-restricted for some p, $5 \leq p \leq n$, and any four consecutive rows of T are linearly independent. Suppose that A is a $n \times 3$ matrix such that $\Delta_n A$ satisfies (5). Then*

$$V(\Delta_m TA) \geq V(\Delta_n A).$$

Proof. Let S be the $(m-1) \times (n-1)$ matrix as in (1). By Lemma 1 S is $(p-1)$-restricted and by Lemma 2 it is TP. By the remarks before Lemma 1, S has rank $(n-1)$ and any three consecutive rows of S are linearly independent. So we may apply Theorem 10 to S and $\Delta_n A$ to give, by (1),

$$V(\Delta_m TA) = V(S\Delta_n A) \geq V(\Delta_n A).$$

In a similar manner to the end of Section 3, we can apply Theorems 10 and 11 to give variation enlarging properties of curves, by taking T to be a collocation matrix or a subdivision matrix. We shall only consider briefly the case of Theorem 11 and collocation matrices.

For any curve $r : [a, b] \to R^3$ we define $\tau(r)$ to be $\sup V(\Delta_m A)$ where the supremum is taken over all matrices $A = [r(t_1), r(t_2), \ldots, r(t_m)]^T$ for increasing sequences $(t_1, \ldots, t_m)$ in $[a, b]$, all $m \geq 4$. Thus $\tau(r)$ can be considered as the angle turned through by the binormal to the curve r.

Now take functions $(\phi_1, \ldots, \phi_n)$ on $[a, b]$ with $\sum_{i=1}^{n} \phi_i = 1$ and suppose that for sufficiently dense points $t_1 < \ldots < t_m$ in $[a, b]$, the collocation matrix $(\phi_j(t_i))_{i=1}^{m}{}_{j=1}^{n}$ satisfies the conditions of Theorem 10. As an example we could take B-splines of degree $\geq p-2$, see Proposition 4.1 of [2]. Then it follows easily from Theorem 10 that if

$$r = \sum_{i=1}^{n} P_i \phi_i, \quad P_i \in R^3, \quad i = 1, \ldots, n,$$

and the matrix $[P_2 - P_1, \ldots, P_n - P_{n-1}]^T$ satisfies (5), then

$$\tau(r) \geq V([P_2 - P_1, \ldots, P_n - P_{n-1}]^T).$$

Essentially this says that the angle turned through by the binormal to the curve r is greater than that turned through by the binormal to the polygonal arc $P_1 \ldots P_n$.

References

[1] C. deBoor and A. Pinkus, The approximation of a totally positive band matrix by a strictly banded totally positive one, Linear Algebra Appl. 42 (1982), 81-98.

[2] J. M. Carnicer, T. N. T. Goodman and J. M. Pena, A generalisation of the variation diminishing property, Adv. Comp. Math. 3 (1995), 375-394.

[3] T. N. T. Goodman, Total positivity and the shape of curves, in "Total Positivity and its Applications" (M. Gasca and C. A. Micchelli, Eds.), pp. 157-186, Kluwer, Dordrecht, 1996.

[4] A. Pinkus, Spectral properties of totally positive kernels and matrices, in "Total Positivity and its Applications" (M. Gasca and C. A. Micchelli, Eds.), pp. 477-511, Kluwer, Dordrecht, 1996.

[5] I. J. Schoenberg, On variation diminishing approximation methods, in "On Numerical Approximation" (R. E. Langer, Ed.), pp. 249-274, University of Wisconsin Press, Madison, 1959.

Inequalities for Maximum Modulus of Rational Functions with Prescribed Poles

N. K. Govil
Department of Mathematics
Auburn University, Auburn, Al 36849

R. N. Mohapatra
Department of Mathematics
University of Central Florida, Orlando, Fl 32816

Abstract

Let $\mathcal{P}_n$ be the set of all complex algebraic polynomials of degree at most n with $\|p\| = \max\limits_{|z|=1} |p(z)|$, and let $\mathcal{R}_n = \mathcal{R}_n(a_1, a_2, \dots, a_n) := \{p(z)/ \prod\limits_{\nu=1}^{n} (z - a_\nu) : p \in \mathcal{P}_n\}$. It is well known that if $p \in \mathcal{P}_n$ then $\max\limits_{|z|=R\geq 1} |p(z)| \leq R^n \|p\|$, and if further $p(z) \neq 0$ for $|z| < 1$ then $\max\limits_{|z|=R\geq 1} |p(z)| \leq \frac{(R^n+1)}{2} \|p\|$. Here we obtain inequalities analogous to these inequalities for rational functions $r \in \mathcal{R}_n$ with $|a_\nu| > 1$ for $\nu = 1, \dots, n$.

1 Introduction and Statement of Results

Let $\mathcal{P}_n$ denote the set of all complex algebraic polynomials $p(z)$ of degree at most n and let $p'(z)$ be the derivative of $p(z)$. For a function f defined on the unit circle $\mathbb{T} := \{z : |z| = 1\}$ in the complex plane $\mathbb{C}$, set $\|f\| = \sup\limits_{z \in \mathbb{T}} |f(z)|$, the Chebyshev norm of f on $\mathbb{T}$.

Let $\mathbb{D}_-$ denote the region inside $\mathbb{T}$, and $\mathbb{D}_+$ the region outside $\mathbb{T}$. For $a_\nu \in \mathbb{C}$, $\nu = 1, 2, \dots, n$, let $w(z) = \prod\limits_{j=1}^{n} (z - a_v)$,

$$B(z) = \prod_{v=1}^{n} (\frac{1 - \bar{a}_v z}{z - a_\nu}),$$

and $\mathcal{R}_n = \mathcal{R}_n(a_1, a_2, \ldots, a_n) := \{\frac{p(z)}{w(z)} : p \in \mathcal{P}_n\}$. Then $\mathcal{R}_n$ is the set of rational functions with possible poles at $a_1, a_2, \ldots, a_n$ and having a finite limit at ∞. Also note that $B(z) \in \mathcal{R}_n$.

Definitions *(i) For $p(z) = \sum_{\nu=0}^{n} b_\nu z^\nu$, the conjugate transpose (reciprocal) p^* of p is defined by*

$$p^*(z) = z^n \overline{\{p(\frac{1}{\bar{z}})\}} = \bar{b}_0 z^n + \bar{b}_1 z^{n-1} + \cdots + \bar{b}_n.$$

(ii) For $r(z) = p(z)/w(z) \in \mathcal{R}_n$, the conjugate transpose, r^*, of r is defined by

$$r^*(z) = B(z)\overline{\{r(\frac{1}{\bar{z}})\}}.$$

(iii) The polynomial $p \in \mathcal{P}_n$ is called self-inversive, if $p^*(z) = \lambda p(z)$ for some $\lambda \in \mathbb{T}$.

(iv) The rational function $r \in \mathcal{R}_n$ is called self-inversive if $r^*(z) = \lambda r(z)$ for some $\lambda \in \mathbb{T}$.

Note that if $r \in \mathcal{R}_n$ and $r = p/w$, then $r^* = p^*/w$ and hence $r^* \in \mathcal{R}_n$. So $r = p/w$ is self-inversive if and only if p is self-inversive.

For a polynomial $p \in \mathcal{P}_n$, the following results are well-known:

Theorem A *If $p \in \mathcal{P}_n$, then*

$$||p'|| \leq n||p|| \tag{1}$$

and

$$\max_{|z|=R\geq 1} |p(z)| \leq R^n ||p||. \tag{2}$$

Both the above inequalities are sharp and become equality for $p(z) = \lambda z^n$, $\lambda \in \mathbb{C}$.

The inequality (1) is the famous Bernstein inequality and (2) is an immediate consequence of the maximum modulus principle.

For polynomials $p \in \mathcal{P}_n$ and $p(z) \neq 0$ in $|z| < 1$, we have

Theorem B *If $p \in \mathcal{P}_n$ and has all its zeros in $\mathbb{T} \cup \mathbb{D}_+$, then*

$$||p'|| \leq \frac{n}{2}||p|| \tag{3}$$

and

$$\max_{|z|=R\geq 1} |p(z)| \leq \frac{(R^n+1)}{2}||p||. \tag{4}$$

Both the above inequalities are best possible and become equality for $p(z) = \alpha z^n + \beta$, $|\alpha| = |\beta|$.

As is well known, the inequality (3) was conjectured by Erdös and proved by Lax [2] and the inequality (4) is due to Ankeny and Rivlin [1].

Bernstein type inequalities for rational functions have appeared in the study of rational approximation problems by Petrushev and Popov [6]. These inequalities contain some constants which are not optimal. Recently Borwein, Erdélyi and Zhang [3] have obtained Bernstein-Markov type inequalities for real rational functions for both algebraic and trigonometric polynomials on a finite interval. Borwein and Erdélyi [2] have obtained Bernstein inequalities for rational spaces of complex algebraic polynomials. Their main result is

Theorem C *If* $z \in \mathbb{T}$, *and* $a_j \in \mathbb{C}\backslash\mathbb{T}$, $j = 1, 2, \ldots, n$, *then for* $r \in \mathcal{R}_n$

$$|r'(z)| \leq \max\left\{\sum_{|a_j|>1} \frac{|a_j|^2-1}{|a_j-z|^2}, \sum_{|a_j|<1} \frac{1-|a_j|^2}{|a_j-z|^2}\right\} ||r||.$$

The above inequality is sharp.

The main point of the above theorem is that the poles are not necessarily restricted to be inside $\mathbb{T}$ or outside $\mathbb{T}$.

Also Li, Mohapatra and Rodriguez [5] have obtained Bernstein type inequalities for rational functions $r \in \mathcal{R}_n$ with all the poles $a_1, \ldots, a_n$ in $\mathbb{D}_+$. In particular for this class of rational functions they obtain inequalities analogous to inequalities (1) and (3). Their results depend upon the following identity for rational functions.

Theorem D *Suppose that* $\lambda \in \mathbb{T}$. *Then the following hold: The equation* $B(z) = \lambda$ *has exactly* n *simple roots, say* $t_1, t_2, \ldots, t_n$, *which lie on the unit circle* $\mathbb{T}$; *and if* $r \in \mathcal{R}_n$ *and* $z \in \mathbb{T}$, *then*

$$B'(z)r(z) - r'(z)[B(z) - \lambda] = \frac{B(z)}{z} \sum_{k=1}^{n} c_k r(t_k) \left|\frac{B(z)-\lambda}{z-t_k}\right|^2,$$

where $c_k = c_k(\lambda)$ *is defined by*

$$c_k^{-1} = \sum_{j=1}^{n} \frac{|a_j|^2-1}{|t_k-a_j|^2} \text{ for } k = 1, 2, \ldots, n.$$

Moreover, for $z \in \mathbb{T}$, *we have*

$$z\frac{B'(z)}{B(z)} = \sum_{k=1}^{n} c_k \left| \frac{B(z) - \lambda}{z - t_k} \right|^2.$$

In this paper we will obtain inequalities analogous to inequalities (2) and (4) for the class of rational functions considered by Li, Mohapatra and Rodriguez [5], that is, for rational functions $r \in \mathcal{R}_n$ with all the poles $a_1, a_2, \ldots, a_n$ in $\mathbb{D}_+$. Our method of proof is different from the methods of Li, Mohapatra and Rodriguez [5], Ankeny and Rivlin [1] and Borwein and Erdélyi [2]. As in Li, Mohapatra and Rodriguez [5] from now on we shall always assume that all poles are in $\mathbb{D}_+$. For the case when all poles are in $\mathbb{D}_-$, analogous results can be obtained by suitable modification of the arguments in the proofs.

Our first result that is presented below provides an inequality analogous to (2) for rational functions.

Theorem 1 *For a rational function*

$$r(z) = \frac{p(z)}{w(z)} = \frac{p(z)}{\prod_{\nu=1}^{n}(z - a_\nu)} \in \mathcal{R}_n, \tag{5}$$

with $|a_\nu| > 1$, $1 \le \nu \le n$, *we have*

$$|r(z)| \le \|r\|\,|B(z)|, \qquad |z| \ge 1. \tag{6}$$

The above result is best possible and the equality holds for

$$r(z) = \lambda \prod_{\nu=1}^{n} \frac{(1 - \bar{a}_\nu z)}{(z - a_\nu)},\ \lambda \in \mathbb{C}.$$

The above theorem is not difficult to prove however for the sake of completeness we will in Section 2 provide brief outlines of its proof. Besides this theorem will be needed to prove Lemma 2, which is essential for the proof of Theorem 2.

Our next result that provides an inequality analogous to (4) for rational functions is given by

Theorem 2 *Let* $r(z) = \dfrac{p(z)}{w(z)} = \dfrac{p(z)}{\prod_{\nu=1}^{n}(z - a_\nu)} \in \mathcal{R}_n$ *with* $|a_\nu| > 1$, $1 \le \nu \le n$. *If all the zeros of* $r(z)$ *lie in* $\mathbb{T} \cup \mathbb{D}_+$, *then for* $|z| \ge 1$,

$$|r(z)| \le \|r\| \left(\frac{|B(z)| + 1}{2} \right). \tag{7}$$

The result is best possible and the equality holds for the rational function $r(z) = \alpha B(z) + \beta, \quad |\alpha| = |\beta|$.

It may be noted that inequalities (2) and (4) can be deduced from inequalities (6) and (7) respectively. To obtain (4) from (7), multiply the two sides of (7) by $\prod_{\nu=1}^{n} a_\nu$ and then let each a_ν go to infinity. The method for deducing inequality (2) from inequality (6) is exactly same.

2 Proofs

We will firstly prove Theorem 1 because it will be needed to prove Lemma 2 given in this section and in order to prove this theorem we will need the following result which is an immediate consequence of the maximum modulus principle for unbounded domains.

Lemma 1 *If $f(z)$ is analytic in $\{z \in \mathbb{C} : |z| \geq 1\}$ and $f(z)$ tends to a finite limit as z tends to infinity, then $|f(z)| \leq \|f\|$ for $|z| \geq 1$.*

Proof of Theorem 1 Since $r \in \mathcal{R}_n$,

$$r(z) = \frac{p(z)}{w(z)} = \frac{\prod_{\nu=1}^{m} (z - z_\nu)}{\prod_{\nu=1}^{n} (z - a_\nu)}, \quad m \leq n$$

and therefore

$$\begin{aligned} \frac{r(z)}{B(z)} &= \left\{ \frac{\prod_{\nu=1}^{m} (z - z_\nu)}{\prod_{\nu=1}^{n} (z - a_\nu)} \right\} \Big/ \left\{ \frac{\prod_{\nu=1}^{n} (1 - \bar{a}_\nu z)}{\prod_{\nu=1}^{n} (z - a_\nu)} \right\} \\ &= \frac{\prod_{\nu=1}^{m} (z - z_v)}{\prod_{\nu=1}^{n} (1 - \bar{a}_\nu z)}, \quad m \leq n. \end{aligned}$$

Furthermore, because $r(z)$ has all its poles in $\mathbb{D}_+$, we have $|a_\nu| > 1$ for $1 \leq \nu \leq n$ and hence the function $r(z)/B(z)$ is analytic for $|z| \geq 1$. Also $|B(z)| = 1$ on $|z| = 1$, and therefore on $|z| = 1$,

$$\frac{|r(z)|}{|B(z)|} = |r(z)| \leq \|r\|.$$

Further, since $m \leq n$, we get, by Lemma 1

$$\frac{|r(z)|}{|B(z)|} \leq \|r\| \text{ for } |z| \geq 1,$$

from which the inequality (6) follows.

For the proof of Theorem 2 besides Lemma 1, we will also need

Lemma 2 *Let $r \in \mathcal{R}_n$ with all its poles in $\mathbb{D}_+$. Then for $|z| \geq 1$,*

$$|r^*(z)| \leq \|r\| \, |B(z)|. \tag{8}$$

Proof. Since $r \in \mathcal{R}_n$,

$$r(z) = \frac{p(z)}{w(z)} = \frac{\prod_{\nu=1}^{m}(z - z_\nu)}{\prod_{\nu=1}^{n}(z - a_\nu)}, \quad m \leq n$$

and therefore

$$r^*(z) = B(z)\{\overline{r(\frac{1}{\bar{z}})}\} = \frac{z^{n-m} z^m \{\overline{p(\frac{1}{\bar{z}})}\}}{\prod_{\nu=1}^{n}(z - a_\nu)} = \frac{z^{n-m} p^*(z)}{\prod_{\nu=1}^{n}(z - a_\nu)}.$$

Clearly the polynomial $z^{n-m}p^*(z) \in \mathcal{P}_n$ and therefore the function $r^* \in \mathcal{R}_n$. Also r^* has all its poles in $\mathbb{D}_+$ and hence by Theorem 1 we get that for $|z| \geq 1$,

$$|r^*(z)| \leq \|r^*\| |B(z)|.$$

Since $\|r^*\| = \|r\|$, the Lemma 2 follows.

Lemma 3 *Let $r \in \mathcal{R}_n$ with all its poles in $\mathbb{D}_+$. If $r(z)$ has all its zeros in $\mathbb{T} \cup \mathbb{D}_+$, then for $|z| \geq 1$,*

$$|r(z)| \leq |r^*(z)|. \tag{9}$$

Proof. Because $r(z) = \dfrac{p(z)}{w(z)} = \dfrac{\prod_{\nu=1}^{m}(z - z_\nu)}{\prod_{\nu=1}^{n}(z - a_\nu)}$, $m \leq n$ has no zeros in $|z| < 1$ and has all poles in $|z| > 1$, the rational function $r^*(z) = B(z)\{\overline{r(\frac{1}{\bar{z}})}\} =$

$z^{n-m}p^*(z)/w(z)$ has all its zeros in $|z| \le 1$. Therefore the function $r(z)/r^*(z) = p(z)/z^{n-m}p^*(z)$ is analytic for $|z| \ge 1$. Also on $|z| = 1$,

$$\left|\frac{r(z)}{r^*(z)}\right| = \frac{|p(z)|}{|z^{n-m}p^*(z)|} = 1,$$

and because $p(z)$ has no zeros in $|z| < 1$, the degree of the polynomial $z^{n-m}p^*(z)$ is at least as big as that of $p(z)$ and therefore by Lemma 1,

$$\frac{|r(z)|}{|r^*(z)|} \le 1 \text{ for } |z| \le 1,$$

from which the inequality (9) follows.

Lemma 4 *Let $r \in \mathcal{R}_n$ with all its poles in $\mathbb{D}_+$. Then for $|z| \ge 1$,*

$$|r(z)| + |r^*(z)| \le ||r||(|B(z)| + 1). \tag{10}$$

Proof. Since $r(z)$ has all its poles in $\mathbb{D}_+$, by the maximum modulus principle $|r(z)| \le ||r||$ for $|z| \le 1$ and hence for every λ with $|\lambda| > 1$, the function $(r(z) - \lambda||r||)$ has no zeros in $|z| \le 1$, implying that the function $(r(z) - \lambda||r||)^* = r^*(z) - \bar{\lambda}||r||B(z)$ has all its zeros in $|z| < 1$, that is, it has no zeros in $|z| \ge 1$. This clearly implies that the function

$$\frac{r(z) - \lambda||r||}{r^*(z) - \bar{\lambda}||r||B(z)}$$

is analytic in $|z| \ge 1$. Further, since on $|z| = 1$

$$\left|\frac{r(z) - \lambda||r||}{r^*(z) - \bar{\lambda}||r||B(z)}\right| = \left|\frac{p(z) - \lambda||r||\,w(z)}{z^{n-m}p^*(z) - \bar{\lambda}||r||\,w^*(z)}\right| = 1,$$

and the polynomial $p(z) - \lambda||r||\,w(z)$ being of degree n with no zeros in $|z| \le 1$, the polynomial $z^{n-m}p^*(z) - \bar{\lambda}||r||\,w^*(z) = (p(z) - \lambda||r||\,w(z))^*$ is also of degree n and therefore again by Lemma 1,

$$\left|\frac{r(z) - \lambda||r||}{r^*(z) - \bar{\lambda}||r||B(z)}\right| \le 1 \text{ for } |z| \ge 1,$$

implying

$$|r(z) - \lambda||r||| \le |r^*(z) - \bar{\lambda}||r||B(z)|$$

for $|z| \ge 1$, which implies that for every λ such that $|\lambda| > 1$, we have for $|z| \ge 1$,

$$|r(z)| - |\lambda|||r|| \le |r^*(z) - \bar{\lambda}||r||B(z)|. \tag{11}$$

Since by Lemma 2, we have

$$|r^*(z)| \le \|r\| |B(z)| \text{ for } |z| \ge 1,$$

we can choose the argument of λ so that the right hand side of (11) is

$$|\lambda| \|r\| |B(z)| - |r^*(z)|. \tag{12}$$

Combining (11) and (12) we get

$$|r(z)| - |\lambda| \|r\| \le |\lambda| \|r\| |B(z)| - |r^*(z)|$$

which is equivalent to, that for $|z| \ge 1$,

$$|r(z)| + |r^*(z)| \le |\lambda| \|r\| (|B(z)| + 1) \tag{13}$$

and if we now make $|\lambda| \to 1$, the inequality (10) follows.

Proof of Theorem 2 The proof of this theorem follows immediately on combining Lemmas 3 and 4.

From Lemma 4, one can immediately obtain

Corollary *Let* $r(z) = \dfrac{p(z)}{w(z)} \in \mathcal{R}_n$ *with all its poles in* $\mathbb{D}_+$. *If* $r(z)$ *is self-inversive then for* $|z| \ge 1$,

$$|r(z)| \le \|r\| \left(\frac{|B(z)| + 1}{2} \right). \tag{14}$$

The result is best possible with equality holding for the rational function $r(z) = \alpha B(z) + \beta$.

Acknowledgment. The authors are extremely grateful to the referee for his many very useful suggestions.

References

[1] N. C. Ankeny and T. J. Rivlin, On a theorem of S. Bernstein, Pacific J. Math. 5 (1955), 849-852.

[2] P. Borwein and T. Erdélyi, Sharp extensions of Bernstein inequality for rational spaces, Mathematika 43 (1996), 413-423.

[3] P. Borwein, T. Erdélyi and J. Zhang, Chebyshev polynomials and Bernstein-Markov type inequalities for rational spaces, J. London Math. Soc. (2) 50 (1994), 501-519.

[4] P. D. Lax, Proof of a conjecture of P. Erdös on the derivative of a polynomial, Bull. Amer. Math. Soc. 50 (1944), 509-513.

[5] Xin Li, R. N. Mohapatra and R. S. Rodriguez, Bernstein-type inequalities for rational functions with prescribed poles, J. London Math. Soc. (2) 51 (1995), 523-531.

[6] P. P. Petrushev and V. A. Popov, Rational approximation of real functions, Encyclopedia of Mathematics and its Applications, vol. 28, Cambridge University Press, Cambridge, 1987.

Recent Progress on Multivariate Splines

Don Hong
Department of Mathematics
East Tennessee State University, Johnson City, TN 37614-0663
E-mail: hong@etsu-tn.edu

Abstract

In many applications, one would like to represent scattered data given at sample points in $\mathbb{R}^s$ using "simple functions". A spline function is one of the most favored choices of such "simple functions". As a result, multivariate splines have been studied in depth in approximation theory. In this survey paper, we discuss the following topics: smoothness conditions of multivariate splines; dimension and local basis problems of multivariate spline spaces; and approximation power by multivariate spline elements. Some most recent results on these topics are sketched. We also mention some of the remaining open questions.

1 Introduction

In many applications, one would like to represent scattered data given at sample points in $\mathbb{R}^s$ using "simple functions". A spline function is one of the most favored choices of such "simple functions". As a result, univariate splines become well-known and there is a very complete constructive theory for univariate splines. In the last decade, multivariate splines have been studied in depth and considerable progress has been made in approximation theory. The aim of this paper is to sketch some of the results which have been obtained in recent years on the study of algebraic structure and approximation properties of multivariate spline spaces over arbitrary triangulations.

Let Δ be an s-dimensional simplicial complex in $\mathbb{R}^s$, i.e., Δ is a finite collection of simplices such that

(a) if $\tau \in \Delta$, then every face of τ also belongs to Δ;
(b) if $\tau, \tau' \in \Delta$ and $\tau \cap \tau'$ is non-empty then this intersection must be a common face of τ and τ'.

We call Δ a triangulation of a polyhedral region of $\mathbb{R}^s$.

As usual, for any nonnegative integers k and r, we define the multivariate spline space $S_k^r(\Delta)$ to be the space of all piecewise polynomial functions on Δ which are of total degree at most k and are smooth of order r. More precisely, $f \in S_k^r(\Delta)$ means that $f \in C^r(\Delta)$ and f is a polynomial with degree at most k on each simplex of Δ.

Multivariate spline functions are useful in representing a mathematical model described by several parameters and in the interpretation of s-dimensional data sets. Multivariate splines also play an important role in Computer Aided Geometric Design (CAGD) for design process in modeling complex parts such as car bodies or airplane fuselages. For these and other reasons, we are interested in the study of the following aspects of the spaces $S_k^r(\Delta)$:

(1) Algebraic structures (dimensions, locally supported bases);

(2) Approximation properties (triangulations and interpolation schemes for optimal approximation order).

This paper is organized as follows. In Section 2, in addition to reviewing the B-net representation of spline functions and some relative properties, we mention a new formulation of the smoothness conditions for splines in terms of the B-net representation which has been applied to construct a stable local basis for the bivariate spline space $S_k^r(\Delta)$ when $k \geq 3r+2$ in [23]. We also include in this section a smoothness condition and a conformality condition for bivariate cubic splines. Section 3 constitutes the discussion of dimension and local basis problems of the multivariate spline spaces. It includes some results of the most recent development on the topics. We also mention the new results on sphere splines and some open problems. In the final section, we sketch the results on the study of the optimal order of approximation from the bivariate spline spaces and optimal triangulations for the full order of approximation.

2 Bézier nets and smoothness conditions

The use of the Bernstein-Bézier net representation (B-form, BB-form) of piecewise polynomial functions (splines) in the study of bivariate splines was formally initiated by G. Farin. In [31], he expressed the C^r continuity conditions for bivariate splines in terms of Bézier coordinates (see also [32]). The general s-dimensional setting is considered in [14] (see also [45]). The B-net method is widely applied in the study of dimension, basis, and approximation properties of multivariate spline spaces (see [5–11], [17- 18], [20—24], [34–42], [49- 51], and [55] for examples). The B-net representation of piecewise polynomial functions is particularly convenient for the construction of various types of bivariate spline approximants on triangulations.

In this section, we begin with the B-net representation of polynomials defined on a simplex and the smoothness conditions for spline functions on triangulations in terms of their B-net ordinates. Then we mention a new formulation of the smoothness conditions for splines in terms of their B-net representations which has played an essential role in the construction of a stable local basis in [23]. Finally, we introduce a smoothness condition and a conformality condition for bivariate cubic splines.

As usual, let $\mathbb{R}$ be the set of all real numbers and $\mathbb{Z}_+$ the set of non-negative integers. Thus $\mathbb{R}^s$ denotes the s-dimensional Euclidean space and $\mathbb{Z}_+^s$ can be used as a multi-index set, while $\pi_k(\mathbb{R}^s)$ is the space of all polynomials of (total) degree $\le k$ in s variables. Let $\delta = [\mathbf{v}_0, \mathbf{v}_1, \cdots, \mathbf{v}_s]$ be a proper s-dimensional simplex with vertices $\mathbf{v}_0$, $\mathbf{v}_1$, $\cdots$, $\mathbf{v}_s \in \mathbb{R}^s$. Then for any $\mathbf{x} \in \mathbb{R}^s$, we have

$$\mathbf{x} = \xi_0 \mathbf{v}_0 + \xi_1 \mathbf{v}_1 + \cdots + \xi_s \mathbf{v}_s \quad \text{with} \quad \xi_0 + \xi_1 + \cdots \xi_s = 1.$$

The $(s+1)$-tuple $\xi = (\xi_0, \xi_1, \cdots, \xi_s)$ is called the barycentric coordinate of $\mathbf{x}$ with respect to the simplex δ. For $\alpha = (\alpha_0, \alpha_1, \cdots, \alpha_s) \in \mathbb{Z}_+^{s+1}$, the length of α is defined by $|\alpha| = \alpha_0 + \alpha_1 + \cdots + \alpha_s$, and the factorial $\alpha!$ is defined as $\alpha_0! \cdots \alpha_s!$. We define the Bernstein Polynomial $B_{\alpha,\delta}$ as

$$B_{\alpha,\delta}(\mathbf{x}) = \binom{|\alpha|}{\alpha} \xi^\alpha,$$

where $\xi^\alpha = \xi_0^{\alpha_0} \xi_1^{\alpha_1} \cdots \xi_s^{\alpha_s}$ and

$$\binom{|\alpha|}{\alpha} = \frac{|\alpha|!}{\alpha_0! \alpha_1! \cdots \alpha_s!}.$$

Moreover, we define the (domain) points

$$\mathbf{x}_{\alpha,\delta} := \frac{\alpha_0 \mathbf{v}_0 + \alpha_1 \mathbf{v}_1 + \cdots + \alpha_s \mathbf{v}_s}{|\alpha|}, \quad |\alpha| = k. \tag{2.1}$$

It is well known that any polynomial $p \in \pi_k$ can be written in a unique way as

$$p = \sum_{|\alpha|=k} b_{\alpha,\delta} B_{\alpha,\delta},$$

where $b_{\alpha,\delta}$ is called the B-net ordinate of p with respect to δ. This gives rise to a mapping $b : \mathbf{x}_{\alpha,\delta} \to b_{\alpha,\delta}$, $|\alpha| = k$. Such a mapping b is called the B-net representation of p with respect to δ.

Now, let us discuss the B-net representation of multivariate splines. Let Δ be a triangulation of a polygonal domain in $\mathbb{R}^s$ and $S_k^0(\Delta)$ the space of all continuous splines of degree k on Δ. Assume $f \in S_k^0(\Delta)$. On each simplex $\delta \in \Delta$, f agrees with some polynomial $p \in \pi_k$. Thus, we have

$$f|_\delta = \sum_{|\alpha|=k} b_{\alpha,\delta} B_{\alpha,\delta}.$$

Let X denote the set of all (domain) points $\mathbf{x}_{\alpha,\delta}$ as defined in (2.1). Then a mapping can be defined as follows:

$$b_f : \mathbf{x}_{\alpha,\delta} \mapsto b_{\alpha,\delta}, \quad |\alpha| = k, \ \delta \in \Delta. \tag{2.2}$$

Such a mapping b_f is called the B-net representation of the spline f.

Now, let us consider the C^r-smoothness conditions in terms of the B-net representation of spline functions. Let $\delta = [\mathbf{v}_0, \mathbf{v}_1, \cdots, \mathbf{v}_s]$ and $\tilde{\delta} = [\mathbf{v}_0, \mathbf{v}_1, \cdots, \tilde{\mathbf{v}}_s]$ be two s-dimensional simplices with a common $(s-1)$-dimensional face $[\mathbf{v}_0, \mathbf{v}_1, \cdots, \mathbf{v}_{s-1}]$ and denote $\mathbf{v}_i = (v_{i,1}, v_{i,2}, \cdots, v_{i,s})$, $i = 0, 1, \cdots, s$, and $\tilde{\mathbf{v}}_s = (\tilde{v}_{s,1}, \tilde{v}_{s,2}, \cdots, \tilde{v}_{s,s})$. Then the oriented volume of the simplex δ is

$$V := vol[\mathbf{v}_0, \mathbf{v}_1, \cdots, \mathbf{v}_s] = \frac{1}{s!} \begin{vmatrix} 1 & v_{0,1} & v_{0,2} & \cdots & v_{0,s} \\ 1 & v_{1,1} & v_{1,2} & \cdots & v_{1,s} \\ \vdots & \vdots & \vdots & \cdots & \vdots \\ 1 & v_{s,1} & v_{s,2} & \cdots & v_{s,s} \end{vmatrix}. \tag{2.3}$$

If we set $\hat{\mathbf{v}}_i = \tilde{\mathbf{v}}_s$, then the oriented volume of the simplex $[\mathbf{v}_0, \mathbf{v}_1, \cdots, \hat{\mathbf{v}}_i, \cdots, \mathbf{v}_s]$ can be denoted by

$$V_i := vol[\mathbf{v}_0, \mathbf{v}_1, \cdots, \hat{\mathbf{v}}_i, \cdots, \mathbf{v}_s] = \frac{1}{s!} \begin{vmatrix} 1 & v_{0,1} & \cdots & v_{0,s} \\ \vdots & \vdots & \cdots & \vdots \\ 1 & v_{i-1,1} & \cdots & v_{i-1,s} \\ 1 & \tilde{v}_{s,1} & \cdots & \tilde{v}_{s,s} \\ 1 & v_{i+1,1} & \cdots & v_{i+1,s} \\ \vdots & \vdots & \cdots & \vdots \\ 1 & v_{s,1} & \cdots & v_{s,s} \end{vmatrix}. \tag{2.4}$$

The following result, which describes C^r-smoothness conditions on a spline function f in terms of its B-net representation, is from [45] (see also [14], [20] etc.).

Theorem 1 *Suppose that the piecewise polynomial function f is defined on $\delta \cup \tilde{\delta}$ by*

$$f|_\delta = \sum_{|\alpha|=k} b_{\alpha,\delta} B_{\alpha,\delta},$$

$$f|_{\tilde{\delta}} = \sum_{|\alpha|=k} b_{\alpha,\tilde{\delta}} B_{\alpha,\tilde{\delta}}.$$

Then the spline f is in $C^r(\delta \cup \tilde{\delta})$ if and only if, for all positive integers $\ell \le r$ and $\alpha = (\alpha_0, \alpha_1, \cdots, \alpha_{s-1}, 0) \in \mathbb{Z}_+^{s+1}$ with $|\alpha| = k - \ell$,

$$b_{\alpha+\ell \mathbf{e}^{s+1},\tilde{\delta}} = \sum_{|\beta|=\ell} \binom{\ell}{\beta} b_{\alpha+\beta,\delta} \frac{V_0^{\beta_0} \cdots V_s^{\beta_s}}{V^{|\beta|}} \tag{2.5}$$

where β and $\mathbf{e}^{s+1} = (0, \cdots, 0, 1)$ are in $\mathbb{Z}_+^{s+1}$.

For example, in the case $s = 2$, Δ is a triangulation of a polygonal plane domain in $\mathbb{R}^2$. We use u, v and w to denote the vertices of a triangle τ in Δ and denote by $\tilde{\tau} = [u, v, \tilde{w}]$ another triangle sharing common edge $e = [u, v]$. Let S, S_u, S_v and S_w denote the oriented areas of the triangles τ, $[\tilde{w}, v, w]$, $[u, \tilde{w}, w]$, and $\tilde{\tau}$, respectively. Then (2.5) can be restated as:

Corollary 2 *Suppose that a bivariate spline function f is defined on the union of two triangles $\tau \cup \tilde{\tau}$ by*

$$f|_\tau = \sum_{|\alpha|=k} b(x_{\alpha,\tau}) B_{\alpha,\tau};$$

$$f|_{\tilde{\tau}} = \sum_{|\alpha|=k} b(x_{\alpha,\tilde{\tau}}) B_{\alpha,\tilde{\tau}}.$$

Then $f \in C^r(\tau \cup \tilde{\tau})$ if and only if for all positive integers $\ell \le r$ and $\gamma = (\gamma_u, \gamma_v, 0) \in \mathbb{Z}_+^3$ with $|\gamma| = k - \ell$,

$$b(x_{\gamma+\ell e^3,\tilde{\tau}}) = \sum_{|\beta|=\ell} \binom{\ell}{\beta} b(x_{\gamma+\beta,\tau}) \left(\frac{S_u}{S}\right)^{\beta_u} \left(\frac{S_v}{S}\right)^{\beta_v} \left(\frac{S_w}{S}\right)^{\beta_w}, \tag{2.6}$$

where $\beta = (\beta_u, \beta_v, \beta_w) \in \mathbb{Z}_+^3$ and $e^3 = (0, 0, 1)$.

Let E_I be the set of interior edges of Δ. For $e \in E_I$ and two triangles $\tau = [u, v, w]$, $\tilde{\tau} = [u, v, \tilde{w}]$ sharing the edge e, and $\alpha = (\alpha_u, \alpha_v, \alpha_{\tilde{w}}) \in \mathbb{Z}_+^3$ with $1 \le \alpha_{\tilde{w}} \le r$, we define the functionals $f_{e,\alpha}$ on $\mathbb{R}^X$ by

$$f_{e,\alpha} b = b(x_{\alpha,\tilde{\tau}}) - \sum_{|\beta|=\alpha_{\tilde{w}}} \binom{\alpha_{\tilde{w}}}{\beta} b(x_{(\alpha_u,\alpha_v,0)+\beta,\tau}) \left(\frac{S_u^{\beta_u} S_v^{\beta_v} S_w^{\beta_w}}{S^{\alpha_{\tilde{w}}}}\right). \tag{2.7}$$

It is clear that the support of the functional $f_{e,\alpha}$ is included in a diamond domain with vertices $\frac{(\alpha_u+\alpha_{\tilde{w}})u+\alpha_v v}{k}$, $\frac{\alpha_u u+(\alpha_v+\alpha_{\tilde{w}})v}{k}$, $x_{\alpha,\tilde{\tau}}$ and $x_{\alpha,\tau}$.

The following new formulation of the smoothness conditions plays an essential role in the constructive proof of the fact that the spaces of bivariate splines over triangulations with smoothness r and total degree $k \geq 3r+2$ achieve the optimal approximation order $k+1$ with estimation constant depending only on k and the smallest angle of the partition (see [23]).

In the following part of this section, we set $\alpha = (\alpha_1, \cdots, \alpha_s) \in \mathbb{Z}_+^s$, and use the standard multi-index notation. For $\alpha, \gamma \in \mathbb{Z}_+^s$, we have

$$\binom{\alpha}{\gamma} = \binom{\alpha_1}{\gamma_1} \cdots \binom{\alpha_s}{\gamma_s} = \frac{\alpha!}{(\alpha-\gamma)!\gamma!}.$$

We will say that $\gamma \leq \alpha$ if and only if $\gamma_i \leq \alpha_i$, for $i = 1, \cdots, s$. Let

$$C_{\alpha,\delta} := \sum_{\gamma \leq \alpha} (-1)^{|\alpha-\gamma|} \binom{\alpha}{\gamma} b(\mathbf{x}_{\gamma,\delta}),$$

where $\mathbf{x}_{\alpha,\delta}$ is defined in (2.1). Then, we have the following new formulation of smoothness conditions (see [39]).

Theorem 3 *A piecewise polynomial function $s \in C^0(\delta \cup \tilde{\delta})$ is of smoothness order r if and only if the corresponding terms $\{C_{\alpha,\delta}\}$ and $\{C_{\alpha,\tilde{\delta}}\}$ satisfy the condition:*

$$C_{\alpha,\tilde{\delta}} = \sum_{|\gamma^-| \leq \alpha_s} C_{(\alpha+\gamma^-, \alpha_s - |\gamma^-|),\delta} \binom{\alpha_s}{|\gamma^-|} \frac{V_1^{\gamma_1} \cdots V_{s-1}^{\gamma_{s-1}} V_s^{\alpha_s - |\gamma^-|}}{V^{\alpha_s}} \tag{2.8}$$

for $1 \leq \alpha_s \leq r$, $\alpha \in \mathbb{Z}_+^s$, where $\gamma^- = (\gamma_1, \cdots, \gamma_{s-1}, 0) \in \mathbb{Z}_+^s$.

For example, in the 2-dimensional setting, let a bivariate spline f be defined on the union $\tau \cup \tilde{\tau}$ of two triangles τ and $\tilde{\tau}$. For $\alpha = (\alpha_u, \alpha_v, \alpha_w)$, $\beta = (\beta_u, \beta_v, \beta_w) \in \mathbb{Z}_+^3$ with $|\alpha| = |\beta| = k$, let

$$C_{\alpha,\tau} := \sum_{\substack{\beta_v \leq \alpha_v \\ \beta_w \leq \alpha_w}} (-1)^{\alpha_v - \beta_v + \alpha_w - \beta_w} \binom{\alpha_v}{\beta_v} \binom{\alpha_w}{\beta_w} b(x_{\beta,\tau}), \tag{2.9}$$

Then we can restate the above lemma for as follows:

Corollary 4 *A spline function $s \in S_k^0(\tau \cup \tilde{\tau})$ is in C^r if and only if the corresponding terms $\{C_{\alpha,\tau}\}$ and $\{C_{\alpha,\tilde{\tau}}\}$ satisfy the condition:*

$$C_{\alpha,\tilde{\tau}} = \sum_{\ell=0}^{\alpha_w} C_{(\alpha_u, \alpha_v+\ell, \alpha_w-\ell),\tau} \binom{\alpha_w}{\ell} \frac{S_v^\ell S_w^{\alpha_w-\ell}}{S^{\alpha_w}}, \tag{2.10}$$

$1 \leq \alpha_w \leq r$, and $\alpha = (\alpha_u, \alpha_v, \alpha_w) \in \mathbb{Z}_+^3$ with $|\alpha| = k$.

For the rest of this section, we restrict ourselves to the two-dimensional setting. Let $\delta = [\mathbf{v}_0, \mathbf{v}_1, \mathbf{v}_2]$ be a planar non-degenerate simplex with vertices $\mathbf{v}_i = (x_i, y_i) \in \mathbb{R}^2$, $i = 0, 1, 2$. Let $\xi = (\xi_0, \xi_1, \xi_2)$ be the barycentric coordinate of $\mathbf{x} \in \mathbb{R}^2$ with respect to the simplex δ. Set

$$\begin{aligned}
\lambda_i &:= y_j - y_k \\
\mu_i &:= -(x_j - x_k) \\
\nu_i &:= x_j y_k - y_j x_k
\end{aligned}$$

with (i, j, k) being a permutation of the subscripts in cyclic order $0 \to 1 \to 2 \to 0$. Then the barycentric coordinate can be expressed as

$$\begin{pmatrix} \xi_0 \\ \xi_1 \\ \xi_2 \end{pmatrix} = \frac{1}{A^{(1)}} \begin{pmatrix} \nu_0 & \lambda_0 & \mu_0 \\ \nu_1 & \lambda_1 & \mu_1 \\ \nu_2 & \lambda_2 & \mu_2 \end{pmatrix} \begin{pmatrix} 1 \\ x \\ y \end{pmatrix} \tag{2.11}$$

where

$$A^{(1)} = \text{area}[\mathbf{v}_0, \mathbf{v}_1, \mathbf{v}_2] = \frac{1}{2!} \begin{vmatrix} 1 & x_0 & y_0 \\ 1 & x_1 & y_1 \\ 1 & x_2 & y_2 \end{vmatrix}$$

is the oriented area of the simplex δ as given in (2.3).

For a bivariate polynomial p with total degree k with B-net representation b_α, i.e.,

$$p = \sum_{|\alpha|=k} b_\alpha B_{\alpha,\delta_1} = \sum_{|\alpha|=k} b_{\alpha_0,\alpha_1,\alpha_2} B_{\alpha,\delta_1},$$

we have the following

$$\begin{aligned}
\begin{pmatrix} \frac{\partial p}{\partial x} \\ \frac{\partial p}{\partial y} \end{pmatrix} &= \begin{pmatrix} \frac{\partial \xi_0}{\partial x} & \frac{\partial \xi_1}{\partial x} \\ \frac{\partial \xi_0}{\partial y} & \frac{\partial \xi_1}{\partial y} \end{pmatrix} \begin{pmatrix} \frac{\partial p}{\partial \xi_0} \\ \frac{\partial p}{\partial \xi_1} \end{pmatrix} \\
&= \frac{1}{A^{(1)}} \begin{pmatrix} \lambda_0 & \lambda_1 \\ \mu_0 & \mu_1 \end{pmatrix} \begin{pmatrix} \frac{\partial p}{\partial \xi_0} \\ \frac{\partial p}{\partial \xi_1} \end{pmatrix} \\
&= \sum_{|\alpha|=k-1} \frac{k}{A^{(1)}} \binom{|\alpha|}{\alpha} \begin{pmatrix} \lambda_0 & \lambda_1 \\ \mu_0 & \mu_1 \end{pmatrix} \begin{pmatrix} \Delta_{1\,3} b_\alpha \\ \Delta_{2\,3} b_\alpha \end{pmatrix} \xi_0^{\alpha_0} \xi_1^{\alpha_1} \xi_2^{\alpha_2},
\end{aligned} \tag{2.12}$$

where

$$\begin{aligned}
\Delta_{1\,3} b_\alpha &= b_{\alpha_0+1,\alpha_1,\alpha_2} - b_{\alpha_0,\alpha_1,\alpha_2+1}, \\
\Delta_{2\,3} b_\alpha &= b_{\alpha_0,\alpha_1+1,\alpha_2} - b_{\alpha_0,\alpha_1,\alpha_2+1}.
\end{aligned} \tag{2.13}$$

Let $f \in C^1(\Omega)$. We consider the following interpolation problem on a simplex $\delta = [\mathbf{v}_0, \mathbf{v}_1, \mathbf{v}_2]$: Find a cubic bivariate polynomial $p \in \pi_3$ such that

$$\begin{aligned} p(\mathbf{v}_i) &= f(\mathbf{v}_i) =: f_i, \\ p(\frac{\mathbf{v}_0+\mathbf{v}_1+\mathbf{v}_2}{3}) &= f(\frac{\mathbf{v}_0+\mathbf{v}_1+\mathbf{v}_2}{3}) =: f_3, \\ \frac{\partial p}{\partial x}|_{\mathbf{v}_i} &= \frac{\partial f}{\partial x}|_{\mathbf{v}_i} =: D_x f_i, \\ \frac{\partial p}{\partial y}|_{\mathbf{v}_i} &= \frac{\partial f}{\partial y}|_{\mathbf{v}_i} =: D_y f_i, \end{aligned} \tag{2.14}$$

for $i = 0, 1, 2$.

We can obtain the interpolation conditions in terms of B-net representation b_α as follows.

$$\begin{aligned} f_0 &= b^{(1)}_{3,0,0}, \quad f_1 = b^{(1)}_{0,3,0}, \quad f_2 = b^{(1)}_{0,0,3}, \\ D_x f_0 &= \frac{3}{A^{(1)}}(\lambda_0 b^{(1)}_{3,0,0} + \lambda_1 b^{(1)}_{2,1,0} + \lambda_2 b^{(1)}_{2,0,1}), \\ D_x f_1 &= \frac{3}{A^{(1)}}(\lambda_1 b^{(1)}_{0,3,0} + \lambda_0 b^{(1)}_{1,2,0} + \lambda_2 b^{(1)}_{0,2,1}), \\ D_x f_2 &= \frac{3}{A^{(1)}}(\lambda_2 b^{(1)}_{0,0,3} + \lambda_0 b^{(1)}_{1,0,2} + \lambda_1 b^{(1)}_{0,1,2}), \\ D_y f_0 &= \frac{-3}{A^{(1)}}(\mu_0 b^{(1)}_{3,0,0} + \mu_1 b^{(1)}_{2,1,0} + \mu_2 b^{(1)}_{2,0,1}), \\ D_y f_1 &= \frac{-3}{A^{(1)}}(\mu_1 b^{(1)}_{0,3,0} + \mu_0 b^{(1)}_{1,2,0} + \mu_2 b^{(1)}_{0,2,1}), \\ D_y f_2 &= \frac{-3}{A^{(1)}}(\mu_2 b^{(1)}_{0,0,3} + \mu_0 b^{(1)}_{1,0,2} + \mu_1 b^{(1)}_{0,1,2}), \\ f_3 &= \frac{1}{27}(b^{(1)}_{3,0,0} + b^{(1)}_{0,3,0} + b^{(1)}_{0,0,3} + 3b^{(1)}_{2,1,0} + 3b^{(1)}_{1,2,0} \\ &\quad + 3b^{(1)}_{2,0,1} + 3b^{(1)}_{1,0,2} + 3b^{(1)}_{0,1,2} + 3b^{(1)}_{0,2,1} + 6b^{(1)}_{1,1,1}). \end{aligned} \tag{2.15}$$

Then we can solve for the B-net representation b_α of p as follows.

$$
\begin{aligned}
b^{(1)}_{3,0,0} &= f_0, \; b^{(1)}_{0,3,0} = f_1, \; b^{(1)}_{0,0,3} = f_2, \\
b^{(1)}_{2,1,0} &= f_0 - \frac{1}{3}\lambda_2 D_x f_0 - \frac{1}{3}\mu_2 D_y f_0, \\
b^{(1)}_{1,2,0} &= f_1 + \frac{1}{3}\lambda_2 D_x f_1 + \frac{1}{3}\mu_2 D_y f_1, \\
b^{(1)}_{2,0,1} &= f_0 + \frac{1}{3}\lambda_1 D_x f_0 + \frac{1}{3}\mu_1 D_y f_0, \\
b^{(1)}_{1,0,2} &= f_2 - \frac{1}{3}\lambda_1 D_x f_2 - \frac{1}{3}\mu_1 D_y f_2, \\
b^{(1)}_{0,2,1} &= f_1 - \frac{1}{3}\lambda_0 D_x f_1 - \frac{1}{3}\mu_0 D_y f_1, \\
b^{(1)}_{0,1,2} &= f_2 + \frac{1}{3}\lambda_0 D_x f_2 + \frac{1}{3}\mu_0 D_y f_2, \\
b^{(1)}_{1,1,1} &= \frac{1}{6}[\, 27 f_3 - 7f_0 - 7f_1 - 7f_2 + (\lambda_2 - \lambda_1) D_x f_0 + (\lambda_0 - \lambda_2) D_x f_1 \\
&\quad + (\lambda_1 - \lambda_0) D_x f_2 + (\mu_2 - \mu_1) D_y f_0 + (\mu_0 - \mu_2) D_y f_1 + (\mu_1 - \mu_0) D_y f_2].
\end{aligned} \tag{2.16}
$$

Thus, there exists a unique solution for the interpolating problem (2.14). Furthermore, we can see that the free parameters of any polynomial p on a simplex can be chosen as the function values and first derivative values at the three vertices as well as the function value at the center of the triangle. In the following, we will see that this group of new free parameters of cubic polynomials can simplify the smoothness conditions and conformality conditions of bivariate cubic splines.

Let $\delta_1 = [\mathbf{v}_0, \mathbf{v}_1, \mathbf{v}_2]$ and $\delta_2 = [\mathbf{v}_0, \mathbf{v}_1, \mathbf{v}_3]$ be two adjacent triangles with a common edge $[\mathbf{v}_0, \mathbf{v}_1]$ and $\mathbf{v}_3 = (x_3, y_3)$. For convenience, let $A^{(1)}$ denote the area of the triangle δ_1 and

$$
\begin{aligned}
A^{(1)}_0 &= \text{area}[\mathbf{v}_3, \mathbf{v}_1, \mathbf{v}_2], \\
A^{(1)}_1 &= \text{area}[\mathbf{v}_1, \mathbf{v}_3, \mathbf{v}_2], \\
A^{(1)}_2 &= \text{area}[\mathbf{v}_0, \mathbf{v}_1, \mathbf{v}_3],
\end{aligned}
$$

and suppose that the piecewise cubic polynomial function $F(x, y)$ is defined on $\delta_1 \cup \delta_2$ by

$$
\begin{aligned}
F|_{\delta_1} &= \sum_{|\alpha|=3} b^{(1)}_\alpha \, B_{\alpha,\delta_1}, \\
F|_{\delta_2} &= \sum_{|\alpha|=3} b^{(2)}_\alpha \, B_{\alpha,\delta_2}.
\end{aligned}
$$

Then, the smoothness conditions in Theorem 1 can be simplified for the bivariate cubic splines as follows.

$$b_\alpha^{(2)} = b_\alpha^{(1)}, \qquad \alpha = (\alpha_0, \alpha_1, 0),\ |\alpha| = 3,$$

$$b_{\alpha+e^3}^{(2)} = \sum_{|\beta|=1} b_{\alpha+\beta}^{(1)} \frac{(A_0^{(1)})^{\beta_0}(A_1^{(1)})^{\beta_1}(A_2^{(1)})^{\beta_2}}{A^{(1)}}, \qquad \alpha = (\alpha_0, \alpha_1, 0),\ |\alpha| = 2.$$

Furthermore, we obtain the following (see [40]).

Theorem 5 *Suppose that the piecewise cubic polynomial function $F(x,y)$ is defined on $\delta_1 \cup \delta_2$ and $F|_{\delta_1} = f$, $F|_{\delta_2} = g$. Let $b_\alpha^{(1)}$, $b_\alpha^{(2)}$ be the B-net representations of F on δ_1 and δ_2, respectively. Then $F(x,y) \in C^1(\delta_1 \cup \delta_2)$ if and only if, for $i = 0, 1$,*

$$\begin{aligned} f_i &= g_i, \\ D_x f_i &= D_x g_i, \\ D_y f_i &= D_y g_i, \end{aligned} \tag{2.17}$$

and

$$b_{1,1,1}^{(2)} = \frac{1}{A^{(1)}}(b_{2,1,0}^{(1)} A_0^{(1)} + b_{1,2,0}^{(1)} A_1^{(1)} + b_{1,1,1}^{(1)} A_2^{(1)}), \tag{2.18}$$

where $b_{2,1,0}^{(1)}$ and $b_{1,2,0}^{(1)}$ are given in (2.16), i.e.,

$$\begin{aligned} b_{2,1,0}^{(1)} &= f_0 - \frac{1}{3}\lambda_2 D_x f_0 - \frac{1}{3}\mu_2 D_y f_0, \\ b_{1,2,0}^{(1)} &= f_1 + \frac{1}{3}\lambda_2 D_x f_1 + \frac{1}{3}\mu_2 D_y f_1. \end{aligned}$$

Therefore, the cubic spline $F(x,y) \in S_3^1(\delta_1 \cup \delta_2)$ can be determined by the given values $F|_{\mathbf{v}_i}$, $D_x F|_{\mathbf{v}_i}$, $D_y F|_{\mathbf{v}_i}$, $i = 0,1,2,3$ and $b_{1,1,1}^{(1)}$(or $b_{1,1,1}^{(2)}$). Thus there are thirteen free parameters in total which uniquely determine a cubic spline on $\delta_1 \cup \delta_2$.

Following the notation in [56], the union of all the triangles with the common vertex $\mathbf{v}$ of a triangulation Δ is called a *standard cell with interior vertex* $\mathbf{v}$ and denoted by $\Delta_{\mathbf{v}}$. The boundary vertices of $\Delta_{\mathbf{v}}$, in the counter clockwise direction, are denoted by $\mathbf{v}_j, j = 1, 2, ..., d$. The number of the edges emanating from $\mathbf{v}$ is called the degree of $\mathbf{v}$ and denoted by $deg(\mathbf{v})$. We call a triangulation Δ an odd- (even-) triangulation if the degree of each interior vertex in Δ is an odd (even) number. For a standard cell $\Delta_{\mathbf{v}}$

with interior vertex $\mathbf{v}$, we define

$$\begin{aligned}\mathbf{v}_0 = \mathbf{v}, \quad e_j &= [\mathbf{v}_0, \mathbf{v}_j],\\ A^{(j)} &= \text{area}[\mathbf{v}_{j+1}, \mathbf{v}_0, \mathbf{v}_j],\\ A_0^{(j)} &= \text{area}[\mathbf{v}_{j+2}, \mathbf{v}_0, \mathbf{v}_j],\\ A_1^{(j)} &= \text{area}[\mathbf{v}_{j+1}, \mathbf{v}_{j+2}, \mathbf{v}_j],\\ A_2^{(j)} &= \text{area}[\mathbf{v}_{j+1}, \mathbf{v}_0, \mathbf{v}_{j+2}],\end{aligned}$$

where $j = 1, ..., d$, $j+1 \bmod(d)$ and $j+2 \bmod(d)$.

Suppose that $\Delta_{\mathbf{v}}$ is a standard cell with an interior vertex $\mathbf{v}$ of the triangulation Δ. Then the conditions (or linear equations) which a spline $s \in S_k^r(\Delta)$ satisfies around the vertex $\mathbf{v}$ are called conformality conditions. Here, we discuss the conformality conditions on bivariate cubic super splines in terms of B-net representation.

The concept of general super splines was introduced by Schumaker in [58]. The subspace of super splines of smoothness r and degree $\leq k$ with enhanced smoothness order $\theta \geq r$ is defined as

$$S_k^{r,\theta}(\Delta) = \{s \in S_k^r(\Delta) : \; s \in C^\theta \text{ at each vertex of } \Delta\}.$$

We consider the conformality conditions for bivariate cubic splines based on the cubic super spline space $S_3^{0,1}(\Delta)$. For this purpose, corresponding to $\Delta_{\mathbf{v}}$ for an interior vertex $\mathbf{v} := \mathbf{v}_0 = (x_0, y_0)$ with $d := deg(\mathbf{v})$ and vertices $\mathbf{v}_j = (x_j, y_j)$, $j = 1, \cdots, d$, identified in the counter-clockwise direction, we define

$$\begin{aligned} s_j &= s(\mathbf{v}_j),\\ D_x s_j &= \frac{\partial s}{\partial x}|_{\mathbf{v}_j},\\ D_y s_j &= \frac{\partial s}{\partial y}|_{\mathbf{v}_j},\end{aligned}$$

for $s \in S_3^{0,1}(\Delta_{\mathbf{v}})$ and $j = 0, \cdots, d$.

We have the following result on conformality conditions of bivariate cubic splines (see [40]).

Theorem 6 *Suppose $s(x,y) \in S_3^{0,1}(\Delta_{\mathbf{v}})$ is a bivariate super cubic spline defined on a standard cell $\Delta_{\mathbf{v}}$ with an interior vertex $\mathbf{v}$. Then a necessary and sufficient condition for $s(x,y) \in S_3^1(\Delta_{\mathbf{v}})$ is that*

i) if d is an even number $d = 2N$, then

$$\sum_{j=1}^{2N} (-1)^j \frac{1}{A^{(j)} A^{(j+1)}} \left[\, b_{2,1,0}^{(j)} \, A_0^{(j)} \; + \; b_{1,2,0}^{(j)} \, A_1^{(j)} \, \right] = 0, \qquad (2.19)$$

ii) if d is an odd number $d = 2N + 1$, then

$$b^{(1)}_{1,1,1} = \frac{1}{2} \sum_{j=1}^{2N+1} (-1)^{j+1} \frac{A^{(1)}}{A^{(j)} A^{(j+1)}} \left[b^{(j)}_{2,1,0} A^{(j)}_0 + b^{(j)}_{1,2,0} A^{(j)}_1 \right], \tag{2.20}$$

where

$$b^{(j)}_{2,1,0} = s_{j+1} - \frac{1}{3}(y_{j+1} - y_0) D_x s_{j+1} - \frac{1}{3}(x_0 - x_{j+1}) D_y s_{j+1},$$
$$b^{(j)}_{1,2,0} = s_0 + \frac{1}{3}(y_{j+1} - y_0) D_x s_0 + \frac{1}{3}(x_0 - x_{j+1}) D_y s_0.$$

3 Dimension and local basis

Though the structure of $S^r_k(\Delta)$ for $s > 1$ is extremely complicated and poorly understood in general, there is a vast literature devoted to the study of the algebraic structure of the spaces $S^r_k(\Delta)$ during the past decade. We begin this section by dealing with the bivariate setting ($s = 2$) and then consider higher dimensional setting ($s > 2$) and sphere splines.

The dimension problem of bivariate spline spaces has a rich history beginning with a conjecture by Strang in 1973 (see [61] and [62]). One of the first results in addressing this problem was due to Morgan and Scott [54], who gave the dimension formula and explicit bases for bivariate spline spaces $S^1_k(\Delta)$, $k \geq 5$. A little later, Schumaker [56] gave a lower bound formula for the dimension of the spaces $S^r_k(\Delta)$. Since then much work has been done on various specific triangulations, particularly, three- or four-directional meshes, or cross-cut partitions (see [20], [25]— [27], [53] and the references therein). The two most important special triangulations are obtained from a uniform rectangular partition. The triangulation which results when all northeast diagonals are drawn in is called a type-1 or three-direction mesh, and it is denoted by $\Delta^{(1)}$. The triangulation which results when all northwest diagonals are also drawn in is called a type-2 or four-direction mesh, and it is denoted by $\Delta^{(2)}$. For a survey of the work on dimension problems of spline spaces over some special partitions, see [28] and also [59].

For $s > 1$, dimension, basis, and interpolation depend on the geometry as well as on the combinatorics of the triangulation. During the past few years, there has been much progress on this surprisingly difficult problem. Let us begin with some notation.

Given a triangulation Δ of $\Omega \subset \mathbb{R}^2$, we denote by E, E_I, V and V_I the sets of edges, interior edges, vertices and interior vertices of Δ respectively.

Let

$$E_b = E \backslash E_I, \text{ and } \quad V_b = V \backslash V_I \tag{3.1}$$

Given a vertex $v \in V$, we use e_v to denote the number of edges with different slopes attached to v. The cardinality of a set A is denoted by $|A|$. Also we let

$$\begin{aligned} \alpha &= \frac{(k+1)(k+2)}{2}, \\ \beta &= \frac{(k-r)(k-r+1)}{2}, \\ \gamma &= \frac{(k+1)(k+2)}{2} - \frac{(r+1)(r+2)}{2}, \end{aligned} \tag{3.2a}$$

and

$$\sigma_v = \sum_{j=1}^{k-r} (r + j + 1 - je_v)_+. \tag{3.2b}$$

Then Schumaker's lower bound formula is given as follows (see [56]).

Theorem 7 *Let Δ be a triangulation of Ω. Then*

$$\dim(S^{\mu}_{k,\Delta}) \geq \alpha + \beta|E_I| - \gamma|V_I| + \sigma \tag{3.3}$$

where α, β, γ and σ_v are given in (3.2a) and (3.2b), while

$$\sigma = \sum_{v \in V_I} \sigma_v. \tag{3.4}$$

Note that Schumaker's lower bound on the dimension of the spaces $S^r_k(\Delta)$ is only valid for simply connected domains (see [37] and [7]). In [47], Jia established a lower bound on the dimension of $S^r_k(\Delta)$ for any polygonal domain possible with "holes". This lower bound is sharp in the sense that it gives the exact dimension for the space $S^r_k(\Delta)$ when $k \geq 3r + 2$.

Alfeld and Schumaker in [8] extended the Morgan-Scott result for all r by showing that Schumaker's lower bound is in fact the dimension of $S^r_k(\Delta)$ when $k \geq 4r + 1$. Together with Piper, they constructed explicit bases for these spaces in [5]. By a clever application of the B-net approach, they also extended the Morgan-Scott results to the space $S^1_4(\Delta)$ (see [6]). By carefully working with the smoothness conditions in terms of B-net representation of spline functions, it was proved in [37] that Schumaker's and Jia's lower bounds indeed give the dimension for the spaces $S^r_k(\Delta)$ when $k \geq 3r + 2$. This is the best result, so far, in this direction.

Theorem 8 *For arbitrary triangulation Δ of $\Omega \subset I\!R^2$ and $k \geq 3r+2$, the dimension of $S_k^r(\Delta)$ is given by*

$$\dim(S^{\mu}_{k,\Delta}) = \alpha(1-c) + \beta|E_I| - \gamma|V_I| + \sigma \tag{3.5}$$

where $c+1$ denotes the number of connected components of $I\!R^2 \backslash \Omega$.

For each $\mu \geq r$, $S_k^{r,\mu}(\Delta)$ is the space of all splines in $S_k^r(\Delta)$ which are in C^μ around each vertex of Δ. Theorem 8 was extended to super spline spaces $S_k^{r,\mu}(\Delta)$ by Ibrahim and Schumaker in [42].

An edge e emanating from a vertex v of Δ is called degenerate if its predecessor and successor are collinear. A triangulation Δ is called non-degenerate provided that it contains no degenerate edges. For a degenerate triangulation Δ, we can always find a non-degenerate triangulation, which is arbitrary close to Δ, obtained by an arbitrary small perturbation of the vertices. Alfeld and Schumaker [9] break the $3r+2$ barrier and extend Theorem 8 to the following.

Theorem 9 *Suppose Δ is a non-degenerate triangulation for a simple connected polygonal domain and $k = 3r+1$. Then*

$$\dim S_k^r(\Delta) = \alpha + \beta|E_I| - \gamma|V_I| + \sigma. \tag{3.6}$$

Furthermore, we can make the following.

Conjecture Suppose Δ is a non-degenerate triangulation and $k \geq 2r+1$. Then

$$\dim(S^{\mu}_{k,\Delta}) \geq \alpha(1-c) + \beta|E_I| - \gamma|V_I| + \sigma. \tag{3.7}$$

There is a gap between the case $k = 2r+1$ and the case $k \geq 3r+2$. For $r = 1$, It has been shown in [6] that the dimension of $S_4^1(\Delta)$ is also given by the lower bound in (3.4). The conjecture for the space $S_3^1(\Delta)$ and $S_2^1(\Delta)$ is still open. Some results are obtained only for some special triangulations (see [52] for an example). For $r = 2$, to date, the dimensions of $S_k^2(\Delta), k = 5, 6, 7$ have not yet been established. In [33], the dimension of the space $S_{3r}^r(\Delta^*)$ is discussed for a triangulation Δ^* obtained by a subdivision of an arbitrary triangulation Δ.

An interior vertex v is said to be singular, if it is the intersection of two lines which are formed by four edges emanating from v. A vertex is called a near-singular vertex if it is an interior vertex and if the sum of two angles of the vertex of two neighboring triangles is near 180^0. Locally supported bases for the bivariate spline spaces were also constructed in [37]. However, the norm of the basis constructed there depends on the smallest angle and the "near-singularity" of the triangulation (see [36]). A local stable basis of the super spline space $S_k^{r,\mu}(\Delta)$ plays an essential role in constructing quasi-interpolation schemes for optimal order approximation from the space $S_k^r(\Delta)$. We now construct a local stable basis for the space $S_k^{r,\mu}(\Delta)$ when $k \geq 3r+2$.

Recall that X denotes the set of domain points as defined in (2.1) and b_s is the B-net representation of the spline function s. By using Theorem 3, the new formulation of smoothness conditions of splines, we can select a subset Y of X so that the linear map: $s \mapsto b_s|_Y$ between the spline s and its B-net representation b_s is one-to-one and onto (see [23]). Such a set Y usually is called a minimally determining set of the space $S_k^{r,\mu}(\Delta)$. For a given point $x \in Y$, there is a unique spline $B_x \in S_k^{r,\mu}(\Delta)$ whose B-net representation b satisfies

$$b(y) = \begin{cases} 1 & y = x, \\ 0 & y \in Y \backslash \{x\}. \end{cases} \tag{3.8}$$

It can be proved that $\{B_x : x \in Y\}$ constitutes a basis of $S_k^{r,\mu}(\Delta)$. Furthermore, we obtain the following (see [23]).

Theorem 10 *The basis $\{B_x : x \in Y\}$ of $S_k^{r,\mu}(\Delta)$ is stable in the sense that there are two positive constants K_1 and K_2, depending only on k and the smallest angle a of the triangulation Δ, such that*

$$K_1 \sup_{x\in Y} |c_x| \leq \left\| \sum_{x\in Y} c_x B_x \right\|_\infty \leq K_2 \sup_{x\in Y} |c_x|. \tag{3.9}$$

This basis is also local in the sense that, for any $x \in Y$, there exists a vertex u such that

$$supp\ B_x \subseteq \overline{St}^{\lfloor r/2 \rfloor + 1}(u), \tag{3.10}$$

where the closed star of a vertex v, denoted by $\overline{St}(v) =: \overline{St}^1(v)$, is the union of all the triangles attached to v, and the m-star of v, denoted by $\overline{St}^m(v)$, is the union of all triangles that intersect $\overline{St}^{m-1}(v)$, $m > 1$.

The generic dimension is such that if $\dim S_k^r(\Delta)$ does not equal it then there is an arbitrarily small perturbation in the location of the vertices that will cause $\dim S_k^r(\Delta)$ to equal the generic value. From a different point of view, Billera in [13] developed a homological approach to this problem. He used techniques from algebraic homology and rigidity theory to obtain certain formulas for the generic dimensions of the spaces $S_k^1(\Delta)$, $k \geq 2$. This work is based on Whiteley's solution of the spline matrix problem (see [63]). For an introduction to this subject, see [64] and the references therein.

Remark 1. Here, we list some unsolved problems for the bivariate spline spaces.

Problem 1. What is the precise dimension of $S_3^1(\Delta)$? The generic dimension is $3|V_b|+2|V_I|+1$, as shown by Billera and Whiteley. The conjecture is

$$\dim S_3^1(\Delta) = 3|V_b| + 2|V_I| + \sigma$$

with σ being the number of singular vertices.
Problem 2. What is the precise dimension of $S_2^1(\Delta)$? The generic dimension is $|V_b|+3$, as shown by Billera and Whiteley. We conjecture that there is no uniform formula for $\dim S_2^1(\Delta)$.
Problem 3. What is the generic dimension of $S_k^r(\Delta)$ for $r>1$ and $k \le 3r$?

It can be seen that the theory of multivariate splines is even more complicated in higher dimensional settings ($s \ge 3$) than in the bivariate case. Recently, some progress has also been made in the study of dimension problems of spline spaces in higher dimensional settings (see [10], [12], [11] and the references therein). By counting the free B-net points in the Bézier net associated with $S_k^0(\Delta)$ we have the following dimension result (see [10]).

Theorem 11 *For $s=3$ and all $k \ge 1$,*

$$\dim(S^0_{k,\Delta}) = |V| + (k-1)|E| + \binom{k-1}{2}|F| + \binom{k-1}{3}T, \qquad (3.11)$$

where V, E, F, and T are the set of vertices, edges, faces, and tetrahedral in the partition, respectively.

Later, the same authors of the paper obtain the following result on generic dimension for trivariate spline spaces.

Theorem 12 *Suppose $k \ge 8$, and let Δ be a generic tetrahedral decomposition. Then*

$$\dim S_k^1(\Delta) = 1 + 5k - 2k^2 + k(k-1)|V_b| + 3(k-1)|V_I| + \frac{k(k-1)(k-5)}{6}T. \qquad (3.12)$$

Remark 2. There are even more unsolved problems in higher dimensional settings. We list some of them as follows.
Problem 4. What is the precise dimension of $S_k^1(\Delta)$ for $k \ge 9$?
Problem 5. What is the generic dimension of $S_k^r(\Delta)$ for $r>1$ and $k \ge 8r+1$?
Problem 6. What is the precise dimension of $S_k^r(\Delta)$?
Problem 7. What are minimal values of k as a function of s for $S_k^r(\Delta)$ to have a local support basis?

Very recently, the spaces of splines defined on triangulations lying on the sphere or on sphere-like surfaces have been discussed in [1– 4]. These spaces arose out of a new kind of Bernstein-Beźier theory on such surfaces. A constructive theory for such spline spaces analogous to the well-known theory of polynomial splines on planar triangulations has been developed. Formulae for the dimension of such spline spaces, and locally supported bases for them, are given in [2]. Some applications of such spline spaces to fit scattered data on sphere-like surfaces are discussed in [4].

4 Approximation from bivariate spline spaces over triangulations

In applications of the spline space $S_k^r(\Delta)$, it is critical to answer the question of how well the splines can approximate classes of smooth functions. In this section, we investigate approximation properties of the bivariate spline space $S_k^r(\Delta)$. Let us begin with a definition of the approximation order of a function space.

Definition 13 *The approximation order of a space S of functions on $I\!R^2$ is defined to be the largest real number ρ for which*

$$dist(f, S) \leq Const\|D^{k+1} f\| \, |\Delta|^\rho \tag{4.1}$$

for any sufficiently smooth function f, with the distance measured in the maximum norm $\| \; \|$, and with the mesh size $|\Delta| := \sup_{\tau\in\Delta} diam\,\tau$.

It is clear that the approximation order from the spline space $S_k^r(\Delta)$ cannot be better than $k+1$ regardless of r and is trivially $k+1$ in the case $r = 0$. A great deal of work has been done on the approximation order of the bivariate spaces $S_k^r(\Delta)$ for a three or four-directional mesh Δ (see [16—19], [29– 30], [43], [44], [46] and the references therein). Most of the results on approximation order of the multivariate spline spaces are in the two dimensional setting. Thus, we restrict ourselves on bivariate spline spaces in this section. In general, it is well known that the approximation order of $S_k^r(\Delta)$ not only depends on k and r, but also on the geometric structure of the partition Δ. According to the results of Ženišek ([65], [66]), it was believed in the past that the full approximation order of $\rho = k+1$ can be obtained from the spline space $S_k^r(\Delta)$ only when the degree of the polynomial k is at least $4r+1$. In a more recent work, however, de Boor and Höllig [18] proved that $S_k^r(\Delta)$ already has full approximation order, provided that $k \geq 3r+2$. To achieve this goal the technique used in [18] was to "disentangle the rings" by carefully working with the smoothness conditions around each vertex using the B-nets and the duality. Consequently, they obtained the approximation order without exhibiting an approximation scheme that attains this order.

For applications, one is required to construct an efficient scheme to achieve the full order $k+1$ of approximation. For this purpose, explicit bases for the spaces $S_k^r(\Delta)$ are set up when $k \geq 3r+2$ and an approximation scheme using such bases to achieve the optimal order of approximation is discussed in [37]. Chui and Lai [24] gave a basis of vertex splines for the super spline spaces $S_k^{r,\mu}(\Delta)$, where $\mu \geq r + \lfloor \frac{k-2r-1}{2} \rfloor$ and $k \geq 3r+2$. They also constructed an approximation scheme by using this super vertex spline basis to prove that the super spline subspace of the space $S_k^r(\Delta)$ already achieves the full order of approximation. Nevertheless, as in [35], the estimation constant in (4. 1) obtained by schemes such as those presented in [24] not only depends on the smallest angle of Δ but also on the measurement of the near-singularity of Δ, i.e., the constant becomes large for near singular vertices (see [36] or [23]). This dependence was also observed by de Boor who pointed out in [15] that the original argument presented in [18] does not fully support the conclusion that the estimation constant depends only on the smallest angle in Δ (see also Schumaker[59, p.547]). Therefore, it becomes even more significant to construct a scheme and to show that the estimation constant depends only on the smallest angle a of the partition Δ. The study of this topic is presented in [23]. Here, we can state the main result in [23] using Theorem 10 as follows.

Theorem 14 *For arbitrary triangulation Δ and $k \geq 3r+2$, the full order of approximation $k+1$ from the space $S_k^r(\Delta)$ can be achieved via a quasi-interpolation scheme using this local super spline basis in Theorem 10 with the estimation constant depending only on the smallest angle of the partition Δ.*

Theorem 14 shows that if k is sufficiently large compared to r, the spline space $S_k^r(\Delta)$ provides the full accuracy expected of piecewise polynomials of degree k. If $k \leq 3r+1$, generally speaking, the space $S_k^r(\Delta)$ does not have full order of approximation. This is shown by some observations on special triangulations. Recall that a type-1 or three-direction mesh is denoted by $\Delta^{(1)}$ and a type-2 or four-direction mesh is denoted by $\Delta^{(2)}$. It was shown in [17] that the approximation order of the space $S_3^1(\Delta^{(1)})$ is only three. In general, de Boor and Jia [19] proved the following.

Theorem 15 *For a three-direction mesh $\Delta^{(1)}$, the bivariate spline space $S_k^r(\Delta^{(1)})$ has approximation order at most k for $k \leq 3r+1$.*

It is well-known that for a type-2 triangulation $\Delta^{(2)}$, Dahmen and Micchelli [30] proved that the space $S_4^1(\Delta^{(2)})$ arrives at the optimal approximation order 5.

For a type-2 triangulation $\Delta^{(2)}$, Jia [48] proved the following general result by considering the local approximation order provided by the box splines in $S_k^r(\Delta^{(2)})$.

Theorem 16 *The approximation order of $S_k^r(\Delta^{(2)})$ for a four-direction mesh $\Delta^{(2)}$ is $k+1$ if $r \leq 1$ and $k \geq r+1$.*

Theorem 16 shows that the space $S_k^1(\Delta^{(2)})$ has full order of approximation if $k \geq 2$. This includes a later result published in [49] on $S_3^1(\Delta^{(2)})$.

It will be very interesting to find some triangulations Δ, which are somewhat specific but more general than the type-2 triangulation, such that the spline space $S_k^r(\Delta)$ still has full approximation order $k+1$. To achieve the optimal (fifth) order approximation from the space $S_4^1(\Delta)$ of C^1 quartic spline functions on a triangulation Δ, a so-called local Clough-Tocher refinement procedure of an arbitrary triangulation Δ is introduced by Chui and Hong in [21]. There some triangles are subdivided, using an interior point (such as the centroid of the triangle), into three subtriangles. To avoid introducing some new data points in addition to the vertex set of the triangulation Δ, as in the local Clough-Tocher refinement, we introduce a so-called mixed three-directional patch in [41] and define a particular three-directional mesh $\Delta^{(3)}$. A rectangle with northeast diagonal is called a *NE-rectangle.* Similarly, a rectangle with northwest diagonal is called a *NW-rectangle.* For a triangulation Δ which consists of NE- and NW-rectangles we may call Δ a mixed three-direction mesh and denote it by $\Delta^{(3)}$. We consider the approximation property of the space $S_4^1(\Delta^{(3)})$. Compactly supported basis functions, which are called mixed three-directional elements, for the space $S_4^1(\Delta^{(3)})$ are constructed, and the fact that the space $S_4^1(\Delta^{(3)})$ has optimal-order of approximation is proved. We also construct an explicit interpolation scheme by using the mixed three-directional elements to achieve this optimal-order of approximation. In [41], we obtain the following.

Theorem 17 *For a mixed three-direction mesh $\Delta^{(3)}$ there is a linear interpolating operator $T : f \in C^1(\Delta^{(3)}) \mapsto s \in S_4^1(\Delta^{(3)})$ such that $Tp = p$ for any polynomial $p \in \pi_4$ and such that T achieves the optimal order of approximation; that is,*

$$||Tg - g|| \leq C||g^{(5)}|||\Delta^{(3)}|^5, \text{ for } g \in C^5(\Delta^{(3)}),$$

where $|\Delta^{(3)}|$ is the mesh size of $\Delta^{(3)}$.

Therefore, the mixed three-directional mesh $\Delta^{(3)}$ is better than the three-directional mesh in the sense that the corresponding spline space has a higher order of approximation. Also the mixed three-directional mesh $\Delta^{(3)}$ is better than local refinements in the sense that the C^1 quartic spline space achieves the optimal approximation order by using a smaller number of data sites in the interpolation. In comparison, the mixed three-directional mesh uses the data only at the intersections of rectangle lines and the optimal-order 5 can also be achieved by the space $S_4^1(\Delta^{(3)})$. Therefore, the mixed three-directional mesh is also better than the four-directional mesh in this regard.

In the representation of scattered data by smooth spline functions, one would like to construct an optimal approximation scheme based on the given sample data set. Therefore, one of the most important problems is to find an optimal triangulation of the given sample sites. Though the notion of optimality depends on the desirable properties in the approximation or modeling problems, here we are concerned with optimal order with respect to the given order r of smoothness and degree k of the polynomial piece of the smooth spline functions. Now, we consider C^1 quartic spline approximation and provide an efficient method for triangulating any finite arbitrarily scattered sample sites, such that these sample sites are the only vertices of the triangulation and such that, for any discrete data given at these sample sites, there is a C^1 quartic polynomial spline on this triangulation that interpolates the given data with the optimal order of approximation. We begin with a notion of type-O triangulation.

A vertex u will be called a *type-O vertex* of a triangulation Δ if u satisfies at least one of the following.
(a) u is a boundary vertex of Δ.
(b) $u \in V_I$ with $deg(u) = 4$.
(c) $u \in V_I$ and $deg(u)$ is an odd integer.
(d) $u \in V_I$ and there exists a vertex v of Δ that satisfies either (i) $v \in V_I$ and $deg(v) = 4$ or $deg(v) =$ an odd integer, or (ii) $v \in V_b$, such that $[u, v]$ is a nondegenerate edge of Δ with respect to u.

We will use V_O to denote the collection of all type-O vertices in V.

Definition 18 *A triangulation of V with only type-O vertices (i.e., $V = V_O$) is called a type-O triangulation.*

The reason for introducing the notion of type-O triangulations is the following (see [22]).

Theorem 19 *Any type-O triangulation Δ admits the 5th order of approximation from $S_4^1(\Delta)$.*

As a consequence of Theorem 19, we have

Corollary 20 *If a triangulation Δ consists only of odd-degree interior vertices, then the spline space $S_4^1(\Delta)$ yields the optimal order of approximation.*

To convert any triangulation to be a type-O triangulation, we introduce a so-called swapping algorithm. Every interior edge e of a triangulation Δ is the diagonal of a quadrilateral Q_e which is the union of two triangles of Δ with common edge e. Following [60], we say that e is a swappable edge if Q_e is convex and no three of its vertices are collinear. If an edge e of a triangulation Δ is swappable, then we can create a new triangulation by swapping the edge. That is, if $v_1, \cdots, v_4$ are the vertices of Q_e ordered in the counterclockwise direction, and if e has endpoints v_1 and v_3, then the swapped edge has endpoints v_2 and v_4. Two vertices in Δ will be called *neighbors* of each other if they are the endpoints of the same edge in Δ. Hence, while v_1 and v_3 are neighbors in the original triangulation Δ, v_2 and v_4 become neighbors in the new triangulation after the edge e is swapped.

For any given set of sample sites, it is clear that, with the exception of those that are collinear, there is a triangulation with these sample sites as its only vertices. Let Δ be a triangulation associated with the given set V, and let V_O be the set of all type-O vertices in Δ. Set

$$\tilde{V} = V \setminus V_O.$$

If $u \in \tilde{V}$, then u and all its neighbors with nondegenerate edges with respect to u must be even-degree vertices with $deg(u) \geq 6$. We can see that, for every interior vertex u with $n := deg(u) \geq 5$, there is a swappable edge $e \in E_u$. Hence, there is at least one vertex u_i such that both $\angle u_{i-1}uu_{i+1}$ and $\angle u_{i-1}u_iu_{i+1}$ are less than π. Therefore, the quadrilateral $Q := [u_{i-1}, u_i, u_{i+1}, u]$ is convex; and hence, the edge $[v, v_i]$ is swappable.

Now we are ready to describe our *Swapping Algorithm* for constructing a type-O triangulation $\widehat{\Delta}$, starting with any triangulation Δ.

Swapping Algorithm

Do while $(\tilde{V} \neq \emptyset)$
Pick any vertex u in $\tilde{V}$ and consider its neighbors.
Pick any neighbor v of u so that the edge $[u, v]$ is swappable.
Swap $[u, v]$, yielding a new edge $[u', v']$.
Form a subset of $\tilde{V}$ by deleting from $\tilde{V}$ all the neighbors w of $w' := u$, v, u', or v', with $[w, w']$ being a nondegenerate edge with respect to w.
Call this subset $\tilde{V}$.
Enddo

The new triangulation obtained by applying this Swapping Algorithm is denoted by $\hat{\Delta}$. It is clear that the triangulations Δ and $\hat{\Delta}$ have the same number of triangles, singular vertices, interior and boundary vertices, and edges. Hence, it follows that

$$\dim S_4^1(\hat{\Delta}) = \dim S_4^1(\Delta).$$

Combining the Swapping Algorithm with Theorem 19, we have the following.

Theorem 21 *Every finite set V of sample sites admits an optimal triangulation Δ, such that the C^1 quartic spline space $S_4^1(\Delta)$ has the fifth order of approximation.*

Remark 3. Some special partitions for optimal order of approximation purposes have been discussed in [50] and [51]. In general, the question of how to find an optimal triangulation for the bivariate spline space $S_k^r(\Delta)$ with $k \leq 3r+1$ is still open. Specially, we may list some of open problems as follows.
Problem 8. Is there any mixed three-direction mesh Δ so that the bivariate cubic spline space over Δ has optimal order of approximation?
Problem 9. What is the relation between k and r so that the spline space $S_k^r(\Delta)$ over a four-direction mesh has optimal order of approximation?

Acknowledgment. This research was supported in part by a Research Development Grant from East Tennessee State University. The author would like to thank the referee's valuable suggestions. He is also grateful to Janice Huang for her carefully reading the paper.

References

[1] P. Alfeld, M. Neamtu, and L. L. Schumaker, Circular Bernstein-B'ezier polynomials, in "Mathematical Methods for Curves and Surfaces", M. æhlen, T. Lyche, and L.L. Schumaker (Eds.), pp.11-20, Vanderbilt University Press, 1995.

[2] P. Alfeld, M. Neamtu, and L. L. Schumaker, Dimension and local bases of homogeneous spline spaces, SIAM J. Math. Anal., to appear.

[3] P. Alfeld, M. Neamtu, and L. L. Schumaker, Bernstein-B'ezier polynomials on spheres and sphere-like surfaces, Compt. Aided Geom. Design, to appear.

[4] P. Alfeld, M. Neamtu, and L. L. Schumaker, Fitting scattered data on sphere-like surfaces using spherical splines, *J. Comp. Appl. Math.*, to appear.

[5] P. Alfeld, B. Piper, and L. L. Schumaker, Minimally supported bases for spaces of bivariate piecewise polynomials of smoothness r and degree $k \geq 4r+1$, *Computer Aided Geometric Design* **4** (1987), 105–123.

[6] P. Alfeld, B. Piper, and L. L. Schumaker, An explicit basis for C^1 quartic bivariate splines, *SIAM J. Numer. Anal.* **24** (1987), 891–911.

[7] P. Alfeld, B. Piper, and L. L. Schumaker, Spaces of bivariate splines on triangulations with holes, *J. Approx. Theory Appl.* **3** (1987), 1-10.

[8] P. Alfeld and L. L. Schumaker, The dimension of bivariate spline spaces of smoothness r for degree $k \geq 4r+1$, *Constr. Approx.* **3** (1987), 189-197.

[9] P. Alfeld and L. L. Schumaker, On the dimension of bivariate spline spaces of smoothness r and degree $d = 3r+1$, *Numer. Math.* **57** (1991), 651–661.

[10] P. Alfeld, L.L. Schumaker, and M. Sirvent, On dimension and existence of local bases for multivariate spline spaces, *J. Approx. Theory* **70** (1992), 243–264.

[11] P. Alfeld and M. Sirvent, The structure of multivariate super spline spaces of high degree, *Math. Comp.* **57** (1991), 299–308.

[12] P. Alfeld, L.L. Schumaker, and W. Whiteley, The generic dimension of the space of C^1 splines of degree $d \geq 8$ on tetrahedral decompositions, *SIAM J. Numer. Anal.* (to appear).

[13] L.J. Billera, Homology of smooth splines: Generic triangulations and a conjecture of Strang, *Trans. Amer. Math. Soc.* **310** (1988), 325-340.

[14] C. de Boor, B-form Basis, in: "Geometric Modeling", (G. Farin, Ed.), pp. 21–28, SIAM, Philadelphia, 1987.

[15] C. de Boor, A Local basis for certain smooth bivariate *pp* spaces, in: "Multivariate Approximation IV", (C. K. Chui, W. Schempp, and K. Zeller, Eds.), pp. 25–30, Birkhäuser, Basel, 1989.

[16] C. de Boor and R. DeVore, Approximation by smooth multivariate splines, *Trans. Amer. Math. Soc.* **276** (1983), 775–788.

[17] C. de Boor and K. Höllig, Approximation order from bivariate C^1-cubics: A counterexample, *Proc. Amer. Math. Soc.* **87** (1983), 649–655.

[18] C. de Boor and K. Höllig, Approximation power of smooth bivariate *pp* functions, *Math. Z.* **197** (1988), 343–363.

[19] C. de Boor and R. Q. Jia, A sharp upper bound on the approximation order of smooth bivariate *pp* functions, *J. Approx. Theory* **72** (1993), 24–33.

[20] C. K. Chui, "Multivariate Splines", CBMS Series in Applied Mathematics, no. **54**, SIAM, Philadelphia, 1988.

[21] C. K. Chui and D. Hong, Construction of local C^1 quartic spline elements for optimal-order approximation, *Math. Comp.* **65** (1996), 85-98. MR **96d:**65023.

[22] C.K. Chui and D. Hong, Swapping edges of arbitrary triangulations to achieve the optimal order of approximation. *SIAM J. Numer. Anal.* **34** (1997), xxx–xxx.

[23] C.K. Chui, D. Hong, and R.Q. Jia, Stability of optimal-order approximation by splines over arbitrary triangulations, *Trans. of Amer. Math. Soc.* **374** (1995), 3301-3318. MR **96d:**41012.

[24] C. K. Chui and M. J. Lai, On bivariate super vertex splines, *Constr. Approx.* **6** (1990), 399–419.

[25] C.K. Chui and R.H. Wang, On smooth multivariate spline functions, *Math. Comp.* **41** (1983), 131–142.

[26] C.K.Chui and R.H. Wang, Multivariate spline spaces, *J. Math. Anal. Appl.* **94** (1983), 197–221.

[27] C.K. Chui and R.H. Wang, On a bivariate B-spline basis, *Scientia Sinica* **27** (1984), 1129–1142.

[28] W. Dahmen and C.A. Micchelli, Recent progress in multivariate splines, in "Approximation Theory IV", (C.K. Chui, L.L. Schumaker, and J.D. Ward, Eds.), pp. 27–121, Academic Press, New York, 1983.

[29] W. Dahmen and C.A. Micchelli, On the approximation order from certain multivariate spline spaces, *J. Austral. Math. Soc. Ser. B* **26** (1984), 233–246.

[30] W. Dahmen and C.A. Micchelli, On the optimal approximation rates from criss-cross finite element spaces, *J. Comp. Appl. Math.* **10** (1984), 255–273.

[31] G. Farin, Bézier polynomials over triangles and the construction of piecewise C^r polynomials, *Technical Report TR/91*, Brunel University, Uxbridge, England, 1980.

[32] G. Farin, Triangular Bernstein-Bézier patches, *Computer Aided Geometric Design* **3** (1986), 87–127.

[33] J.B. Gao, On the dimension of the bivariate spline spaces $S^r_{3r}(\Delta^*)$, *J. Math. Res. & Exp.* **14** (1994), 367-378.

[34] Z.R. Guo and R.Q. Jia, A B-net approach to the study of multivariate splines, (Chinese), *Adv. Math.* **19** (1990), 189–198. MR **91c:**41009.

[35] D. Hong, "On bivariate spline spaces over arbitrary triangulations", Master's Thesis, Zhejiang University, Hangzhou, Zhejiang, China, 1987.

[36] D. Hong, "Construction of stable local spline bases over arbitrary triangulations for optimal order approximation", Ph.D. Dissertation, Texas A&M University, College Station, TX, 1993.

[37] D. Hong, Spaces of bivariate spline functions over triangulations, *Approx. Theory and Appl.* **7** (1991), 56–75. MR **92f:**65016.

[38] D. Hong, A new formulation of Bernstein-Bezier based smoothness conditions for *pp* functions, *Approx. Theory and Appl.* **11** (1995), 67–75.

[39] D. Hong, Optimal triangulations for the best C^1 quartic spline approximation, in "Approximation Theory VIII, Vol 1: Approximation and Interpolation," (C.K. Chui and L.L. Schumaker, Eds.), pp.249-256, World Scientific Publishing Co., Inc., 1995.

[40] D. Hong and H.-W. Liu, Some new formulations of smoothness conditions and conformality conditions for bivariate splines, *Computers and Mathematics with Applications (1998).*

[41] D. Hong and R. N. Mohapatra, Optimal-order approximation by mixed three-directional spline elements, *Computers and Mathematics with Applications* **xx** (1998), xx-xx.

[42] A. K. Ibrahim and L. L. Schumaker, Super spline spaces of smoothness r and degree $d \geq 3r + 2$, *Constr. Approx.* **7** (1991), 401–423.

[43] R.Q. Jia, Approximation by smooth bivariate splines on a three-direction mesh, in "Approximation Theory IV", (C.K. Chui, L.L. Schumaker, and J.D. Ward, Eds.), pp. 539–545, Academic Press, New York, 1983.

[44] R.Q. Jia, Approximation order from certain spaces of smooth bivariate splines on a three direction mesh, *Trans. Amer. Math. Soc.* **295** (1986), 199–212.

[45] R. Q. Jia, B-net Representation of Multivariate Splines, *Ke Xue Tong Bao* (*A Monthly Journal of Science*) **11** (1987), 804–807.

[46] R. Q. Jia, Local approximation order of box splines, *Scientia Sinica* **31** (1988), 274–285.

[47] R. Q. Jia, Lower bounds on the dimension of spaces of bivariate splines, in "Multivariate Approximation and Interpolation", (W. Haussmann and K. Jetter Eds.), pp. 155–165, Birkhäuser Verlag, Berlin, 1990.

[48] R.Q. Jia, Lecture Notes on Multivariate Splines, Department of Mathematics, University of Alberta, Edmonton, Canada. 1990.

[49] M.J. Lai, Approximation order from bivariate C^1-cubic on a four-directional mesh is full, *Comput. Aided Geometic Design* **11** (1994), 215-223.

[50] M.J. Lai and L.L. Schumaker, Scattered data interpolation using C^2 piecewise polynomials of degree six, *SIAM Numer. Anal.* **34** (1997), 905-921.

[51] M.J. Lai and L.L. Schumaker, Quadrangulating scattered data points, manuscript.

[52] H.-W. Liu, ntegral representation of bivariate splines and the dimension of quadratic spline spaces over stratified triangulations, *Acta Math. Sinica* **4** (1994), 534-543.

[53] C. Manni, On the dimension of bivariate spline spaces on generalized quasi-cross-cut partitions, *J. Approx. Theory* **69** (1992), 141–155.

[54] J. Morgan and R. Scott, A nodal basis for C^1 piecewise polynomials of degree ≥ 5, *Math. Comp.* **29** (1975), 736–740.

[55] P. Sablonniére, Bernstein-Bézier methods for the construction of bivariate spline approximants, *Computer Aided Geometric Design* **2** (1985), 29–36.

[56] L. L. Schumaker, On the dimension of spaces of piecewise polynomials in two variables, in "Multivariable Approximation Theory", (W. Schempp and K. Zeller, Eds.), pp. 396–412, Birkhäuser, Basel, 1979.

[57] L.L. Schumaker, Bounds on the dimension of spaces of multivariate piecewise polynomials, *Rocky Mountain J. Math.* **14** (1984), 251–264.

[58] L.L. Schumaker, On super splines and finite elements, *SIAM J. Numer. Anal.* **26** (1989), 997–1005.

[59] L.L. Schumaker, Recent progress on multivariate splines, in "Mathematics of Finite Elements VII", (J. Whiteman, Ed.), pp. 535–562, Academic Press, London, 1991.

[60] L. L. Schumaker, Computing optimal triangulations using simulated annealing, *Computer Aided Geometric Design* **10** (1993), 329–345.

[61] G. Strang, Piecewise polynomials and the finite elements method, *Bull. Amer. Math. Soc.* **79** (1973), 736–740.

[62] G. Strang, The dimension of piecewise polynomials, and one-sided approximation, "Proc. Conf. Numerical Solution of Differential Equations", Dundee (1973), pp. 144–152, Lecture Notes in Mathematics, no. **365**, Springer-Verlag, New York, 1974.

[63] W. Whiteley, A matrix for splines, in "Progress in Approximation Theory", (P. Nevai and A. Pinkus, Eds.), pp. 821–828, Academic Press, Boston, 1991.

[64] W. Whiteley, The combinatorics of bivariate splines, in "Applied Geometry and Discrete Mathematics", (P. Gritzmann and B. Sturmfels, Eds.), pp. 587–608, DIMACS Series, AMS Press, Providence, 1991.

[65] A. Ženišek, Interpolation polynomials on the triangle, *Numer. Math.* **15** (1970), 283-296.

[66] A. Ženišek, Polynomial approximation on tetrahedron in the finite element method, *J. Approx. Theory* **7** (1973), 334-351.

Hermite Interpolation on Chebyshev Nodes and Walsh Equiconvergence

A. Jakimovski
School of Mathematical Sciences
Tel-Aviv University, Tel-Aviv, Israel

A. Sharma
Department of Mathematics
University of Alberta, Edmonton, Alberta, Canada T6G 2G1

Abstract

While Brück et al. [1] have studied Walsh equiconvergence on Faber nodes with respect to an arbitrary compact set E for which $C_\infty \backslash E$ is simply connected, their results when specialized to the case of Chebyshev nodes (which are Faber nodes with respect to $E = [-1, 1]$) are not sharp. Here using elementary methods we first find the explicit representation of Hermite interpolant on Chebyshev nodes with multiplicity $p \geq 1$ to a function $\in A(C_\rho)$, obtain the region of Walsh equiconvergence and show that it is sharp. An explicit representation of the Hermite interpolant to a function on Chebyshev nodes with multiplicity $p \geq 1$ is not known to the best of our knowledge.

1 Introduction

Let C_R, $R \geq 1$, be the ellipse in the complex z plane which is the image of the circle $|w| = R$ in the complex w plane by the mapping $z = \frac{1}{2}(w + w^{-1})$. The ellipse C_R has foci ± 1 and the axes $a = \frac{1}{2}(R + \frac{1}{R})$, $b = \frac{1}{2}(R - \frac{1}{R})$. Notice that an ellipse with foci ± 1 and half axes (in the direction of the $x, y - axes$ whose sum is $R > 1$ is necessarily C_R . For $\rho > 1$, $A(C_\rho)$ denotes the set of functions analytic inside C_ρ but having a singularity on C_ρ. Let $T_k(z)$ denote the Chebyshev polynomial (of the first kind) of degree k. Each $z \in C_R$, $R > 1$, has a unique representation $z = \frac{1}{2}(w + \frac{1}{w})$ for a unique $|w| = R > 1$. For such a pair z, w we have $T_k(z) = \frac{1}{2}(w^k + \frac{1}{w^k})$. A function

$f(z) \in A(C_R)$ has a unique Chebyshev expansion inside the ellipse C_R of the form

$$f(z) = \frac{1}{2}A_0 + \sum_{k=1}^{\infty} A_k T_k(z). \tag{1.1}$$

Set $S_n(f,z) := \frac{1}{2}A_0 + \sum_{k=0}^{n} A_k T_k(z)$. It has been shown by Rivlin [6] that the Lagrange interpolant to $f(z)$ in the zeros of $T_m(z) = \cos m\theta$, $z = e^{i\theta}$ is given by

$$L_{m-1}(f,z) = \sum_{j=0}^{m-1} a_j^{(m)} T_j(z), \tag{1.2}$$

where

$$\begin{cases} a_0^{(m)} &= \frac{1}{2}A_0 + \sum\limits_{k=1}^{\infty} (-1)^k A_{2mk} \\ a_j^{(m)} &= A_j + \sum\limits_{k=1}^{\infty} (-1)^k (A_{2mk-j} + A_{2mk+j}), \quad (j = 1, \ldots, m-1). \end{cases} \tag{1.3}$$

His method used properties of Chebyshev-polynomials and the Gauss quadrature formula [7]. It seems to be difficult to obtain the coefficients of the Hermite interpolant to $f(z)$, given by (1.1), in the zeros of $\left(T_m(z)\right)^p$ by using Rivlin's method. Here we obtain by elementary methods an explicit formula for the Hermite interpolant to $f(z)$ on the zeros of $(T_m(z))^p$. Rivlin used the Lagrange interpolant (1.2) to obtain a Walsh type equiconvergence result for such interpolants for functions in $A(C_\rho)$. Since Walsh equiconvergence has been widely studied ([2], [3], [9], [10]) for functions analytic in $|z| < \rho$ with respect to their Lagrange and Hermite interpolants in the roots of unity, it is desirable to examine if Rivlin's result can be extended for Hermite interpolants to functions of the type (1.1) which are in $A(C_\rho)$. For a historical survey of Walsh equiconvergence, we refer to [8].

Recently explicit formulae for Hermite interpolation in the zeros of $(z^n - 1)^r$ to a function given by a power series $\sum_{\nu=0}^{\infty} a_\nu z^\nu$ have been obtained by Goodman et al. [4]. However the methods used there and the results obtained therein do not seem to be related to the present results. Brück et al. [1] have considered the problem of Walsh equi-convergence on Faber nodes with respect to a general set E. When $E = [-1,1]$, Brück et al. give a region for Walsh equiconvergence but do not give the explicit representation of the Hermite interpolant. Here we give an explicit formula for the Hermite interpolant and make the results of Brück et al. [1] more precise. In fact, we get here the best region of Walsh equiconvergence.

The object of this paper is twofold: first to obtain an explicit formula for the Hermite interpolant to $f(z)$ of the form (1.1) in the zeros of $(T_n(z))^p$ for a given integer $p > 1$ and then to apply it to obtain an extension of Rivlin's result [6] on Walsh equiconvergence to Hermite interpolation. When $p = 2$, and 4, the explicit expressions are easier but for general p, the problem is difficult but our approach leads to relatively simple expressions.

The plan of the paper is as follows. In Sec.2, we give the notation and a statement of the main results, with some examples of the coefficients of the Hermite interpolants for $1 \leq p \leq 4$. Sec.3 deals with finding the Hermite interpolant of Chebyshev monomials like $\cos((ps + j)n + r)\theta$, $0 \leq j \leq p - 1$, $1 \leq r \leq n - 1$. The case $r = 0$ is the subject of Sec.4 where we find the Hermite interpolant to Chebyshev monomials of the form $\cos(ps+j)n\theta$. In Sec.5 we apply the result of Sec.4 to a linear combination of Chebyshev monomials of the form $\cos(ps + j)n\theta$.. This leads to the proof of Theorem 2.1(ii) and Theorem 2.4(ii) when $\sigma = \rho n$, ρ even, and of Theorem 2.4(i) when $\sigma = \rho n$, ρ odd.

In Sec.6, we make some remarks on Theorem 3.4 which prepare the groundwork for the proof of Theorem 2.4 and Theorem 2.1 in the subsections 5.1 and 5.2 to follow. Lastly in Sec.7, we apply Theorems 2.1 and 2.4 to the study of Walsh equiconvergence. Rivlin considered the difference $\Delta_n(f, z) := S_{n-1}(L_{m-1}(f, \cdot), z) - S_{n-1}(f, z)$, $m = nq + c$, c a constant, and showed that $\limsup_{n\to\infty} |\Delta_n(f, z)|^{1/n} = 0$ for $z = \frac{1}{2}(w + \frac{1}{w})$, $\rho < |w| < \rho^{2q-1}$. A slight generalization is in Juneja and Dua [5]. We replace the Lagrange interpolant by Hermite interpolant $H^{\{p\}}_{pm-1}(f, z)$ in the zeros of $(T_m(z))^p$. We consider the difference $\Delta_{l,n,m,p}(f, z)$ defined in Sec.2 and obtain the region of equiconvergence of this difference. For $p = 1$, $l = 1$, it gives the result of Rivlin. For $p = 1$, $l \geq 1$, it is the result of Juneja and Dua [5]. It is interesting that the region of equiconvergence depends on the parity of p -- a phenomenon which seems to be new and is not present in Walsh equiconvergence for functions analytic in a circle of radius ρ.

Rivlin [7] has shown that the Lagrange interpolation polynomial to $f(z)$ given by (1.1) on the extrema of Chebyshev polynomial $T_m(z)$ (i.e., at the zeros of $\frac{\sin m\theta}{\sin\theta}$) is given by $\sum_{j=0}^{m-1} b_j^{(m)} T_j(z)$, where

$$b_0^{(m)} = \tfrac{1}{2}A_0 + \textstyle\sum_{k=1}^{\infty} A_{2mk}$$

$$b_j^{(m)} = A_j + \textstyle\sum_{k=1}^{\infty}(A_{2mk-j} + A_{2mk+j}),\ j = 1, \dots, m-1.$$

It can be easily seen that the Hermite interpolant to $f(z)$ in the zeros of $\left(\frac{\sin m\theta}{\sin\theta}\right)^2$ is given by $\sum_{j=0}^{2m-1} b_j^{(m)} T_j(z)$, where

$$b_0^{(m)} = \tfrac{1}{2}A_0 + \textstyle\sum_{\lambda=1}^{\infty} A_{2m\lambda}$$

$$b_j^{(m)} = A_j + \textstyle\sum_{\lambda=1}^{\infty}\left\{(\lambda+1)A_{2m\lambda+j} - \lambda A_{2m(\lambda+1)-j}\right\},\ j = 1, \dots m.$$

Following the method used in the present paper it is possible to find the Hermite interpolants on the zeros of $(\sin n\theta)^p$. A corresponding result on Walsh equiconvergence will also follow naturally. We propose to return to this and other related problems later.

2 Notation and statements of main results

The symbol $\sum_{k=0}^{\infty}{}' a_k$ is defined as the sum $\frac{1}{2}a_0 + a_1 + a_2 + a_3 + \cdots$ and a'_k denotes a_k when $k \geq 1$ and it denotes $\frac{1}{2}a_0$ when $k = 0$. Empty sums, that is, sums $\sum_p^q$ with $p > q$ are defined to be zero. Write $I\!N := \{1, 2, 3, \ldots\}$, $I\!N^+ := \{0, 1, 2, 3, \ldots\}$, $Z := \{0, \pm 1, \pm 2, \ldots\}$ and $e_n := 1$ when n is odd and $e_n := 0$ when n is even. The following relations are easily proved and will be used often:

$$\begin{cases} e_p + e_{p+1} = 1\ ,\ e_{ps} = e_p e_s\ ,\ e_{(e_p)} = e_p\ ,\ e_{ps+2u-e_j} = (-1)^j e_p e_s + e_j\ , \\ e_{e_p e_s - 1} = 1 - e_p e_s = e_{s-1} + e_s e_{p-1}\ ,\ (-1)^p = (-1)^{e_p}\ ,\ e_{-p} = e_p\ , \\ e_{ps+j} = (-1)^j e_p e_s + e_j\ ,\ (-1)^{ps} e_p e_s + e_p e_s = 0\ . \end{cases} \tag{2.0}$$

Let $p, n \in I\!N^+$. Given two real functions f, g defined on $[-1, 1]$ we write $f \overset{p,n}{\sim} g$ if $f(\theta) - g(\theta)$ and its derivatives up to, and including, the order $p-1$ vanish at the zeros of the function $\cos^p n\theta = 0$. For the sake of brevity we say that f is congruent to g modulo $(\cos n\theta)^p$. Let η denote the sequence $\eta_0 := 1$, $\eta_k := 2$ for $k \in I\!N$.

The following is the first of the main results. It gives the explicit form of the Hermite interpolant of order p to a function belonging to $A(C_\rho)$ when p is an even integer.

Theorem 2.1 *Assume that p is a positive even integer and s, n are positive integers. For a function $f(z) \in A(C_R)$ given by (1.1), let $H_{pn-1}^{\{p\}}(z, f) = \sum_{\sigma=0}^{pn-1} h_\sigma^{(p)} T_\sigma(z)$ where $f(z) \overset{p,n}{\sim} H_{pn-1}^{\{p\}}(z, f)$. Then we have:*
(i) *For each integer σ, $0 < \sigma < pn$, $2\lambda n < \sigma < (2\lambda + 2)n$, $0 \leq \lambda \leq \frac{1}{2}p - 1$, we have*

$$\begin{aligned} h_\sigma^{(p)} &= A_\sigma - \binom{p-1}{\frac{1}{2}p+\lambda} \sum_{s=1}^{\infty} (-1)^{\frac{1}{2}p(s+1)} \sum_{\tau=0}^{\frac{1}{2}p-1} (-1)^\tau \binom{\frac{1}{2}p(s+1)+\tau}{p} \times \\ &\quad \times \left(\frac{p}{\frac{1}{2}ps+\tau-\lambda} A_{(ps+2\tau-2\lambda)n+\sigma} + \frac{p}{\frac{1}{2}ps+\tau+\lambda+1} A_{(ps+2\tau+2(\lambda+1))n-\sigma} \right); \end{aligned} \tag{2.1}$$

(ii) *For each* even *integer* ρ, $\rho \in \{0, 2, 4, \ldots, p-2\}$, *we have*

$$\begin{aligned} h_{\rho n}^{(p)} &= A'_{\rho n} - \tfrac{1}{2}\eta_\rho \tbinom{p}{\frac{1}{2}p+\frac{1}{2}\rho} \textstyle\sum_{s=1}^{\infty} (-1)^{\frac{1}{2}p(s+1)} \sum_{\tau=0}^{\frac{1}{2}p-1} (-1)^\tau \frac{1}{p} \times \\ &\quad \times \tbinom{\frac{1}{2}p(s+1)+\tau-1}{p-1} \tfrac{(p^2-\rho^2)(ps+2\tau)}{(ps+2\tau)^2-\rho^2} A_{(ps+2\tau)n} \end{aligned} \tag{2.2}$$

Theorem 2.1 is proved in sections 5 and 6.

Remark It should be observed that in case (i) of the above theorem when $\sigma = (2\lambda+1)n$, we have

$$A_{(ps+2\tau-2\lambda)n+\sigma} = A_{(ps+2\tau+1)n} = A_{(ps+2\tau+2(\lambda+1))n-\sigma}$$

and formula (2.1) becomes a bit simpler.

Example 2.2 *Applying Theorem 2.1 to the case* $p = 2$, *we get the following:*

(i) *For* $\sigma = 0$

$$h_0^{(2)} = \frac{1}{2}A_0 + \sum_{s=1}^{\infty}(-1)^s A_{2sn}.$$

(ii) *For each* σ, $0 < \sigma < 2n$ *we have*

$$h_\sigma^{(2)} = A_\sigma + \sum_{s=1}^{\infty}(-1)^s \left((s+1)A_{2sn+\sigma} + sA_{2(s+1)n-\sigma}\right).$$

When $\sigma = n$ *we have the simpler expression*

$$h_n^{(2)} = A_n + \sum_{s=1}^{\infty}(-1)^s(2s+1)A_{(2s+1)n}.$$

Example 2.3 *Applying Theorem 2.1 to the case* $p = 4$, *we obtain the following:*

(i) *For* $\sigma = 0$ *we have*

$$h_0^{(4)} = \frac{1}{2}A_0 - \sum_{s=1}^{\infty}\left\{(2s+1)(2s-1)A_{4sn} - (2s+2)2sA_{(4s+2)n}\right\},$$

(ii) *For* $0 < \sigma < 2n$

$$h_\sigma^{(4)} = A_\sigma - \tfrac{1}{2}\sum_{s=1}^{\infty} \{(2s+2)(2s+1)(2s-1)A_{4sn+\sigma} +(2s+2)2s(2s-1)A_{(4s+2)n-\sigma} -(2s+3)(2s+2)2sA_{(4s+2)n+\sigma} -(2s+3)(2s+1)2sA_{(4s+4)n-\sigma}\} ,$$

(iii) *For* $\sigma = 2n$ *we have*

$$h_{2n}^{(4)} = A_{2n} - \frac{1}{4}\sum_{s=1}^{\infty} \left(4s^2 A_{4sn} - (2s+1)^2 A_{(4s+2)n}\right) ,$$

and

(iv) *For* $2n < \sigma < 4n$ *we have*

$$h_\sigma^{(4)} = A_\sigma - \tfrac{1}{6}\sum_{s=1}^{\infty} \{(2s+2)(2s+1)2sA_{(4s-2)n+\sigma} +(2s+1)2s(2s-1)A_{(4s+4)n-\sigma} -(2s+3)(2s+2)(2s+1)A_{4sn+\sigma} -(2s+2)(2s+1)2sA_{(4s+6)n-\sigma}\} .$$

For $\sigma = n$ and $\sigma = 3n$, the expressions in (ii) and (iv) become simpler.

The following is the second main result. It gives a formula for the coefficients of the Hermite interpolant of order p, when p is an odd positive integer.

Theorem 2.4 *Assume that* p *is a positive odd integer. For a function* $f(z) \in A(C_R)$ *satisfying (1.1), set* $H_{pn-1}^{\{p\}}(z, f) = \sum_{\sigma=0}^{pn-1} h_\sigma^{(p)} T_\sigma(z)$ *where* $f(z) \overset{p,n}{\sim} H_{pn-1}^{\{p\}}(z, f)$. *Then:*

(i) *For each* σ *satisfying* $\rho n < \sigma < (\rho+1)n$, $(\rho = 0, \ldots, p-1)$ *or for*

$\sigma = \rho n$ where ρ is an even integer, we have

$$h_\sigma^{(p)} = A'_\sigma + \binom{p-1}{\frac{1}{2}(p-1)-(-1)^\rho\frac{1}{2}(\rho-e_\rho)+e_\rho} \sum_{s=1}^{\infty}(-1)^s \sum_{\tau=0}^{p-1}(-1)^\tau \binom{ps+\tau}{p} \times$$

$$\times \frac{p}{ps+\tau-\frac{1}{2}(p-1)+\frac{1}{2}(\rho+e_\rho)} \left\{ A_{2psn+\left((\rho+e_\rho)-(p-1)+2\tau\right)n-\sigma} \right.$$

$$\left. + A_{2psn+\left(-(\rho+e_\rho)-(p-1)+2\tau)\right)n+\sigma} \right\}; \tag{2.3}$$

(ii) *For each $\sigma = \rho n$ where ρ is an* odd *integer $0 \le \rho \le p-1$, we have*

$$h_\sigma^{(p)} = h_{\rho n}^{(p)} = A_{\rho n} + \binom{p}{\frac{1}{2}(p-\rho)} \sum_{s=1}^{\infty}(-1)^s \sum_{\tau=0}^{p-1}(-1)^\tau \binom{ps+\tau}{p} \times$$
$$\times \left\{ \frac{\frac{1}{2}(p+\rho)}{ps-\frac{1}{2}(p-\rho)+\tau} + \frac{\frac{1}{2}(p-\rho)}{ps-\frac{1}{2}(p+\rho)+\tau} - \frac{p}{ps+\tau} \right\} A_{(2ps+2\tau-p)n}. \tag{2.4}$$

Theorem 2.4(ii) is proved in Sec. 5.2 and Theorem 2.4(i) in Sec. 6.2.
Remark It should be observed that in case (i) of the above theorem when $\sigma = \rho n$ and ρ is even we have

$$\begin{aligned} A_{2psn+((\rho+e_\rho)-(p-1)+2\tau)n-\sigma} &= A_{2psn+(-(p-1)+2\tau)n} \\ &= A_{2psn+(-(\rho+e_\rho)-(p-1)+2\tau)n+\sigma}, \end{aligned}$$

and (2.3) becomes simpler.

Example 2.5 *Applying Theorem 2.4 to the case $p = 1$ (since $\binom{0}{0} = 1$), we have*

$$h_\sigma^{(1)} = \frac{1}{2}\eta_\sigma \left(A_\sigma + \sum_{s=1}^{\infty}(-1)^s (A_{2sn-\sigma} + A_{2sn+\sigma}) \right) \quad \text{for } \ 0 \le \sigma < n.$$

Example 2.6 *Applying Theorem 2.4 to the case $p = 3$ we get the following:*
(i) *For $\sigma = 0$*

$$h_0^{(3)} = \tfrac{1}{2}A_0 + 2\sum_{s=1}^{\infty} \quad (-1)^s \left\{ 3s(3s-2)A_{(6s-2)n} - (3s+1)(3s-1)A_{6sn} \right.$$
$$\left. +(3s+2)(3s)A_{(6s+2)n} \right\},$$

(ii) *For* $0 < \sigma < n$,

$$h_\sigma^{(3)} = A_\sigma + \sum_{s=1}^{\infty} (-1)^s \{3s(3s-2)(A_{(6s-2)n-\sigma} + A_{(6s-2)n+\sigma}) - (3s+1)(3s-1)(A_{6sn-\sigma} + A_{6sn+\sigma}) + (3s+2)3s(A_{(6s+2)n-\sigma} + A_{(6s+2)n+\sigma})\},$$

(iii) *For* $\sigma = n$,

$$h_n^{\{3\}} = A_n + \sum_{s=1}^{\infty} (-1)^s \{(6s-3)A_{(6s-3)n} - (6s-1)A_{(6s-1)n} + (6s+1)A_{(6s+1)n}\}.$$

(iv) *For* $n < \sigma < 2n$,

$$h_\sigma^{(3)} = A_\sigma + \sum_{s=1}^{\infty} (-1)^s \{\tbinom{3s-1}{2} A_{6sn-\sigma} + \tbinom{3s}{2} A_{(6s-4)n+\sigma} - - \tbinom{3s}{2} A_{(6s+2)n-\sigma} - \tbinom{3s+1}{2} A_{(6s-2)n+\sigma} + \tbinom{3s+1}{2} A_{(6s+4)n-\sigma} + \tbinom{3s+2}{2} A_{6sn+\sigma}\}.$$

(v) *For* $\sigma = 2n$,

$$h_{2n}^{\{3\}} = A_{2n} + \sum_{s=1}^{\infty} (-1)^s \{(3s-1)^2 A_{(6s-2)n} - (3s)^2 A_{6sn} + (3s+1)^2 A_{(6s+2)n}\}.$$

(vi) *For* $2n < \sigma < 3n$,

$$h_\sigma^{(3)} = A_\sigma + \sum_{s=1}^{\infty} (-1)^s \{\tbinom{3s-1}{2} A_{6sn-\sigma} + \tbinom{3s}{2} A_{(6s-4)n+\sigma} - \tbinom{3s}{2} A_{(6s+2)n-\sigma} - \tbinom{3s+1}{2} A_{(6s-2)n+\sigma} + \tbinom{3s+1}{2} A_{(6s+4)n-\sigma} + \tbinom{3s+2}{2} A_{6sn+\sigma}\}.$$

Rivlin was the first in [6] to obtain an equiconvergence theorem for functions $f(z) \in A(C_R)$ given by (1.1). He considered the difference

$$\Delta_{n,m}(f,z) := S_n(L_m(f,\cdot),z) - S_n(f,z), \tag{2.5}$$

where $S_n(f,z) := \sum_{\nu=0}^{n}{}' A_\nu T_\nu(z)$ and $L_m(f,z)$ is the Lagrange interpolant of $f(z)$ in the zeros of $T_m(z)$, where $m = nq+c$, c a positive constant and q is a constant greater than 1. He showed that $\limsup_{n\to\infty} \Delta_{n,m}(f,z) = 0$ for z in the interior of the ellipse $C_{\rho^{2q-1}}$. Here we consider the Hermite interpolant to $f(z) \in A(C_\rho)$, $\rho > 1$, in the zeros of $(T_n(z))^p$. For any positive integer $l \geq 1$, we consider the difference

$$\Delta_{l-1,n,m,p}(f,z) := S_{n-1}(H^{\{p\}}_{pm-1}(f,\cdot),z) - \begin{cases} S_{n-1}(H^{\{p\}}_{pm-1}(S_{plm-1}(f,\cdot)),z) \text{ if } p \text{ is even} \\ S_{n-1}(H^{\{p\}}_{pm-1}(S_{(p(2l-1)-1)m+n-1}(f,\cdot)),z) \text{ if } p \text{ is odd.} \end{cases} \tag{2.6}$$

The motivation for the definition (2.6) when p is even, is based on formula (7.2a) in Section 7.1. When $l = 1,\ p = 1,$ the Hermite interpolant becomes Lagrange interpolant and the above difference reduces to the case treated by Rivlin. In order to state our result we need

$$f_{l,q,p}(R) := \begin{cases} \frac{R}{\rho^{plq+1}} \text{ if } p \text{ is even and } R \geq \rho \\ \frac{1}{\rho^{plq}} \text{ if } p \text{ is even and } 1 < R < \rho \\ \frac{R}{\rho^{(p(2l-1)+1)q-1}} \text{ if } p \text{ is odd.} \end{cases}$$

We can now state our theorem on equiconvergence:

Theorem 2.7 *Let $m = m(n)$ be a sequence of positive integers such that $m/n \to q > 1$ as $n \to \infty$. If $f \in A(C_\rho)$, $\rho > 1$ and l,p are positive integers, then for each $R > 1$*

$$\limsup_{n\to\infty} \max_{z\in C_R} |\Delta_{l-1,n,m,p}(f,z)|^{1/n} \leq f_{l,q,p}(R). \tag{2.6a}$$

When p is odd and $R > 1$, or p is even and $R > \rho$, the following holds:

$$\max_{f\in A(C_R)} \limsup_{n\to\infty} \max_{z\in C_R} |\Delta_{l-1,n,m,p}(f,z)|^{1/n} = f_{l,q,p}(R). \tag{2.7}$$

If

$$f(z) = f_*(z) := \frac{\rho(\rho - z)}{1 - 2\rho z + \rho^2} \in A(C_\rho) \tag{2.8}$$

then when p is odd or when p is even and $R > \rho$ we have

$$\lim_{n\to\infty}\min_{z\in C_R}|\Delta_{l-1,n,m,p}(f_*,z)|^{1/n} = \lim_{n\to\infty}\max_{z\in C_R}|\Delta_{l-1,n,m,p}(f_*,z)|^{1/n} = f_{l,q,p}(R). \tag{2.9}$$

This shows that the bound in (2.7) is sharp when $R > \rho$ and that

$$\Delta_{l-1,n,m,p}(f,z) \to 0 \text{ as } n \to \infty$$

for each z in the interior of the ellipse C_P where

$$P = \begin{cases} \rho^{plq+1} & \text{if } p \text{ is even,} \\ \rho^{(p(2l-1)+1)q-1} & \text{if } p \text{ is odd.} \end{cases}$$

3 Hermite interpolants of Chebyshev monomials: $\cos((ps+j)n+r)\theta, 0 < r < n$

Let p, s, n be positive integers and let r be an integer with $0 < r < n$ and let j be an integer satisfying $0 \le j < p$. We want to find the cosine polynomial which interpolates $\cos(psn + jn + r)\theta$ in the zeros of $(\cos n\theta)^p$. We set $z = e^{i\theta}$, so that

$$\cos(psn + jn + r)\theta = \frac{1}{2}\left(z^{psn+jn+r} + z^{-psn-jn-r}\right).$$

The Hermite interpolant for this monomial will be given in Theorem 3.4. We shall need some preliminary lemmas.

Given a positive integer p and a non-negative integer λ we define the following polynomial

$$C_{p,\lambda}(t) := \sum_{\nu=0}^{p-1}(-1)^\nu\binom{\lambda}{\nu}t^\nu.$$

Lemma 3.1 *Given a positive integer p and a non-negative integer λ, $\lambda \ge p$, then for each μ, $0 \le \mu < p$ we have*

$$\frac{1}{\mu!}C^{(\mu)}_{p,\lambda}(t) = (-1)^\mu\binom{\lambda}{\mu}C_{p-\mu,\lambda-\mu}(t) \tag{3.1}$$

and

$$\frac{1}{\mu!}C^{(\mu)}_{p,\lambda}(1) = (-1)^{p-1}\binom{\lambda}{\mu}\binom{\lambda-\mu-1}{p-\mu-1} = (-1)^{p-1}\frac{p}{\lambda-\mu}\binom{\lambda}{p}\binom{p-1}{\mu}. \tag{3.2}$$

Proof. It is easy to see that $C'_{p,\lambda}(t) = -\binom{\lambda}{1}C_{p-1,\lambda-1}(t)$. Suppose we have

$$\frac{1}{\mu!}C^{(\mu)}_{p,\lambda}(t) = (-1)^{\mu}\binom{\lambda}{\mu}C_{p-\mu,\lambda-\mu}(t)$$

for some $\mu < p-1$. Then by the previous identity we get that it is also true for $\mu + 1$. This proves (3.1).
It is clear that $C_{1,\lambda}(1) = 1$ for all values of λ. Assume that for some p, $C_{p,\lambda}(1) = (-1)^{p-1}\binom{\lambda-1}{p-1}$. ¿From the definition of $C_{p,\lambda}(t)$ we see that

$$C_{p+1,\lambda}(t) = C_{p,\lambda}(t) + (-1)^{p}\binom{\lambda}{p}t^{p}.$$

Using the induction assumption we get

$$\begin{aligned} C_{p+1,\lambda}(1) &= C_{p,\lambda}(1) + (-1)^{p}\tbinom{\lambda}{p} = (-1)^{p-1}\tbinom{\lambda-1}{p-1} + (-1)^{p}\tbinom{\lambda}{p} \\ &= (-1)^{(p+1)-1}\tbinom{\lambda-1}{(p+1)-1}. \end{aligned}$$

This implies the truth of (3.2). □

Remark. It is clear that $C_{p,\lambda}(1) = (-1)^{p+1}$ if $\lambda = p$ and $C_{p,\lambda}(1) = 0$ if $\lambda < p$.

Set $\Lambda \equiv \Lambda_{p,s,j} := \frac{1}{2}(ps + j + p - e_{ps+j+p})$. We can now prove

Lemma 3.2 *For integers p, s, j, n such that $0 \le j \le p-1$ we have*
(i) *For $e_p \le \lambda \le \frac{1}{2}(p + e_p) - 1$, we have*

$$\frac{1}{(\frac{1}{2}(p+e_p)-\lambda-1)!}C^{(\frac{1}{2}(p+e_p)-\lambda-1)}_{p,\Lambda}(1) + \frac{1}{(\frac{1}{2}(p-e_p)+\lambda)!}C^{(\frac{1}{2}(p-e_p)+\lambda)}_{p,\Lambda}(1)$$

$$= (-1)^{p-1}\tbinom{\Lambda}{p}\tbinom{p-1}{\frac{1}{2}(p-e_p)+\lambda}\left(\frac{p}{\Lambda-\frac{1}{2}(p+e_p)+\lambda+1} + \frac{p}{\Lambda-\frac{1}{2}(p-e_p)-\lambda}\right).$$

(ii) *For $1 - e_p \le \lambda \le \frac{1}{2}(p - e_p) - 1$, we have*

$$\frac{1}{(\frac{1}{2}(p-e_p)-\lambda)!}C^{(\frac{1}{2}(p-e_p)-\lambda)}_{p,\Lambda}(1) + \frac{1}{(\frac{1}{2}(p+e_p)+\lambda)!}C^{(\frac{1}{2}(p+e_p)+\lambda)}_{p,\Lambda}(1) -$$

$$-\tbinom{p}{\frac{1}{2}(p+e_p)+\lambda}C_{p,\Lambda}(1)$$

$$= (-1)^{p-1}\tbinom{\Lambda}{p}\tbinom{p}{\frac{1}{2}(p-e_p)-\lambda}\left(\frac{\frac{1}{2}(p+e_p)+\lambda}{\Lambda-\frac{1}{2}(p-e_p)+\lambda} - \frac{p}{\Lambda} + \frac{\frac{1}{2}(p-e_p)-\lambda}{\Lambda-\frac{1}{2}(p+e_p)-\lambda}\right).$$

(iii) *The following identity holds:*

$$\frac{1}{(\frac{1}{2}(p-e_p))!}C_{p,\Lambda}^{(\frac{1}{2}(p-e_p))}(1) - \tfrac{1}{2}\tbinom{p}{\frac{1}{2}(p-e_p)}C_{p,\Lambda}(1)$$

$$= (-1)^{p-1}\tbinom{\Lambda}{p}\tbinom{p-1}{\frac{1}{2}(p-e_p)}\left(\frac{p}{\Lambda-\frac{1}{2}(p-e_p)} - \frac{p}{\Lambda}\cdot\frac{p}{(p+e_p)}\right).$$

Lemma 3.3 *Given an integer β, a positive integer p and a non-negative integer $\lambda \geq p$, we have*

$$\cos(\beta+2n\lambda)\theta \overset{p,n}{\sim} (-1)^\lambda \sum_{\mu=0}^{p-1} \frac{1}{\mu!}C_{p,\lambda}^{(\mu)}(1)\cos(\beta+2n\mu)\theta. \tag{3.3}$$

Equivalently,

$$\cos(\beta+2n\lambda)\theta \overset{p,n}{\sim} (-1)^{\lambda+p-1}\binom{\lambda}{p}\sum_{\mu=0}^{p-1}\binom{p-1}{\mu}\frac{p}{\lambda-\mu}\cos(\beta+2n\mu)\theta. \tag{3.4}$$

Proof. We have

$$\begin{aligned}
z^{\beta+2n\lambda} &= z^\beta(z^{2n})^\lambda = z^\beta((z^{2n}+1)+(-1))^\lambda \\
&= z^\beta \textstyle\sum_{\nu=0}^{\lambda}(-1)^{\lambda-\nu}\binom{\lambda}{\nu}(z^{2n}+1)^\nu \\
&\overset{p,n}{\sim} (-1)^\lambda z^\beta \textstyle\sum_{\nu=0}^{p-1}(-1)^\nu\binom{\lambda}{\nu}(z^{2n}+1)^\nu \\
&= (-1)^\lambda \textstyle\sum_{\nu=0}^{p-1}(-1)^\nu\binom{\lambda}{\nu}\sum_{\mu=0}^{\nu}\binom{\nu}{\mu}z^{2n\mu+\beta} \\
&= (-1)^\lambda \textstyle\sum_{\mu=0}^{p-1}\left(\sum_{\nu=\mu}^{p-1}(-1)^\nu\binom{\lambda}{\nu}\binom{\nu}{\mu}\right)z^{\beta+2n\mu} \\
&= (-1)^\lambda \textstyle\sum_{\mu=0}^{p-1}\frac{1}{\mu!}C_{p,\lambda}^{(\mu)}(1)z^{\beta+2n\mu}.
\end{aligned}$$

Replacing z by $\frac{1}{z}$ we get, since $((1/z)^{2n}+1)^p = 0$ is equivalent to $(z^{2n}+1)^p = 0$, that

$$(1/z)^{\beta+2n\lambda} \overset{p,n}{\sim} (-1)^\lambda \sum_{\mu=0}^{p-1}\frac{1}{\mu!}C_{p,\lambda}^{(\mu)}(1)(1/z)^{\beta+2n\mu}.$$

Since $z = e^{i\theta}$ we get by adding the last two relations

$$\cos(\beta+2n\lambda)\theta \overset{p,n}{\sim} (-1)^\lambda \sum_{\mu=0}^{p-1}\frac{1}{\mu!}C_{p,\lambda}^{(\mu)}(1)\cos(\beta+2n\mu)\theta.$$

Applying now (3.2) when $\lambda \geq p$ we get (3.4). □

Theorem 3.4 *p, s, n be positive integers. Let j be an integer satisfying $0 \leq j < p$ and let r be an integer satisfying $0 < r < n$. Then we have*

$$\cos(psn + jn + r)\theta \overset{p,n}{\sim} (-1)^{\Lambda} \sum_{\mu=0}^{p-1} \frac{1}{\mu!} C_{p,\Lambda}^{(\mu)}(1) \cos(r + n(2\mu - p + e_{ps+j+p}))\theta, \tag{3.5}$$

where Λ is a constant defined after the Remark above.

Remark. *When $0 < r < n$ or when $r = 0$ and $e_{ps+j+p} = 1$ we have*

$$\begin{aligned} -pn &< \min_{0 \leq \mu < p} \left(r + n(2\mu - p + e_{ps+j+p})\right) \\ &\leq \max_{0 \leq \mu < p} \left(r + n(2\mu - p + e_{ps+j+p})\right) < pn. \end{aligned}$$

Since $0 < r < n$, it is clear that $r + (2\mu - p + e_{ps+j+p})n$ cannot be zero.

Proof. The proof is split into the following four cases: *Case 1* Here ps is even and j is even. *Case 2* Here ps is even and j is odd. *Case 3* Here ps is odd and j is even. *Case 4* Here ps is odd and j is odd. In all the four cases we write $ps = 2q - 1$ when ps is odd and $ps = 2q$ when ps is even. Similarly we write $j = 2k - 1$ when j is odd and $j = 2k$ when j is even. The integer w will be chosen in each of the four cases differently.
Case 1 In this case we have $e_{ps+j+p} = e_p$. We have

$$psn + jn + r = (r - 2wn) + 2n(q + k + w) \equiv \beta + 2n\lambda.$$

Applying the Lemma 3.3 for these values of β, λ when $w = \frac{1}{2}(p - e_p)$ we get (3.5).
Case 2 In this case we have $e_{ps+j+p} = 1 - e_p$. We have

$$psn + jn + r = (r - (2w + 1)n) + 2n(q + k + w) \equiv \beta + 2n\lambda.$$

Applying the Lemma 3.3 for these values of β, λ when $w = \frac{1}{2}(p + e_p - 2)$ we get the first conclusion of the theorem. Applying now (3.2) the second conclusion follows.
Case 3 In this case we have $e_{ps+j+p} = 0$. We have

$$psn + jn + r = (r - (2w + 1)n) + 2n(q + k + w) \equiv \beta + 2n\lambda.$$

Applying the Lemma 3.3 for these values of β, λ when $w = \frac{1}{2}(p - e_p)$ (in this case p is necessarily odd) we get the first conclusion of the theorem.

Applying now (3.2) the second conclusion follows.

Case 4 In this case we have $e_{ps+j+p} = 1$. We have

$$psn + jn + r = (r - 2wn) + 2n(q + k + w - 1) \equiv \beta + 2n\lambda.$$

Applying the Lemma 3.3 for these values of β, λ when $w = \frac{1}{2}(p - e_p)$ (in this case p is necessarily odd) we get the first conclusion of the theorem. Applying now (3.2) the second conclusion follows. □

4 Hermite interpolants of Chebyshev monomials: $\cos(ps + j)n\theta$.

The following two lemmas are easily proved and their proofs are left out.

Lemma 4.1 *Let p, s, j be integers. Then the numbers $ps + j + p$ and $p \pm e_{ps+j}$ are either both even or both odd.*

Lemma 4.2 *For any integers p, s, j we have $e_p - e_{ps+j+p} = -(-1)^p e_{ps+j}$.*

Lemma 4.3 *For any integer $p \geq 1$, we have*

$$\cos pn\theta \overset{p,n}{\sim} (e_p - 1)\frac{1}{2}\binom{p}{\frac{1}{2}(p - e_p)} - \sum_{\lambda=1-e_p}^{\frac{1}{2}(p-e_p)-1} \binom{p}{\frac{1}{2}(p + e_p) + \lambda} \cos(2\lambda + e_p)n\theta, \tag{4.1}$$

where empty sums are defined to be zero.

Proof. Write $z := e^{i\theta}$. We have

$$\begin{aligned}
(2\cos n\theta)^p &= (z^n + z^{-n})^p = \textstyle\sum_{k=0}^{p} \binom{p}{k} z^{(p-2k)n} \\
&= \textstyle\sum_{k=0}^{\frac{1}{2}(p-e_p)} \binom{p}{k} z^{(p-2k)n} + \sum_{k=\frac{1}{2}(p-e_p)+1}^{p} \binom{p}{k} z^{(p-2k)n} \\
&= \sum_{k\in\{p,p-2,\cdots,e_p\}} \binom{p}{\frac{1}{2}(p-k)} z^{kn} + \sum_{k\in\{p,p-2,\cdots,2-e_p\}} \binom{p}{\frac{1}{2}(p+k)} z^{-kn} \\
&= \begin{cases} \binom{p}{\frac{1}{2}p} + \sum_{k\in\{2,4,\cdots,p-2\}} \binom{p}{\frac{1}{2}(p+k)} 2\cos kn\theta + 2\cos pn\theta, \\ \qquad \text{if } p \text{ is even,} \\ \sum_{k\in\{1,3,\cdots,p-2\}} \binom{p}{\frac{1}{2}(p-k)} 2\cos kn\theta + 2\cos pn\theta, \\ \qquad \text{if } p \text{ is odd.} \end{cases}
\end{aligned}$$

The last relation is equivalent to (4.1). □

Theorem 4.4 *Let p, s, j, n be integers such that $0 \le j \le p-1$. Suppose $ps+j+p$ is odd. Then*

$$\cos(ps+j)n\theta \overset{p,n}{\sim} (-1)^{\Lambda+p-1}\binom{\Lambda}{p}\times$$

$$\times \sum_{\lambda=0}^{\frac{1}{2}(p+e_p)-1}{}' \binom{p-1}{\frac{1}{2}(p-e_p)+\lambda} B_\lambda(\Lambda,p)\cos(2\lambda+1-e_p)n\theta,$$

where $\Lambda := \frac{1}{2}(ps+j+p-e_{ps+j+p}) = \frac{1}{2}(ps+j+p-1)$ and

$$B_\lambda(\Lambda,p) := \frac{p}{\Lambda-\frac{1}{2}(p+e_p)+\lambda+1} + \frac{p}{\Lambda-\frac{1}{2}(p-e_p)-\lambda}.$$

Theorem 4.5 *Let p, s, j, n be integers such that $0 \le j \le p-1$. Suppose $ps+j+p$ is even. Then*

$$\cos(ps+j)n\theta \overset{p,n}{\sim} (-1)^{\Lambda+p-1}\binom{\Lambda}{p}\times$$

$$\times \sum_{\lambda=0}^{\frac{1}{2}(p-e_p)-1} \binom{p}{\frac{1}{2}(p-e_p)-\lambda} B^*_\lambda(\Lambda,p)\cos(2\lambda+e_p)n\theta,$$

where $\Lambda := \frac{1}{2}(ps+j+p-e_{ps+j+p}) = \frac{1}{2}(ps+j+p)$ and

$$B^*_\lambda(\Lambda,p) := \frac{\frac{1}{2}(p+e_p)+\lambda}{\Lambda-\frac{1}{2}(p-e_p)+\lambda} - \frac{p}{\Lambda} + \frac{\frac{1}{2}(p-e_p)-\lambda}{\Lambda-\frac{1}{2}(p+e_p)-\lambda}.$$

Remark: Lemma 4.3 is the special case $s=1$ and $j=0$ of Theorem 4.5 and is used in the proof of Theorem 4.5.

Proof of Theorem 4.4 Assume that $ps+j+p$ is odd. By Lemma 4.1 $p+e_{ps+j}$ is odd. Since $z^{psn+jn} = z^{\beta+2\lambda n}$ where $\beta := ne_{ps+j} - 2wn$, $\lambda := \frac{1}{2}(ps+j-e_{ps+j})+w$, it follows from Lemma 3.3 that $\cos(ps+j)n\theta \overset{p,n}{\sim} I(\theta)$ where

$$I(\theta) = (-1)^{\frac{1}{2}(ps+j-e_{ps+j})+w}\sum_{\mu=0}^{p-1} \frac{1}{\mu!}C^{(\mu)}_{p,\frac{1}{2}(ps+j-e_{ps+j})+w}(1)\times$$

$$\times\cos(2\mu-2w+e_{ps+j})n\theta.$$

We choose w such that

$$-pn \le \min_{0\le\mu\le p-1}(2\mu-2w+e_{ps+j})n \le \max_{0\le\mu\le p-1}(2\mu-2w+e_{ps+j})n < pn.$$

This yields $w = \frac{1}{2}(p - e_p)$. Since $p + e_{ps+j}$ is odd, we have $e_p + e_{ps+j} = 1$, so that $w = \frac{1}{2}(p + e_{ps+j} - 1)$. For this value of w we get

$$\cos(ps+j)n\theta \overset{p,n}{\sim} (-1)^\Lambda \left\{ \sum_{\mu=0}^{\mu^*} + \sum_{\mu=\mu^*+1}^{p-1} \right\} \frac{1}{\mu!} C_{p,\Lambda}^{(\mu)}(1) \cos(2\mu - p + 1)n\theta$$

$$\equiv (-1)^\Lambda (S_1 + S_2),$$

where $\mu^* := w + e_p - 1 = \frac{1}{2}(p + e_p) - 1$. We replace μ by $\mu^* - \mu$ in S_1 and by $\mu^* + 1 + \mu$ in S_2. Then

$$\begin{aligned} S_1 &= \sum_{\mu=0}^{\mu^*} \tfrac{1}{(\mu^*-\mu)!} C_{p,\Lambda}^{(\mu^*-\mu)}(1) \cos(2(\mu^* - \mu) - p + 1)n\theta \\ &= \sum_{\mu=0}^{\frac{1}{2}(p-e_p)-1+e_p} \tfrac{1}{(\mu^*-\mu)!} C_{p,\Lambda}^{(\mu^*-\mu)}(1) \cos(-(2\mu + 1 - e_p))n\theta. \end{aligned}$$

Similarly we have

$$S_2 = \sum_{\mu=0}^{\frac{1}{2}(p-e_p)-1} \frac{1}{(\mu^* + 1 + \mu)!} C_{p,\Lambda}^{(\mu^*+1+\mu)}(1) \cos(2\mu + 1 + e_p)n\theta.$$

Now in S_1 we make the change of variable $\mu = \lambda$ and in S_2 we make the change of variable $\mu = \lambda - e_p$. Then from $S_1 + S_2$ we see that

$$\begin{aligned} \cos(ps+j)n\theta \quad &\overset{p,n}{\sim} (-1)^\Lambda \Big\{ e_p \tfrac{1}{(\frac{1}{2}(p-e_p))!} C_{p,\Lambda}^{(\frac{1}{2}(p-e_p))}(1) \cos(1 - e_p)n\theta \\ &\quad + \sum_{\lambda=e_p}^{\frac{1}{2}(p+e_p)-1} \Big(\tfrac{1}{(\frac{1}{2}(p+e_p)-\lambda-1)!} C_{p,\Lambda}^{(\frac{1}{2}(p+e_p)-\lambda-1)}(1) \\ &\quad + \tfrac{1}{(\frac{1}{2}(p-e_p)+\lambda)!} C_{p,\Lambda}^{(\frac{1}{2}(p-e_p)+\lambda)}(1) \Big) \cos(2\lambda + 1 - e_p)n\theta \Big\}. \end{aligned}$$

An application of Lemma 3.2(i) completes the proof. □

Proof of Theorem 4.5 We recall that empty sums are defined to be equal to zero. Assume $ps+j+p$ is even. Write $\beta := -pn$ and $\lambda := \frac{1}{2}(ps+j+p) = \Lambda$. We have $z^{psn+jn} = z^{\beta+2n\lambda}$. For these values of β, λ we get from Lemma 3.3 that

$$\begin{aligned} \cos(ps+j)n\theta \quad &\overset{p,n}{\sim} (-1)^\Lambda \left\{ \sum_{\mu=0}^{\mu^*} + \sum_{\mu=\mu^*+1}^{p-1} \right\} \tfrac{1}{\mu!} C_{p,\Lambda}^{(\mu)}(1) \cos(2\mu - p)n\theta \\ &\equiv (-1)^\Lambda (S_1 + S_2) \quad \text{where} \quad \mu^* := \tfrac{1}{2}(p - e_p). \end{aligned} \tag{4.2}$$

We replace μ by $\mu^* - \mu$ in S_1 and by $\mu^* + 1 + \mu$ in S_2. Then on simplifying we get

$$S_1 = \sum_{\mu=0}^{\frac{1}{2}(p-e_p)-1} \frac{1}{(\frac{1}{2}(p-e_p)-\mu)!} C_{p,\Lambda}^{(\frac{1}{2}(p-e_p)-\mu)}(1)\cos(-(2\mu+e_p))n\theta \\ +C_{p,\Lambda}(1)\cos pn\theta. \tag{4.3}$$

Similarly by simplifying we have

$$S_2 = \sum_{\mu=0}^{\frac{1}{2}(p+e_p)-2} \frac{1}{(\frac{1}{2}(p-e_p)+1+\mu)!} C_{p,\Lambda}^{(\frac{1}{2}(p-e_p)+1+\mu)}(1)\cos(2(\mu+1)-e_p)n\theta.$$

In the last formula we replace $\mu + 1$ by μ only when p is even and when p is odd we leave the last formula as it is. Then we see that for both even and odd values of p we have

$$S_2 = \sum_{\mu=1-e_p}^{\frac{1}{2}(p-e_p)-1} \frac{1}{(\frac{1}{2}(p+e_p)+\mu)!} C_{p,\Lambda}^{(\frac{1}{2}(p+e_p)+\mu)}(1)\cos(2\mu+e_p)n\theta. \tag{4.4}$$

Combining (4.2),(4.3) and (4.4) and applying Lemma 4.3 to the term $\cos pn\theta$, we get

$$\begin{aligned}\cos(psn+jn)\theta \overset{p,n}{\sim} (-1)^\Lambda \Big\{ \Big(&(1-e_p)\tfrac{1}{(\frac{1}{2}(p-e_p))!} C_{p,\Lambda}^{(\frac{1}{2}(p-e_p))}(1) \\ &+ (e_p-1)\tfrac{1}{2}\tbinom{p}{\frac{1}{2}(p-e_p)} C_{p,\Lambda}(1)\Big) \\ &+ \textstyle\sum_{\lambda=1-e_p}^{\frac{1}{2}(p-e_p)-1} \Big(\tfrac{1}{(\frac{1}{2}(p-e_p)-\lambda)!} C_{p,\Lambda}^{(\frac{1}{2}(p-e_p)-\lambda)}(1) \\ &+ \tfrac{1}{\frac{1}{2}(p+e_p)+\lambda)!} C_{p,\Lambda}^{(\frac{1}{2}(p+e_p)+\lambda)}(1) \\ &- \tbinom{p}{\frac{1}{2}(p+e_p)+\lambda} C_{p,\Lambda}(1)\Big)\cos(2\lambda+e_p)n\theta\Big\}.\end{aligned}$$

Applying now Lemma 3.2(ii),(iii) the proof is complete. □

5 Applications of Sec.4 to linear sums of Chebyshev monomials: $\cos(ps+j)n\theta$

5.1 Proof of Theorem 2.1(ii).

We assume here that p is even. We split the sum over j, $0 \le j \le p-1$, into a sum over even values of j say $j = 2\tau$, $0 \le \tau \le \frac{1}{2}p - 1$, and a sum over odd values of j say $j = 2\tau + 1$, $0 \le \tau \le \frac{1}{2}p - 1$. When j is even then $ps + j + p$ is even, so for the sum over even j we use Theorem 4.5. When j is odd then $ps + j + p$ is odd, so in the sum over odd j we use Theorem 4.4. Thus applying Theorem 4.5 to the first sum and Theorem 4.4 to the second sum, we have

$$\sum_{s=1}^{\infty} \sum_{j=0}^{p-1} A_{(ps+j)n} \cos(ps+j)n\theta = \sum_{s=1}^{\infty} \sum_{\tau=0}^{\frac{1}{2}p-1} A_{(ps+2\tau)n} \cos(ps+2\tau)n\theta$$

$$+\sum_{s=1}^{\infty} \sum_{\tau=0}^{\frac{1}{2}p-1} A_{(ps+2\tau+1)n} \cos(ps+2\tau+1)n\theta \overset{p,n}{\sim} I_1 + I_2,$$

where

$$I_1 = -\sum_{s=1}^{\infty} (-1)^{\frac{1}{2}p(s+1)} \binom{\frac{1}{2}p(s+1)+\tau}{p} \sum_{\tau=0}^{\frac{1}{2}p-1} (-1)^{\tau} {\sum_{\lambda=0}^{\frac{1}{2}p-1}}' \binom{p}{\frac{1}{2}p+\lambda} \times$$

$$\times \left(\frac{\frac{1}{2}p+\lambda}{\frac{1}{2}ps+\tau+\lambda} - \frac{p}{\frac{1}{2}p(s+1)+\tau} + \frac{\frac{1}{2}p-\lambda}{\frac{1}{2}ps-\tau-\lambda} \right) A_{(ps+2\tau)n} \cos 2\lambda n\theta \,,$$

$$I_2 = -\sum_{s=1}^{\infty} (-1)^{\frac{1}{2}p(s+1)} \binom{\frac{1}{2}p(s+1)+\tau}{p} \sum_{\tau=0}^{\frac{1}{2}p-1} (-1)^{\tau} \times$$

$$\times \sum_{\lambda=0}^{\frac{1}{2}p-1} \binom{p-1}{\frac{1}{2}p+\lambda} \left(\frac{p}{\frac{1}{2}ps+\tau+\lambda+1} + \frac{p}{\frac{1}{2}ps+\tau-\lambda} \right) A_{(ps+2\tau+1)n} \times$$

$$\times \cos(2\lambda+1)n\theta.$$

Writing $\rho := 2\lambda$ in I_1 and $\rho = 2\lambda + 1$ in I_2 and applying the identity

$$\binom{\frac{1}{2}p(s+1)+\tau}{p} \left(\frac{\frac{1}{2}(p+\rho)}{\frac{1}{2}ps+\tau+\frac{1}{2}\rho} - \frac{p}{\frac{1}{2}p(s+1)+\tau} + \frac{\frac{1}{2}(p-\rho)}{\frac{1}{2}ps+\tau-\frac{1}{2}\rho} \right)$$
$$= \frac{1}{p} \binom{\frac{1}{2}p(s+1)+\tau-1}{p-1} \frac{\frac{1}{2}(p^2-\rho^2)(\frac{1}{2}ps+\tau)}{(\frac{1}{2}ps+\tau)^2-(\frac{1}{2}\rho)^2} \tag{5.0}$$

in I_1 yields formula (2.2) and completes the proof of Theorem 2.1(ii). Also for $\rho = (2\lambda+1)n$, I_2 yields formula (2.1) when $\sigma = (2\lambda+1)$. This completes the proof of Theorem 2.1 when $\sigma = (2\lambda + 1)n$. □

5.2 Proof of Theorem 2.4(ii).

Here p is an odd positive integer. We shall break the double sum over s and j on the left hand side of (5.1) into four sums according to the parity of s and j and find the Hermite interpolant to each sum. When s is even we replace it by $2s$ and when s is odd we replace it by $2s-1$. When j is even we replace it by 2τ and when j is odd we replace it in one sum by $2\tau-1$ and in the other sum by $2\tau+1$. Denote $A_{(ps+j)n}\cos(ps+j)n\theta$ by $I_\theta(s,j)$. Then we have

$$\begin{aligned}\sum_{s=1}^{\infty}\sum_{j=0}^{p-1} A_{(ps+j)n}\cos(ps+j)n\theta &= \sum_{s=1}^{\infty}\sum_{\tau=0}^{\frac{1}{2}(p-1)} I_\theta(2s,2\tau)\\ &+\sum_{s=1}^{\infty}\sum_{\tau=0}^{\frac{1}{2}(p-1)} I_\theta(2s-1,2\tau)+\sum_{s=1}^{\infty}\sum_{\tau=1}^{\frac{1}{2}(p-1)} I_\theta(2s,2\tau-1)\\ &+\sum_{s=1}^{\infty}\sum_{\tau=0}^{\frac{1}{2}(p-1)-1} I_\theta(2s-1,2\tau+1)\\ &\equiv I_1+I_2+I_3+I_4.\end{aligned} \tag{5.1}$$

Interpolant for I_1. The number $p2s+2\tau+p=2(ps+\tau)+p$ is odd. Therefore applying Theorem 4.4 we get

$$\begin{aligned}\cos(2ps+2\tau)n\theta \quad &\overset{p,n}{\sim} (-1)^s(-1)^{\tau+\frac{1}{2}(p-1)}\binom{ps+\tau+\frac{1}{2}(p-1)}{p}\times\\ &\times\left\{\binom{p-1}{\frac{1}{2}(p-1)}\frac{p}{ps+\tau}+\sum_{\lambda=1}^{\frac{1}{2}(p-1)}\binom{p-1}{\frac{1}{2}(p-1)+\lambda}\times\right.\\ &\left.\times\left(\frac{p}{ps+\tau+\lambda}+\frac{p}{ps+\tau-\lambda}\right)\cos 2\lambda n\theta\right\}.\end{aligned}$$

Changing the summation index from λ to ρ where $\rho:=2\lambda$, and summing over ρ and τ , and making the change of variable $\widetilde{\tau}=\tau+\frac{1}{2}(p-1)$ in the sum over τ on the righthand side and then replacing $\widetilde{\tau}$ by τ, we get

$$\begin{aligned}I_1 \quad &\overset{p,n}{\sim} \sum_{s=1}^{\infty}(-1)^s\sum_{\tau=\frac{1}{2}(p-1)}^{p-1}(-1)^\tau\binom{ps+\tau}{p}\sum_{\rho\in\{0,2,\ldots,p-1\}}\tfrac{1}{2}\eta_\rho\binom{p-1}{\frac{1}{2}(p-1)+\frac{1}{2}\rho}\times\\ &\times\left(\frac{p}{ps+\tau-\frac{1}{2}(p-1)+\frac{1}{2}\rho}+\frac{p}{ps+\tau-\frac{1}{2}(p-1)-\frac{1}{2}\rho}\right)A_{(p(2s-1)+(2\tau+1))n}\cos\rho n\theta.\end{aligned}$$

Interpolant for I_2. The number $p(2s-1)+2\tau+p=2(ps+\tau)$ is even. Therefore applying Theorem 4.5 we get

$$\cos(p(2s-1)+2\tau)n\theta \overset{p,n}{\sim} (-1)^s(-1)^\tau \binom{ps+\tau}{p} \sum_{\lambda=0}^{\frac{1}{2}(p-1)-1} \binom{p}{\frac{1}{2}(p-1)-\lambda} \times$$
$$\times \left(\frac{\frac{1}{2}(p+1)+\lambda}{ps+\tau-\frac{1}{2}(p-1)+\lambda} - \frac{p}{ps+\tau} + \frac{\frac{1}{2}(p-1)-\lambda}{ps+\tau-\frac{1}{2}(p+1)-\lambda} \right) \times$$
$$\times \cos(2\lambda+1)n\theta \,.$$

Changing the summation index from λ to ρ where $\rho := 2\lambda+1$, and summing over s and τ we get

$$I_2 \overset{p,n}{\sim} \sum_{s=1}^{\infty}(-1)^s \sum_{\tau=1}^{\frac{1}{2}(p-1)}(-1)^\tau \binom{ps+\tau}{p} \times$$
$$\times \sum_{\rho\in\{1,3,\ldots,p-2\}} \binom{p}{\frac{1}{2}(p+\rho)} \left(\frac{\frac{1}{2}(p+\rho)}{ps+\tau-\frac{1}{2}(p-\rho)} - \frac{p}{ps+\tau} + \frac{\frac{1}{2}(p-\rho)}{ps+\tau-\frac{1}{2}(p+\rho)} \right) \times$$
$$\times A_{(p(2s-1)+2\tau)n} \cos \rho n\theta.$$

Interpolant for I_3. The number $p2s+(2\tau-1)+p=2(ps+\tau)+p-1$ is even. Therefore applying Theorem 4.5 we get

$$\cos(p2s \;\; +(2\tau-1))n\theta \overset{p,n}{\sim} (-1)^s(-1)^{\tau+\frac{1}{2}(p-1)} \binom{ps+\tau+\frac{1}{2}(p-1)}{p} \times$$
$$\times \sum_{\lambda=0}^{\frac{1}{2}(p-1)-1} \binom{p}{\frac{1}{2}(p-1)-\lambda} \times$$
$$\times \left(\frac{\frac{1}{2}(p+1)+\lambda}{ps+\tau+\lambda} - \frac{p}{ps+\tau+\frac{1}{2}(p-1)} + \frac{\frac{1}{2}(p-1)-\lambda}{ps+\tau-\lambda-1} \right) \cos(2\lambda+1)n\theta \,.$$

Changing the summation index from λ to ρ where $\rho := 2\lambda+1$, and then summing over s and τ and making the change of variable $\widetilde{\tau} = \tau + \frac{1}{2}(p-1)$ in the sum over τ on the righthand side and then replacing $\widetilde{\tau}$ by τ, we get

$$I_3 \overset{p,n}{\sim} \sum_{s=1}^{\infty}(-1)^s \sum_{\tau=\frac{1}{2}(p-1)+1}^{p-1}(-1)^\tau \binom{ps+\tau}{p} \sum_{\rho\in\{1,3,\ldots,p-2\}} \binom{p}{\frac{1}{2}(p-\rho)} \times$$
$$\times \left(\frac{\frac{1}{2}(p+\rho)}{ps+\tau-\frac{1}{2}(p-\rho)} - \frac{p}{ps+\tau} + \frac{\frac{1}{2}(p-\rho)}{ps+\tau-\frac{1}{2}(p+\rho)} \right) A_{(p(2s-1)+2\tau)n} \cos \rho n\theta.$$

Interpolant for I_4. The number $p(2s-1)+(2\tau+1)+p=2(ps+\tau)+1$ is odd. Therefore applying Theorem 4.4 we get

$$\cos(p(2s-1)+(2\tau+1))n\theta \overset{p,n}{\sim} (-1)^s(-1)^\tau \binom{ps+\tau}{p} \left\{ \binom{p-1}{\frac{1}{2}(p-1)} \frac{p}{ps+\tau-\frac{1}{2}(p-1)} \right.$$
$$\left. + \sum_{\lambda=1}^{\frac{1}{2}(p-1)} \binom{p-1}{\frac{1}{2}(p-1)+\lambda} \left(\frac{p}{ps+\tau-\frac{1}{2}(p-1)+\lambda} + \frac{p}{ps+\tau-\frac{1}{2}(p-1)-\lambda} \right) \cos 2\lambda n\theta \right\}.$$

Changing the summation index from λ to ρ where $\rho := 2\lambda$ and summing over s and τ we get

$$I_4 \overset{p,n}{\sim} \sum_{s=1}^{\infty}(-1)^s \sum_{\tau=0}^{\frac{1}{2}(p-1)-1} (-1)^\tau \binom{ps+\tau}{p} \sum_{\rho\in\{0,2,\dots,p-1\}} \tfrac{1}{2}\eta_\rho \binom{p-1}{\frac{1}{2}(p-1)+\frac{1}{2}\rho} \times$$

$$\times \left(\frac{p}{ps+\tau-\frac{1}{2}(p-1)+\frac{1}{2}\rho} + \frac{p}{ps+\tau-\frac{1}{2}(p-1)-\frac{1}{2}\rho}\right) \times$$

$$\times A_{(p(2s-1)+(2\tau+1))n} \cos \rho n\theta \ .$$

Finally adding together the Hermite interpolants for I_1, I_2, I_3, I_4 we get

$$\sum_{s=1}^{\infty} \sum_{j=0}^{p-1} A_{(ps+j)n} \cos(ps+j)n\theta \overset{p,n}{\sim} (I_2+I_3)+(I_1+I_4)$$

$$= \sum_{s=1}^{\infty}(-1)^s \sum_{\tau=0}^{p-1} \binom{ps+\tau}{p} \sum_{\rho\in\{1,3,\dots,p-2\}} \binom{p}{\frac{1}{2}(p-\rho)} \times$$

$$\times \left(\frac{\frac{1}{2}(p+\rho)}{ps+\tau-\frac{1}{2}(p-\rho)} - \frac{p}{ps+\tau} + \frac{\frac{1}{2}(p-\rho)}{ps+\tau-\frac{1}{2}(p+\rho)}\right) A_{(p(2s-1)+2\tau)n} \cos \rho n\theta$$

$$+ \sum_{s=1}^{\infty}(-1)^s \sum_{\tau=0}^{p-1} (-1)^\tau \binom{ps+\tau}{p} \sum_{\rho\in\{0,2,\dots,p-1\}} \tfrac{1}{2}\eta_\rho \binom{p-1}{\frac{1}{2}(p-1)+\frac{1}{2}\rho} \times$$

$$\times \left(\frac{p}{ps+\tau-\frac{1}{2}(p-1)+\frac{1}{2}\rho} + \frac{p}{ps+\tau-\frac{1}{2}(p-1)-\frac{1}{2}\rho}\right) A_{(p(2s-1)+(2\tau+1))n} \cos \rho n\theta \ .$$

When ρ is odd, the coefficient of $\cos \rho n\theta$ gives $h_{\rho n}^{(p)}$ in (2.4) in Theorem 2.4(ii). When ρ is even, the coefficient of $\cos \rho n\theta$ gives $h_{\rho n}^{(p)}$ in (2.3) in Theorem 2.4(i). This completes the proof. □

6 Proof of Theorem 2.4(i) and Theorem 2.1(i)

6.1 Some remarks on Theorem 3.4

Set

$$\begin{cases} E_{p,s,j} & := \{2-e_{ps+j}, 4-e_{ps+j}, \dots, p-e_{ps+j+p}\}, \\ E^*_{p,s,j} & := \{e_{ps+j}, 2+e_{ps+j}, \dots, p-2+e_{ps+j+p}\}. \end{cases} \tag{6.1}$$

By Theorem 3.4 we have for $0 \le j \le p-1$, $0 < r < n$,

$$A_{(ps+j)n+r} \cos((ps+j)n+r)\theta \overset{p,n}{\sim} A_{(ps+j)n+r} I_1 + A_{(ps+j)n+r} I_2, \tag{6.2}$$

where using formula (3.4), we have

$$\begin{cases} I_1 &= (-1)^{\Lambda+p-1}\binom{\Lambda}{p}\sum_{\mu=0}^{\mu^*}\binom{p-1}{\mu}\frac{p}{\Lambda-\mu}\cos(-r+(p-2\mu-e_{ps+j+p})n)\theta\ , \\ I_2 &= (-1)^{\Lambda+p-1}\binom{\Lambda}{p}\sum_{\mu=\mu^*+1}^{p-1}\binom{p-1}{\mu}\frac{p}{\Lambda-\mu}\cos(r+(2\mu-p+e_{ps+j+p})n)\theta, \end{cases} \tag{6.3}$$

and

$$\Lambda := \frac{1}{2}(ps+j+p-e_{ps+j+p})\ ,\ \mu^* := \frac{1}{2}(p-e_p)-1+e_pe_{ps+j}\ . \tag{6.4}$$

¿From the easily verified identity

$$-e_p+2e_pe_{ps+j}+e_{ps+j+p} = e_{ps+j}, \tag{6.5}$$

we see that

$$-r+n(p-2\mu^*-e_{ps+j+p}) = (2-e_{ps+j})n-r \geq n-r \geq 1.$$

Therefore the coefficient of θ in each angle appearing in the cosines in I_1 is greater than or equal to 1. Similarly we get $r+n(2(\mu^*+1)-p+e_{ps+j+p}) = ne_{ps+j}+r \geq 1$, so that the coefficient of θ in each angle appearing in the cosines in I_2 is greater then or equal to 1. We now show that

$$\begin{aligned} A_{(ps+j)n+r}I_1 &= (-1)^{\Lambda+p-1}\binom{\Lambda}{p}\sum_{\sigma\in -r+n\cdot E_{p,s,j}}\binom{p-1}{\frac{1}{2}(p-e_p)-1+e_pe_{ps+j}-\lambda}\times \\ &\times\frac{p}{\frac{1}{2}(ps+j-e_{ps+j})+\lambda+1}A_{(ps+j)n-\sigma+(2\lambda+2-e_{ps+j})n}\cos\sigma\theta \\ &\text{where } \lambda := \left[\tfrac{\sigma}{2n}\right] \quad\text{and}\quad 2\lambda+1-e_{ps+j} = \left[\tfrac{\sigma}{n}\right], \end{aligned} \tag{6.6}$$

and

$$\begin{aligned} A_{(ps+j)n+r}\quad I_2 &= (-1)^{\Lambda+p-1}\binom{\Lambda}{p}\sum_{\sigma\in r+n\cdot E^*_{p,s,j}}\binom{p-1}{\frac{1}{2}(p-e_p)+e_pe_{ps+j}+\lambda}\times \\ &\times\frac{p}{\frac{1}{2}(ps+j-e_{ps+j})+\lambda+1}A_{(ps+j)n+\sigma-(2\lambda+e_{ps+j})n}\cos\sigma\theta \\ &\text{where } \lambda := \left[\tfrac{\sigma}{2n}\right] \quad\text{and}\quad 2\lambda+e_{ps+j} = \left[\tfrac{\sigma}{n}\right]. \end{aligned} \tag{6.7}$$

Proof of (6.6) We change the summation over μ in $A_{(ps+j)n+r}I_1$ to a summation over σ given by

$$\sigma := -r+(p-2\mu-e_{ps+j+p})n. \tag{6.8}$$

Putting $\lambda := \mu^* - \mu$ we get

$$\mu = \mu^* - \lambda = \frac{1}{2}(p - e_p) - 1 + e_p e_{ps+j} - \lambda\,, \tag{6.9}$$

and applying (6.5) we have

$$\Lambda - \mu = \frac{1}{2}(ps + j - e_{ps+j}) + \lambda + 1. \tag{6.10}$$

Then from (6.8), using (6.9) and simplifying, we get

$$\sigma = n - r + ((2\lambda + 1) - e_{ps+j})n\,. \tag{6.11}$$

so that

$$\sigma \in -r + n \cdot E_{p,s,j}. \tag{6.12}$$

By (6.11) we have for $\rho := \left[\frac{\sigma}{n}\right]$,

$$\lambda = \left[\frac{\sigma}{2n}\right], \ \rho := \left[\frac{\sigma}{n}\right] = 2\lambda + 1 - e_{ps+j}, \quad \text{and} \ \lambda = \frac{1}{2}(\rho - e_\rho). \tag{6.13}$$

¿From (6.11), we also get

$$(ps + j)n - \sigma + (2\lambda + 2 - e_{ps+j})n = (ps + j)n + r\,, \tag{6.14}$$

so that for $1 \le r \le n - 1$,

$$A_{(ps+j)n+r} \cos(-r + (p - 2\mu - e_{ps+j+p}))\theta = A_{(ps+j)n-\sigma+(2\lambda+2-e_{ps+j})n} \cos \sigma\theta, \tag{6.15}$$

where σ satisfies (6.12). Substituting the above expressions in the expression for $A_{(ps+j)n+r} I_1$ and using (6.9) and (6.10), we get (6.6). □

Proof of (6.7). We change the summation over μ in $A_{(ps+j)n+r} I_2$ to summation over σ where

$$\sigma := r + (2\mu - p + e_{ps+j+p})n, \quad \text{and} \ \ \lambda := \mu - \mu^* - 1. \tag{6.16}$$

Then using the value of μ^* we get

$$0 \le \lambda \le \lambda^* := p - 1 - (\mu^* + 1) = \frac{1}{2}(p + e_p) - e_p e_{ps+j} - 1, \tag{6.17}$$

$$\mu = \lambda + \mu^* + 1 = \lambda + \frac{1}{2}(p - e_p) + e_p e_{ps+j}\,. \tag{6.18}$$

Applying (6.5) we have

$$\sigma = r + (2\lambda - e_p + 2e_p e_{ps+j} + e_{ps+j+p})n = r + n(2\lambda + e_{ps+j}), \tag{6.19}$$

and

$$\Lambda - \mu = \frac{1}{2}(ps + j - e_{ps+j}) - \lambda. \tag{6.19a}$$

Set $\rho := \left[\frac{\sigma}{n}\right]$. By (6.19) we have

$$\rho := \left[\frac{\sigma}{n}\right] = 2\lambda + e_{ps+j},\ \lambda = \frac{1}{2}(\rho - e_\rho), \tag{6.20}$$

and

$$\sigma \in r + n \cdot E^*_{p,s,j}, \tag{6.21}$$

and for $1 \le r \le n-1$, we have

$$A_{(ps+j)n+r}\cos(r + (2\mu - p + e_{ps+j+p})n)\theta = A_{(ps+j)n+\sigma-(2\lambda+e_{ps+j})n}\cos\sigma\theta\ . \tag{6.22}$$

Substituting the above expressions in the expression for $A_{(ps+j)n+r}I_2$ and using (6.18) and (6.19a) we get (6.7). □

We shall use (6.6) and (6.7) in the following two sections.

6.2 Proof of Theorem 2.4(i).

We assume now that p is odd. We have $\Omega = \Omega_0 \bigcup \Omega_1$ where

$$\Omega := \left(\bigcup_{\nu=0}^{p-1}[\nu n + 1, (\nu+1)n - 1]\right) \bigcap Z\ ,$$

$$\Omega_0 := \left(\bigcup_{\nu\in\{0,2,\ldots,p-1\}}[\nu n + 1, (\nu+1)n - 1]\right) \bigcap Z\ ,$$

$$\Omega_1 := \left(\bigcup_{\nu\in\{1,3,\ldots,p-2\}}[\nu n + 1, (\nu+1)n - 1]\right) \bigcap Z\ .$$

We sum both sides of (6.6) and (6.7) over j, $0 \le j \le p-1$. This sum is split into two sums. One over even j and the other sum over odd j. We treat these two sums separately when s is even and separately when s is odd. Altogether we have the following four cases: **(i)** s is even and j is even. In this case we replace s by $2s$ and j by 2τ where $s = 1, 2, \ldots,\ ;\ 0 \le \tau \le \frac{1}{2}(p-1)$. Since $p2s + 2\tau + p = 2(ps+\tau) + p$ is odd we have $\Lambda = ps + \tau + \frac{1}{2}(p-1)$. **(ii)** s is odd and j is odd. In this case we replace s by $2s-1$ and j by $2\tau+1$ where $s = 1, 2, \ldots\ ;\ 0 \le \tau \le \frac{1}{2}(p-1) - 1$. Then $p(2s-1) + (2\tau+1) + p = 2(ps+\tau) + 1$ is odd so that $\Lambda = ps + \tau$. **(iii)** s is even and j is odd. In this case we replace s by $2s$ and j by $2\tau - 1$ where $s = 1, 2, \ldots\ ;\ 1 \le \tau \le \frac{1}{2}(p-1)$. Then $p2s + (2\tau-1) + p = 2(ps+\tau) + (p-1)$ is even so that $\Lambda = ps + \tau + \frac{1}{2}(p-1)$. **(iv)** s is odd and j is even. In this case we replace s by $2s-1$ and j by 2τ where $s = 1, 2, \ldots; 0 \le \tau \le \frac{1}{2}(p-1)$. Here $p(2s-1) + 2\tau + p = 2(ps+\tau)$ is even and $\Lambda = ps + \tau$. We consider the above four possibilities separately.

The case (i) In this case p is odd, s is even and j is even. Then $E_{p,s,j} = \{2,4,\dots,p-1\}$, $E^*_{p,s,j} = \{0,2,\dots,p-1\}$. We sum both sides of (6.6) and (6.7) over $j = 2\tau$, $0 \le \tau \le \frac{1}{2}(p-1)$ and use (6.2) and we get

$$\sum_{\tau=0}^{\frac{1}{2}(p-1)} A_{(2ps+2\tau)n+r} \cos((2ps+2\tau)n+r)\theta \overset{p,n}{\sim} I_1^* + I_2^*,$$

where, after making in the right hand side the change the summation variable from τ to $\widetilde{\tau} = \tau + \frac{1}{2}(p-1)$ and then writing τ for $\widetilde{\tau}$, we obtain

$$I_1^* = (-1)^{ps} \sum_{\tau=\frac{1}{2}(p-1)}^{p-1} (-1)^\tau \binom{ps+\tau}{p} \sum_{\sigma\in -r+n\{2,4,\dots,p-1\}} \binom{p-1}{\frac{1}{2}(p-1)+\lambda+1} \times$$

$$\times \frac{p}{ps+\tau-\frac{1}{2}(p-1)+\lambda+1} A_{(2ps+2\tau-(p-1))n-\sigma+(2\lambda+2)n} \cos\sigma\theta$$

$$(\text{here } \ 2\lambda+1 = \left[\tfrac{\sigma}{n}\right] =: \rho),$$

and

$$I_2^* = (-1)^{ps} \sum_{\tau=\frac{1}{2}(p-1)}^{p-1} (-1)^\tau \binom{ps+\tau}{p} \sum_{\sigma\in r+n\{0,2,\dots,p-1\}} \binom{p-1}{\frac{1}{2}(p-1)+\lambda} \times$$

$$\times \frac{p}{ps+\tau-\frac{1}{2}(p-1)-\lambda} A_{(2ps+2\tau-(p-1))n+\sigma-2\lambda n}$$

$$(\text{here } \ 2\lambda = \left[\tfrac{\sigma}{n}\right] := \rho).$$

We observe that in both I_1^* and I_2^* we have $\lambda = \frac{1}{2}(\rho - e_\rho)$. Replacing in the right hand side λ by $\frac{1}{2}(\rho - e_\rho)$ and then summing both sides over r, $1 \le r \le n-1$ we get

$$\sum_{r=1}^{n-1} \sum_{\tau=0}^{\frac{1}{2}(p-1)} A_{(2ps+2\tau)n+r} \cos((2ps+2\tau)n+r)\theta \overset{p,n}{\sim} S_1(s) + S_2(s)$$

where

$$S_1(s) = (-1)^{ps} \sum_{\tau=\frac{1}{2}(p-1)}^{p-1} (-1)^\tau \binom{ps+\tau}{p} \sum_{\sigma\in\Omega_1} \binom{p-1}{\frac{1}{2}(p-1)+\frac{1}{2}(\rho-e_\rho)+1} \times$$

$$\times \frac{p}{ps+\tau-\frac{1}{2}(p-1)+\frac{1}{2}(\rho-e_\rho)+1} A_{(2ps+2\tau-(p-1))n-\sigma+((\rho-e_\rho)+2)n} \cos\sigma\theta,$$

and

$$S_2(s) = (-1)^{ps} \sum_{\tau=\frac{1}{2}(p-1)}^{p-1} (-1)^\tau \binom{ps+\tau}{p} \sum_{\sigma\in\Omega_0} \binom{p-1}{\frac{1}{2}(p-1)+\frac{1}{2}(\rho-e_\rho)} \times$$

$$\times \frac{p}{ps+\tau-\frac{1}{2}(p-1)-\frac{1}{2}(\rho-e_\rho)} A_{(2ps+2\tau-(p-1))n+\sigma-(\rho-e_\rho)n} \cos\sigma\theta.$$

The case (iii). In this case p is odd, s is even and j is odd. Here $E_{p,s,j} = \{1, 3, \ldots, p\}$, $E^*_{p,s,j} = \{1, 3, \ldots, p-2\}$. Repeating the same argument as in the case (i) we get

$$\sum_{r=1}^{n-1} \sum_{\tau=1}^{\frac{1}{2}(p-1)} A_{(2ps+(2\tau-1))n+r} \cos((2ps + (2\tau - 1))n + r)\theta \overset{p,n}{\sim} S_3(s) + S_4(s)$$

where

$$\begin{aligned} S_3(s) &= (-1)^{ps} \sum_{\tau=\frac{1}{2}(p+1)}^{p-1} (-1)^\tau \binom{ps+\tau}{p} \sum_{\sigma\in\Omega_0} \binom{p-1}{\frac{1}{2}(p-1)+\frac{1}{2}(\rho-e_\rho)} \times \\ &\times \frac{p}{ps+\tau-\frac{1}{2}(p-1)+\frac{1}{2}(\rho-e_\rho)} A_{(2ps+2\tau-(p-1))n-\sigma+((\rho-e_\rho))n} \cos\sigma\theta, \end{aligned}$$

and

$$\begin{aligned} S_4(s) &= (-1)^{ps} \sum_{\tau=\frac{1}{2}(p+1)}^{p-1} (-1)^\tau \binom{ps+\tau}{p} \sum_{\sigma\in\Omega_1} \binom{p-1}{\frac{1}{2}(p-1)+\frac{1}{2}(\rho-e_\rho)+1} \times \\ &\times \frac{p}{ps+\tau-\frac{1}{2}(p-1)-\frac{1}{2}(\rho-e_\rho)-1} A_{(2ps+2\tau-(p-1))n+\sigma-(\rho-e_\rho+2)n} \cos\sigma\theta. \end{aligned}$$

The case (ii). In this case p is odd, s is odd and j is odd. Here $E_{p,s,j} = \{2, 4, \ldots, p-1\}$, $E^*_{p,s,j} = \{0, 2, \ldots, p-1\}$. The same argument used in the case (i), but without the change of variable τ, gives

$$\begin{aligned} &\sum_{r=1}^{n-1} \sum_{\tau=0}^{\frac{1}{2}(p-1)-1} A_{(p(2s-1)+(2\tau+1))n+r} \cos((p(2s-1) + (2\tau+1))n + r)\theta \\ &\overset{p,n}{\sim} S_5(s) + S_6(s), \end{aligned}$$

where

$$\begin{aligned} S_5(s) &= (-1)^{ps} \sum_{\tau=0}^{\frac{1}{2}(p-1)-1} (-1)^\tau \binom{ps+\tau}{p} \sum_{\sigma\in\Omega_1} \binom{p-1}{\frac{1}{2}(p-1)+\frac{1}{2}(\rho-e_\rho)+1} \times \\ &\times \frac{p}{ps+\tau-\frac{1}{2}(p-1)+\frac{1}{2}(\rho-e_\rho)+1} A_{(2ps+2\tau-(p-1))n-\sigma+((\rho-e_\rho)+2)n} \cos\sigma\theta, \end{aligned}$$

and

$$S_6(s) = (-1)^{ps} \sum_{\tau=0}^{\frac{1}{2}(p-1)-1} (-1)^{\tau} \binom{ps+\tau}{p} \sum_{\sigma\in\Omega_0} \binom{p-1}{\frac{1}{2}(p-1)+\frac{1}{2}(\rho-e_\rho)} \times$$

$$\times \frac{p}{ps+\tau-\frac{1}{2}(p-1)-\frac{1}{2}(\rho-e_\rho)} A_{(2ps+2\tau-(p-1))n+\sigma-(\rho-e_\rho)n} \cos\sigma\theta.$$

The case (iv). In this case p is odd, s is odd and j is even. So that $E_{p,s.j} = \{1,3,\dots,p\}$, $E^*_{p,s,j} = \{1,3,\dots,p-2\}$. The same argument used in the case (i), but without the change of variable τ, we get

$$\sum_{r=1}^{n-1} \sum_{\tau=0}^{\frac{1}{2}(p-1)} A_{(p(2s-1)+2\tau)n+r} \cos((p(2s-1)+2\tau)n+r)\theta \overset{p,n}{\sim} S_7(s) + S_8(s),$$

where

$$S_7(s) = (-1)^{ps} \sum_{\tau=0}^{\frac{1}{2}(p-1)} (-1)^{\tau} \binom{ps+\tau}{p} \sum_{\sigma\in\Omega_0} \binom{p-1}{\frac{1}{2}(p-1)+\frac{1}{2}(\rho-e_\rho)} \times$$

$$\times \frac{p}{ps+\tau-\frac{1}{2}(p-1)+\frac{1}{2}(\rho-e_\rho)} A_{(2ps+2\tau-(p-1))n-\sigma+((\rho-e_\rho))n} \cos\sigma\theta,$$

and

$$S_8(s) = (-1)^{ps} \sum_{\tau=0}^{\frac{1}{2}(p-1)} (-1)^{\tau} \binom{ps+\tau}{p} \sum_{\sigma\in\Omega_1} \binom{p-1}{\frac{1}{2}(p-1)+\frac{1}{2}(\rho-e_\rho)+1} \times$$

$$\times \frac{p}{ps+\tau-\frac{1}{2}(p-1)-\frac{1}{2}(\rho-e_\rho)-1} A_{(2ps+2\tau-(p-1))n+\sigma-(\rho-e_\rho+2)n} \cos\sigma\theta.$$

Finally summing up the expressions we obtained we get

$$\sum_{s=1}^{\infty} \sum_{r=1}^{n-1}\sum_{j=0}^{p-1} A_{(ps+j)n+r)} \cos((ps+j)n+r)\theta \overset{p,n}{\sim} \sum_{s=1}^{\infty}\left(\sum_{w=1}^{8} S_w(s)\right)$$

$$= \sum_{s=1}^{\infty}\sum_{w=1}^{4}(S_w(s)+S_{w+4}(s))$$

$$= \sum_{s=1}^{\infty}(-1)^s\sum_{\tau=0}^{p-1}(-1)^{\tau}\binom{ps+\tau}{p}\sum_{\sigma\in\Omega_1}\binom{p-1}{\frac{1}{2}(p-1)+\frac{1}{2}(\rho-e_\rho)+1}\times$$

$$\times\left\{\frac{p}{ps+\tau-\frac{1}{2}(p-1)+\frac{1}{2}(\rho-e_\rho)+1}A_{(2ps+2\tau-(p-1)n-\sigma+(\rho-e_\rho+2)n}\right.$$

$$\left.+\frac{p}{ps+\tau-\frac{1}{2}(p-1)-\frac{1}{2}(\rho-e_\rho)-1}A_{(2ps+2\tau-(p-1)n+\sigma-(\rho-e_\rho+2)n}\right\}\cos\sigma\theta$$

$$+\sum_{s=1}^{\infty}(-1)^s\sum_{\tau=0}^{p-1}(-1)^{\tau}\binom{ps+\tau}{p}\sum_{\sigma\in\Omega_0}\binom{p-1}{\frac{1}{2}(p-1)+\frac{1}{2}(\rho-e_\rho)}\times$$

$$\times\left\{\frac{p}{ps+\tau-\frac{1}{2}(p-1)-\frac{1}{2}(\rho-e_\rho)}A_{(2ps+2\tau-(p-1)n+\sigma-(\rho-e_\rho)n}\right.$$

$$\left.+\frac{p}{ps+\tau-\frac{1}{2}(p-1)+\frac{1}{2}(\rho-e_\rho)}A_{(2ps+2\tau-(p-1)n-\sigma+(\rho-e_\rho)n}\right\}\cos\sigma\theta$$

$$= \sum_{s=1}^{\infty}(-1)^s\sum_{\tau=0}^{p-1}(-1)^{\tau}\binom{ps+\tau}{p}\sum_{\sigma\in\Omega}\binom{p-1}{\frac{1}{2}(p-1)+\frac{1}{2}(\rho-e_\rho)+e_\rho}\times$$

$$\times\left\{\frac{p}{ps+\tau-\frac{1}{2}(p-1)+\frac{1}{2}(\rho-e_\rho)+e_\rho}A_{(2ps+2\tau-(p-1)n-\sigma+(\rho-e_\rho+2e_\rho)n}\right.$$

$$\left.+\frac{p}{ps+\tau-\frac{1}{2}(p-1)-\frac{1}{2}(\rho-e_\rho)-e_\rho}A_{(2ps+2\tau-(p-1)n+\sigma-(\rho-e_\rho+2e_\rho)n}\right\}\cos\sigma\theta.$$

Observe that when $\sigma\in\Omega_1$ then $\rho := \left[\frac{\sigma}{n}\right] = \nu$ is odd and when $\sigma\in\Omega_0$ then $\rho := \left[\frac{\sigma}{n}\right] = \nu$ is even. This gives us the last equality and completes the proof of Theorem 2.4(i). □

6.3 Proof of Theorem 2.1(i).

Assume now that p is even. We have $\Omega = \Omega_2 \bigcup \Omega_3$ where

$$\Omega_2 := \left(\bigcup_{\nu\in\{1,3,\dots,p-1\}}[\nu n+1,(\nu+1)n-1]\right)\bigcap Z,$$

$$\Omega_3 := \left(\bigcup_{\nu\in\{0,2,\dots,p-2\}}[\nu n+1,(\nu+1)n-1]\right)\bigcap Z.$$

We sum both sides of (6.6) and (6.7) over j, $0 \le j \le p-1$. This sum is split into two sums: one over even j and the other over odd j.

We first consider *case* **(i)** when p is even and j is even. We replace j by 2τ, $0 \le \tau \le \frac{1}{2}p - 1$. We have $ps + 2\tau + p = p(s+1) + 2\tau$, so that $\Lambda = \frac{1}{2}p(s+1) + \tau$. Here $E_{p,s,j} = \{2,4,\dots,p\}$, $E^*_{p,s,j} = \{0,2,\dots,p-2\}$. We sum both sides of (6.6) and (6.7) over $j = 2\tau$, $0 \le \tau \le \frac{1}{2}p - 1$ and use (6.2). Then we get

$$\sum_{\tau=0}^{\frac{1}{2}p-1} A_{(ps+2\tau)n+r}\cos((ps+2\tau)n+r)\theta \overset{p,n}{\sim}$$

$$\overset{p,n}{\sim} -(-1)^{\frac{1}{2}p(s+1)}\sum_{\tau=0}^{\frac{1}{2}p-1}(-1)^{\tau}\binom{\frac{1}{2}p(s+1)+\tau}{p}\sum_{\sigma\in -r+n\{2,4,\dots,p\}}\binom{p-1}{\frac{1}{2}p+\lambda}\times$$

$$\times\frac{p}{\frac{1}{2}ps+\tau+\lambda+1}A_{(ps+2\tau)n-\sigma+(2\lambda+2)n}\cos\sigma\theta$$

$$-(-1)^{\frac{1}{2}p(s+1)}\sum_{\tau=0}^{\frac{1}{2}p-1}(-1)^{\tau}\binom{\frac{1}{2}p(s+1)+\tau}{p}\sum_{\sigma\in r+n\{0,2,\dots,p-2\}}\binom{p-1}{\frac{1}{2}p+\lambda}\times$$

$$\times\frac{p}{\frac{1}{2}ps+\tau-\lambda}A_{(ps+2\tau)n+\sigma-2\lambda n}\cos\sigma\theta,$$

where $\lambda = \left[\frac{\sigma}{2n}\right]$.

Summing both sides over r, $1 \le r \le n-1$ we get

$$\sum_{r=1}^{n-1}\sum_{\tau=0}^{\frac{1}{2}p-1} A_{(ps+2\tau)n+r}\cos((ps+2\tau)n+r)\theta \overset{p,n}{\sim} S_1(s) + S_2(s) \qquad (6.23)$$

where

$$S_1(s) = -(-1)^{\frac{1}{2}p(s+1)}\sum_{\tau=0}^{\frac{1}{2}p-1}(-1)^{\tau}\binom{\frac{1}{2}p(s+1)+\tau}{p}\sum_{\sigma\in\Omega_2}\binom{p-1}{\frac{1}{2}p+\lambda}\times$$

$$\times\frac{p}{\frac{1}{2}ps+\tau+\lambda+1}A_{(ps+2\tau)n-\sigma+(2\lambda+2)n}\cos\sigma\theta,$$

and

$$S_2(s) = -(-1)^{\frac{1}{2}p(s+1)} \sum_{\tau=0}^{\frac{1}{2}p-1} (-1)^\tau \binom{\frac{1}{2}p(s+1)+\tau}{p} \sum_{\sigma\in\Omega_3} \binom{p-1}{\frac{1}{2}p+\lambda} \times$$

$$\times \frac{p}{\frac{1}{2}ps+\tau-\lambda} A_{(ps+2\tau)n+\sigma-2\lambda n} \cos\sigma\theta$$

where $\lambda = \left[\frac{\sigma}{2n}\right]$.

Case **(ii).** In this case p is even and j is odd. Here $E_{p,s,j} = \{1,3,\ldots,p-1\}$, $E^*_{p,s,j} = \{1,3,\ldots,p-1\}$. We replace j by $2\tau+1$ $0 \le \tau \le \frac{1}{2}p-1$. We have $ps+(2\tau+1)+p = p(s+1)+2\tau+1$ so that $\Lambda = \frac{1}{2}p(s+1)+\tau$. We sum both sides of (6.6) and (6.7) over $j = 2\tau+1$, $0 \le \tau \le \frac{1}{2}p-1$ and use (6.2) and obtain

$$\sum_{\tau=0}^{\frac{1}{2}p-1} A_{(ps+2\tau+1)n+r} \cos((ps+2\tau+1)n+r)\theta \overset{p,n}{\sim} -(-1)^{\frac{1}{2}p(s+1)} \times$$

$$\times \sum_{\tau=0}^{\frac{1}{2}p-1} (-1)^\tau \binom{\frac{1}{2}p(s+1)+\tau}{p} \sum_{\sigma\in -r+n\{1,3,\ldots,p-1\}} \binom{p-1}{\frac{1}{2}p+\lambda} \times$$

$$\times \frac{p}{\frac{1}{2}ps+\tau+\lambda+1} A_{(ps+2\tau+1)n-\sigma+(2\lambda+1)n} \cos\sigma\theta \quad \left(\text{here } \lambda = \left[\tfrac{\sigma}{2n}\right]\right)$$

$$-(-1)^{\frac{1}{2}p(s+1)} \sum_{\tau=0}^{\frac{1}{2}p-1} (-1)^\tau \binom{\frac{1}{2}p(s+1)+\tau}{p} \sum_{\sigma\in r+n\{1,3,\ldots,p-1\}} \binom{p-1}{\frac{1}{2}p+\lambda} \times$$

$$\times \frac{p}{\frac{1}{2}ps+\tau-\lambda} A_{(ps+2\tau+1)n+\sigma-(2\lambda+1)n} \cos\sigma\theta \quad \left(\text{where } \lambda = \left[\tfrac{\sigma}{2n}\right]\right).$$

We observe that in both I_1 and I_2 we have $\lambda = \left[\frac{\sigma}{2n}\right]$. Summing both sides over r, $1 \le r \le n-1$ we get

$$\sum_{r=1}^{n-1} \sum_{\tau=0}^{\frac{1}{2}p-1} A_{(ps+2\tau+1)n+r} \cos((ps+2\tau+1)n+r)\theta \overset{p,n}{\sim} S_3(s)+S_4(s), \quad (6.24)$$

where

$$S_3(s) = -(-1)^{\frac{1}{2}p(s+1)} \sum_{\tau=0}^{\frac{1}{2}p-1} (-1)^\tau \binom{\frac{1}{2}p(s+1)+\tau}{p} \sum_{\sigma\in\Omega_3} \binom{p-1}{\frac{1}{2}p+\lambda} \times$$

$$\times \frac{p}{\frac{1}{2}ps+\tau+\lambda+1} A_{(ps+2\tau+1)n-\sigma+(2\lambda+1)n} \cos\sigma\theta \quad \left(\text{here } \lambda = \left[\tfrac{\sigma}{2n}\right]\right)$$

and

$$S_4(s) = -(-1)^{\frac{1}{2}p(s+1)} \sum_{\tau=0}^{\frac{1}{2}p-1} (-1)^\tau \binom{\frac{1}{2}p(s+1)+\tau}{p} \sum_{\sigma\in\Omega_2} \binom{p-1}{\frac{1}{2}p+\lambda} \times$$

$$\times \frac{p}{\frac{1}{2}ps+\tau-\lambda} A_{(ps+2\tau+1)n+\sigma-(2\lambda+1)n} \cos\sigma\theta \quad \left(\text{where } \lambda = \left[\tfrac{\sigma}{2n}\right]\right).$$

Combining (6.23) and (6.24), we obtain

$$\sum_{s=1}^{\infty}\sum_{r=1}^{n-1}\sum_{j=0}^{p-1} A_{(ps+j)n+r)} \cos((ps+j)n+r)\theta \overset{p,n}{\sim} \sum_{s=1}^{\infty}(S_1(s)+S_3(s))$$

$$+\sum_{s=1}^{\infty}(S_2(s)+S_4(s))$$

$$=\sum_{s=1}^{\infty}(-1)^{\frac{1}{2}p(s+1)}\sum_{\tau=0}^{\frac{1}{2}p-1}(-1)^{\tau}\binom{\frac{1}{2}p(s+1)+\tau}{p}\sum_{\sigma\in\Omega_3\bigcup\Omega_2}\binom{p-1}{\frac{1}{2}p+\lambda}\times$$

$$\times\left\{\frac{p}{\frac{1}{2}ps+\tau+\lambda+1}A_{(ps+2\tau)n-\sigma+(2\lambda+2)n}+\frac{p}{ps+\tau-\lambda}A_{(ps+2\tau)n+\sigma-2\lambda n}\right\}\cos\sigma\theta,$$

which completes the proof since $\Omega = \Omega_2 \bigcup \Omega_3$, once we observe that for $2\lambda n < \sigma < (2\lambda+2)n$, we have $\lambda = \left[\frac{\sigma}{2n}\right]$. □

7 Walsh equiconvergence theorems for Hermite interpolants

We shall later need the special function $f_*(z) \in A(C_\rho)$, given by (2.8) and having the following Chebyshev expansion

$$f_*(z) = \sum_{\sigma=0}^{\infty} \frac{1}{\rho^\sigma} T_\sigma(z) \,, \tag{7.1}$$

inside the ellipse C_ρ $(\rho > 1)$. The following three lemmas are straightforward and their proofs are left out:

Lemma 7.1 *Let* $(a_n)_{n\geq 0}, (a_n^{(1)})_{n\geq 0}, \ldots (a_n^{(j)})_{n\geq 0}$, *be given sequences. Then:*
(i) If the sequences satisfy $a := \limsup_{n\to\infty} |a_n|^{1/n} > \max_{1\leq k\leq j} \limsup_{n\to\infty} |a_n^{(k)}|^{1/n}$ *then*

$\limsup_{n\to\infty} \left|a_n + a_n^{(1)} + \ldots a_n^{(j)}\right|^{1/n} = a$,

(ii) If the sequences satisfy $\exists a := \lim_{n\to\infty} |a_n|^{1/n} > \max_{1\leq k\leq j} \limsup_{n\to\infty} |a_n^{(k)}|^{1/n}$ *then*

$\exists \lim_{n\to\infty} \left|a_n + a_n^{(1)} + \ldots a_n^{(j)}\right|^{1/n} = a$.

Lemma 7.2 *Suppose the two sequences* (a_n), (b_n) *satisfy* $\limsup |a_n|^{1/n} = a$ *and* $\lim_{n\to\infty} |b_n|^{1/n} = b$. *Then we have* $\limsup_{n\to\infty} |a_n b_n|^{1/n} = ab$.

Lemma 7.3 *For numbers ρ, R satisfying $R > 1$, $\rho > 1$, we have*

$$\lim_{n\to\infty}\min_{|w|=R}\left|\sum_{\sigma=1}^{n-1}\left(w^{\sigma}+w^{-\sigma}\right)\right|^{1/n}=\lim_{n\to\infty}\max_{|w|=R}\left|\sum_{\sigma=1}^{n-1}\left(w^{\sigma}+w^{-\sigma}\right)\right|^{1/n}=R,$$

and

$$\lim_{n\to\infty}\min_{|w|=R}\quad\sum_{\sigma=1}^{n-1}\left|\left(\rho^{\sigma}+\rho^{-\sigma}\right)\left(w^{\sigma}+w^{-\sigma}\right)\right|^{1/n}$$

$$=\lim_{n\to\infty}\max_{|w|=R}\sum_{\sigma=1}^{n-1}\left|\left(\rho^{\sigma}+\rho^{-\sigma}\right)\left(w^{\sigma}+w^{-\sigma}\right)\right|^{1/n}=\rho R\,.$$

Lemma 7.4 *Assume $\rho > 1$. Then we have:*
(i) If p is a positive even integer then

$$\lim_{n\to\infty}\left|\sum_{\tau=0}^{\frac{1}{2}p-1}(-1)^{\tau}\binom{\frac{1}{2}p(j+1)+\tau-1}{p-1}\frac{p}{pj+2\tau}\frac{1}{\rho^{2\tau m}}\right|^{1/n}=1\,,$$

and

$$\lim_{n\to\infty}\left|\sum_{\tau=0}^{\frac{1}{2}p-1}(-1)^{\tau}\binom{\frac{1}{2}p(j+1)+\tau}{p}\frac{p}{\frac{1}{2}pj+\tau}\frac{1}{\rho^{2\tau m}}\right|^{1/n}=1\,;$$

(ii) If p is a positive integer then

$$\lim_{n\to\infty}\left|\sum_{\tau=0}^{p-1}(-1)^{\tau}\binom{pj+\tau}{p}\frac{p}{pj+\tau-(p-1)}\frac{1}{\rho^{2\tau m}}\right|^{1/n}=1.$$

Proof of (i). We have

$$\sum_{\tau=0}^{\frac{1}{2}p-1}(-1)^{\tau}\binom{\frac{1}{2}p(j+1)+\tau-1}{p-1}\frac{p}{pj+2\tau}\cdot\frac{1}{\rho^{2m\tau}}=\binom{\frac{1}{2}p(j+1)-1}{p-1}\frac{p}{pj}+A$$

where

$$A=\frac{1}{\rho^{2m}}\sum_{\tau=1}^{\frac{1}{2}p-1}(-1)^{\tau}\binom{\frac{1}{2}p(j+1)+\tau-1}{p-1}\frac{p}{pj+2\tau}\frac{1}{\rho^{2m(\tau-1)}}.$$

Now

$$|A|\le\frac{c}{\rho^{2m}},$$

where c depends on p and j only. Hence

$$\limsup_{n\to\infty} |A|^{1/n} \le \frac{1}{\rho^{2q}} < 1.$$

The proof of the first conclusion of (i) follows now by applying Lemmas 7.1 and 7.2. The proof of (ii) and of the second conclusion of (i) is similar. □

Lemma 7.5 *Let l be a positive integer. Then:*
(i) If p is a positive even integer then there exist constants c_1, c_2, depending on p and l only, satisfying

$$0 < c_1 \equiv c_1(p,l) \le c_2 \equiv c_2(p,l) < \infty,$$

such that

$$0 < c_1 \binom{s-l+p-1}{s-l} \le \binom{\frac{1}{2}p(s+1)+\tau}{p} \max\left(\frac{p}{\frac{1}{2}ps+\tau-\lambda}, \frac{p}{\frac{1}{2}ps+\tau+\lambda+1}\right)$$
$$\le c_2 \binom{s-l+p-1}{s-l}$$

for the integers $0 \le \tau$, $\lambda \le \frac{1}{2}p - 1$ and $s \ge l$.

(ii) When p is a positive odd integer then there exist constants c_1, c_2, depending on p and l only, satisfying

$$0 < c_1 \equiv c_1(p,l) \le c_2 \equiv c_2(p,l) < \infty,$$

such that

$$0 < c_1 \binom{s-l+p-1}{s-l} \le \binom{ps+\tau}{p} \frac{p}{ps+\tau-\frac{1}{2}(p-1)}$$
$$\le c_2 \binom{s-l+p-1}{s-l} < \infty,$$

for the integers τ, s , $0 \le \tau \le p-1$ and $s \ge l$,.

Proof of (i). We shall prove (i) only, since the proof of (ii) is similar. We have for $0 \le \tau, \lambda \le \frac{1}{2}p - 1$ and $x \ge 1$

$$\frac{1}{2}p(x+1)+\tau \ge \frac{1}{2}px+\tau-\lambda \ge \frac{1}{2}px+\tau-\frac{1}{2}p+1 = \frac{1}{2}p(x+1)+\tau-(p-1).$$

Also

$$\frac{1}{2}p(x+1)+\tau = \frac{1}{2}px+\tau+\frac{1}{2}p-1+1 \ge \frac{1}{2}px+\tau+\lambda+1 \ge \frac{1}{2}px+\tau.$$

Hence for $x \ge 1$,

$$\binom{\frac{1}{2}p(x+1)+\tau}{p}\frac{p}{\frac{1}{2}px+\tau-\lambda} \quad \text{and} \quad \binom{\frac{1}{2}p(x+1)+\tau}{p}\frac{p}{\frac{1}{2}px+\tau+\lambda+1}$$

are polynomials of degree $p-1$ which are positive for $x \geq 1,\ 0 \leq \tau, \lambda \leq \frac{1}{2}p - 1$. The leading power in both these polynomials is $\frac{(\frac{1}{2}p)^{p-1}}{(p-1)!}x^{p-1}$. Also, for each $x \geq l$, $\binom{x-l+p-1}{p-1}$ is a positive polynomial of degree $p-1$ for $x \geq l$. Its leading term is $\frac{1}{(p-1)!}x^{p-1}$. Therefore the continuous rational functions

$$\frac{p\binom{\frac{1}{2}p(x+1)+\tau}{p}/(\frac{1}{2}px + \tau - \lambda)}{\binom{x-l+p-1}{x-l}}, \frac{p\binom{\frac{1}{2}p(x+1)+\tau}{p}/(\frac{1}{2}px + \tau + \lambda + 1)}{\binom{x-l+p-1}{x-l}}$$

are positive for $x \geq l$ and tend to a positive limit as $x \to \infty$. Therefore to each pair $0 \leq \tau, \lambda \leq \frac{1}{2}p - 1$ there correspond positive and finite constants $0 < c_1(\tau, \lambda) \leq c_2(\tau, \lambda) < \infty$ such that for each $x \geq l$ we have

$$\begin{aligned} 0 < c_1(\tau, \lambda) \quad &\leq \frac{p\binom{\frac{1}{2}p(x+1)+\tau}{p}/(\frac{1}{2}px+\tau-\lambda)}{\binom{x-l+p-1}{x-l}}, \frac{p\binom{\frac{1}{2}p(x+1)+\tau}{p}/(\frac{1}{2}px+\tau+\lambda+1)}{\binom{x-l+p-1}{x-l}} \\ &\leq c_2(\tau, \lambda) < \infty, \quad \text{for } x \geq l. \end{aligned}$$

Since the number of pairs τ, λ that satisfy $0 \leq \tau, \lambda \leq \frac{1}{2}p - 1$ is finite, the proof of the lemma is complete. □

7.1 Walsh equiconvergence when p is even.

We assume here that $m > n \geq 1$ are integers. Set

$$s_{n-1,m,0}(f, z) := \frac{1}{2}A_0 + \sum_{\sigma=1}^{n-1} A_\sigma T_\sigma(z),$$

and when $p \geq 2$ is even then for $j \geq 1$ set

$$\begin{aligned} s_{n-1,m,j} \quad &(f, z) := -\tfrac{1}{2}\tbinom{p}{\frac{1}{2}p}(-1)^{\frac{1}{2}p(j+1)} \times \\ &\times \sum_{\tau=0}^{\frac{1}{2}p-1} (-1)^\tau \tbinom{\frac{1}{2}p(j+1)+\tau-1}{p-1} \tfrac{p}{pj+2\tau} A_{(pj+2\tau)m} \\ &- \tbinom{p-1}{\frac{1}{2}p}(-1)^{\frac{1}{2}p(j+1)} \sum_{\tau=0}^{\frac{1}{2}p-1} (-1)^\tau \tbinom{\frac{1}{2}p(j+1)+\tau}{p} \times \\ &\times \sum_{\sigma=1}^{n-1} \left(\tfrac{p}{\frac{1}{2}pj+\tau} A_{(pj+2\tau)m+\sigma} + \tfrac{p}{\frac{1}{2}pj+\tau+1} A_{(pj+2\tau+2)m-\sigma} \right) T_\sigma(z). \end{aligned}$$

It is easy to verify that for positive integers $m > n$, an even integer $p \geq 2$ and a positive integer l, we have

$$S_{n-1}\left(H^{\{p\}}_{pm-1}(S_{plm-1}(f,\cdot),z)\right) = \sum_{j=0}^{l-1} s_{n-1,m,j}(f,z). \tag{7.2a}$$

Thus the difference $\Delta_{l-1,n,m,p}(f,z)$ of (2.6) can also be written as

$$S_{n-1}(H^{\{p\}}_{pm-1}(f,\cdot),z) - \sum_{j=0}^{l-1} s_{n-1,m,j}(f,z).$$

Lemma 7.6 *Assume that $p \geq 2$ is an even integer. Assume $m := m(n)$ is such that $m/n \to q > 1$ as $n \to \infty$ and $R > \rho > 1$. Then for each integer $j \geq 1$, we have*

$$\lim_{n\to\infty} \min_{z\in C_R} |s_{n-1,m,j}(f_*,z)|^{1/n} = \lim_{n\to\infty} \max_{z\in C_R} |s_{n-1,m,j}(f_*,z)|^{1/n} = \frac{R}{\rho^{pjq+1}},$$

where f_* is given by (2.8).

Proof. For each $z \in C_R$ we have $w = w(z)$ where $|w| = R$ and $T_\sigma(z) = \frac{1}{2}(w^\sigma + w^{-\sigma})$. We have

$$s_{n-1,m,j}(f_*,z) = S_{1,j} + S_{2,j} + S_{3,j},$$

where (since $A_m = \rho^{-m}$)

$$S_{1,j} = -\tfrac{1}{2}\tbinom{p}{\frac{1}{2}p}(-1)^{\frac{1}{2}p(j+1)}\frac{1}{\rho^{pjm}}\sum_{\tau=0}^{\frac{1}{2}p-1}(-1)^\tau\tbinom{\frac{1}{2}p(j+1)+\tau-1}{p-1}\times$$
$$\times\frac{p}{pj+2\tau}\frac{1}{\rho^{2m\tau}},$$

$$S_{2,j}(w) = -\tbinom{p-1}{\frac{1}{2}p}(-1)^{\frac{1}{2}p(j+1)}\frac{1}{\rho^{pjm}}\sum_{\tau=0}^{\frac{1}{2}p-1}(-1)^\tau\tbinom{\frac{1}{2}p(j+1)+\tau}{p}\times$$
$$\times\frac{p}{\frac{1}{2}pj+\tau}\frac{1}{\rho^{2\tau m}}\sum_{\sigma=1}^{n-1}\frac{1}{2}\left((\tfrac{w}{\rho})^\sigma + (\tfrac{1}{\rho w})^\sigma\right),$$

and

$$S_{3,j}(w) = -\tbinom{p-1}{\frac{1}{2}p}\quad(-1)^{\frac{1}{2}p(j+1)}\frac{1}{\rho^{2m}}\frac{1}{\rho^{pjm}}\sum_{\tau=0}^{\frac{1}{2}p-1}(-1)^\tau\tbinom{\frac{1}{2}p(j+1)+\tau}{p}\times$$
$$\times\frac{p}{\frac{1}{2}pj+\tau+1}\frac{1}{\rho^{2\tau m}}\sum_{\sigma=1}^{n-1}\frac{1}{2}\left((w\rho)^\sigma + (\tfrac{\rho}{w})^\sigma\right).$$

Applying Lemmas 7.1, 7.2, 7.3 and 7.4 we get

$$\lim_{n\to\infty} |S_{1,j}|^{1/n} = \frac{1}{\rho^{pjq}},$$

$$\lim_{n\to\infty} \min_{|w|=r} |S_{2,j}(w)|^{1/n} = \lim_{n\to\infty} \max_{|w|=r} |S_{2,j}(w)|^{1/n} = \frac{R}{\rho^{pjq+1}},$$

$$\lim_{n\to\infty} \min_{|R|=R} |S_{3,j}(w)|^{1/n} = \lim_{n\to\infty} \max_{|R|=R} |S_{3,j}(w)|^{1/n} = \frac{R\rho}{\rho^{2q}\rho^{pjq}}.$$

Hence

$$\begin{aligned}\lim_{n\to\infty} \min_{|w|=R} |s_{n-1,m,j}(f_*, z)|^{1/n} &= \lim_{n\to\infty} \max_{|w|=R} |s_{n-1,m,j}(f_*, z)|^{1/n} \\ &= \max\left(\tfrac{1}{\rho^{pjq}}, \tfrac{R}{\rho^{pjq+1}}, \tfrac{R}{\rho^{2q+pjq-1}}\right) = \tfrac{R}{\rho^{pjq+1}},\end{aligned}$$

which completes the proof. □

Lemma 7.7 *Assume $f \in C_\rho$, p is even, $m := m(n)$ satisfies $m/n \to q > 1$ and $R > 1$, $\rho > 1$. Then for each $z \in C_\rho$ and each positive integer l, we have*

$$\limsup_{n\to\infty} \max_{z\in C_R} |\Delta_{l-1,n,m,p}(f, z)|^{1/n} \le f_{l,q,p}(R)$$

Proof. We assume here that $z \in C_\rho$. Then $z = \frac{1}{2}(w + \frac{1}{w})$ where $|w| = R$, and we have $T_\sigma(z) = \frac{1}{2}(w^\sigma + w^{-\sigma})$. Hence $|T_\sigma(z)| \le \frac{1}{2}(R^\sigma + \frac{1}{R^\sigma})$. We have $\Delta_{l-1,n,m,p}(f, z) = J_1 + J_2$ where

$$\begin{aligned}J_1(z) &= -\tfrac{1}{2}\tbinom{p}{\frac{1}{2}p} \textstyle\sum_{s=l}^{\infty}(-1)^{\frac{1}{2}p(s+1)} \sum_{\tau=0}^{\frac{1}{2}p-1} \binom{\frac{1}{2}p(s+1)+\tau}{p} \times \\ &\quad \times \left(\tfrac{\frac{1}{2}p}{\frac{1}{2}ps+\tau} + \tfrac{\frac{1}{2}p}{\frac{1}{2}ps+\tau} - \tfrac{p}{\frac{1}{2}p(s+1)+\tau}\right) A_{(ps+2\tau)m},\end{aligned}$$

$$\begin{aligned}J_2(z) &= -\tbinom{p-1}{\frac{1}{2}p} \textstyle\sum_{\sigma=1}^{n-1} \sum_{s=l}^{\infty}(-1)^{\frac{1}{2}p(s+1)} \sum_{\tau=0}^{\frac{1}{2}p-1}(-1)^\tau \binom{\frac{1}{2}p(s+1)+\tau}{p} \times \\ &\quad \times \left(\tfrac{p}{\frac{1}{2}ps+\tau} A_{(ps+2\tau)m+\sigma} + \tfrac{p}{\frac{1}{2}ps+\tau+1} A_{(ps+2\tau+2)m-\sigma}\right) T_\sigma(z).\end{aligned}$$

The expression for J_1 above is obtained from (2.2) for $\rho = 0$ or equivalently from the value of I_1 in Section 5.1 for $\rho = 0$. See also the identity (5.0), for $\rho = 0$. Since $f \in C_\rho$, therefore for each ϵ, with $1 < \rho - \epsilon < \rho$, there exists a finite constant $c \equiv c(\epsilon) > 0$ such that $|A_k| \le c(\rho-\epsilon)^{-k}$, $\quad k = 0, 1, \dots$. For $w = w(z)$ we have $|w| = R > \rho$. Hence using Lemma 7.5 (i), we have

$$\begin{aligned} |J_1| &= \mathcal{O}(1) \sum_{s=l}^{\infty} \sum_{\tau=0}^{\frac{1}{2}p-1} \binom{s-l+p-1}{s-l} \frac{1}{(\rho-\epsilon)^{(ps+2\tau)m}} \\ &= \mathcal{O}(1) \sum_{\tau=0}^{\frac{1}{2}p-1} \frac{1}{((\rho-\epsilon)^{2m})^{\tau}} \sum_{s=l}^{\infty} \binom{s-l+p-1}{s-l} \frac{1}{((\rho-\epsilon)^{pm})^{s}} \\ &= \mathcal{O}(1) \frac{1}{(\rho-\epsilon)^{pml}} \frac{1}{(1-(\rho-\epsilon)^{-pm})^{p}} \sum_{\tau=0}^{\frac{1}{2}p-1} \frac{1}{((\rho-\epsilon)^{2m})^{\tau}} \\ &= \mathcal{O}(1) \frac{1}{(\rho-\epsilon)^{pml}}. \end{aligned}$$

Therefore we have

$$\limsup_{n\to\infty} |J_1|^{1/n} \le \frac{1}{(\rho-\epsilon)^{pql}}.$$

We also have

$$\begin{aligned} |J_2(z)| &= \mathcal{O}(1) \sum_{\sigma=1}^{n-1} \sum_{s=l}^{\infty} \sum_{\tau=0}^{\frac{1}{2}p-1} \binom{\frac{1}{2}p(s+1)+\tau}{p} \times \\ &\times \left(\frac{p}{\frac{1}{2}ps+\tau} \frac{1}{(\rho-\epsilon)^{(ps+2\tau)m+\sigma}} + \frac{p}{\frac{1}{2}ps+\tau+1} \frac{1}{(\rho-\epsilon)^{(ps+2\tau+2)m-\sigma}} \right) \tfrac{1}{2}(|w|^{\sigma} + |w|^{-\sigma}) \\ &= \mathcal{O}(1) \sum_{s=l}^{\infty} \sum_{\tau=0}^{\frac{1}{2}p-1} \binom{\frac{1}{2}p(s+1)+\tau}{p} \frac{p}{\frac{1}{2}ps+\tau} \frac{1}{(\rho-\epsilon)^{(ps+2\tau)m}} \times \\ &\times \left\{ \max\left(\frac{R}{\rho-\epsilon}, \left(\frac{R}{\rho-\epsilon}\right)^{n} \right) + \frac{1}{R(\rho-\epsilon)} + \frac{(R(\rho-\epsilon))^{n}}{(\rho-\epsilon)^{2m}} + \right. \\ &\left. + \frac{1}{(\rho-\epsilon)^{2m}} \max\left(\frac{\rho-\epsilon}{R}, \left(\frac{\rho-\epsilon}{R}\right)^{n} \right) \right\} \\ &= \mathcal{O}(1) \sum_{\tau=0}^{\frac{1}{2}p-1} \frac{1}{(\rho-\epsilon)^{2m\tau}} \max\left(\frac{R}{\rho-\epsilon}, \left(\frac{R}{\rho-\epsilon}\right)^{n} \right) \times \\ &\times \sum_{s=l}^{\infty} \binom{s-l+p-1}{s-l} \frac{1}{((\rho-\epsilon)^{pm})^{s}} \\ &= \mathcal{O}(1) \frac{1}{(\rho-\epsilon)^{pml}} \max\left(\frac{R}{\rho-\epsilon}, \left(\frac{R}{\rho-\epsilon}\right)^{n} \right). \end{aligned}$$

Therefore we have

$$\limsup_{n\to\infty} \max_{z\in R} |J_2(z)|^{1/n} \le \frac{R}{(\rho-\epsilon)^{pql}} \max\left(\frac{1}{R}, \frac{1}{\rho-\epsilon}\right).$$

Combining the last result with the analogous result for J_1 and letting $\epsilon \searrow 0$ we get

$$\limsup_{n\to\infty} \max_{z\in C_R} |\Delta_{l-1,n,m,p}(f,z)|^{1/n} \le f_{\ell,q,p}(R).$$

So the proof is completed. □

Proof of (2.9) of Theorem 2.7 when p is even. For $z \in C_R$ we have $z = \frac{1}{2}(w + \frac{1}{w})$ where $|w| = R$ and $T_\sigma(z) = \frac{1}{2}(w^\sigma + w^{-\sigma})$. The proof follows now from the identity

$$\Delta_{l-1,n,m,p}(f_*, z) = s_{n-1,m,l}(f_*, z) + \Delta_{l,n,m,p}(f_*, z),$$

and by applying Lemma 7.1(ii), Lemma 7.6 and Lemma 7.7. □

Proof of (2.6a) and(2.7) of Theorem 2.7 when p is even. (2.6a) is an immediate consequence of Lemma 7.7 and (2.7) is a consequence of (2.6a) and (2.9). □

7.2 Walsh Equiconvergence when p is odd.

We assume here that $m > n \geq 1$ are integers. We set as in Sec. 7.1

$$s_{n-1,m,0}(f, z) := \frac{1}{2}A_0 + \sum_{\sigma=1}^{n-1} A_\sigma T_\sigma(z),$$

but for $j \geq 1$ when $p \geq 1$ is odd, set

$$s_{n-1,m,j}(f, z) := \binom{p-1}{\frac{1}{2}(p-1)}(-1)^j \sum_{\tau=0}^{p-1}(-1)^\tau \binom{pj+\tau}{p} \times$$

$$\times \frac{p}{pj+\tau-\frac{1}{2}(p-1)} \sum_{\sigma=1}^{n-1} \left(A_{(2pj+2\tau-(p-1))m-\sigma} + A_{(2pj+2\tau-(p-1))m+\sigma}\right) T_\sigma(z).$$

It is easy to verify that when p is odd, then for any three positive integers l, $m > n$, we have

$$S_{n-1}\left(H_{pm-1}^{\{p\}}(S_{(p(2l-1)-1)m+n-1}(f, \cdot), z)\right) = \sum_{j=0}^{l-1} s_{n-1.m,j}(f, z). \tag{7.2}$$

Lemma 7.8 *Assume $p \geq 1$ is an odd integer. Assume $m := m(n)$ satisfies $m/n \to q > 1$ as $n \to \infty$. Then for each $\rho > 1$, each $z \in C_R$, $R > 1$, $w = w(z)$, $|w| = R$ and each integer $j \geq 1$, we have*

$$\begin{aligned}\lim_{n\to\infty} \min_{z\in C_R} |s_{n-1,m,j}(f_*, z)|^{1/n} &= \lim_{n\to\infty} \max_{z\in C_R} |s_{n-1,m,j}(f_*, z)|^{1/n} \\ &= \frac{R}{\rho^{(2pj-(p-1))q-1}}.\end{aligned}$$

Proof. Assume $z \in C_R$. We have

$$\begin{aligned} s_{n-1,m,j}(f_*,z) &= \binom{p-1}{\frac{1}{2}(p-1)}(-1)^j \frac{1}{\rho^{(2pj-(p-1))m}} \times \\ &\times \left(\sum_{\tau=0}^{p-1}(-1)^\tau \binom{pj+\tau}{p} \frac{p}{pj+\tau-\frac{1}{2}(p-1)}\right) \frac{1}{\rho^{2\tau m}} \times \\ &\times \frac{1}{2}\left(\sum_{\sigma=0}^{n-1}(\rho^\sigma+\rho^{-\sigma})(w^\sigma+w^{-\sigma})\right). \end{aligned}$$

Applying Lemma 7.1, Lemma 7.3 and Lemma 7.4(ii) we get

$$\begin{aligned} \lim_{n\to\infty} \min_{z\in C_R} |s_{n-1,m,j}(f_*,z)|^{1/n} &= \lim_{n\to\infty} \max_{z\in C_R} |s_{n-1,m,j}(f_*,z)|^{1/n} \\ &= \frac{1}{\rho^{(2pj-(p-1))q}} \cdot 1 \cdot \rho R = \frac{R}{\rho^{(2pj-(p-1))q-1}}. \end{aligned}$$

□

Recall that from (2.6) when p is odd, we have

$$\begin{aligned} \Delta_{l-1,n,m,p}(f,z) := & -S_{n-1}\left(H^{\{p\}}_{pm-1}(S_{(p(2l-1)-1)m+n-1)}(f,\cdot),z)\right) \\ & + S_{n-1}(H^{\{p\}}_{pm-1}(f,z). \end{aligned} \tag{7.3}$$

Lemma 7.9 *Assume $f \in C_\rho$, p is odd, $m := m(n)$ satisfies $m/n \to q > 1$ and $R > 1$. Then we have for each positive l*

$$\limsup_{n\to\infty} \max_{z\in C_R} |\Delta_{l-1,n,m,p}(f,z)|^{1/n} \leq \frac{R}{\rho^{(2pl-p+1)q-1}}.$$

Proof. For $z \in C_R$ we have $z = \frac{1}{2}(w + \frac{1}{w})$ where $|w| = R$ and $T_\sigma(z) = \frac{1}{2}(w^\sigma + w^{-\sigma})$. We have

$$\begin{aligned} \Delta_{l-1,n,m,p}(f,z) &= \binom{p-1}{\frac{1}{2}(p-1)} \sum_{s=l}^{\infty}(-1)^s \sum_{\tau=0}^{p-1}(-1)^\tau \binom{ps+\tau}{p} \frac{p}{ps+\tau-\frac{1}{2}(p-1)} \times \\ &\times \left(\sum_{\sigma=0}^{n-1} A_{(2ps+2\tau-(p-1))m-\sigma} + \sum_{\sigma=0}^{n-1} A_{(2ps+2\tau-(p-1))m+\sigma}\right) T_\sigma(z) \\ &\equiv J_1(z) + J_2(z). \end{aligned}$$

Since $f \in C_\rho$, to each ϵ, with $1 < \rho - \epsilon < \rho$, there exists a finite constant $c \equiv c(\epsilon) > 0$ such that $|A_k| \leq c(\rho-\epsilon)^{-k}, \quad k = 0, 1, \ldots$. For w such that

$|w| = R > 1,\ \rho > 1$, we have

$$|J_1(z)| = \mathcal{O}(1) \sum_{s=l}^{\infty} \sum_{\tau=0}^{p-1} \binom{s-l+p-1}{s-l} (\rho-\epsilon)^{(p-1-2\tau)m} \tfrac{1}{(\rho-\epsilon)^{2pm(s-l)}} \tfrac{1}{(\rho-\epsilon)^{2pml}} \times$$

$$\times \sum_{\sigma=0}^{n-1} \left(((\rho-\epsilon)R)^{\sigma} + ((\rho-\epsilon)/R)^{\sigma}\right)$$

$$= \mathcal{O}(1) \tfrac{(\rho-\epsilon)^{(p-1)m}}{(\rho-\epsilon)^{2pml}} \sum_{\tau=0}^{p-1} \tfrac{1}{(\rho-\epsilon)^{2\tau m}} ((\rho-\epsilon)R)^n.$$

Therefore we have

$$\limsup_{n\to\infty} \max_{z\in C_R} |J_1(z)|^{1/n} \le \frac{(\rho-\epsilon)^{(p-1)q}}{(\rho-\epsilon)^{2plq}} (\rho-\epsilon)R = \frac{R}{(\rho-\epsilon)^{(2pl-p+1)q-1}}\ .$$

Similarly,

$$|J_2(z)| = \mathcal{O}(1) \textstyle\sum_{s=l}^{\infty} \sum_{\tau=0}^{p-1} \binom{s-l+p-1}{s-l} (\rho-\epsilon)^{(p-1)m} \tfrac{1}{(\rho-\epsilon)^{2\tau m}} \tfrac{1}{(\rho-\epsilon)^{2spm}} \times$$

$$\times \textstyle\sum_{\sigma=0}^{n-1} \left(\left(\tfrac{R}{\rho-\epsilon}\right)^{\sigma} + ((\rho-\epsilon)R)^{-\sigma}\right)$$

$$= \mathcal{O}(1) \tfrac{(\rho-\epsilon)^{(p-1)m}}{(\rho-\epsilon)^{2plm}} \max\left(1, \left(\tfrac{R}{\rho-\epsilon}\right)^n\right).$$

Therefore we have

$$\limsup_{n\to\infty} \max_{z\in C_R} |J_2(z)|^{1/n} \le \frac{(\rho-\epsilon)^{(p-1)q}}{(\rho-\epsilon)^{2plq}} R \max\left(\frac{1}{R}, \frac{1}{\rho-\epsilon}\right)$$

Combining the last result with the analogous result for $J_1(z)$ we get, using Lemma 7.1,

$$\limsup_{n\to\infty} \max_{z\in C_R} |J_1(z) + J_2(z)|^{1/n} =$$

$$= \max\left(\limsup_{n\to\infty} |J_1(z)|^{1/n}, \limsup_{n\to\infty} |J_2(z)|^{1/n}\right)$$

$$= \tfrac{R}{(\rho-\epsilon)^{(2pl-p+1)q-1}}.$$

Hence we have

$$\limsup_{n\to\infty} \max_{z\in C_R} |\Delta_{l-1,n,m,p}(f,z)|^{1/n} \le \frac{R}{(\rho-\epsilon)^{(2pl-p+1)q-1}}.$$

Letting $\epsilon \searrow 0$, the proof follows. □

Proof of (2.9) of Theorem 2.7 when p is odd. We have

$$\Delta_{l-1,n,m,p}(f_*, z) = s_{n-1,m,l}(f_*, z) + \Delta_{l,n,m,p}(f_*, z).$$

The proof follows now by Lemma 7.1 and Lemma 7.8. □

Proof of (2.7) of Theorem 2.7 when p is odd. The proof of (2.7) follows now from (2.9) and Lemma 7.9. □

References

[1] R. Brück, A. Sharma and R.S. Varga, An extension of a result of Rivlin on Walsh equiconvergence (Faber Nodes), ISBN Brkhauser Verlag, Basel **119**(1994), 41-66.

[2] A.S. Cavaretta, A. Sharma, and R.S. Varga, Interpolation in the roots of unity: An extension of a theorem of J.L. Walsh, Resultate Math. **3**(1980), 155-191.

[3] H.P. Dikshit, A. Sharma, V.Singh and F. Stenger, Rivlin's theorem on Walsh equiconvergence, J. Approximation Th. **52(3)**(1988), 339-349.

[4] T.N.T. Goodman, K.G. Ivanov and A. Sharma, Hermite Interpolation in the roots of unity, J. Approximation Th. **84**(1996), 41-60.

[5] O.P. Juneja, and Pratibha Dua, On Walsh equiconvergence by Chebyshev polynomials, (Preprint).

[6] T.J. Rivlin, On Walsh equiconvergence, J. Approx. Th. **36**(1982), 334-345.

[7] T.J. Rivlin, Chebyshev polynomials, 2nd edn., Wiley & Sons. Inc. Toronto, 1990

[8] A. Sharma, Some recent results on Walsh theory of equiconvergence Approximation Theory V, (Chui C.K., Schumaker L.L, Ward J.D, Eds), 173-190, Academic Press, New York, 1986.

[9] J.L. Walsh, Approximation in the complex Domain, 5th ed., Amer. Math. Soc. Providence, R.I., 1969.

[10] L. Yuanren, Extension of a theorem of J.L. Walsh on overconvergence, Approx. Theory and its Applications, **2(3)**(1986), 19-32.

Continuous Functions Which Change Sign Without Properly Crossing the x-Axis

Peter D. Johnson Jr.
Department of Discrete and Statistical Sciences
120 Math Annex
Auburn University, Alabama 36849

Abstract

For a real interval $[a, b]$ and a perfect subset S of $[a, b]$ without interior, it is shown that continuous real-valued functions on $[a, b]$ can be constructed which "change sign around" each $x \in S$, but whose graphs never "properly" cross the x-axis. (The words and phrases in quotation marks will be defined below.) It is also shown that for every continuous function f on $[a, b]$ that changes sign, never properly crosses the x-axis, and is not zero throughout any subinterval of $[a, b]$, the set of points around which f changes sign is a perfect subset of $[a, b]$ without interior.

This note is inspired by an interesting oversimplification in the wonderful book of Rivlin [2]. The oversimplification occurs in the hint for problem 2.4, section 2.4, p. 62. The result of problem 2.4 is a little shaky, for reasons of definition that will be mentioned below, but the result that is really the object of the exercise is true. This result is as follows.

Proposition A. *Suppose that $[a, b]$ is a real interval, w is a continuous function on $[a, b]$, strictly positive on (a, b), $f \in C[a, b]$, n is a non-negative integer, and q_n^* is the polynomial of degree less than or equal to n which best approximates f in the least-squares sense with respect to the weight function w.* [This means that q_n^* is the unique polynomial among polynomials q of degree $\leq n$ which minimizes $\int_a^b (f(x) - q(x))^2 w(x)dx$.] *Suppose that f is not itself a polynomial of degree $\leq n$. Then there exist $x_0 < \cdots < x_{n+1}$ in (a, b) at which $g = f - q_n^*$ alternates in sign.* [This means that $g(x_i)g(x_{i+1}) < 0$, $i = 0, \ldots, n$.]

Here is the gist of Rivlin's hint. If there were $n+1$ or more points in (a, b) *at which g changes sign*, then you could choose the $n+2$ points $x_0, \ldots, x_{n+1}$

in the intervals between those points and a and b. So suppose that g changes sign *only* at $y_1, \dots, y_r \in (a,b)$, with $r \le n$. Set $p(x) = \prod_{j=1}^{r}(x - y_j)$. Then $p(x)$ is a polynomial of degree $r \le n$, and $p(x)g(x)$ does not change sign on (a,b), and is not identically zero. Therefore, $\int_a^b p(x)g(x)w(x)dx \ne 0$, so $g(x) = f - q_n^*$ is not orthogonal (in $L^2([a,b], wdx))$ to the set of polynomials of degree $\le n$, a contradiction.

Very neat; the oversimplification resides in the meaning of the phrase "g changes sign at y". For instance, at which point or points would you say that the function h defined by

$$h(x) = \begin{cases} x, & x < 0 \\ 0, & 0 \le x \le 1 \\ x - 1, & x > 1 \end{cases}$$

changes sign?

However, it is not my purpose to quibble with Rivlin's suggested proof of Proposition A, which can be easily fixed, with some damage to the brisk and memorable simplicity of the original.[1] The purpose of this note is to administer hopefully a salutary admonition concerning the usual mental picture conjured by the words "changing sign". The pesky behavior exhibited by the function h above is not really of interest. Functions which take the value zero on intervals can be reduced to ones that do not by collapsing those intervals to points.

Definition *A real-valued function g* changes sign properly *[non-strictly] at y if its values are one sign in an open interval to the left of y [non-strictly] and the opposite sign [non-strictly] in an open interval to the right of y. Equivalently, for some $\epsilon > 0$, we have $g(x_1)g(x_2) < 0$ $[g(x_1)g(x_2) \le 0]$ whenever $y - \epsilon < x_1 < y < x_2 < y + \epsilon$.*

The function g is jittery on the left *of y if it takes both positive and negative values in every interval $(y - \epsilon, y)$, $\epsilon > 0$.* Jittery on the right *is defined similarly. If g is either jittery on the right or on the left of y, then it will be called* jittery *at y. We will say that g* changes sign around *y if either g is jittery at y, or if g changes sign properly, possibly non-strictly,*

[1] Assume that $x_0 < \cdots < x_r$, $r \le n$, is a longest sequence in (a,b) satisfying $g(x_i)g(x_{i+1}) < 0$, $i = 0, \dots, r-1$, and take it from there. It may fail to be the case that there is a point y_i between x_{i-1} and x_i at which g changes sign, in the proper sense implicit in Rivlin's hint, but there is some y_i between x_{i-1} and x_i such that g is, possibly non-strictly, one sign between x_{i-1} and y_i, and the other between y_i and x_i, and these y_i will serve. Incidentally, Cheney [1] proves a result similar to that in Rivlin's problem 2.4, and thus to Proposition A (see [1], Chapter 4, Theorem 5 and its Corollaries), using the same argument as in Rivlin's hint.

at y and g is not identically zero on any interval of the form $(y-\epsilon, y)$ or $(y, y+\epsilon)$.

For example, if $f_1(x) = x \sin 1/x$, and

$$f_2(x) = \begin{cases} |f_1(x)|, & x < 0 \\ -|f_1(x)|, & x > 0, \end{cases}$$

then f_1 is jittery at 0, and f_2 is not; both functions change sign around 0. The function $f_3 = \max(f_2, 0)$ does not change sign around 0, but $f_1 f_3$, which is jittery on the left of 0, does.

The function h defined earlier does not change sign around any point.

The following may seem self-evident, but we include it anyway, to quiet certain logical qualms.

Lemma 1 *If f is a continuous real-valued function on a real interval, $f(x_1) < 0$ and $f(x_2) > 0$, and f is not zero on any subinterval of the interval with endpoints x_1, x_2, then f changes sign around some point x between x_1 and x_2.*

Proof. Without loss of generality, assume that $x_1 < x_2$. Set $x = \sup\{t \geq x_1; f(s) \leq 0 \text{ for all } s \in [x_1, t]\}$. Clearly the hypotheses on f, x_1 and x_2 guarantee that $x_1 < x < x_2$. The definition of x and the continuity of f guarantee that $f(t) \leq 0$ on $[x_1, x]$. If for some $\epsilon > 0$, $f(t) \geq 0$ for all $t \in (x, x+\epsilon)$, then f changes sign properly, possibly non-strictly, at x, and the hypothesis about f not being zero on any subinterval of $[x_1, x_2]$ then implies that f changes sign around x. Otherwise, f takes negative values in every interval of the form $(x, x+\epsilon)$. By the definition of x, f also takes positive values in every such interval. Thus, if f doesn't change sign properly at x, then f is jittery at x on the right, and so changes sign around x.

The function f_1 above changes sign around 0 due to being jittery there. However, f_1 also changes sign properly at many points. Is it possible for a continuous function to change sign *only* around points where it is jittery, and never properly? Well, sure: for instance, one could define a function to be zero on the Cantor set in $[0,1]$ and then define it on each interval in the complement of the Cantor set so that the resulting function is (a) continuous, (b) zero only on the Cantor set, and (c) jittery at each point of the Cantor set. Achieving (a) and (c) is not completely trivial; for instance, to achieve continuity, it is not sufficient to arrange for the function to be zero at each endpoint of each maximal interval in the complement of the Cantor set. However, the arrangements to be made are not difficult. It is of some interest that making these arrangements with the Cantor set is essentially the only way to produce continuous functions that change sign only around points where they are jittery.

A perfect set is a non-empty closed set with no isolated points. The lemma following, which is well known in various forms, at least folklorically, says that every bounded perfect set without interior can be obtained in the same way as the Cantor set usually is. In what follows, if I, J are disjoint open intervals, $I < J$ means that $x < y$ for all $x \in I$, $y \in J$.

Lemma 2 *Let D denote the set of dyadic fractions in $(0,1)$. Suppose that $[a,b]$ is a real interval, $S \subseteq [a,b]$ is perfect, with no interior, and a, $b \in S$. Then the maximal subintervals of $[a,b] \backslash S$ can be indexed by D in such a way that if r, $s \in D$ and $r < s$ then $I(r) < I(s)$.*

Proof. The maximal intervals of $[a,b]\backslash S$ are open, and no two of them share an endpoint, because such an endpoint would be an isolated point of S. For the same reason, since a, $b \in S$, neither a nor b is an endpoint of one of those intervals. Therefore, if I and J are maximal intervals of $[a,b]\backslash S$, and $I < J$, then the distances from a to the left hand endpoint of I, from the right hand endpoint of I to the left hand endpoint of J, and from the right hand endpoint of J to b are all positive. Since S has no interior, there are points of $[a,b]\backslash S$ in those interstices, and therefore there are maximal intervals K_1, K_2, K_3 of $[a,b]\backslash S$ satisfying $K_1 < I < K_2 < J < K_3$.

With this in mind, let $r_1, r_2, \ldots$ be an ordering of D and $J_1, J_2, \ldots$ an ordering of the maximal intervals of $[a,b]\backslash S$. Set $I(r_1) = J_1$. Having found $I(r_1), \ldots, I(r_n)$, ordered as $r_1, \ldots, r_n$ are ordered with respect to the usual ordering on the real line, pick $I(r_{n+1})$ from $\{J_1, J_2, \ldots\}\backslash\{I(r_1), \ldots, I(r_n)\}$ by picking the J_k with the smallest index k, among those intervals that lie in relation to $I(r_1), \ldots, I(r_n)$ as r_{n+1} lies in relation to $r_1, \ldots, r_n$. Clearly the sequence $I(r_1), I(r_2), \ldots$ defined thus satisfies the order requirement, that for r, $s \in D$, $r < s$ implies $I(r) < I(s)$. That every J_i appears in the sequence $I(r_1), I(r_2), \ldots$ is a straightforward consequence of the fact that between any two distinct dyadic fractions lies an infinity of dyadic fractions.

Theorem 3 *Suppose that $[a,b]$ is a real interval, and $S \subseteq [a,b]$ is a perfect set with no interior. Then there is a continuous real-valued function g on $[a,b]$ such that $g(x) = 0$ if and only if $x \in S$, g is jittery at each $x \in S$, and g never changes sign properly (even non-strictly).*

Conversely, if $f \in C[a,b]$ takes both positive and negative values, $f^{-1}(\{0\})$ contains no interval, and f never changes sign properly (even non-strictly) then $\{x \in [a,b]; f$ changes sign around $x\}$ is a perfect subset of $[a,b]$ with no interior.

Proof. Yes, there is a redundancy in the claims for g, above. If $S = \{x; g(x) = 0\}$ and g is jittery at each $x \in S$, and S is perfect with no interior, then of course g never changes sign properly. But it is the 90's and redundancies in a conclusion are not considered terribly sinful.

Suppose that $S \subseteq [a, b]$ is perfect, with no interior. Let $\alpha = \inf S$ and $\beta = \sup S$. Then α, $\beta \in S$, and if we can find $g \in C[\alpha, \beta]$ as called for, then we can easily extend g to all of $[a, b]$ so that the conclusion of the theorem holds. So assume that a, $b \in S$.

We will define $g \equiv 0$ on S, and finish the job by defining g on each maximal interval of $[a, b]\backslash S$. On each such interval I, we can easily arrange for g to be non-zero, continuous, tending to zero at the endpoints, and, for $t \in I$, for $|g(t)|$ to be no greater than the distance from t to the nearest endpoint of I. If g satisfies these requirements, then we will have $S = \{x; g(x) = 0\}$ and also that g is continuous: if $x \in S$, $\epsilon > 0$, $t \in [a, b]\backslash S$, and $|x - t| < \epsilon$, then the distance from t to the nearest endpoint of the maximal interval I of $[a, b]\backslash S$ in which t lies is less than ϵ, so $|g(t) - g(x)| = |g(t)| < \epsilon$.

Therefore, we will have proven the first part of the theorem if we can distribute signs, ± 1, to the maximal intervals of $[a, b]\backslash S$ so that if g is required to have these signs on those intervals, then g will be jittery at each point of S.

For this purpose, we order D lexicographically with respect to the (denominator, numerator) pairs; in plain English, the ordering is $1/2, 1/4, 3/4, 1/8, 3/8, 5/8, 7/8$, etc. For $m \geq 2$, $1 \leq j \leq 2^{m-1} - 1$, let $pr((2j-1)/2^m) = 2j/2^m = j/2^{m-1}$, which is the dyadic fraction among those preceding $(2j-1)/2^m$ in the ordering just defined which lies closest to $(2j-1)/2^m$ on the right with respect to the usual real order. For $2 \leq j \leq 2^{m-1}$, let $p\ell((2j-1)/2^m) = (2j-2)/2^m = (j-1)/2^{m-1}$, which is the predecessor of $(2j-1)/2^m$ immediately to the left of $(2j-1)/2^m$. Now, we define $\sigma : D \to \{-1, 1\}$ recursively as follows: $\sigma(1/2) = 1$, and for $m \geq 2$, to define $\sigma((2j-1)/2^m)$, $j = 1, \ldots, 2^{m-1}$, we break into cases.

$$\begin{gathered} m \text{ even: set } \sigma(1/2^m) = -\sigma(1/2^{m-1}) \\ \sigma(s) = -\sigma(pr(p\ell(s))), s = (2j-1)/2^m, 2 \leq j \leq 2^{m-1} - 1, \\ \text{and } \sigma((2^m - 1)/2^m) = 1. \end{gathered}$$

m odd: set $\sigma(1/2^m) = 1$,

$$\sigma(s) = -\sigma(p\ell(pr(s))), s = (2j-1)/2^m, 2 \leq j \leq 2^{m-1} - 1,$$

and $\sigma((2^m - 1)/2^m) = -\sigma((2^{m-1} - 1)/2^{m-1})$.

Now let $r \to I(r)$ be an indexing of the maximal intervals of $[a, b]\backslash S$ by D as in Lemma 2, and require g to have sign $\sigma(r)$ on $I(r)$ for each $r \in D$. We claim that this makes g jittery at each $x \in S$. To see this, we need to demystify the definition of σ above. Since $r \to I(r)$ preserves the usual real order, for $m \geq 2$ the intervals $I((2j-1)/2^m)$, $j = 1, \ldots, 2^{m-1}$ fall into the interstices between a, b, and the intervals $I(k/2^i)$, $1 \leq i < m$, k odd. Imagine that these intervals have already had their signs assigned. When m is even, the idea behind the assignment of signs to $I(1/2^m)$, $I(3/2^m), \ldots, I((2^m-3)/2^m)$ is that for a and each right hand endpoint of each interval that has already had its sign assigned, whatever the sign was on the closest interval to the right of such a point, the point now has a new closest right hand neighbor with a different sign. (Of course, $I((2^{m-1}-1)/2^{m-1})$ had no neighbor on its right, so we arbitrarily assign 1 to $I((2^m-1)/2^m)$, just to keep things simple.) The assignment when m is odd serves the same neighbor's-sign-switching purpose for b and the left hand endpoints.

If $x = a$ or if x is a right hand endpoint of some maximal interval in $[a,b]\backslash S$, then for $\epsilon > 0$, $(x, x+\epsilon)$ contains infinitely many $I(r)$, $r \in D$; thus $(x, x+\epsilon)$ contains some $I(r)$ which was a closest neighbor on the right of x, with respect to the ordering of D we are considering, and $(x, x+\epsilon)$ will then contain all subsequent closest neighbors of x on the right, as we go through the intervals by this ordering, because those later intervals fall between x and $I(r)$. Thus g will take both positive and negative values in $(x, x+\epsilon)$. Consequently, g is jittery at x on the right. Similarly, if $x = b$ or x is a left hand endpoint of a maximal subinterval of $[a,b]\backslash S$, then g is jittery on the left of x. Finally, if $x \in S$, and $x \neq a, b$, and x is not an endpoint of an interval in S's complement, then x is the limit, on either side, of such endpoints, so g takes both positive and negative values in all intervals of the form $(x-\epsilon, x)$ or $(x, x+\epsilon)$; so g is jittery at x, on both sides.

Now suppose that f is as described in the second part of the theorem. Since f takes positive and negative values on $[a,b]$, $S = \{x; f$ changes sign around $x\}$ is non-empty, and since f does not change sign properly, f is jittery at each point of S, by Lemma 1 (using also the hypothesis that f is not zero on any interval). Any limit point of points at which f is jittery will clearly be a point at which f is jittery, so S is closed. By Lemma 1, and the hypothesis on f, a point x at which f is jittery could not possibly be an isolated point of S (since f has to change sign around points other than x in any neighborhood of x). Thus S is perfect.

Finally, since f is continuous, $f = 0$ on S, so the fact that f is never zero throughout an interval implies that S has no interior.

Given $S \subseteq [a,b]$, perfect without interior, can you find $g \in C[a,b]$ satisfying (i) $S = \{x \in [a,b]; g(x) = 0\}$ and (ii) g is jittery at each $x \in S$, which is

(iii) infinitely differentiable?

(iv) infinitely differentiable and/or orthogonal to the polynomials of degree $\leq n$, with respect to some previously specified weight function on $[a, b]$?

Surely the answers are all yes, but the proofs will be a chore.

References

[1] Cheney, E. W., *Introduction to Approximation Theory*, New York, McGraw-Hill, 1966.

[2] Rivlin, T. J., *An Introduction to the Approximation of Functions*, New York, Dover, 1981 (originally Waltham, Mass., Blaisdell, 1969).

Some Remarks on Weighted Interpolation

Theodore Kilgore
Department of Mathematics, Auburn University,
Auburn, Alabama 36849 USA

The Bernstein-Erdös conditions, which characterize interpolation of minimal norm, are valid in the spaces of polynomials with a Jacobi weight, provided that the exponents in the Jacobi weight are non-negative. The Bernstein-Erdös characterization also provides some good estimates concerning the magnitude of the norm of the related interpolation operators.

1 Introduction

Let $Y_{n,\alpha,\beta}$ denote the space of polynomials of degree n or less on the interval $[-1,1]$, weighted by the Jacobi weight $w_{\alpha,\beta}(x) := (1-t)^{\alpha}(1+t)^{\beta}$, where $\alpha \geq 0$ and $\beta \geq 0$. In these spaces it is possible to interpolate a function $f \in C[-1,1]$ on any set of nodes $t_0, \ldots, t_n$ satisfying

$$-1 < t_0 < \ldots < t_n < 1.$$

However, if $\alpha = 0$ it will be assumed that $t_n = 1$, and if $\beta = 0$ it will be assumed that $t_0 = -1$.

Aside from intrinsic interest, interpolation in the weighted polynomial spaces $Y_{n,\alpha,\beta}$ is important in the study of simultaneous approximation of derivatives. For example, in the articles [1] and [2] of Balázs and Kilgore, some results are expressed in terms of interpolation into the spaces $Y_{n,\frac{1}{2},\frac{1}{2}}$ using as nodes the zeroes of the Chebyshev polynomials. Here, it will be shown that for the spaces $Y_{n,\alpha,\beta}$ the interpolation of minimal norm is characterized by conditions which are a slight adaptation of the conditions stated in the conjectures of Bernstein and Erdös concerning the (unweighted) Lagrange interpolation of minimal norm. As a result of this, certain estimates concerning the norm of interpolation into $Y_{n,\alpha,\beta}$ will be seen to follow as well.

2 Notation

To describe the weighted interpolation into $Y_{n,\alpha,\beta}$, nodes $t_0, \ldots, t_n$ will be used, as already described. For notational convenience, when $\alpha \neq 0$, we will also define $t_{-1} := -1$, and, when $\beta \neq 0$, we will define $t_{n+1} := 1$, though the symbols t_{-1} and t_{n+1} are not used to denote nodes of interpolation. The fundamental functions (basis functions) for the interpolation $y_0, \ldots, y_n$ in $Y_{n,\alpha,\beta}$ will satisfy

$$y_i(t_j) = \delta_{ij} \text{ (Kronecker delta) for } i, j = 0, \ldots, n$$

and, when $\alpha > 0$ or respectively $\beta > 0$,

$$y_i(1) = 0 \text{ or respectively } y_i(-1) = 0, \text{ for } i = 0 \ldots, n.$$

Explicitly,

$$y_j(t) = \frac{(1-t)^\alpha (1+t)^\beta}{(1-t_j)^\alpha (1+t_j)^\beta} l_j(t) = \frac{w_{\alpha,\beta}(t)}{w_{\alpha,\beta}(t_j)} l_j(t), \tag{1}$$

where $w_{\alpha,\beta}$ signifies the weight function, and l_j is the corresponding fundamental polynomial for ordinary Lagrange interpolation. That is,

$$l_j(t) = \prod_{k=0,\ k\neq j}^{n} \frac{t - t_k}{t_j - t_k}.$$

The weighted interpolation operator with range $Y_{n,\alpha,\beta}$ is $P_{n,\alpha,\beta}$, defined by

$$P_{n,\alpha,\beta} f = \sum_{j=0}^{n} f(t_j) y_j,$$

and it is easily seen that

$$||P_{n,\alpha,\beta}|| = ||\sum_{j=0}^{n} |y_j|||,$$

in which the sum which is normed on the right is called the *Lebesgue function* of $P_{n,\alpha,\beta}$ and is denoted by $\Lambda_{n,\alpha,\beta}(t)$.

3 The Bernstein-Erdös conditions applied to interpolation in $Y_{n,\alpha,\beta}$

The Bernstein-Erdös conditions, which have been shown in the succession of papers Kilgore [8] and [9] and in de Boor and Pinkus [4] to characterize the unweighted Lagrange interpolation of minimal norm, were originally given as conjectures in Bernstein [3] and in Erdös [6] and [7]. In order to adapt the essential content of the Bernstein-Erdös conditions to the present context, a closer description of $\Lambda_{n,\alpha,\beta}$ is needed. This function is a piecewise weighted polynomial with continuous splicing at the nodes $t_0, \ldots, t_n$, and $\Lambda_{n,\alpha,\beta}$ has value 1 at each of these nodes. If $\alpha > 0$, then $\Lambda_{n,\alpha,\beta}(1) = 0$, and if $\beta > 0$, then $\Lambda_{n,\alpha,\beta}(-1) = 0$. If either $\alpha > 0$ or $\beta > 0$, then $\Lambda_{n,\alpha,\beta}(t) > 1$ for every point in $[t_0, t_n]$ which is not a node, and if $\alpha = \beta = 0$, then this holds for all $n \geq 2$. In all cases, the function $\Lambda_{n,\alpha,\beta}$ has a unique local maximum T_k in each of the subintervals (t_{k-1}, t_k), for $1 \leq k \leq n$, with $\Lambda_{n,\alpha,\beta}(T_k) := \lambda_k$. That this local maximum must occur and be unique may be seen by noting that there is a unique weighted polynomial X_k in $Y_{n,\alpha,\beta}$ whose restriction to $[t_{k-1}, t_k]$ agrees with $\Lambda_{n,\alpha,\beta}$ on that interval, having on the nodes the sign pattern

$$X_k(t_j) = (-1)^{j-k-1} \text{ for } j \leq k-1$$

and

$$X_k(t_j) = (-1)^{j-k} \text{ for } j \leq k-1.$$

The existence and unicity of the point T_k can now be determined by counting the zeroes of X'_k which occur at the local maxima and minima determined by the sign alternation. In particular, $X'_k(T_k) = 0$.

Now, if $\beta > 0$, then $-1 = t_{-1} < t_0$, and the weighted polynomial X_0 shall be defined as

$$X_0 = \sum_{j=0}^{n} (-1)^j y_j,$$

and T_0 is taken to be the leftmost local maximum of X_0, with $X_0(T_0) = \lambda_0$ and $X'_0(T_0) = 0$. Similarly, if $\alpha > 0$, then $t_n < t_{n+1} = 1$, and

$$X_0 = \sum_{j=0}^{n} (-1)^{n-j} y_j,$$

with T_{n+1} the rightmost local maximum of X_{n+1}. Then $X_{n+1}(T_{n+1}) = \lambda_{n+1}$, and $X'_{n+1}(T_{n+1}) = 0$. Note that these quantities may or may not, depending on the placement of the nodes, represent actual local maxima of $\Lambda_{n,\alpha,\beta}$. Also, there is no predetermined order relation between T_0 and t_0, nor between T_{n+1} and t_n. It is obvious, however, that $T_0 < t_1$ and $t_{n-1} < T_{n+1}$.

The natural reformulations of the two conjectures of Bernstein and Erdös for the weighted space $Y_{n,\alpha,\beta}$ will involve all of the local maxima $T_0, \ldots T_{n+1}$. The precise statements of the two conjectures may be seen in Theorem 1.

4 Results

Theorem 1 *For $\alpha \geq 0$ and $\beta \geq 0$, not both zero, and for $n \geq 1$ let $P_{n,\alpha,\beta}$ denote the weighted interpolation into the space $Y_{n,\alpha,\beta}$, on nodes $t_0, \ldots, t_n$ with $-1 < t_0 < \ldots < t_n < 1$ if α and β are both positive; $t_n = 1$ if $\alpha = 0$; $t_0 = 1$ if $\beta = 0$. Then*

(i) the following conditions must hold if $P_{n,\alpha,\beta}$ is of minimal norm

$$\lambda_0 = \ldots = \lambda_{n+1} \text{ if } \alpha > 0 \text{ and } \beta > 0$$
$$\lambda_0 = \ldots = \lambda_n \text{ if } \alpha = 0$$
$$\lambda_1 = \ldots = \lambda_{n+1} \text{ if } \beta = 0$$

(ii) the nodes which give the conditions in (i) are in each case uniquely determined (implying because of this unicity that the condition given in (i) for each case is sufficient as well as necessary)

(iii) if the configuration of the nodes is not such as to give the minimal norm, then at least one of the quantities λ_i is strictly less than the value of the minimal norm.

Theorem 2 *The norm of the minimal $P_{n,\alpha,\beta}$ is bounded below by the norm of $P_{n,0,0}$, that is, by the norm of the unweighted Lagrange interpolation of minimal norm.*

5 The Proofs

Some preliminary analysis is helpful.

First, one must note that the function which maps the nodes to the λ's is differentiable, and the derivative is given by the matrix with entries

$$\frac{\partial \lambda_i}{\partial t_j} = -y_j(T_i)X_i'(t_j), \tag{2}$$

which formula originated in a work of Morris and Cheney [5]. The restrictions on the indices i and j differ of course in our three cases. If $\alpha > 0$ and $\beta > 0$, then the formula (2) is valid for $i = 0, \ldots, n+1$ and $j = 0, \ldots, n$. If $\alpha = 0$, then λ_{n+1} is not defined, and the node $t_n = 1$ is immovable. Therefore, the formula holds for $i = 0, \ldots, n$ and $j = 0, \ldots, n-1$ in this case. Similarly, if $\beta = 0$, then $t_0 = -1$ is immovable, and λ_0 is not defined, Then the formula holds for $i = 1, \ldots, n+1$ and $j = 1, \ldots, n$.

Now, consider the matrix

$$\left(\frac{\partial \lambda_i}{\partial t_j}\right)_{ij} = \left(-y_j(T_i)X_i'(t_j)\right)_{ij}, \tag{3}$$

which in each of the three cases under consideration has one more row than the number of its columns. If any single row is removed, the matrix which remains is square.

The results in the theorems can be directly related to the properties of the matrix given in (3):

If the matrices obtained from (3) by removal of a single row are always nonsingular, then Theorem 1 part (i) follows. For, let the removed row be indexed by k, where k is any given one of the indices. Then the nonsingularity of the resulting square matrix implies that the function which maps the nodes to all of the λ's except λ_k is locally one-to-one, and thus the nodes may be perturbed in such a way that all of the λ's except λ_k decrease. This says that if k is chosen so that λ_k is the least of the λ's (assuming that they are not all equal), then the maximum of all the λ's can always be decreased. This argument and the proof of the requisite nonsingularity conditions to show Theorem 1 part (i) for the case $\alpha, \beta = 0$ were done in Kilgore [8], and all other work related to the Bernstein-Erdös conjectures depends on this result.

Here, however, the situation is somewhat more complicated. If $\alpha > 0$ or $\beta > 0$ or both, then it does not follow always that a matrix obtained from (3) by removal of a single row is nonsingular. However, the statement is almost true. The exceptions must be noted and analysed, and the general line of the argument follows, with some modification.

If the determinants of the matrices obtained by striking the kth row alternate in sign, then the unicity condition, Theorem 1 part (ii) follows. This may be seen in Kilgore [9]. The sign alternation was shown in de Boor and Pinkus [4], where a topological argument, also based on this sign alternation, establishes Theorem 1 part (iii) as well. Again, this sign alternation does not hold here unconditionally.

Now, it is possible to present the proofs of the theorems. For the sake of clarity and definiteness in indexing the entries, it will be assumed that both $\alpha > 0$ and $\beta > 0$. The proofs of the other two cases are entirely analogous.

Proof of Theorem 1

One begins with a sequence of reduction steps on the matrix (3) which reduce it to an equivalent matrix, which is an "evaluation matrix" of polynomials. That is, the entries in the ith row of the resulting matrix will have the form of a polynomial q_i of degree at most n which is evaluated at the successive points $t_0, \dots, t_n$. This sequence of reduction steps performed on the matrix thus reduces the question of whether the square submatrix obtained by removing the kth row, any k, to the question of whether the set of polynomials $q_0, \dots, q_{k-1}, q_{k+1}, \dots, q_n$ is linearly independent.

The first reduction step is to remove from the jth column of the matrix (3) the factor

$$w_{\alpha,\beta}(t_j) \prod_{\ell=0,\ \ell\neq j}^{n} (t_j - t_\ell),$$

which, in accordance with (1), occurs in the denominator of each entry in the jth column. This should be done for $j = 0, \ldots, n$.

The second reduction step is to factor out from the ith row of the matrix the expression

$$w_{\alpha,\beta}(T_i) \prod_{\ell=0}^{n} (T_i - t_\ell).$$

Note that, if $1 \leq i \leq n$, this factor is never zero. However, there is the actual possibility that $T_0 = t_0$ or that $T_{n+1} = t_n$, giving rise to singularities, or that $T_0 > t_0$ or that $T_{n+1} < t_n$, leading to sign changes in the determinants of the square submatrices under consideration. However, it is clear that if $t_0 \leq T_0$, then $\lambda_0 = X_0(T_0) < X_1(T_1) = \lambda_1$. Under these circumstances, the value λ_0 cannot be equal to $\|\Lambda_{n,\alpha,\beta}\|$, and its actual behavior (aside from the basic consideration that its behavior is an analytic function of the nodes) is not important. Similar observations apply on the other end of the domain; the precise behavior of λ_{n+1} only becomes of interest if $t_n < T_{n+1}$. The possibility that $t_0 = T_0$ or that $T_{n+1} = t_n$ can also be handled, as will be seen shortly.

Assuming for the moment that $T_0 \neq t_0$ and that $T_{n+1} \neq t_n$, the cancellation just noted can be done for $i = 0, \ldots, n+1$. The matrix which remains, equivalent except under the already-noted exceptional circumstances to (3), is

$$\left(\frac{X_i'(t_j)}{t_j - T_i} \right)_{ij}. \tag{4}$$

In this matrix, it may now be assumed that the functions X_i and the points T_i are fixed, and then the points t_j may be released from their role as nodes and be viewed as an arbitrary ordered list of points in $(-1, 1)$. Furthermore, for each index i the function X_i has zero derivative at T_i, and thus for each i the singularity of the function

$$\frac{X_i'(t)}{t - T_i}$$

which occurs at the point T_i is in fact removable. Thus, while not forgetting that the matrix (4) is not equivalent to the matrix (3) in the cases that $T_0 = t_0$ or $T_{n+1} = t_n$, it is still possible even in these exceptional cases to regard the matrix (4) as originating from the matrix (3), invoking a continuity argument.

Now, a further reduction step may be carried out upon (4). If the jth column of the matrix (4) is multiplied by the quantity $(1-t)^{-1}(1+t)^{-1}w_{\alpha,\beta}(t_j)$, then the entries in the matrix in fact become polynomials of degree at most n. With these considerations, the matrix (4) becomes the matrix

$$(q_i(t_j))_{ij}, \tag{5}$$

in which for $i = 0, \ldots, n+1$

$$q_i(t) = \frac{w_{\alpha,\beta}(t)X_i'(t)}{(1-t)(1+t)(t-T_i)}. \tag{6}$$

The proof of Theorem 1 now follows from the observation that each square submatrix of (5) obtained by striking a single row is globally nonsingular, in combination with the observation that the signs of the determinants of these square submatrices are alternating when the points T_0 and T_{n+1} are in their "proper locations," that is, when $-1 < T_0 < t_0$ and $t_n < T_{n+1} < 1$. In particular, the global nonsingularity conditions are established by noting that they hold if and only if any $n+1$ of the polynomials $q_0, \ldots, q_{n+1}$ comprise a linearly independent set of polynomials of degree at most n. The argument is tedious and is essentially a repetition of known work. While the details will be omitted, the proof of this depends upon the fact that the zeroes of the functions $X_0', \ldots, X_n'$ must interlace strictly, while, as the functions X_0' and X_{n+1}' are scalar multiples of one another, their zeroes coincide. This pattern of the zeroes of $X_0', \ldots, X_{n+1}'$ gives rise to a similar interlacing pattern of the zeroes of the polynomials $q_0, \ldots, q_{n+1}$, in which the points $T_o, \ldots, T_{n+1}$ are no longer in the list. The sign pattern of the polynomials $q_0, \ldots, q_{n+1}$ upon the points $T_0, \ldots T_{n+1}$ now gives the result. The details of this argument may be seen in Kilgore [10].

Proof of Theorem 2

The proof of this result is analogous to the proof given in Kilgore [11]. Briefly, one may assume that the nodes $t_0, \ldots, t_n$ are so strategically placed as to guarantee the minimality of $\|\Lambda_{n,\alpha,\beta}\|$. Then one may, leaving the node t_0 fixed, move the node t_n toward it, at the same time that the values $\lambda_1, \ldots, \lambda_{n-1}$ retain their initial (equal) values. This can be carried out by appeal to the Implicit Function Theorem, provided that the matrix

$$\left(\frac{\partial \lambda_i}{\partial t_j}\right)_{i=1}^{n-1}{}_{j=1}^{n-1}$$

is nonsingular. That this matrix is nonsingular follows from Proposition 2 of Kilgore [11]. The contents of Theorem 1 guarantee that, while t_n is being moved toward t_1, the value of λ_n must decrease. Now, the norm of the Lebesgue function restricted to the (shrinking) interval $[t_0, t_n]$ does not increase but rather remains constant, and on this shrinking interval the weighted polynomial functions $X_1, \ldots, X_n$ uniformly approximate with increasing exactitude the unweighted polynomials which pass through the same values at the nodes $t_0, \ldots, t_n$. Thus, the limiting situation as the interval $[t_0, \ldots, t_n]$ shrinks to zero length is a placement of nodes inside of the interval which gives unweighted polynomial interpolation of norm equal to the norm of the optimal $\Lambda_{n,\alpha,\beta}$. And the unweighted polynomial approximation is not of optimal norm, as the value of λ_n has decreased.

References

[1] K. Balázs and T. Kilgore, On the simultaneous approximation of derivatives by Lagrange and Hermite interpolation, *Journal of Approximation Theory*, 60 (1990), 231-244.

[2] K. Balázs and T. Kilgore, A discussion of simultaneous approximation of derivatives by Lagrange interpolation, *Numerical Functional Analysis and Optimization* 11 (1990), 225-237.

[3] S. Bernstein, Sur la limitation des valeurs d'un polynôme $P(x)$ de degré n sur tout un segment par ses valeurs en $(n + 1)$ points du segment, *Isvestia Akademiia Nauk SSSR* 7 (1931), 1025-1050.

[4] C. de Boor and A. Pinkus, Proof of the conjectures of Bernstein and Erdös concerning the optimal nodes for polynomial interpolation, *Journal of Approximation Theory* 24 (1978), 289-303.

[5] E. Cheney and P. Morris, On the existence and characterization of minimal projections, *Journal für die reine und angewandte Mathematik*, 270 (1974), 61-76.

[6] P. Erdös, Some remarks on polynomials, *Bulletin of the American Mathematical Society* (1947), 1169-1176.

[7] P. Erdös, Problems and results on the theory of interpolation. I, *Acta Mathematica Hungarica* 9 (1958), 381-388.

[8] T. Kilgore, Optimization of the norm of the Lagrange interpolation operator, *Bulletin of the American Mathematical Society*, 83 (1977), 1069-1071.

[9] T. Kilgore, A characterization of the Lagrange Interpolating projection with minimal Tchebycheff norm, *Journal of Approximation Theory*, 24 (1978), 273-288.

[10] T. Kilgore, A lower bound on the norm of interpolation with an extended Tchebycheff system, *Journal of Approximation Theory* 47 (1986), 240-245.

[11] T. Kilgore, Some recent results in optimal interpolation, in "Constructive Theory of Functions '87" (B. Sendov et al., Eds.), pp,. 251-259, Publishing House of the Bulgarian Academy of Sciences, Sofia, 1988.

A Note on Chebyshev's Inequality

Xin Li
Department of Mathematics, University of Central Florida
Orlando, FL 32816

Abstract

Starting with Chebyshev's inequality, we introduce improvements and variations that lead to new problems and conjectures on polynomials and logarithmic potentials.

1 Introduction

A result of Chebyshev on monic polynomials with minimum deviation from 0 on the interval $[-1,1]$ can be described as follows (cf. [3, Theorem 2.1]): For $n \geq 1$,

$$\max_{x\in[-1,1]} |x^n + a_1x^{n-1} + a_2x^{n-2} + \cdots + a_n| \geq \frac{1}{2^{n-1}},$$

for all possible choices of $a_1, a_2, ..., a_n$, and the equality holds only if

$$x^n + a_1x^{n-1} + a_2x^{n-2} + \cdots + a_n = \frac{1}{2^{n-1}} \cos(n \arccos x) =: \tilde{T}_n(x).$$

We call $\tilde{T}_n$ the (monic) Chebyshev polynomial (of the first kind) of degree n. Let us use $\|f\|$ for $\max_{x\in[-1,1]} |f(x)|$. Then, by factoring the monic polynomial, the above inequality takes a new form:

$$\|(x - x_1)(x - x_2)\cdots(x - x_n)\| \geq \frac{1}{2^{n-1}}, \tag{1}$$

for all possible choices of $x_1, x_2, ..., x_n$. We will call inequality (1) Chebyshev's inequality. Improving Chebyshev's inequality Turán established in a short paper [6] (see also [7, Lemma 5.1] and [3, p. 89]) that:

$$\|(x - x_1)(x - x_2)\cdots(x - x_n)\| \geq \frac{1}{2^{n-1}} \prod_{|x_j|>1} |x_j|. \tag{2}$$

Note that the Chebyshev polynomial $\tilde{T}_n$ has its zeros all contained in the interval $[-1, 1]$:

$$\tilde{T}_n\left(\cos\frac{(2k+1)\pi}{2n}\right) = 0, \; k = 0, 1, ..., n-1.$$

It is easy to see that when

$$(x - x_1)(x - x_2)\cdots(x - x_n) = \tilde{T}_n(x),$$

the equality in (2) holds, and Turán showed that $\tilde{T}_n$ is the only such polynomial. Turán's inequality (2) improves Chebyshev's (1) only if some zeros of the polynomials lie outside of the unit circle in the complex plane. For example, if we know the monic polynomial $P_n(x) = x^n + ...$(lower degree terms) has all its zeros outside of the circle with radius 2, then Chebyshev's inequality gives us only

$$||P_n|| \geq \frac{1}{2^{n-1}},$$

but Turán's inequality tells us more:

$$||P_n|| \geq 2.$$

Using his improved inequality, Turán established some one-sided theorems on extremal problems for polynomials, which played important roles in his book [7]. Since Turán's inequality takes into account only zeros outside of the unit circle, we ask: Can we also use the information of zeros that lie in the unit circle? In this note, we will propose an improvement of Chebyshev's inequality that strengthens Turán's inequality as well. Here is the organization of the rest of this note: We state our new results in Section 2 and prove these results in Section 3. Then we discuss the problems brought out by our results and formulate some questions in Section 4.

2 New Results

We have the following version of improvement for Chebyshev's inequality.

Theorem 1 *Let $n \geq 1$. Then, for any $x_1, x_2, ..., x_n \in \mathbf{C}$, we have*

$$\begin{aligned} &||(x - x_1)(x - x_2)\cdots(x - x_n)|| \\ \geq\; & \frac{1}{2^n}\left\{\prod_{i=1}^{n}\left|x_i + \sqrt{x_i^2 - 1}\right| + \frac{1}{\prod_{i=1}^{n}\left|x_i + \sqrt{x_i^2 - 1}\right|}\right\}, \end{aligned}$$

where we take that branch of $\sqrt{z^2-1}$ that has branch cut $[-1,1]$ and is such that $\left|z+\sqrt{z^2-1}\right|>1$ for $z\in\mathbf{C}\setminus[-1,1]$. Furthermore, the equality holds if and only if

$$(x-x_1)(x-x_2)\cdots(x-x_n)=\tilde{T}_n(x)+ir \quad \textit{for some } r\in\mathbf{R},$$

after some suitable replacements of x_j by $\overline{x}_j$ for some complex x_j's.

The next result shows that the inequality in Theorem 1 improves Turán's inequality as well.

Theorem 2 *For any $x_1,x_2,...,x_n\in\mathbf{C}$, we have*

$$\prod_{i=1}^{n}\left|x_i+\sqrt{x_i^2-1}\right|+\frac{1}{\prod_{i=1}^{n}\left|x_i+\sqrt{x_i^2-1}\right|}\geq 2\prod_{|x_i|>1}|x_i|.$$

Our proof of Theorem 1 leads us to the study of the following minimization problem on the unit circle: *Find the minimum value of*

$$\max_{|z|=1}|Az^n+c_1z^{n-1}+\cdots+c_{n-1}z+B|$$

among all possible choices of $c_1,c_2,...,c_{n-1}$. It turns out that this problem was solved completely by a result of Smirnov and Lebedev, [4, §5.1, Theorem 2], which can also be viewed as an attempt to improve and extend Chebyshev's inequality.

Theorem of Smirnov and Lebedev *For every $n\geq 2$, we have*

$$\max_{|z|=1}|Az^n+c_1z^{n-1}+\cdots+c_{n-1}z+B|\geq|A|+|B|.$$

Furthermore, the equality holds only when $c_1=...=c_{n-1}=0$.

We will also prove the following extension of the Theorem of Smirnov and Lebedev.

Theorem 3 *For $n\geq 2$ and $p\geq 1$, we have*

$$\int_{|z|=1}|Az^n+c_1z^{n-1}+\cdots+c_{n-1}z+B|^p|dz|\geq\int_{|z|=1}|Az^n+B|^p|dz|.$$

Furthermore, the equality holds only when $c_1=...=c_{n-1}=0$.

The integral value of $\int_{|z|=1}|Az^n+B|^p|dz|$ can be calculated using Hypergeometric function:

$$\begin{aligned}&\int_{|z|=1}|Az^n+B|^p|dz|\\ =\;&2\pi(|A|^2+|B|^2)^{p/2}\,{}_2F_1\left(\frac{2-p}{4},-\frac{p}{4},1,\frac{4|AB|^2}{(|A|^2+|B|^2)^2}\right).\end{aligned}\tag{3}$$

3 Proofs

We need the Joukowski transformation:

$$J(w) = \frac{1}{2}\left(w + \frac{1}{w}\right),$$

which maps the unit circle $|w| = 1$ onto the interval $[-1, 1]$ and the exterior of the unit circle conformally onto the exterior of $[-1, 1]$. The Joukowski transformation has an inverse: For $|w| > 1$ and $z \in \mathbf{C} \setminus [-1, 1]$,

$$w = z + \sqrt{z^2 - 1} \text{ if and only if } z = J(w).$$

(The branch of $\sqrt{\cdots}$ is chosen such that $\left|z + \sqrt{z^2 - 1}\right| > 1$ for $z \in \mathbf{C} \setminus [-1, 1]$.)

Proof of Theorem 1. Let

$$p(x) = (x - x_1)(x - x_2)\cdots(x - x_n)$$

and

$$x_j = J(w_j) \text{ for } |w_j| \geq 1,\ j = 1, 2, ..., n.$$

So, using the inverse of the Joukowski transformation, we can write the w_j in terms of x_j:

$$w_j = x_j + \sqrt{x_j^2 - 1},\ j = 1, 2, ..., n.$$

Now, we have

$$\begin{aligned}
\|p(x)\| &= \max_{|w|=1} |w^n p(J(w))| = \max_{|w|=1} \left| \frac{1}{2^n} w^{2n} + \cdots + \frac{1}{2^n} \right| \\
&= \max_{|w|=1} \left| \frac{1}{2^n}\left(w^{2n} + \cdots + 1\right) \prod_{|w_j|>1} \frac{1 - \overline{w}_j w}{w - w_j} \right| \\
&= \frac{1}{2^n} \max_{|w|=1} \left| \prod_{j=1}^{n} \left(w - \frac{1}{w_j}\right) \prod_{|w_j|=1} (w - w_j) \prod_{|w_j|>1} (1 - \overline{w}_j w) \right| \\
&= \frac{1}{2^n} \max_{|w|=1} \left| \prod_{|w_j|>1} (-\overline{w}_j) w^{2n} + \cdots + \prod_{|w_j|=1} (-w_j) \prod_{j=1}^{n} \left(-\frac{1}{w_j}\right) \right| \\
&\geq \frac{1}{2^n} \left(\prod_{|w_j|>1} |w_j| + \frac{1}{\prod_{|w_j|>1} |w_j|} \right).
\end{aligned}$$

Here, we have used the decomposition $w^{2n}+\cdots+1=\prod_{j=1}^{n}(w-w_j)\prod_{j=1}^{n}(w-\frac{1}{w_j})$, and in the last inequality, we have used the Theorem of Smirnov and Lebedev. Changing back from w_j to x_j, we obtain

$$\|p\| \geq \frac{1}{2^n}\left(\prod_{\left|x_j+\sqrt{x_j^2-1}\right|>1}\left|x_j+\sqrt{x_j^2-1}\right| + \frac{1}{\prod_{\left|x_j+\sqrt{x_j^2-1}\right|>1}\left|x_j+\sqrt{x_j^2-1}\right|}\right).$$

The equality holds if and only if

$$(w^{2n}+\cdots+1)\prod_{|w_j|>1}\frac{1-\overline{w}_j w}{w-w_j} = \prod_{|w_j|>1}(-\overline{w}_j)w^{2n} + \prod_{|w_j|>1}\left(-\frac{1}{w_j}\right), \quad (4)$$

by the necessary and sufficient condition on the case when the equality holds in the Theorem of Smirnov and Lebedev. This implies that either $|w_j|=1$ for all $j=1,2,...,n$ or $|w_j|>1$ for all $j=1,2,...,n$. In the former case, we obtain $w^{2n}+\cdots+1=w^{2n}+1$, i.e., $w^n p(J(w))=\frac{1}{2^n}(w^{2n}+1)$, which gives $p(x)=\tilde{T}_n(x)$.

Now we consider the case when $|w_j|>1$ for all $j=1,2,...,n$. If $|w_j|>1$ for all $j=1,2,...,n$, then (4) would imply that $\frac{1}{w_j},\frac{1}{\overline{w}_j}$, $j=1,2,...,n$, are the solutions of

$$w^{2n}+\left(\prod_{j=1}^{n}\frac{1}{|w_j|}\right)^2=0. \quad (5)$$

Let

$$R:=\prod_{j=1}^{n}\frac{1}{|w_j|}.$$

Then $0<R<1$ and we must have

$$\prod_{j=1}(w-\frac{1}{w_j})(w-\frac{1}{\overline{w}_j}) = w^{2n}+R^2=(w^n-iR)(w^n+iR).$$

If we choose w_j such that $\prod_{j=1}^{n}(w-1/w_j)=w^n-iR$, then $\prod_{j=1}^{n}(w-w_j)=w^n+i/R$, and so

$$\begin{aligned} p(x) &= \frac{1}{2^n w^n}\prod_{j=1}^{n}(w-w_j)(w-\frac{1}{w_j}) \\ &= \frac{1}{2^n}(w^n-iR)(w^n+\frac{i}{R}) = \tilde{T}_n(x)-\frac{i}{2^n}(R-\frac{1}{R}). \end{aligned}$$

If, instead, we take w_j such that $\prod_{j=1}^n (w - 1/w_j) = w^n + iR$, then $\prod_{j=1}^n (w - w_j) = w^n - i/R$, which leads to

$$p(x) = \tilde{T}_n(x) + \frac{i}{2^n}(R - \frac{1}{R}).$$

All the other choices of w_j will give us the alterations in the zeros of $p(x)$ in the form of replacing some x_j by their conjugates. Finally, note that any nonzero real number r can be written as either $\frac{1}{2^n}(R - \frac{1}{R})$ or $-\frac{1}{2^n}(R - \frac{1}{R})$ for some $R \in (0, 1)$. Therefore, the conditions for equality in the theorem is verified. □

Next, we prove Theorem 2. The following lemma is needed.

Lemma 4 *If $a_j \geq 1$, $j = 1, 2, ..., m$, then*

$$\prod_{j=1}^m a_j + \frac{1}{\prod_{j=1}^m a_j} \geq \frac{1}{2^{m-1}} \prod_{j=1}^m \left(a_j + \frac{1}{a_j}\right).$$

Proof. The proof follows from the repeated use of

$$ab + \frac{1}{ab} \geq \frac{1}{2}\left(a + \frac{1}{a}\right)\left(b + \frac{1}{b}\right) \quad \text{for } a \geq 1 \text{ and } b \geq 1. \tag{6}$$

The proof of inequality (6) can be obtained from

$$2ab + \frac{2}{ab} - \left(a + \frac{1}{a}\right)\left(b + \frac{1}{b}\right) = \left(a - \frac{1}{a}\right)\left(b - \frac{1}{b}\right).$$

This completes the proof of the lemma. □

Proof of Theorem 2. Let $a_j = \left|x_j + \sqrt{x_j^2 - 1}\right|$ for $j = 1, 2, .., n$. Then $a_j \geq 1$ $(j = 1, 2, ..., n)$ and the proof is completed by applying the lemma. □

Proof of Theorem 3. We apply the trick of Smirnov and Lebedev of using the roots of unity (cf. [4, §5.1]). Let ζ_k denote the n-th roots of unity, i.e.,

$$\zeta_k = \exp\left\{\frac{2\pi i k}{n}\right\}, \quad k = 1, 2, ..., n.$$

Then

$$\sum_{k=1}^n \zeta_k^j = \begin{cases} 0 & \text{if } j = 1, 2, ..., n-1, \\ n & \text{if } j = 0, n. \end{cases}$$

It follows that for any polynomial

$$p(z) = Az^n + c_1 z^{n-1} + \cdots + c_n z + B,$$

we have

$$\frac{1}{n}\sum_{k=1}^{n} p(\zeta_k z) = Az^n + B.$$

Therefore,

$$\begin{aligned}
\int_{|z|=1} |Az^n + B|^p |dz| &= \int_{|z|=1} \left|\frac{1}{n}\sum_{k=1}^{n} p(\zeta_k z)\right|^p |dz| \\
\leq & \left\{\frac{1}{n}\sum_{k=1}^{n}\left(\int_{|z|=1} |p(\zeta_k z)|^p \, |dz|\right)^{1/p}\right\}^p \\
= & \left\{\frac{1}{n}\sum_{k=1}^{n}\left(\int_{|z|=1} |p(z)|^p \, |dz|\right)^{1/p}\right\}^p = \int_{|z|=1} |p(z)|^p \, |dz|,
\end{aligned}$$

where, the inequality follows from an application of Minkowski's inequality. The equality holds if and only if

$$p(\zeta_k z) = Az^n + B, \; k = 1, 2, ..., n,$$

which implies $p(z) = p(\zeta_1(\zeta_1^{-1} z)) = A(\zeta_1^{-1} z)^n + B = Az^n + B$. This completes the proof. □

Proof of (3). It is easy to verify that

$$I := \int_{|z|=1} |Az^n + B|^p |dz| = |A|^p \int_0^{2\pi} \left|e^{i\varphi} + \frac{B}{A}\right|^p d\varphi.$$

Writing $B/A = re^{i\varphi_0}$ for some $r, \varphi_0 \in \mathbf{R}$, we have

$$I = |A|^p \int_0^{2\pi} \left|e^{i\varphi} + re^{i\varphi_0}\right|^p d\varphi = |A|^p \int_0^{2\pi} \left|e^{i(\varphi-\varphi_0)} + r\right|^p d\varphi.$$

By the periodicity of the integrand, we can write

$$I = |A|^p \int_0^{2\pi} \left|e^{i\varphi} + r\right|^p d\varphi = |A|^p \int_0^{2\pi} (1 + 2r\cos\varphi + r^2)^{p/2} d\varphi.$$

Now, using tables of integration or Mathematica to calculate the integral and then combining the results we get the equality (3). □

4 Concluding Remarks

In this section, we collect some remarks and problems related to the results discussed in this note. First, we mention that Turán's inequality was originally stated in the equivalent form

$$||p|| \geq \frac{1}{2^{n-1}} \exp\left\{\frac{1}{2\pi}\int_{|z|=1} \log|p(z)||dz|\right\},$$

for $p(z) = z^n + ...$(lower degree terms). This is indeed equivalent to the Turán's inequality because

$$\exp\left\{\frac{1}{2\pi}\int_{|z|=1} \log|p(z)||dz|\right\} = \prod_{|x_i|>1} |x_i|$$

if $p(z) = (z - x_1)(z - x_2)\cdots(z - x_n)$. Next, note that this quantity is the geometric mean of $p(z)$ on the unit circle. In view of the fact that the norm in the inequality is taken on the interval $[-1, 1]$, it is natural to ask if we can replace the geometric mean on the unit circle by that on the interval which equals

$$G(p, [-1, 1]) := \exp\left\{\frac{1}{\pi}\int_{-1}^{1} \frac{\log|p(x)|}{\sqrt{1-x^2}}dx\right\}.$$

To check out whether this replacement works is actually one of the motivations of this note. In fact, using the current notation, the inequality in Theorem 1 can be put in the form

$$||p|| \geq G(p, [-1, 1]) + \frac{1}{2^{2n}G(p, [-1, 1])}. \tag{7}$$

Indeed, we can use (cf. [2, Lemma 4.1])

$$\frac{1}{\pi}\int_{-1}^{1} \frac{\log|x - x_0|}{\sqrt{1-x^2}}dx = -\log 2 + \log\left|x_0 + \sqrt{x_0^2 - 1}\right|$$

to verify that

$$\exp\left\{\frac{1}{\pi}\int_{-1}^{1} \frac{\log|(x - x_1)(x - x_2)\cdots(x - x_n)|}{\sqrt{1-x^2}}dx\right\} = \frac{1}{2^n}\prod_{i=1}^{n}\left|x_i + \sqrt{x_i^2 - 1}\right|.$$

Now, we employ some terminologies and notations from potential theory to rewrite the inequalities we have seen. First, note that $\text{cap}([-1, 1]) = 1/2$

and $dx/(\pi\sqrt{1-x^2})$ is the extremal measure, $d\mu_{[-1,1]}$, for the set $[-1,1]$. So,

$$G(p,[-1,1]) = \exp\left\{\int_{-1}^{1} \log|p(x)|d\mu_{[-1,1]}(x)\right\},$$

and we can put the inequality (7) into the form:

$$\begin{aligned} \|p\| \;\geq\; & \exp\left\{\int_{-1}^{1} \log|p(x)|d\mu_{[-1,1]}(x)\right\} \\ & +\text{cap}([-1,1])^{2n} \exp\left\{-\int_{-1}^{1} \log|p(x)|d\mu_{[-1,1]}(x)\right\}. \end{aligned} \tag{8}$$

For a Jordan arc J in the complex plane, we use $\|\cdot\|_J$ for the supremum norm taken on J, cap(J) for the capacity of J, and $d\mu_J$ for the equilibrium (extremal) measure associated with J (cf. [5]). To generalize Theorem 1 to the case when the norm is taken on Jordan arcs in the complex plane, we formulate the following question:

Problem 1. *Let J be a Jordan arc with positive logarithmic capacity. Is it true that, for any monic polynomial of degree n, $p(x) = x^n + \cdots$, we have*

$$\|p\|_J \geq \exp\left\{\int_J \log|p(x)|d\mu_J(x)\right\} + \text{cap}(J)^{2n} \exp\left\{-\int_J \log|p(x)|d\mu_J(x)\right\}?$$

It is also interesting to find the corresponding inequality for more than one piece of Jordan arcs like $J = [-1,a]\cup[b,1]$ for some $-1 < a < 0 < b < 1$.

We write the inequality (8) in a weaker form:

$$\log\|p\| > \int_{-1}^{1} \log|p(x)|d\mu_{[-1,1]}(x). \tag{9}$$

It seems that the strict inequality is caused by the end points of the interval – for a closed curved like the unit circle $J = \{z : |z| = 1\}$, the following inequality is sharp:

$$\|(z-z_1)\cdots(z-z_n)\|_J \geq \exp\frac{1}{2\pi}\int_J |(z-z_1)\cdots(z-z_n)||dz| = \prod_{|z_j|>1} |z_j|;$$

the equality holds only if $(z-z_1)\cdots(z-z_n) = z^n$.

Let us write (9) in a more general form. We have

$$\sup_{x\in[-1,1]} \int_{\mathbf{C}} \log|x-t|d\mu(t) > \int_{-1}^{1}\int_{\mathbf{C}} \log|x-t|d\mu(t)d\mu_{[-1,1]}(x),$$

for $d\mu(t) = \frac{1}{n}\sum_{j=1}^{n} d\delta_{x_j}(t)$, where $d\delta_z$ denotes the probability measure concentrated on the single point z. The expression $U(\mu;x) := \int_{\mathbf{C}} \log|x-t|d\mu(t)$ is called the logarithmic potential of μ. Is the above inequality holds for all probability measures μ with compact support in $\mathbf{C}$?

Another direction for generalization of Theorem 1 is to replace the supremum norm by L_p norm. In this case, Theorem 3 is not quite enough to help us in L_p norm for any sharp results. Finally, we ask the following question.

Problem 2. *Find the minimum value of*

$$\frac{\left\{\int_{-1}^{1} |(x-x_1)(x-x_2)\cdots(x-x_n)|^p dx\right\}^{1/p}}{G(x_1, x_2, ..., x_n)}$$

among all possible choices of $x_1, x_2, ..., x_n$, *where*

$$G(x_1, x_2, ..., x_n) = \prod_{|x_i|>1} |x_i|.$$

It is worth mentioning that Ya.L. Geronimus [1] considered related extremal problems. He used L_σ^p-norms taken on a contour in $\mathbf{C}$ for a general measure σ on the contour, but he was mainly concerned about the asymptotic behavior of the minimum values.

References

[1] Ya. L. Geronimus, On some extremal problems in the space $L_\sigma^{(p)}$, Mat. Sbornik N.S. 31(1952), 3-26. (Russian)

[2] H.N. Mhaskar and E.B. Saff, Extremal problems for polynomials with exponential weights, Trans. Amer. Math. Soc., 285(1984), 203-234.

[3] T.J. Rivlin, "Chebyshev Polynomials", 2nd edn., John Wiley & Sons, New York, 1990.

[4] V.I. Smirnov and N.A. Lebedev, "Functions of a Complex Variable (Constructive Theory)", English Edition, Iliffe Books, London, 1968.

[5] M. Tsuji, "Potential Theory in Modern Function Theory", 2nd edn., Chelsea, New York, 1958.

[6] P. Turán, On an inequality of Chebyshev (Cebysev), Ann. Univ. Sci. Budapest, Eötvős, Sect. Math. 11(1968), 15-16.

[7] P. Turán, "On a New Method of Analysis and its Applications", John Wiley & Sons, New York, 1984.

Smooth Maclaurin Series Coefficients in Padé and Rational Approximation

Dedicated to the Memory of A.K. Varma

D.S. Lubinsky
Department of Mathematics, University of the Witwatersrand
P.O. Wits 2050, South Africa

Abstract

It is folklore amongst Padé approximators that when the Maclaurin series coefficients $\{a_j\}$ of a power series

$$f(z) = \sum_{j=0}^{\infty} a_j z^j$$

behave "smoothly", then the Padé approximants of f "behave well". We survey some of the notions of smoothness and results that lend credence to this belief. We focus especially on results involving the behaviour of $a_{j-1}a_{j+1}/a_j^2$ as $j \to \infty$.

1 Introduction

Let

$$f(z) = \sum_{j=0}^{\infty} a_j z^j \tag{1}$$

be a formal power series with complex coefficients and $m, n \geq 0$ be integers. The m, n Padé approximant to f is a rational function

$$[m/n](z) = P(z)/Q(z) \tag{2}$$

where P, Q are polynomials of degree at most m, n respectively with Q not identically zero, and

$$(fQ - P)(z) = O(z^{m+n+1}). \tag{3}$$

The Padé table of f is the array

$$\begin{array}{cccc} [0/0] & [0/1] & [0/2] & \dots \\ [1/0] & [1/1] & [1/2] & \cdots \\ [2/0] & [2/1] & [2/2] & \cdots . \\ \vdots & \vdots & \vdots & \ddots \end{array} \tag{4}$$

The problem of convergence of $[m/n]$ to f as m or n or both approach ∞, is one of the most important problems associated with Padé approximation. A deep understanding of convergence theory provides justification to the application of the Padé technique and also gives insight as to when and how they should be applied [1], [2].

It it is now well known that the chance of proving uniform or even pointwise convergence is very rare, if we place only restrictions on the analytic nature of f, such as analyticity or meromorphicity. The most widely applicable theorems generally establish convergence in planar or linear Lebesgue measure, or in some other set function such as logarithmic capacity [2]. This lack of pointwise convergence is due to the phenomenon of "spurious poles": for infinitely many m, n, $[m/n]$ may have poles that do not reflect the analytic properties of f. For example, f may be entire, but infinitely of the diagonal approximants $[n/n]$ may have a pole at 1 (or any other fixed point) [24].

Despite these pathologies, uniform and very rapid convergence of the Padé aproximants (under suitable restrictions on m, n as $m + n \to \infty$) has been established for very large classes of special functions [1], [2]. For many of these special functions, their Maclaurin series coefficients tend to behave in a regular or definite pattern. Indeed, folklore amongst Padé approximators asserts that when the Maclaurin series coefficients of f are "smooth" then the approximants should behave well, and there should be rapid convergence.

This belief arises not only from numerical experience but also is intuitively reasonable from a determinantal representation for the denominator in $[m/n](z)$. Let us write

$$[m/n](z) = P_{mn}(z)/Q_{mn}(z) \tag{5}$$

where P_{mn}, Q_{mn} are normalized by the condition

$$Q_{mn}(0) = 1. \tag{6}$$

If we define the n by n Toeplitz determinant

$$D(m/n) := \det\left(a_{m-j+k}\right)_{j,k=1}^{n} = \det\begin{bmatrix} a_m & a_{m+1} & \cdots & a_{m+n-1} \\ a_{m-1} & a_m & \cdots & a_{m+n-2} \\ \vdots & \vdots & \ddots & \vdots \\ a_{m-n+1} & a_{m-n+2} & \cdots & a_m \end{bmatrix} \tag{7}$$

then

$$Q_{mn}(z) = \frac{1}{D(m/n)} \det\begin{bmatrix} a_m & a_{m+1} & \cdots & a_{m+n} \\ a_{m-1} & a_m & \cdots & a_{m+n-1} \\ \vdots & \vdots & \ddots & \vdots \\ a_{m-n+1} & a_{m-n+2} & \cdots & a_{m+1} \\ z^n & z^{n-1} & \cdots & 1 \end{bmatrix}. \tag{8}$$

This representation may be derived as follows: equate coefficients of powers of $z^j, j = m+1, m+2, \ldots, m+n$ on both sides of (3). From the resulting n linear equations, solve for the coefficients of Q_{mn} by Cramer's rule, and then derive (8).

Since determinants involve signed sums of products of their entries, it seems reasonable that when the $\{a_j\}$ are smooth in some sense, then Q_{mn} should behave well. Conversely, a single irregular coefficient a_j could dramatically affect the behaviour of Q_{mn} and the location of its zeros. Of course justifying such intuition is extremely difficult, especially if the size of the determinant grows to ∞, and especially if the a_j decay rapidly in size as $j \to \infty$.

It is then not surprising that most of the general theorems on convergence of Padé approximants avoid using the determinantal representation, concentrating instead on contour integrals derived from Cauchy's integral formula. There is one, namely de Montessus' theorem, which deals with $m \to \infty$, n fixed, that was first proved by analysing the behaviour of the determinants above, but subsequently a more attractive proof was found that avoids determinants. de Montessus' Theorem asserts that if a function f is analytic in $|z| < r$, except for poles of total multiplicity n, none at 0, then $[m/n]$ converges to f as $m \to \infty$ uniformly in compact subsets of $|z| < r$ omitting poles of f. In a sense, such a function does have smooth Maclaurin series coefficients. Let us suppose, for simplicity, that f has n simple poles $u_1, u_2, \ldots u_n$. Then on subtraction of the principal part of f, we see that

$$f(z) = \sum_{k=1}^{n} \frac{c_k}{1 - z/u_k} + g(z)$$

where g is analytic in $|z| < r$. Then given $0 < s < r$

$$a_j = \sum_{k=1}^{n} c_k u_k^{-j} + O(s^{-j}),\ j \to \infty.$$

The sum is the dominant part, since each $|u_k| < r$ and also the sum is "smooth" in j, especially if the $|u_k|$ are distinct.

This is an example of linear combinations of geometric sequences $\{u_1^{-k}\}_{k=1}^{\infty}$, $\{u_2^{-k}\}_{k=1}^{\infty}$, $\ldots \{u_n^{-k}\}_{k=1}^{\infty}$, which intuitively are very smooth. Other examples of obviously smooth sequences are $\{1/k!\}_{k=1}^{\infty}$ or even $\{1/k!^{\theta}\}_{k=1}^{\infty}$ for any real θ.

Is it possible to make precise this notion of smoothness? Much depends on what the goal is: one can give implicit definitions of smoothness, for example involving high order differences of the coefficients, or probably more appropriately involving the asymptotic behaviour of the determinants $D(m/n)$. These might allow one to prove nice theorems, but unfortunately verifying the hypotheses of the theorem would be perhaps more difficult than the proofs of the theorems. Most of those who have worked on notions of smoothness have preferred to start with some fairly explicit conditions on the Maclaurin series coefficients and then prove directly applicable theorems. This of course to some extent restricts the generality. It seems unlikely that even in the case where we focus on n fixed, $m \to \infty$, that a completely general notion of smoothness of $\{a_j\}$ can ever be formulated. One of the problems in this direction is dealing simultaneously with sequences that exhibit geometric decay, or much faster than geometric decay, or have very complicated asymptotic behaviour.

We shall formulate one of Wilson's theorems, as extended by G.A. Baker, in Section 2. In Section 3, we begin our discussion of smoothness in terms of hypotheses on $a_{j-1}a_{j+1}/a_j^2$. We examine there what can be achieved for columns $\{[m/n]\}_{m=1}^{\infty}$of the Padé table and corresponding sequences of best approximants. In Section 4, we look at diagonals $\{[n/n]\}_{n=1}^{\infty}$, and in Section 5, examine extensions to simultaneous approximants and further results.

2 Wilson's Theorems

In the late 1920's and early 1930's, R. Wilson investigated extensions of de Montessus' theorem. For example if f has poles of total multiplicity n in $|z| < r$, and specified singularities such as poles or branchpoints on $|z| = r$, what can we say about the convergence or divergence of $\{[m/n + \mu]\}_{m=1}^{\infty}$ in $|z| < r$ for non-negative μ? This led him also to study convergence of $\{[m/n+\mu]\}_{m=1}^{\infty}$ to "smooth" entire or meromorphic functions with at most

n poles in $\mathbb{C}$. G.A. Baker extended these in his book [1]. We quote only that dealing with functions meromorphic in a disc of finite radius; there is an analogue for functions meromorphic in $\mathbb{C}$ [1,p.149].

Theorem 1 *Let f be analytic in $|z| < r$, except for poles z_j of multiplicity $r_j, j = 1, 2, \ldots l$, and of total multiplicity*

$$n = \sum_{j=1}^{l} r_j.$$

Let $\mu \geq 0$. Consider

$$g(z) := f(z) \prod_{j=1}^{l} (1 - z/z_j)^{r_j} = \sum_{j=0}^{\infty} g_j z^j.$$

Assume that the $\{g_k\}$ satisfy for $k = m - n - \mu + 1$ to $m + n + \mu$,

$$g_k = \Gamma^{-k} \beta(m) \left[\sum_{j=0}^{2\mu-2} \alpha_j(m) \left(\frac{k}{m} - 1 \right)^j + \gamma(k, m) \left(\frac{n + \mu}{m} \right)^{2\mu - 1} \right]$$

where $|\Gamma| = R$, $\beta(m) \neq 0$, $\alpha_j(m)$ and $\gamma(k, m)$ are uniformly bounded as $m \to \infty$ and

$$\lim_{m \to \infty} \det \begin{bmatrix} (2\mu - 2)!\alpha_{2\mu-2}(m) & \cdots & (\mu - 1)!\alpha_{\mu-1}(m) \\ \vdots & & \vdots \\ (\mu - 1)!\alpha_{\mu-1}(m) & \cdots & \alpha_0(m) \end{bmatrix}$$

exists and is non-zero. Then uniformly in compact subsets of $|z| < r$ omitting poles of f,

$$\lim_{m \to \infty} [m/n + \mu](z) = f(z).$$

The proof [1,pp.144-146] involves asymptotic analysis of the determinants in (7), (8).

3 The use of $a_{j-1}a_{j+1}/a_j^2$

One of the difficulties in describing smoothness is dealing with different rates of decay of $a_j, j \to \infty$. How can we simultaneously handle sequences such as $\{2^{-j}\}_{j=0}^{\infty}$ or for $\lambda > 0, \{j!^{-1/\lambda}\}_{j=0}^{\infty}$ or $\{1/\Gamma(1 + j/\lambda)\}_{j=0}^{\infty}$

or $\{2^{-j^2}\}_{j=0}^{\infty}$? A single division of successive coefficients to form a_{j+1}/a_j already dramatically reduces the variability and a second division to form

$$q_j := a_{j-1}a_{j+1}/a_j^2 \tag{9}$$

gives a quantity q_j that in all the above cases approaches a finite non-zero limit as $j \to \infty$ (namely 1 for the first three sequences and $\frac{1}{4}$ for the fourth).

Not only does q_j approach a finite limit, but it does so very smoothly: We see that

$$a_j = 2^{-j} \Rightarrow q_j = 1;$$

$$a_j = j!^{-1/\lambda} \Rightarrow q_j = \left(1 + \frac{1}{j}\right)^{-1/\lambda}; \tag{10}$$

For each fixed $m \geq 1$, there exist $c_1, c_2, \ldots c_m$ such that

$$a_j = 1/\Gamma(1 + j/\lambda) \Rightarrow q_j = 1 + \sum_{k=1}^{m} c_k j^{-k} + o(j^{-m}); \tag{11}$$

$$a_j = 2^{-j^2} \Rightarrow q_j = \frac{1}{4}.$$

The only non-trivial one is the one involving the gamma function, and that follows from the well known asymptotic expansion for $\Gamma(x), x \to \infty$:

$$\Gamma(x) \sim \left(\frac{2\pi}{x}\right)^{1/2} \left(\frac{x}{e}\right)^x \left(1 + \frac{1}{12x} + \frac{1}{288x^2} + \cdots\right).$$

These simple observations become of interest when one notices that the determinant $D(m/n)$ can be transformed by suitably scaling its rows and columns, thereby obtaining a determinant of a Toeplitz matrix whose entries involve the q_j :

Lemma 2 *Let $a_j \neq 0, j \geq N$ and q_j be defined by (9). Then for $m-n+1 \geq N$,*

$$D(m/n)/(a_m)^n = \det\left(b_{k-j}\right)_{j,k=1}^{n} \tag{12}$$

where

$$b_l := \begin{cases} q_m^{l/2} \cdot q_{m+1}^{l-1} q_{m+2}^{l-2} \cdots q_{m+l-1}, & l > 0 \\ 1, & l = 0 \\ q_m^{-l/2} \cdot q_m^{-l} q_{m-1}^{-l-1} \cdots q_{m+l+1}, & l < 0 \end{cases}. \tag{13}$$

Here any value of the $l/2$th power may be taken.

Proof. We multiply the jth row of $D(m/n)$ by $q_m^{-j/2}a_m^{-1}(a_{m+1}/a_m)^j$ and multiply the kth column of $D(m/n)$ by $q_m^{k/2}(a_{m+1}/a_m)^{-k}, j, k = 1, 2, \ldots n$. These operations taken together effectively multiply the determinant by a_m^{-n}. Thus

$$D(m/n)/a_m^n = \det\left(q_m^{(k-j)/2}(a_{m+1}/a_m)^{j-k}a_{m-j+k}/a_m\right)_{j,k=1}^n.$$

Here for $l > 0$

$$\begin{aligned} a_{m+l}/a_m &= \prod_{t=0}^{l-1}(a_{m+t+1}/a_{m+t}) \\ &= \prod_{t=0}^{l-1}(q_{m+t}q_{m+t-1}\cdots q_{m+1}a_{m+1}/a_m) \\ &= (a_{m+1}/a_m)^l q_{m+1}^{l-1}q_{m+2}^{l-2}\cdots q_{m+l-1} \end{aligned}$$

and hence for $k - j = l > 0$,

$$q_m^{(k-j)/2}(a_{m+1}/a_m)^{j-k}a_{m-j+k}/a_m = b_l = b_{k-j}.$$

The case $k - j = l \le 0$ is similar.□

Note that the quantities b_l depend on m, n, but the number of factors q_j appearing in b_l is

$$l/2 + (l-1) + (l-2) + \ldots + 1 = l^2/2,$$

independent of m, n. This suggests:

Theorem 3 *Let f of (1) be a formal power series with $a_j \neq 0$, j large enough, and assume that for some $q \in \mathbb{C}$,*

$$\lim_{j\to\infty} a_{j-1}a_{j+1}/a_j^2 = q. \tag{14}$$

Then for each fixed $n \ge 1$,

$$\lim_{m\to\infty} D(m/n)/(a_m)^n = \det\left(q^{(k-j)^2/2}\right)_{j,k=1}^n = \prod_{j=1}^{n-1}(1-q^j)^{n-j}. \tag{15}$$

Proof. The first equality in (15) follows directly from the lemma and the limit (14). When $q = 0$, the determinant in the middle of (15) is that of the identity matrix, so equals 1 and the second equality follows. It remains to evaluate

$$\det\left(q^{(k-j)^2/2}\right)_{j,k=1}^{n}$$

when $q \neq 0$. There are several ways to do this. Perhaps the simplest is to note that the hypothesis (14) holds when we choose

$$a_j := q^{j^2/2},\ j = 0, 1, 2, \ldots .$$

We now evaluate $D(m/n)$ for this special case. Observe that

$$a_{l+k-1}/a_l = q^{l(k-1)} \cdot q^{(k-1)^2/2}.$$

Thus if we extract a_m from the first row, a_{m-1} from the second row, ..., a_{m-n+1} from the nth row in (7), we obtain a determinant in which the factor $q^{(k-1)^2/2}$ appears in each entry in the kth column, $k = 1, 2, \ldots n$. Then extracting these factors from each column, we obtain for this special case,

$$D(m/n) = a_m a_{m-1} \ldots a_{m-n+1}\left[\prod_{k=1}^{n} q^{(k-1)^2/2}\right]\det\left(q^{(m+1-k)(j-1)}\right)_{j,k=1}^{n}.$$

The last determinant is a Vandermonde determinant involving $z_k = q^{m+1-k}$, $k = 1, 2, \ldots n$, and evaluating it gives

$$D(m/n) = a_m a_{m-1} \ldots a_{m-n+1}\left[\prod_{k=1}^{n} q^{(k-1)^2/2}\right]\prod_{1\leq k<l\leq n}\left(q^{m+1-l} - q^{m+1-k}\right).$$

After some straightforward manipulations, we deduce that for this special case

$$D(m/n)/a_m^n = \prod_{j=1}^{n-1}(1 - q^j)^{n-j}.$$

Since the right-hand side depends only on q and n, we obtain the result in the general case.□

This is an attractive result, but of far more interest to Padé approximators is the behaviour of the Padé denominators Q_{mn} :

Theorem 4 *Let $n \geq 1$. Under the hypothesis of Theorem 3, we have*

$$\lim_{m\to\infty} Q_{mn}(ua_m/a_{m+1})D(m/n)/a_m^n$$

$$= \det \begin{bmatrix} 1 & q^{1/2} & q^{2^2/2} & \cdots & q^{n^2/2} \\ q^{1/2} & 1 & q^{1/2} & \cdots & q^{(n-1)^2/2} \\ \vdots & \vdots & & \ddots & \vdots \\ q^{(n-1)^2/2} & q^{(n-2)^2/2} & q^{(n-3)^2/2} & \cdots & q^{1/2} \\ (q^{-1/2}u)^n & (q^{-1/2}u)^{n-1} & (q^{-1/2}u)^{n-2} & \cdots & 1 \end{bmatrix} =: V_n(u), \tag{16}$$

uniformly for u in compact subsets of $\mathbb{C}$. (When $q = 0, V_n(u) \equiv 1$). In particular, if $q^j \neq 1, j = 1, 2, \ldots, n-1$, then uniformly in compact subsets of $\mathbb{C}$,

$$\lim_{m\to\infty} Q_{mn}(ua_m/a_{m+1}) = B_n(u) \tag{17}$$

where

$$B_n(u) := V_n(u) \Big/ \prod_{j=1}^{n-1} (1-q^j)^{n-j} \tag{18}$$

is a polynomial of degree $\leq n$ with $B_n(0) = 1$.

Proof. We apply the same operations to the determinant in (8) defining Q_{mn} as we did to that defining $D(m/n)$. Thus, we multiply the jth row by $q_m^{-j/2}a_m^{-1}(a_{m+1}/a_m)^j$, $j = 1, 2, \ldots n$, and multiply the kth column by $q_m^{k/2}(a_{m+1}/a_m)^{-k}, k = 1, 2, \ldots, n+1$. Moreover we multiply the $(n+1)$st row by $q_m^{-(n+1)/2}(a_{m+1}/a_m)^{n+1}$.These operations taken together effectively multiply the determinant by a_m^{-n}. If we define, for a fixed m, b_l by (13), then we see as in Lemma 2 that

$$Q_{mn}(ua_m/a_{m+1})D(m/n)/a_m^n$$

$$= \det \begin{bmatrix} 1 & b_1 & b_2 & \cdots & b_n \\ b_{-1} & 1 & b_1 & \cdots & b_{n-1} \\ \vdots & \vdots & & \ddots & \vdots \\ b_{-(n-1)} & b_{-(n-2)} & b_{-(n-3)} & \cdots & b_1 \\ (q_m^{-1/2}u)^n & (q_m^{-1/2}u)^{n-1} & (q_m^{-1/2}u)^{n-2} & \cdots & 1 \end{bmatrix}. \tag{19}$$

Now letting $m \to \infty$ gives the result if $q \neq 0$. The case $q = 0$ requires a slightly different approach. When $q^j \neq 1$ for all j, we can use the fact that the right-handside of (15) is non-zero and divide by the limit there.□

The surprising thing about this result is the simplicity of its hypothesis and the powerful conclusion. When for example

$$\lim_{m\to\infty} a_{m+1}/a_m = 0$$

so that f is entire, the conclusion is that all the poles of $[m/n]$ approach ∞ as $m \to \infty$ with rate $|a_m/a_{m+1}|$. More precisely, if $z_1, z_2, \ldots z_n$ are the zeros of $B_n(u)$, then for large m, $Q_{mn}(z)$ has zeros $z_{m1}, z_{m2}, \ldots z_{mn}$ that satisfy

$$\lim_{m\to\infty} z_{mj}a_{m+1}/a_m = z_j.$$

Since all poles of $[m/n]$ leave each compact set as $m \to \infty$, it is then easy to prove uniform convergence [17].

There is an interesting negative converse to this: If $q^j \neq 1, 1 \leq j \leq n-1$, one can show that $B_n(u)$ has at least one zero z_1, say, in $|u| < 1$ [19]. Suppose now that both (14) holds with q not a root of unity and

$$\lim_{j\to\infty} a_{j+1}/a_j = 1/R > 0.$$

Then f is analytic in $|z| < R$, but the considerations of the previous paragraph show that $[m/n]$ has a pole approaching $z_1 R$ as $m \to \infty$ and so $\{[m/n]\}_{m=1}^{\infty}$ (nor any subsequence) can converge uniformly to f in $\{z : S < |z| < R\}$ for any $S < R|z_1|$. This phenomenon and its relation to the Baker-Graves-Morris Conjecture (resolved in 1984 by Buslaev, Goncar and Suetin [4]) is examined in [19].

The polynomials $B_n(-u)$ are in fact the Rogers-Szegö polynomials, well known in the theory of $q-$ or basic, hypergeometric series. One can show that

$$B_n(-u) = \sum_{j=0}^{n} \begin{bmatrix} n \\ j \end{bmatrix} u^j \tag{20}$$

where

$$\begin{bmatrix} n \\ j \end{bmatrix} = \frac{(1-q^n)(1-q^{n-1})\ldots(1-q^{n+1-j})}{(1-q)(1-q^2)\ldots(1-q^j)} \tag{21}$$

is the Gaussian, or basic, binomial coefficient. It is well defined even when q is a root of unity, for example as $q \to 1$, it converges to $\begin{pmatrix} n \\ j \end{pmatrix}$. The $\{B_n\}_{n=0}^{\infty}$ satisfy the three term recurrence relation

$$B_n(u) = B_{n-1}(u) - uq^{n-1}B_{n-1}(u/q), n = 1, 2, \ldots. \tag{22}$$

The latter may be proved by applying Sylvester's determinant identity to the right-hand side of (8) and then (22) can be proved by induction on n [19].

Despite the elegance of Theorems 3 and 4, they do not give useful results when q is a root of unity. Indeed the most likely value of q in applications is $q = 1$ and here the right-hand sides of (15), (16) are zero. However the above results are at least suggestive of what might be possible under additional assumptions. For many examples, including those in (10), (11), q_j has a complete asymptotic expansion in negative powers of j. This is treated in

Theorem 5 *Let q be a primitive lth root of unity. Assume that for some $N \geq 1$,*

$$q_j = q\left[1 + \sum_{k=1}^{N} c_k j^{-k} + o(j^{-N})\right], \; j \to \infty, \tag{23}$$

where $c_1 \neq 0$. Then for $n = 0, 1, 2, \ldots, lN + 1$,

$$\lim_{m\to\infty} D(m/n) \Big/ \left\{ a_m^n \prod_{k=1}^{n-1} \left(1 - q_m^j\right)^{n-j} \right\} = 1 \tag{24}$$

and uniformly in compact subsets of $\mathbb{C}$,

$$\lim_{m\to\infty} Q_{mn}(ua_m/a_{m+1}) = B_n(u). \tag{25}$$

Sketch of Proof One has to be far more careful than in the proofs of Theorems 3 and 4. One uses induction on n, and the following identities, which may be proved using Sylvester's determinant identity [17]:

$$D(m/n+1)D(m/n-1) = D(m/n)^2 - D(m-1/n)D(m+1/n);$$

$$Q_{mn}(z) = Q_{m,n-1}(z) - zQ_{m-1,n-1}(z)\frac{D(m-1/n-1)D(m+1/n)}{D(m/n-1)D(m/n)}.$$

In the induction step, one has to use a more precise asymptotic than (24), carrying those terms of the asymptotic expansion (23) that are still available. Full details appear in [17].□

So the only difference when q is a root of unity is that q is replaced by q_m in the asymptotic for $D(m/n)$. We may rewrite (24) in a slightly more explicit form: As $m \to \infty$,

$$D(m/n) = a_m^n \left\{ \prod_{\substack{1 \leq j < n \\ q^j \neq 1}} (1 - q^j)^{n-j} \right\} \left\{ \prod_{\substack{1 \leq j < n \\ q^j = 1}} (-c_1 j/m)^{n-j} \right\} (1 + o(1)). \tag{26}$$

There are many sequences of smooth coefficients that do not satisfy (23), but have more complicated asymptotic behaviour, for example

$$a_j = 1 \Big/ \prod_{k=0}^{j} \log(k+2) \tag{27}$$

or

$$a_j = (\log j)^{-j\theta} \tag{28}$$

where $\theta \in \mathbb{R}$. For both these sequences,

$$q_j = h\left(\frac{1}{j}, \frac{1}{\log j}\right)$$

where h is a bivariate function analytic at $(0,0)$ with $h(0,0) = 1$. The following theorem [17] deals with both these examples with $l = 2$ and $\psi_1(x) = x; \psi_2(x) = \log x$. For an l-tuple of non-negative integers $J = (j_1, j_2, \dots j_l)$, we set

$$\Psi^J := \psi_1^{j_1} \psi_2^{j_2} \dots \psi_l^{j_l}.$$

Theorem 6 *Let f of (1) be a formal power series that is not a rational function with $a_j \neq 0$, j large enough. Let l be a positive integer, and $\psi_j(x), j = 1, 2, \dots l$ be complex valued functions defined for large enough x, such that*

(I)

$$\lim_{x \to \infty} |\psi_j(x)| = \infty;$$

(II) For any l tuple of integers J, either $\Psi^J(x)$ is identically constant for large x, or

$$\lim_{x \to \infty} \left|\Psi^J(x)\right| = 0 \text{ or } \infty;$$

(III) There exist for $k = 1, 2, \dots l$ functions $g_{k\pm}$ of l complex variables, analytic near $(0, 0, \dots, 0)$ such that for large enough x,

$$\frac{1}{\psi_k(x \pm 1)} = g_{k\pm}\left(\frac{1}{\psi_1(x)}, \frac{1}{\psi_2(x)}, \dots, \frac{1}{\psi_l(x)}\right).$$

(IV) There is a function h of l complex variables, analytic near $(0, 0, \dots, 0)$ with $q := h(0, 0, \dots, 0) \neq 0$ and for large enough j,

$$a_{j-1}a_{j+1}/a_j^2 = h\left(\frac{1}{\psi_1(j)}, \frac{1}{\psi_2(j)}, \dots, \frac{1}{\psi_l(j)}\right).$$

Let $n \geq 1$. Then there exists an l-tuple J of non-positive integers, and $C \neq 0$ such that

$$\lim_{m\to\infty} D(m/n)/\{a_m^n \Psi^J(m)\} = C.$$

(V) *If moreover, there exist $\gamma_k \neq 0$ such that*

$$\lim_{x\to\infty} \psi_k(x+1)/\psi_k(x) = \gamma_k,\ k = 1,2,\ldots,l$$

then there exists a polynomial $B_n^*(u)$ of degree $\leq n$ with $B_n^*(0) = 1$ and

$$\lim_{m\to\infty} Q_{mn}(ua_m/a_{m+1}) = B_n^*(u), \tag{29}$$

uniformly in compact subsets of $\mathbb{C}$. In the special case where all $\gamma_k = 1$, we may take $B_n^*(u) = B_n(u)$.

The proof of this proceeds along much the same lines as that of Theorem 5. Note that we require analyticity of h rather than just a multivariate asymptotic expansion. One result which does weaken the analyticity condition was proved in [14]. It involves forward differences

$$\Delta q_m := q_{m+1} - q_m;$$

$$\Delta^l q_m = \Delta(\Delta^{l-1} q_m),\ l \geq 2.$$

Theorem 7 *Let f be a formal power series with $a_j \neq 0$ for j large enough, and let (14) hold for some root of unity. Assume that there is a non-decreasing sequence of numbers $\{\lambda_m\}_{m=1}^{\infty}$ such that for each $\beta \geq 0$,*

$$\lim_{m\to\infty} \lambda_m^\beta \left|q_m - q\right| = 0.$$

Assume moreover that for each $p, l \geq 0$, there exists $C > 0$ and m_0 such that for $m \geq m_0$ and $i \geq -p$,

$$\left|\Delta^l (q - q_{m+i})\right| \leq C \left|q - q_m\right|^{l+1} \lambda_m^{l+1}.$$

Assume also that $|q_m| < 1$ for m large enough. Then the conclusions (24) and (25) are valid.

The choice $\lambda_m = \log m$ in this last result allows treatment of both examples (27) and (28). We have concentrated on asymptotic behaviour of the Padé denominators Q_{mn} as $m \to \infty$. Now we may turn to the asymptotic behaviour of $f - [m/n]$:

Theorem 8 *Let f of (1) be entire with $a_j \neq 0$, j large enough satisfying (14). Assume either that q is not a root of unity or that q is a root of unity, and the hypotheses of Theorem 5 or 6 or 7 hold. Then*

$$\lim_{m\to\infty} \{f - [m/n]\}(z) \Big/ \left\{ (-1)^n \frac{D(m+1/n+1)}{D(m/n)} z^{m+n+1} \right\} = 1 \tag{30}$$

uniformly in compact subsets of $\mathbb{C}$.

Sketch of Proof The Padé condition (3) can be put into the more precise form

$$(fQ_{mn} - P_{mn})(z) = (-1)^n \frac{D(m+1/n+1)}{D(m/n)} z^{m+n+1} + \sum_{j=m+n+2}^{\infty} e_j z^j \tag{31}$$

where the $\{e_j\}$ may also be expressed as a determinant of an $(n+1) \times (n+1)$ matrix divided by $D(m/n)$. (This is easily derived by taking the determinantal expression for Q_{mn} and its analogue for P_{mn} and forming $fQ_{mn} - P_{mn}$). The e_j are estimated by a similar sort of analysis to that for $D(m+1/n+1)$ or $D(m/n)$ and are found to be much smaller. The details appear in [14].□

Note that on the circle $|z| = r$, $|f - [m/n]|(z)$ is more or less constant:

$$|f - [m/n]|(z) = \left| \frac{D(m+1/n+1)}{D(m/n)} \right| r^{m+n+1} (1 + o(1)).$$

This means that as z winds around $|z| = r$ once, the error function $(f - [m/n])(z)$ winds around the circle $|u| = \left| \frac{D(m+1/n+1)}{D(m/n)} \right| r^{m+n+1}$ with approximately the same magnitude $m+n+1$ times. This is a so called "near circularity" phenomenon [15], [23]. It implies that $[m/n]$ behaves almost as well as a best uniform approximant $R_{mn}(z) = R_{mn}(z; r)$ of type (m, n) to f on $|z| \leq r$. The latter is a rational function with numerator, denominator of degree $\leq m, n$ respectively, that satisfies

$$\| f - R_{mn} \|_{L_\infty(|z|=r)} = \min_{\substack{\deg(P)\leq m, \\ \deg(Q)\leq n, Q\neq 0}} \| f - P/Q \|_{L_\infty(|z|=r)} =: E_{mn}(f; r).$$

The first entire function for which very precise asymptotics were given for E_{mn} is the exponential function, treated by E.B. Saff [21], [22] and later by Braess [3], Trefethen [23]. There are a few other functions that had previously been treated, such as $\sin z$ or $\cos z$, by Dzadyk and Filosof [6], [7]. A.L. Levin and the author [14] used Theorem 8 to greatly extend the scope of previous work:

Theorem 9 *Let f of (1) be entire with $a_j \neq 0$, j large enough, satisfying (14). Assume either that q is not a root of unity or that q is a root of unity, and the hypotheses of Theorem 5 or 6 or 7 hold. Let $r > 0$. Then*

(I)

$$\lim_{m\to\infty} E_{mn}(f;r) \Big/ \left\{ \left| a_{m+1} \left(\frac{a_{m+1}}{a_m} \right)^n \prod_{j=1}^{n} (1 - q_m^j) \right| r^{m+n+1} \right\} = 1. \tag{32}$$

(II) Let $R_{mn}(z;r)$ be the best rational approximant to f of type (m,n) on $|z| \leq r$. Then

$$\lim_{m\to\infty} R_{mn}(z;r) = f(z)$$

uniformly in compact subsets of $\mathbb{C}$, and uniformly in compact subsets of $|z| > r$,

$$\lim_{m\to\infty} \{f(z) - R_{mn}(z;r)\} \Big/ \left\{ (-1)^n \frac{D(m+1/n+1)}{D(m/n)} z^{m+n+1} \right\} = 1. \tag{33}$$

4 Diagonals

The reader will observe that the dimension n of the determinant $D(m/n)$ was kept constant throughout the previous section. The sensitivity of determinants to perturbation grows with their size, and so we should expect somewhat less general results if we want to estimate $D(n/n)$ as $n \to \infty$, for example. Certainly, the simple approach of letting $m \to \infty$ in individual entries in (12) is no longer applicable.

There is one tool that can be applied successfully: diagonal dominance. Recall that a matrix is diagonally dominant if each diagonal entry exceeds the sum of the absolute values of the off-diagonal entries in its row. Diagonal dominance guarantees invertibility, but far more can be said. For our purposes, some inequalities of Ostrowski, Oeder and Price (see for example, the delightful short paper [20]) were very useful: Let

$$F := (f_{jk})_{j,k=1}^{n}$$

be an n by n matrix with complex entries, non-zero diagonal elements and define the normalized row sums

$$\sigma_j := \left[\sum_{\substack{k=1\\k\neq j}}^{n} |f_{jk}| \right] \Big/ |f_{jj}|,\ j = 1,2,\ldots n. \tag{34}$$

Assume that

$$\sigma := \max_{1 \le j \le n} \sigma_j < 1. \tag{35}$$

Then

$$\prod_{j=1}^{n} |f_{jj}| (1 - \sigma_j) \le |\det(F)| \le \prod_{j=1}^{n} |f_{jj}| (1 + \sigma_j). \tag{36}$$

Moreover, if we introduce the sums to the left and right of the diagonal

$$\lambda_j := \sum_{k=1}^{j-1} |f_{jk}|\,; \; \rho_j := \sum_{k=j+1}^{n} |f_{jk}|$$

then

$$|f_{nn}| \prod_{j=1}^{n-1} (|f_{jj}| - \sigma\rho_j) \le |\det(F)| \le |f_{nn}| \prod_{j=1}^{n-1} (|f_{jj}| + \sigma\rho_j)\,. \tag{37}$$

Similar inequalities hold with ρ_j replaced by λ_j or with combinations of them [20]. Still more impressive are Ostrowski's estimates for the elements $(F^{-1})_{jk}$ of the inverse of F: For $1 \le j \le n$,

$$(F^{-1})_{jj} = \frac{1}{f_{jj}(1 + \theta_j \sigma_j \sigma)} \tag{38}$$

where $|\theta_j| \le 1$; and for $1 \le k \le n, k \ne j$,

$$\left|(F^{-1})_{jk}\right| \le \sigma_j \left|(F^{-1})_{jj}\right|. \tag{39}$$

At first sight the quantitative estimate (37) leads one to expect that one could replace σ_j in (35) by ρ_j or λ_j, but this is not possible, and it is really (35) that is the most stringent condition in applying diagonal dominance. By applying these considerations to the matrix $(b_{k-j})_{j,k=1}^{n}$, we obtain:

Theorem 10 *Let f of (1) have $a_j \ne 0$, $j = 0, 1, 2, \ldots$. Further assume that*

$$\left|a_{j-1}a_{j+1}/a_j^2\right| \le \rho_0^2, \; j = 1, 2, 3, \ldots \tag{40}$$

where $\rho_0 = 0.4559\ldots$ is the positive root of the equation

$$2\sum_{j=1}^{\infty} \rho^{j^2} = 1. \tag{41}$$

Then the Padé table of f is normal, that is

$$D(m/n) \ne 0, \; m, n = 0, 1, 2, 3, \ldots . \tag{42}$$

Moreover, given any sequence of non-negative integers $\{n_m\}_{m=1}^{\infty}$, *we have*

$$\lim_{m\to\infty}[m/n_m](z) = f(z) \tag{43}$$

uniformly in compact subsets of $\mathbb{C}$. *If further*

$$\lim_{m\to\infty} n_m \rho_0^{2m} = 0, \tag{44}$$

then uniformly in compact subsets of $\mathbb{C}$,

$$\lim_{m\to\infty} Q_{mn_m}(z) = 1; \ \lim_{m\to\infty} P_{mn_m}(z) = f(z).$$

In particular, this is true for the diagonal sequence $\{[m/m]\}_{m=1}^{\infty}$.

Sketch of Proof Consider the determinant in (12),

$$D(m/n)/a_m^n = \det\,(b_{k-j})_{j,k=1}^n$$

where for the given m, n, the $\{b_l\}$ are defined by (13). Our condition (40) ensures that

$$|q_j| \le \rho_0^2,\ j = 1, 2, 3, \dots .$$

Then from (13), for $l = \pm 1, \pm 2, \pm 3, \dots$,

$$|b_l| \le (\rho_0^2)^{|l|/2+(|l|-1)+(|l|-2)+\dots+1} = \rho_0^{l^2}.$$

Then the sum of the off-diagonal elements in any row of $(b_{k-j})_{j,k=1}^n$ is at most

$$\sum_{\substack{k=-n\\k\neq 0}}^{n} |b_k| < 2\sum_{k=1}^{n} \rho_0^{j^2} < 1.$$

Since $b_0 = 1$, the matrix $(b_{k-j})_{j,k=1}^n$ is diagonally dominant, and moreover (37) gives

$$\left(1 - \sum_{j=1}^{n} \rho_0^{j^2}\right)^{n-1} \le |D(m/n)/a_m^n| \le \left(1 + \sum_{j=1}^{n} \rho_0^{j^2}\right)^{n-1}.$$

If we apply similar considerations to the matrix in (19), we see that for $\left|q_m^{-1/2}u\right| \le \frac{1}{4}$,

$$\left(1 - \sum_{j=1}^{n} \rho_0^{j^2}\right)^{n} \le |Q_{mn}(ua_m/a_{m+1})D(m/n)/a_m^n| \le \left(1 + \sum_{j=1}^{n} \rho_0^{j^2}\right)^{n}.$$

We deduce that $Q_{mn}(v)$ has no zeros in

$$|v| \leq \left| \frac{a_m}{a_{m+1}} \right| |q_m|^{1/2} /4.$$

Since

$$a_m / a_{m+1} \to \infty$$

rapidly, (in fact far more rapidly than q_m may decay to 0, if that is the case) one deduces that given $r > 0$, Q_{mn} has for large m and uniformly in $n \geq 1$, no zeros in $|z| \leq r$. The rest of the proof follows fairly standard lines, see [16].□

It was noted in [16] that the constant ρ_0 is best possible: For each $\rho > \rho_0$, there exists $\rho_1 \in (\rho_0, \rho)$, an entire function f with

$$\left| a_{j-1} a_{j+1} / a_j^2 \right| = \rho_1^2 \; \forall j$$

and an integer n such that $D(m/n) = 0$ for infinitely many m, while $\{[m/n]\}_{m=1}^{\infty}$ has 0 as a limit point of poles, and hence does not converge uniformly in any neighbourhood of 0. The proof of this given in [16] contained an oversight, but it was repaired in Section 6 of [17].

The functions satisfying (40) have

$$\limsup_{j \to \infty} |a_j|^{1/j^2} \leq \rho_0$$

and so are a limited class of entire functions of order 0. If we replace the condition $|q_j| \leq \rho_0^2$ by an asymptotic one, we can allow slightly more rapid growth of the coefficients:

Theorem 11 *Let f of (1) have $a_j \neq 0, j$ large enough, and*

$$\lim_{j \to \infty} q_j = q : \; |q| < 1. \tag{45}$$

(I) Given $0 < \varepsilon < 1$, there exists m_0 such that for $m \geq m_0$ and $n \geq 1$,

$$(1-\varepsilon)^n \leq \left| D(m/n) \Big/ \left\{ a_m^n \prod_{j=1}^{n-1} (1 - q_m^j)^{n-j} \right\} \right| \leq (1+\varepsilon)^n. \tag{46}$$

(II) Moreover, there exists $C > 0$ such that for any sequence $\{n_m\}_{m=1}^{\infty}$ of positive integers with

$$\lim_{m \to \infty} \left\{ n_m \max_{|j| \leq C\sqrt{\log n_m}} |q_{m+j}/q_m - 1| \right\} = 0, \tag{47}$$

we have

$$\lim_{m\to\infty} D(m/n_m) \Big/ \left\{ a_m^{n_m} \prod_{j=1}^{n_m-1} (1-q_m^j)^{n_m-j} \right\} = 1. \tag{48}$$

(III) For any sequence of non-negative integers $\{n_m\}_{m=1}^{\infty}$, *we have*

$$\lim_{m\to\infty} [m/n_m](z) = f(z)$$

uniformly in compact subsets of $\mathbb{C}$. *In particular this is true of* $\{[m/m]\}_{m=1}^{\infty}$.

Sketch of Proof We use the formula

$$D(m/n)/a_m^n = \det(B)$$

where

$$B := (b_{k-j})_{j,k=1}^{n}$$

and the b_l are given by (13). Define

$$A_n(q) := (q^{(k-j)^2/2})_{j,k=1}^{n}.$$

The basic idea is to show that $BA_n(q_m)^{-1}$ is diagonally dominant, and more precisely, is approximately the identity matrix, so that $\det(B)$ is approximately $\det\left(A_n(q_m)\right)$. We present a few more of the details below:

Step 1: Estimates for $B - A_n(q_m)$
Define for $m, j \geq 0$.

$$\eta_{m,j} := q_{m+j}/q_m - 1$$

For $l > 0$, (13) shows that

$$\begin{aligned} b_l/q_m^{l^2/2} &= \prod_{j=1}^{l-1} \left(\frac{q_{m+j}}{q_m}\right)^{l-j} = \prod_{j=1}^{l-1} (1+\eta_{m,j})^{l-j} \\ &= \exp\left(\sum_{j=1}^{l-1} (l-j)\log(1+\eta_{m,j})\right). \end{aligned}$$

Using the inequalities

$$|e^u - 1| \leq 2|u|; \ |\log(1+u)| \leq 2|u|, \ |u| \leq \frac{1}{2},$$

we see that

$$\left|b_l/q_m^{l^2/2}-1\right|\leq 2l^2\max_{|j|<l}|\eta_{m,j}|,$$

provided the quantity on the right-hand side does not exceed $\frac{1}{2}$. The same inequality holds for $l<0$, and we deduce that

$$\left|\left(B-A_n(q_m)\right)_{j,k}\right|\leq 2(k-j)^2\,|q_m|^{(k-j)^2/2}\max_{|s|<|k-j|}|\eta_{m,s}|. \tag{49}$$

at least for $|j-k|\leq p$ and $m\geq m_0(p)$, any fixed positive integer p. For $|j-k|>p$, one uses cruder estimates.

Step 2: Estimates for $A_n(q_m)^{-1}$

It is well known that the inverse of a Vandermonde matrix is given in terms of Lagrange interpolation polynomials at the points appearing in the Vandermonde matrix. Since $A_n(q)$ is apart from some multiplications by diagonal matrices, a Vandermonde matrix, (recall the proof of Theorem 3) we can explicitly compute $A_n(q)^{-1}$. Let

$$H_n(x):=\prod_{j=1}^{n}(1-q^jx)$$

and let $[P]_j$ denote the coefficient of x^j in a polynomial P. Then [18,p.326]

$$\left(A_n(q)^{-1}\right)_{kl}=\frac{(-1)^{l-1}q^{(l-k)^2/2}\left[H_n(x)/(1-q^lx)\right]_{k-1}}{H_{l-1}(1)H_{n-l}(1)}.$$

Using this formula one can show that

$$\left|\left(A_n(q)^{-1}\right)_{kl}\right|\leq S(q)\,|q|^{|l-k|/2}$$

where

$$S(q):=2\prod_{j=1}^{\infty}\left\{\frac{1+|q|^j}{1-|q|^j}\right\}^2.$$

Since $q_m\to q$, we deduce that given $r>|q|$, we have uniformly in $m\geq m_1$, $n\geq 1$,

$$\left|\left(A_n(q_m)^{-1}\right)_{kl}\right|\leq S(r)\,|r|^{|l-k|/2}. \tag{50}$$

Step 3: Apply Diagonal Dominance

Let

$$F:=\left(B-A_n(q_m)\right)A_n(q_m)^{-1}=BA_n(q_m)^{-1}-I.$$

Using the bounds (49), (50), and straightforward matrix multiplication, one obtains that

$$\sum_{k=1}^{n} |(F)_{jk}| \leq \varepsilon_m < \frac{1}{2}$$

uniformly for $m \geq m_2, n \geq 1$. Here $\varepsilon_m \to 0, m \to \infty$. Then diagonal dominance gives

$$(1-\varepsilon_m)^{n-1} \leq |\det(I+F)| \leq (1+\varepsilon_m)^{n-1}.$$

Since

$$\begin{aligned}\det(I+F) &= \det\left(BA_n(q_m)^{-1}\right) \\ &= \det(B)/\det\left(A_n(q_m)\right) = D(m/n)\Big/\left\{a_m^n \prod_{j=1}^{n-1}(1-q_m^j)^{n-1}\right\}\end{aligned}$$

we obtain (46) and the remaining parts follow similarly.□

As was the case for rows, one can deduce the rates of decay of $E_{mn}(f;r)$ [14]. However, we emphasize that the details are quite technical, so we shall only state:

Theorem 12 *Let f satisfy either the hypotheses of Theorem 10 with ρ_0 in (40) replaced by any $\rho < \rho_0$; or let f satisfy the hypotheses of Theorem 11. Let $\{n_m\}_{m=1}^{\infty}$ be an infinite sequence of positive integers such that for some $B > 0$,*

$$n_m \leq Bm, m \text{ large enough.}$$

Under the hypotheses of Theorem 10, we need $B = 1.4$. Let $r > 0$. Then

(I)

$$\lim_{m\to\infty} E_{m,n_m}(f;r)\Big/\left\{\left|\frac{D(m+1/n_m+1)}{D(m/n_m)}\right| r^{m+n_m+1}\right\} = 1; \qquad (51)$$

(II) Uniformly in compact subsets of $|z| > r$,

$$\lim_{m\to\infty} \left(f(z) - R_{mn_m}(z;r)\right)\Big/\left\{\frac{(-1)^{n_m} D(m+1/n_m+1)}{D(m/n_m)} z^{m+n_m+1}\right\} = 1$$

and the same is true if we replace $R_{mn_m}(z;r)$ by $[m/n_m](z)$. Moreover, $\{R_{m,n_m}\}_{m=1}^{\infty}$ and $\{[m/n_m]\}_{m=1}^{\infty}$ converge uniformly in compact subsets of $\mathbb{C}$ to f.

When f satisfies the hypotheses of Theorem 11 and also (47) holds, we note that

$$\frac{D(m+1/n_m+1)}{D(m/n_m)} = a_{m+1}\left(\frac{a_{m+1}}{a_m}\right)^{n_m}\prod_{j=1}^{\infty}(1-q^j)\big(1+o(1)\big),\ m\to\infty.$$

An example to which Theorems 10, 11 may be applied is

$$f(z) = \sum_{j=0}^{\infty}(j!)^{\beta}q^{j^2}z^j$$

where $\beta \in \mathbb{C}$, |ıı| < ⊮.

5 Further results

A natural generalisation of Padé approximants are Hermite-Padé approximants. Let

$$f(z) = \sum_{j=0}^{\infty}a_j z^j;\ g(z) = \sum_{j=0}^{\infty}b_j z^j$$

be formal power series, and let (l, m, n) be non-negative integers whose sum is even. The vector $(P_1/Q, P_2/Q)$ of rational functions is called a simultaneous rational approximant of type (l, m, n) to (f, g) if

$$\deg(P_1) \le l;\ \deg(P_2) \le m;\ \deg(Q) \le n$$

and

$$(fQ - P_1)(z) = O\left(z^{1+\frac{1}{2}(l+m+n)}\right); \tag{52}$$

$$(gQ - P_2)(z) = O\left(z^{1+\frac{1}{2}(l+m+n)}\right).$$

One can show that if

$$r := (n+m-l)/2;\ s := (n+l-m)/2, \tag{53}$$

then

$$Q(z) = Q_{lmn}(z) = \frac{1}{D(l,m,n)}\det\begin{bmatrix} 1 & z & z^2 & \cdots & z^n \\ a_{l+1} & a_l & a_{l-1} & \cdots & a_{l-n+1} \\ \vdots & \vdots & \vdots & & \vdots \\ a_{l+r} & a_{l+r-1} & a_{l+r-2} & \cdots & a_{l-s} \\ b_{m+1} & b_m & b_{m-1} & \cdots & b_{m-n+1} \\ \vdots & \vdots & \vdots & & \vdots \\ b_{m+s} & b_{m+s-1} & b_{m+s-2} & \cdots & b_{m-r} \end{bmatrix},$$

where

$$D(l,m,n) := \det \begin{bmatrix} a_l & a_{l-1} & a_{l-2} & \cdots & a_{l-n+1} \\ \vdots & \vdots & \vdots & & \vdots \\ a_{l+r-1} & a_{l+r-2} & a_{l+r-3} & \cdots & a_{l-s} \\ b_m & b_{m-1} & b_{m-1} & \cdots & b_{m-n+1} \\ \vdots & \vdots & \vdots & & \vdots \\ b_{m+s} & b_{m+s-1} & b_{m+s-2} & \cdots & b_{m-r} \end{bmatrix}.$$

The reader may find it strange that we require $l+m+n$ even so that the exponent in (52) is integral. It is partly for this reason that simultaneous approximants are normally defined with l, m, n replaced by $\sigma-\rho_0, \sigma-\rho_1, \sigma-\rho_2$ respectively, for suitable integers ρ_j, σ. We chose the above formulation to facilitate statement of the following theorem, proved by K.A. Driver [5]:

Theorem 13 *Assume that the power series f, g have $a_j b_j \neq 0$ for large enough j, and assume that (14) holds with q not a root of unity. Assume moreover that*

$$\lim_{j\to\infty} \frac{b_j}{b_{j+1}} \Big/ \frac{a_j}{a_{j+1}} = \lambda$$

where $\lambda q \neq 0$ and $\lambda \neq 1$. Let k be an integer and n be a non-negative integer with $n+k$ even. Then with $r = r(l), s = s(l)$ given by (53), and with $m = m(l) = l+k$

$$\lim_{l\to\infty} D(l,m,n)/\{a_l^r b_{m-r}^s\}$$

$$= \left[\prod_{j=1}^{r-1}(1-q^j)^{r-j} \prod_{j=1}^{s-1}(1-q^j)^{s-j}\right] \left[\prod_{j=1}^{r}\prod_{i=1}^{s}(1-\lambda^{-1}q^{k+i-j})\right].$$

Moreover, uniformly in compact subsets of $\mathbb{C}$,

$$\lim_{l\to\infty} Q_{lmn}(ua_l/a_{l+1}) = W(u)$$

where W is a polynomial of degree $\leq n$ with $W(0) = 1$.

Explicit expressions were given for W in [5]. The case where q is a root of unity was also treated, under the assumption that q_j has an asymptotic expansion in negative powers of j, with the same true of

$$\lambda_j := \frac{b_j}{b_{j+1}} \Big/ \frac{a_j}{a_{j+1}}.$$

In the 1930's, Szegö studied the distribution of zeros of the partial sums of the Maclaurin series of e^z. He proved that after a suitable normalization, these zeros approach what is now called the Szegö curve. There have been numerous extensions of this to other classes of entire functions [10] and also to the zeros of the Padé numerators P_{mn}, with n fixed, $m \to \infty$ [8], [9]. R. Kovacheva and E.B. Saff [12] have undertaken such a study for entire functions satisfying (14). Subsequently Kovacheva [11] studied the zeros of $f - [m/n]$ under (14). We mention here just a small sample of the results in [11], [12]:

Theorem 14 *Let f of (1) have $a_j \neq 0$, j large enough, and let the hypotheses of Theorem 5 hold with $q = 1$. Let the coefficient c_1 in the asymptotic expansion (23) be real and negative. Fix $n \geq 1$. Then*

(I) uniformly in compact subsets of $|u| > 1$,

$$\lim_{m\to\infty} \frac{P_{mn}(ua_m/a_{m+1})}{(ua_m/a_{m+1})^m D(m/n+1)/D(m/n)} = \frac{u^{n+1}}{(u-1)^{n+1}}. \tag{54}$$

(II) Given $0 < \varepsilon < 1$, $[m/n](z)$ has for large enough m, no zeros in $|z| > |a_m/a_{m+1}|(1+\varepsilon)$ and at least one zero in $|a_m/a_{m+1}|(1-\varepsilon) < |z| < |a_m/a_{m+1}|(1+\varepsilon)$. Moreover, 1 is a limit point of zeros of $P_{mn}(ua_m/a_{m+1}), m \to \infty$.

The case where q is not a root of unity is also treated in [11], [12].

References

[1] G.A. Baker, "Essentials of Padé Approximants", Academic Press, New York, 1975.

[2] G.A. Baker and P.R. Graves-Morris, "Padé Approximants", 2nd Edition, Encyclopaedia of Mathematics and Applications, Vol. 59, Cambridge University Press, Cambridge, 1996.

[3] D. Braess, On the Conjecture of Meinardus to Rational Approximation to e^x, II, J. Approx. Theory, 40(1984), 375-379.

[4] V.I. Buslaev, A.A. Goncar, S.P. Suetin, On the Convergence of Subsequences of the mth Row of the Padé Table, Math. USSR Sbornik, 48(1984), 535-540.

[5] K.A. Driver, Simultaneous Rational Approximants for a Pair of Functions with Smooth Maclaurin Series Coefficients, J. Approx. Theory, 83(1995), 308-329.

[6] V.K. Dzadyk, On the Asymptotics of Diagonal Padé Approximants of the Functions $\sin z, \cos z, \sinh z$ and $\cosh z$, Math. USSR Sbornik, 36(1980), 231-249.

[7] V.K.Dzadyk and L.I.Filozof, The Rate of Convergence of Padé Approximants for Some Elementary Functions, Math. USSR. Sbornik, 35(1979), 615-629.

[8] A. Edrei, Angular Distribution of the Zeros of Padé Polynomials, J. Approx. Theory, 24(1978), 251-265.

[9] A. Edrei, The Padé Tables of Entire Functions, J. Approx. Theory, 28(1980), 54-82.

[10] A. Edrei, E.B. Saff, R.S. Varga, "Zeros of Sections of Power Series", Springer Lecture Notes in Maths., Vol. 1002, Springer, Berlin, 1983.

[11] R.K. Kovacheva, Zeros of Padé Error Functions for Functions with Smooth Maclaurin Coefficients, to appear.

[12] R.K. Kovacheva, E.B. Saff, Zeros of Padé Approximants for Entire Functions with Smooth Maclaurin Coefficients, J. Approx. Theory, 79(1994), 347-384.

[13] A.L.Levin and D.S.Lubinsky, Best Rational Approximants of Entire Functions whose Maclaurin Series Coefficients Decrease Rapidly and Smoothly, Trans. Amer. Math. Soc., 93(1986), 533-545.

[14] A.L. Levin and D.S. Lubinsky, Rows and Diagonals of the Walsh Array for Entire Functions with Smooth Maclaurin Series Coeficients, Constr. Approx, 6(1990), 257-286.

[15] A.L. Levin, V.M. Tikhomirov, About One Theorem of Erohin, Russian Math. Surveys, 23(1968), 122-135.

[16] D.S. Lubinsky, Padé Tables of Entire Functions of Very Slow and Smooth Growth, Constr. Approx., 1(1985), 349-358.

[17] D.S. Lubinsky, Uniform Convergence of Rows of the Padé Table for Functions with Smooth Maclaurin Series Coefficients, Constr. Approx., 3(1987), 307-330.

[18] D.S. Lubinsky, Padé Tables of Entire Functions of Very Slow and Smooth Growth II, Constr. Approx., 4(1988), 321-339.

[19] D.S. Lubinsky and E.B. Saff, Convergence of Padé Approximants of the Partial Theta Functions and the Rogers-Szegö Polynomials, 3(1987), 331-361.

[20] A.M. Ostrowski, Note on Bounds for Determinants with Dominant Principal Diagonal, Proc. Amer. Math. Soc., 3(1952), 26-30.

[21] E.B. Saff, The Convergence of Rational Functions of Best Approximation to the Exponential Function II, Proc. Amer. Math. Soc., 32(1972), 187-194.

[22] E.B. Saff, On the Degree of Best Rational Approximation to the Exponential Function, J. Approx. Theory, 9(1973), 97-101.

[23] L.N. Trefethen, The Asymptotic Accuracy of Rational Best Approximations to e^z on a Disk, J. Approx. Theory, 40(1980), 380-383.

[24] H. Wallin, The Convergence of Padé Approximants and the Size of the Power Series Coefficients, Applicable Anal., 4(1974), 235-251.

On Marcinkiewicz-Zygmund-Type Inequalities

H. N. Mhaskar *
Department of Mathematics,
California State University, Los Angeles, California, 90032

J. Prestin
FB Mathematatik
Universität Rostock, Universitätsplatz 1, 18051 Rostock, Germany

Abstract

We investigate the relationships between the Marcinkiewicz-Zygmund-type inequalities and certain shifted average operators. Applications to the mean boundedness of a quasi-interpolatory operator in the case of trigonometric polynomials, Jacobi polynomials, and Freud polynomials are presented.

1 Introduction

The Marcinkiewicz-Zygmund inequalities assert the following ([26], Theorem X.7.5). If $n \geq 1$ is an integer, $1 < p < \infty$, and S is a trigonometric polynomial of order at most n, then

$$\int_{-\pi}^{\pi} |S(t)|^p dt \leq \frac{c_1}{2n+1} \sum_{j=1}^{2n+1} \left| S\left(\frac{2\pi j}{2n+1}\right)\right|^p \leq c_2 \int_{-\pi}^{\pi} |S(t)|^p dt, \tag{1}$$

where c_1 and c_2 are positive constants depending only on p, but not on S or n. An important application of these inequalities is to deduce the mean boundedness of the operator of trigonometric polynomial interpolation from that of the partial sum operator of the trigonometric Fourier series ([26], Theorem X.7.14). There is a very close connection between the existence of such inequalities, the boundedness of interpolatory operators, and that

*This research was supported, in part, by National Science Foundation Grant DMS 9404513, the Air Force Office of Scientific Research, and the Alexander von Humboldt Foundation.

of the partial sum operators (cf. [3], [4].) Similar inequalities have been studied by many authors in connection with orthogonal polynomial expansions and Lagrange interpolation at the zeros of orthogonal polynomials, both in the case of generalized Jacobi weights on $[-1,1]$ and the Freud-type and more general weights on the whole real axis ([10], [24], [25]). In the context of spline spaces, analogous questions have been studied in [18]. The mean convergence of interpolatory processes have been studied also by Varma and his collaborators. For example, Erdös-Feldheim-type results are obtained in [22] and [23].

This theory is usually not applicable in the "extreme cases"; i.e., in the case when one is interested in either the uniform convergence or convergence in the L^1 norm. It is well known that Lagrange interpolation polynomials, and in general also the Fourier operators are not convergent either uniformly or in the L^1 sense. In the case of trigonometric series, a useful substitute for the partial sums of the Fourier series is the shifted arithmetic mean $(1/n)\sum_{k=n+1}^{2n} s_k$ where s_k is the operator of taking the k-th partial sum of the Fourier expansion. These operators are linear, and provide a "near best" order of approximation both in the uniform and L^1 sense. They have many other interesting properties as well (cf. [8]). These operators have been studied also in more general situations ([14] and references therein). In the context of interpolation at the zeros of generalized Jacobi and Freud-type weight functions, we have studied certain quasi-interpolatory operators in [15].

In this paper, our main purpose is to investigate the connections between these operators and inequalities similar to (1). We formulate a few abstract results in the next section. These abstract results will be illustrated in the context of trigonometric polynomials in Section 3, generalized Jacobi polynomials in Section 4, and Freud polynomials in Section 5.

2 Abstract results

Let μ be a (positive) measure on a measure space Σ. For a μ-measurable function f, we write

$$\|f\|_{p,\mu} := \begin{cases} \left(\int |f|^p d\mu\right)^{1/p}, & \text{if } 1 \le p < \infty, \\ \mu - \text{ess sup}\,|f|, & \text{if } p = \infty. \end{cases}$$

The class L^p_μ consists of all functions f for which $\|f\|_{p,\mu} < \infty$, where two functions are considered equal if they are equal μ-almost everywhere.

We may interprete the inequalities (1) as the statement

$$\|S\|_{p,\mu} \le c_1 \|S\|_{p,\nu} \le c_2 \|S\|_{p,\mu},$$

where μ is the Lebesgue measure and ν is a discrete measure. The partial sum operator for the Fourier series and the Lagrange interpolation operator at the jump points of ν can be expressed in the form $\int f(t)D_n(\cdot - t)d\mu(t)$ and $\int f(t)D_n(\cdot - t)d\nu(t)$ respectively, where D_n is the Dirichlet kernel. Similar expressions hold also for orthogonal expansions and Lagrange interpolation polynomials at the zeros of orthogonal polynomials.

Motivated by these examples, we consider two σ-finite measures μ and ν on a measure space Σ, and corresponding operators (when defined)

$$T(\tau; f, x) := \int f(t)K(x,t)d\tau(t), \qquad \tau = \mu, \nu,$$

where K is a symmetric function. We assume that K is essentially bounded as well as integrable with respect to all the product measures $\mu \times \mu$, $\nu \times \nu$, and $\mu \times \nu$. We also consider two weight functions, w and W, which are both measurable and positive almost everywhere with respect to both μ and ν. We will adopt the convention that the symbols $c, c_1, c_2, \ldots$ will denote positive constants depending only on μ and other fixed constants of the problem under consideration, such as the norms and other explicitly indicated quantities. Their values may be different at different occurrences, even within a single formula. In this context, ν is not considered to be a fixed parameter. We are intentionally vague regarding the weight functions and the norms. In applications, it will be clearly specified if the constants depend upon these paramters or not.

Let $1 \leq p \leq \infty$. By condition $I(p)$, we denote the statement

$$\|w(wW)^{-1/p}T(\nu; f)\|_{p,\mu} \leq c\|W(wW)^{-1/p}f\|_{p,\nu}, \quad f \in L^1_\nu.$$

By condition $O(p)$, we denote the statement

$$\|w(wW)^{-1/p}T(\mu; f)\|_{p,\nu} \leq c\|W(wW)^{-1/p}f\|_{p,\mu}, \quad f \in L^1_\mu,$$

where we note that the constants in both the above inequalities depend upon μ, and not upon ν. We are also interested in the following variation of the condition $O(p)$, to be denoted by $C(p)$:

$$\|w(wW)^{-1/p}T(\mu; f)\|_{p,\mu} \leq c\|W(wW)^{-1/p}f\|_{p,\mu}, \quad f \in L^1_\mu.$$

In typical applications, ν will be a discrete measure, and μ will be a continuous measure. The condition $O(p)$ then appears to be weaker than both $C(p)$ and $I(p)$, as we are estimating only a few values of the continuous operator in terms of the integrals over the whole domain of μ. Nevertheless, the following theorem shows a close connection between these conditions.

Theorem 1 *Let μ, ν be σ-finite measures on a measure space Σ, $1 \le p \le q \le r \le \infty$, and p' be the conjugate index for p. Further suppose that w and W^{-1} are in both L^1_μ and L^1_ν.*
(a) *Let $f, g : \Sigma \to I\!R$, and $f(x)g(t)K(x,t)$ be integrable with respect to the product measures $\mu(t) \times \nu(x)$. Then the following* reciprocity law *holds.*

$$\int fT(\mu; g)d\nu = \int T(\nu; f)gd\mu.$$

(b) *The conditions $I(p)$ and $O(p')$ are equivalent. The conditions $C(p)$ and $C(p')$ are equivalent.*
(c) *The conditions $I(p)$ and $I(r)$ together imply $I(q)$. The conditions $O(p)$ and $O(r)$ together imply $O(q)$. The conditions $C(p)$ and $C(r)$ together imply $C(q)$.*

Proof. By Fubini's theorem and the fact that $K(x,t) = K(t,x)$, we have

$$\begin{aligned}\int f(x)T(\mu; g, x)d\nu(x) &= \int f(x) \int g(t)K(x,t)d\mu(t)d\nu(x) \\ &= \int g(t) \int f(x)K(x,t)d\nu(x)d\mu(t) \\ &= \int g(t)T(\nu; f, t)d\mu(t).\end{aligned}$$

This proves part (a).

To verify part (b), we first prove that $O(p')$ implies $I(p)$. Let $g : \Sigma \to I\!R$ be an arbitrary, μ-measurable, simple function such that $\|g\|_{p',\mu} \le 1$. An application of Hölder's inequality,

$$\|w(wW)^{-1/p}\|_{1,\mu} \le \|w^{1/p'}\|_{p',\mu}\|W^{-1/p}\|_{p,\mu} = \|w\|_{1,\mu}^{1/p'}\|W^{-1}\|_{1,\mu}^{1/p},$$

shows that $w(wW)^{-1/p} \in L^1_\mu$. Since g is μ-essentially bounded, the function $T(\mu; w(wW)^{-1/p}g)$ is well defined, and the conditions of part (a) are also satisfied. Then using the reciprocity law and the facts that $W^{-1}(wW)^{1/p} = w(wW)^{-1/p'}$, $O(p')$, and $w(wW)^{-1/p} = W^{-1}(wW)^{1/p'}$, we obtain

$$\begin{aligned}\left|\int w(wW)^{-1/p}T(\nu; f)gd\mu\right| &= \left|\int fT(\mu; w(wW)^{-1/p}g)d\nu\right| \\ &\le \|W(wW)^{-1/p}f\|_{p,\nu}\|w(wW)^{-1/p'}T(\mu; w(wW)^{-1/p}g)\|_{p',\nu} \\ &\le c\|W(wW)^{-1/p}f\|_{p,\nu}\|W(wW)^{-1/p'}w(wW)^{-1/p}g\|_{p',\mu} \\ &\le c\|W(wW)^{-1/p}f\|_{p,\nu}.\end{aligned}$$

Since

$$\|w(wW)^{-1/p}T(\nu;f)\|_{p,\mu} = \sup\left|\int w(wW)^{-1/p}T(\nu;f)gd\mu\right|,$$

where the supremum is taken over all μ-measurable simple functions g with $\|g\|_{p',\mu} \le 1$, we have proved $I(p)$. Similarly, the condition $I(p)$ implies $O(p')$. Finally, we observe that when $\mu = \nu$, the conditions $I(p)$, $C(p)$ and $O(p)$ are the same. This completes the proof of part (b).

If $I(p)$ and $I(r)$ both hold, then the Stein-Weiss interpolation theorem ([2], Corollary 5.5.2) implies $I(q)$. The other assertions are proved in the same way.

In order to relate Theorem 1 with the Marcinkiewicz-Zygmund-type (M-Z) inequalities, we restrict ourselves to the case when $w = W$. We consider an increasing sequence of sets $\{\Pi_k\}$, $\Pi_k \subseteq \Pi_{k+1}$, $k = 0, 1, \ldots$, which may be thought of as subsets of each of the spaces $L^1_\mu \cap L^\infty_\mu$ and $L^1_\nu \cap L^\infty_\nu$. We next consider a sequence of symmetric kernel functions K_k, and operators T_k defined (if possible) by

$$T_k(\tau; f, x) := \int f(t)K_k(x,t)d\tau(t), \qquad \tau = \mu, \nu, \quad k = 1, 2, \ldots.$$

In applications, the measure ν will also depend upon n, but we do not need this fact in the current abstract formulation of our results. We assume that there exist integers $a \ge 1$ and b such that

$$T_k(\tau; f) \in \Pi_{ak+b}, \quad \tau = \mu, \nu, \quad k = 1, 2, \ldots.$$

Sometimes, we also assume the following *reproducing property,* referred to as R_n:

$$T_k(\tau; P) = P, \quad P \in \Pi_k, \quad \tau = \mu, \nu, \quad k = 1, 2, \ldots, n.$$

In applications, we will choose ν, depending upon n, so that the condition R_n will be satisfied. The condition $I_n(p)$ (respectively $O_n(p)$, $C_n(p)$) denotes the fact that each of the operators T_k, $1 \le k \le n$, satisfies the condition $I(p)$ (respectively $O(p)$, $C(p)$). Clearly, the conditions $O_n(p)$ (with $w = W$) and the reproducing property R_n imply the "simpler" M-Z inequality

$$\|w^{(p-2)/p}P\|_{p,\nu} \le c\|w^{(p-2)/p}P\|_{p,\mu}, \qquad P \in \Pi_n .$$

This condition will be referred to as $SMZ_n(p)$. By the full M-Z inequality, to be denoted by $MZ_n(p)$, we mean the estimates

$$\|w^{(p-2)/p}P\|_{p,\mu} \le c_1\|w^{(p-2)/p}P\|_{p,\nu} \le c_2\|w^{(p-2)/p}P\|_{p,\mu}, \qquad P \in \Pi_n .$$

The following theorem gives several connections between the various conditions defined above.

Theorem 2 *Let μ, ν be as in Theorem 1, $n \geq 1$ be an integer, $w = W$, and w, w^{-1} be in both L^1_μ and L^1_ν.*
(a) *Let R_n hold and let $1 \leq p \leq 2 \leq r \leq \infty$. The conditions $O_n(p)$, $O_n(r)$ (or $I_n(p)$, $I_n(r)$) together imply the full M-Z inequalities $MZ_n(q)$ for each q with $\max(p, r') \leq q \leq \min(p', r)$ where p' be the conjugate index of p.*
(b) *Let $1 \leq p \leq \infty$, and $SMZ_{an+b}(p')$ hold. The condition $C_n(p)$ implies $I_n(p)$.*
(c) *Let p, p' be as above, and $MZ_{an+b}(p')$ hold. The condition $I_n(p)$ implies $C_n(p)$.*

Proof. In view of Theorem 1(c), the conditions $O_n(p)$ and $O_n(r)$ together imply each of the conditions $O_n(s)$, $p \leq s \leq r$. Theorem 1(b) then yields the conditions $I_n(s')$, $r' \leq s' \leq p'$. Thus, each of the conditions $O_n(q)$ and $I_n(q)$ hold for $\max(p, r') \leq q \leq \min(p', r)$. The part (a) then follows easily from the reproduction property R_n.

Next, we prove part (b). Let $1 \leq k \leq n$ be an integer, and $C_n(p)$ hold. Theorem 1(b) then implies $C_n(p')$. Let $w_{p'} := w^{(p'-2)/p'}$. Since $T_k(\mu; f) \in \Pi_{an+b}$ for all f for which it is defined, the conditions $SMZ_{an+b}(p')$ and $C_n(p')$ together imply

$$||w_{p'}T_k(\mu; f)||_{p',\nu} \leq c||w_{p'}T_k(\mu; f)||_{p',\mu} \leq c||w_{p'}f||_{p',\mu},$$

which is the condition $O_n(p')$. In turn, this implies $I_n(p)$.

Finally, we prove part (c). The condition $I_n(p)$ implies $O_n(p')$. Together with $MZ_{an+b}(p')$, this gives for integer $1 \leq k \leq n$

$$||w_{p'}T_k(\mu; f)||_{p',\mu} \leq c||w_{p'}T_k(\mu; f)||_{p',\nu} \leq c||w_{p'}f||_{p',\mu}.$$

Thus, $C_n(p')$ holds, and therefore, $C_n(p)$ holds.

3 Trigonometric approximation

In this section, let μ be the Lebesgue measure on $[-\pi, \pi]$, normalized to be 1. All functions will be assumed to be 2π-periodic. If $f \in L^1_\mu$, and $n \geq 1$ is an integer, the n-th partial sum of its Fourier series is given by

$$s^*_n(f, x) = \int_{-\pi}^{\pi} f(t)D_n(x - t)d\mu(t),$$

where the *Dirichlet kernel* is defined by

$$D_n(t) := \frac{\sin((n + 1/2)t)}{\sin(t/2)}.$$

Let $\ell \geq 0$ be an integer. The *de la Vallée Poussin* operators are defined by

$$P^*_{n,\ell}(f,x) := \int_{-\pi}^{\pi} f(t)V^*_{n,\ell}(x,t)d\mu(t),$$

where the kernel function $V^*_{n,\ell}$ is defined by

$$V^*_{n,\ell}(x,t) := \frac{1}{\ell+1}\sum_{k=n}^{n+\ell} D_k(x-t).$$

Let $m \geq 1$ be an integer, $-\pi < x_{1,m} < \cdots < x_{2m+1,m} \leq \pi$ be any distinct points, and ν_m be the measure that associates the mass $1/(2m+1)$ with each of these points. As an application of our results in Section 2, we study the convergence behavior of the discretized de la Vallée Poussin operators, defined by

$$L^*_{n,\ell,m}(f,x) = \int_{-\pi}^{\pi} f(t)V^*_{n,\ell}(x,t)d\nu_m(t).$$

The class of all trigonometric polynomials of order at most n will be denoted by $\mathbb{H}_n$. In the notation of Section 2, $\Pi_n = \mathbb{H}_n$, $K_n = V^*_{n,\ell}$, $T_n(\mu; f) = P^*_{n,\ell}(f)$, $T_n(\nu_m; f) = L^*_{n,\ell,m}(f)$, $a = 1$, $b = \ell$, and $w = W = 1$. For $1 \leq p < \infty$, let $\mathcal{R}_p$ denote the class of all 2π-periodic, Riemann integrable functions, and let $\mathcal{R}_\infty$ denote the class of all continuous 2π-periodic functions. We wish to estimate $\|f - L^*_{n,\ell,m}(f)\|_{p,\mu}$ for all $f \in \mathcal{R}_p$. This estimate will be given in terms of the degree of best one-sided trigonometric approximation, defined for $f \in \mathcal{R}_p$ by

$$\tilde{E}_{n,p}(f) := \inf\{\|Q - R\|_{p,\mu}\}$$

where the infimum is over all $Q, R \in \mathbb{H}_n$ for which $Q(x) \leq f(x) \leq R(x)$ for all $x \in [-\pi, \pi]$. There are several known Jackson-type estimates for $\tilde{E}_{n,p}(f)$ (cf. [19]). These imply that $\tilde{E}_{n,p}(f) \to 0$ as $n \to \infty$ for every $f \in \mathcal{R}_p$.

Theorem 3 *Let $n \leq \ell \leq \kappa n$, $m \geq n + \ell$, and*

$$\min_{1\leq k\leq 2m+1} |x_{k+1,m} - x_{k,m}| \geq c/m, \tag{2}$$

where $x_{2m+2,m} := x_{1,m}$. Then for any $f \in \mathcal{R}_p$,

$$\|L^*_{n,\ell,m}(f)\|_{p,\mu} \leq \begin{cases} c(\kappa)\|f\|_{p,\nu_m}, & \text{if } \kappa > 1,\ 1 \leq p \leq \infty, \\ c(p)\|f\|_{p,\nu_m}, & \text{if } 1 < p < \infty. \end{cases} \tag{3}$$

Further, suppose that the nodes $\{x_{k,m}\}$ are chosen so that $L^*_{n,\ell,m}(T) = T$ for every $T \in \mathbb{H}_n$. If $f \in \mathcal{R}_p$, we have

$$\|f - L^*_{n,\ell,m}(f)\|_{p,\mu} \leq \begin{cases} c(\kappa)\tilde{E}_{n,p}(f), & \text{if } \kappa > 1,\ 1 \leq p \leq \infty, \\ c(p)\tilde{E}_{n,p}(f), & \text{if } 1 < p < \infty. \end{cases}$$

Proof. It is known [10] that the condition (2) implies $SMZ_{n+\ell}(p)$ for every p, $1 \le p \le \infty$. It is also known ([8], [21]) that for any $f \in L^p_\mu$,

$$\|P^*_{n,\ell}(f)\|_{p,\mu} \le \begin{cases} c(\kappa)\|f\|_{p,\mu}, & \text{if } \kappa > 1,\ 1 \le p \le \infty, \\ c(p)\|f\|_{p,\mu}, & \text{if } 1 < p < \infty. \end{cases} \tag{4}$$

Thus, the condition $C_n(p)$ is satisfied for all p, $1 < p < \infty$, and even for $p = 1, \infty$ if $\kappa > 1$. Since $SMZ_{n+\ell}(p')$ holds as well, Theorem 2(b) implies $I_n(p)$ for the same values of p. This is (3).

Next, let $1 < p < \infty$, $f \in \mathcal{R}_p$, $Q, R \in I\!H_n$ be chosen so that $Q(x) \le f(x) \le R(x)$ for all $x \in [-\pi, \pi]$ and $\|Q - R\|_{p,\mu} \le 2\tilde{E}_{n,p}(f)$. Using the reproduction property, the bounds (3), and the condition $SMZ_{n+\ell}(p)$, we deduce that

$$\begin{aligned} \|f - L^*_{n,\ell,m}(f)\|_{p,\mu} &= \|f - Q - L^*_{n,\ell,m}(f - Q)\|_{p,\mu} \\ &\le \|f - Q\|_{p,\mu} + c(p)\|f - Q\|_{p,\nu_m} \\ &\le \|R - Q\|_{p,\mu} + c(p)\|R - Q\|_{p,\nu_m} \\ &\le c_1(p)\|R - Q\|_{p,\mu} \le c_2(p)\tilde{E}_{n,p}(f). \end{aligned}$$

The remaining assertion is proved in the same way.

We end this section with a few remarks comparing the quantity $\tilde{E}_{n,p}(f)$ with the degree of best approximation of f, defined by

$$E_{n,p}(f) := \inf_{Q \in I\!H_n} \|Q - f\|_{p,\mu}.$$

In view of (4), one easily concludes that

$$\|f - P^*_{n,\ell}(f)\|_{p,\mu} \le cE_{n,p}(f).$$

The integral operator $P^*_{n,\ell}$ neglects the behavior of its argument f on a set of measure zero, whereas the estimates on the discrete operator $L^*_{n,\ell,m}$ depend upon countably many points through the best **one-sided** approximation (cf. also [6]). However, if f is of bounded variation on $[-\pi, \pi]$, and possesses at least one jump, then it is known [19] that

$$cn^{-1/p} \le E_{n,p}(f) \le \tilde{E}_{n,p}(f) \le c_1 n^{-1/p}.$$

Our results imply in this case that

$$cn^{-1/p} \le \|f - L^*_{n,\ell,m}(f)\|_{p,\mu} \le cn^{-1/p}.$$

If $f \notin \mathcal{R}_p$, then the sequence $L^*_{n,\ell,m}(f)$, even if well defined, may not converge.

4 Generalized Jacobi polynomials

A *generalized Jacobi weight* is a function of the form

$$w(x) := \begin{cases} \prod_{k=1}^{\rho} (x - \xi_k)^{\beta_k}, & x \in [-1,1], \\ 0, & \text{otherwise}, \end{cases} \tag{5}$$

where $\rho \geq 2$ is an integer, $-1 =: \xi_\rho < \cdots < \xi_1 := 1$, and $\beta_k > -1$ for $k = 1, \ldots, \rho$. The class of generalized Jacobi weights will be denoted by GJ. In this and the next section, $\mathcal{P}_n$ denotes the class of all polynomials of degree at most n. Associated with a weight $w \in GJ$, there is a system of orthonormalized polynomials $\{p_k \in \mathcal{P}_k\}$, called generalized Jacobi polynomials. For each integer $m \geq 1$, the polynomial p_m has m distinct zeros $\{x_{k,m}\}$ on the interval $[-1,1]$. Further, there exist positive numbers $\lambda_{k,m}$ such that

$$\int_{-1}^{1} P(t)w(t)dt = \sum_{k=1}^{m} \lambda_{k,m} P(x_{k,m}), \qquad P \in \mathcal{P}_{2m-1}.$$

Generalized Jacobi polynomials have been studied extensively by Nevai [17]. Following [17], if $w \in GJ$ is of the form (5), and $m \geq 1$ is an integer, we write for $x \in [-1,1]$,

$$\overline{w}_m(x) := \left(\sqrt{1-x} + \frac{1}{m}\right)^{2\beta_1+1} \prod_{k=2}^{\rho-1} \left(|x - \xi_k| + \frac{1}{m}\right)^{\beta_k} \left(\sqrt{1+x} + \frac{1}{m}\right)^{2\beta_\rho+1}.$$

If w is the Legendre weight, i.e., $\beta_k = 0$, $k = 1, \ldots, \rho$, then it is easy to see that

$$c_1 \overline{w}_m(x) \leq \Delta_m(x) := \sqrt{1-x^2} + \frac{1}{m} \leq c_2 \overline{w}_m(x).$$

In the rest of this section, the constants $c, c_1, \ldots$ may depend upon w.

We let μ the measure $w(t)dt$, and ν_m be the measure that associates the mass $\lambda_{k,m}$ with the zero $x_{k,m}$. The de la Vallée Poussin kernel is defined by

$$V_n(x,t) := \sum_{k=0}^{n} p_k(x)p_k(t) + \sum_{k=n+1}^{2n-1} \left(2 - \frac{k}{n}\right) p_k(x)p_k(t).$$

The de la Vallée Poussin-type operator is defined by

$$v_n(f,x) := \int_{-1}^{1} f(t)V_n(x,t)w(t)dt,$$

and its discrete analogue is defined for $f : [-1,1] \to I\!R$ by

$$\tau_{n,m}(f,x) := \int_{-1}^{1} f(t)V_n(x,t)d\nu_m(t).$$

If $m \geq 2n$, the operators satisfy $\tau_{n,m}(P) = v_n(P) = P$ for every $P \in \mathcal{P}_n$.

In the language of Section 2, $T(\mu; f) = v_n(f)$, $T(\nu_m; f) = \tau_{n,m}(f)$, $a = 2$, $b = -1$. We let $\gamma \in (-1,1)$, and take $g_n := \Delta^{\gamma}_{\sqrt{n}}\sqrt{w_n}$ in place of the weight w of Section 2. In place of the weight W of Section 2, we take $G(x) := (1-x^2)^{\gamma/2}\sqrt{w(x)}$, where $w \in GJ$ is the fixed weight function considered in this section. It is easy to verify that g_n and G^{-1} are both in L^1_μ. In [15], we have proved that if $2n \leq m \leq Ln$ for some constant L, then

$$\begin{aligned} \|g_n v_n(f)\|_{\infty,\mu} &\leq c\|Gf\|_{\infty,\mu}, \\ \|g_n \tau_{n,m}(f)\|_{\infty,\mu} &\leq c\|Gf\|_{\infty,\nu_m}, \end{aligned}$$

where the constants may depend upon w, γ, and L only. Thus, we have proved $C(\infty)$ (and hence, $O(\infty)$) and $I(\infty)$ in the language of Section 2. Since the weight functions are not the same on both sides of these inequalities, we may not apply Theorem 2 directly in this context. We may apply Theorem 1. Together with the reproducing property R_n, this gives the following theorem.

Theorem 4 *Let $w \in GJ$, $-1 < \gamma < 1$, $1 \leq p \leq \infty$, $L \geq 2$, $n \geq 1$, $2n \leq m \leq Ln$ be integers. Let*

$$g_n := \Delta^{\gamma}_{\sqrt{n}}\sqrt{w_n}, \qquad G(x) := (1-x^2)^{\gamma/2}\sqrt{w(x)}.$$

Then

$$\begin{aligned} \|g_n(g_nG)^{-1/p}v_n(f)\|_{p,\mu} &\leq c\|G(g_nG)^{-1/p}f\|_{p,\mu}, \\ \|g_n(g_nG)^{-1/p}\tau_{n,m}(f)\|_{p,\mu} &\leq c\|G(g_nG)^{-1/p}f\|_{p,\nu_m}, \\ \|g_n(g_nG)^{-1/p}v_n(f)\|_{p,\nu_m} &\leq c\|G(g_nG)^{-1/p}f\|_{p,\mu}. \end{aligned}$$

For any polynomial $P \in \mathcal{P}_n$,

$$\|g_n(g_nG)^{-1/p}P\|_{p,\nu_m} \leq c\|G(g_nG)^{-1/p}P\|_{p,\mu}\ , \tag{6}$$

$$\|g_n(g_nG)^{-1/p}P\|_{p,\mu} \leq c\|G(g_nG)^{-1/p}P\|_{p,\nu_m}\ . \tag{7}$$

In the same manner as described for the trigonometric polynomial case, the approximation error for $v_n(f)$ and $\tau_{n,m}(f)$ could be bounded by a weighted best one-sided approximation. The convergence order of such weighted best one-sided approximation has been investigated in e.g. [5], [16], [20], [13].

In (6), (7), the number of nodes at which the polynomials are evaluated is greater than the degree of the polynomials, and we do not have a complete equivalence required by the condition $MZ_n(p)$. The conditions $SMZ_n(p)$ are proved in [17] and [10] with different discrete measures ν_n, where the number of nodes is less than the degree of the polynomials involved. In light of the results of Y. Xu [24] regarding the mean boundedness of the Fourier sums and of Kallaev [7] regarding the uniform boundedness of the de la Vallée Poussin operators for certain ultraspherical polynomials, results analogous to Theorem 3 can also be obtained. However, since this does not add any new insight to the problem, we do not present the details. In conclusion, we note that Y. Xu [24], [25] has studied the conditions $MZ_n(p)$, again with different discrete measures. These results require more stringent conditions on the weights involved than those in Theorem 4. Moreover, these do not apply in the extreme cases $p = 1, \infty$.

5 Freud-type weight functions

Let $w : \mathbb{R} \to (0, \infty)$, and $Q := \log(1/w)$. The function w is called a Freud-type weight function if each of the following conditions is satisfied. The function Q is an even, convex function on $\mathbb{R}$, Q is twice continuously differentiable on $(0, \infty)$ and there are constants c_1 and c_2 such that

$$0 < c_1 \le \frac{xQ''(x)}{Q'(x)} \le c_2 < \infty, \qquad 0 < x < \infty .$$

The most commonly discussed examples include $\exp(-|x|^\alpha)$, $\alpha > 1$. Associated with the Freud-type weight function w, there is a system of numbers $\{a_x\}$, called MRS-numbers, defined for $x > 0$ by the equations

$$x = \frac{2}{\pi} \int_0^1 \frac{a_x t Q'(a_x t)}{\sqrt{1-t^2}} dt .$$

One of the most important properties of a_x is the following: For every integer $n \ge 1$ and $P \in \mathcal{P}_n$,

$$\max_{x \in \mathbb{R}} |P(x)w(x)| = \max_{|x| \le a_n} |P(x)w(x)| .$$

For a detailed discussion of approximation theory involving these weight functions, we refer the reader to [14].

In the remainder of this section, w will denote a fixed Freud-type weight function, the measure μ will be the measure $w^2(x)dx$ on $\mathbb{R}$, $\{p_k\}$ will denote the sequence of polynomials orthonormal on $\mathbb{R}$ with respect to the measure

μ. For each integer $n \geq 1$, the polynomial p_n has n real and simple zeros, $\{x_{k,n}\}$. We have the following quadrature formula for all $P \in \mathcal{P}_{2n-1}$:

$$\int_{\mathbb{R}} P(t)d\mu(t) = \sum_{k=1}^{n} \lambda_{k,n} P(x_{k,n}),$$

where the *Cotes' numbers* $\lambda_{k,n}$ are all positive. The measure ν_n associates the mass $\lambda_{k,n}$ with the point $x_{k,n}$. In the remainder of this section, all constants $c, c_1, \ldots$ will depend upon w.

The de la Vallée Poussin kernel for the Freud-type weight function is defined by

$$V_n(x,t) := \sum_{k=0}^{n} p_k(x)p_k(t) + \sum_{k=n+1}^{2n-1} \left(2 - \frac{k}{n}\right) p_k(x)p_k(t).$$

If $1 \leq p \leq \infty$, the de la Vallée Poussin operator is defined for $w^{(p-2)/p} f \in L^p_\mu$ by

$$v_n(f,x) := \int_{\mathbb{R}} f(t)V_n(x,t)d\mu(t),$$

and its discretized version is defined for all $f : \mathbb{R} \to \mathbb{R}$ by

$$\tau_{n,m}(f,x) := \int_{\mathbb{R}} f(t)V_n(x,t)d\nu_m(t).$$

In the notation of Section 2, $\Pi_n = \mathcal{P}_n$, $K_n = V_n$, $T_n(\mu; f) = v_n(f)$, $T_n(\nu_m; f) = \tau_{n,m}(f)$, the weights w and W are both equal to the fixed Freud-type weight function w, $a = 2$, and $b = -1$. It is not difficult to verify that both w and w^{-1} are in L^1_μ. It is known [14] that the conditions $C_n(p)$ are satisfied for all p, $1 \leq p \leq \infty$. Moreover, if $m \geq 2n$ then the reproducing property R_n also holds.

Theorem 5 *Let $1 \leq p \leq \infty$. Let w be a Freud-type weight function, and $Q := \log(1/w)$ satisfy the following Lipschitz condition:*

$$|Q'(a_y \cos\theta) - Q'(a_y \cos\phi)| \leq c\frac{y}{a_y}|\theta - \phi|^\lambda, \qquad y > 0,$$

where c and λ are positive constants independent of y. Let $L, \delta > 0$, $n \geq 1$ be an integer, and $(2+\delta)n \leq m \leq Ln$. Then for each $P \in \mathcal{P}_n$,

$$\|w^{(p-2)/p}P\|_{p,\mu} \leq c_1\|w^{(p-2)/p}P\|_{p,\nu_m} \leq c_2\|w^{(p-2)/p}P\|_{p,\mu}. \tag{8}$$

Further, for any $f : \mathbb{R} \to \mathbb{R}$,

$$\|w^{(p-2)/p}\tau_{n,m}(f)\|_{p,\mu} \leq c\|w^{(p-2)/p}f\|_{p,\nu_m}. \tag{9}$$

Proof. It is proved in [15] that (9) is satisfied if $p = \infty$. In the language of Section 2, this means that the condition $I_n(\infty)$ is satisfied. Clearly, the condition $C_n(\infty)$, which is known to hold [14], implies the condition $O_n(\infty)$. Theorem 1 then implies that the condition $I_n(1)$ holds as well, and hence, all the conditions $I_n(p)$ are satisfied; i.e., (9) holds. Since the reproduction property R_n also holds, Theorem 2 implies all the conditions $MZ_n(p)$; i.e., the estimates (8).

The conditions $MZ_n(p)$ in this context have been studied by Lubinsky and Matjila [9], [11] with a discrete measure depending upon less nodes than the degree of the polynomials involved. Again, they do not hold for $p = 1, \infty$. Lubinsky and Matjila [12] have also studied the mean convergence of Lagrange interpolation operators for the Freud weights. Typically, these results use very deep techniques. One sided approximation by polynomials is also studied in this context by Nevai (see [16], [14] for further references). In particular, we may obtain analogues of Theorem 3 in this case as well, but we do not find this worth pursuing at this time.

References

[1] V. M. Badkov, Convergence in the mean and almost everywhere of Fourier series in orthogonal polynomials, Mat. Sbornik 35 (1974), 229-262.

[2] J. Bergh and J. Löfström, "Interpolation spaces, an introduction", Springer Verlag, Berlin, 1976.

[3] C. K. Chui, X. C. Shen and L. Zhong, On Lagrange interpolation at disturbed roots of unity, Trans. Amer. Math. Soc. 336 (1993), 817-830.

[4] C. K. Chui, and L. Zhong, Polynomial interpolation and Marcinkiewicz-Zygmund inequalities on the unit circle, CAT Report 365 (1995).

[5] G. Freud, Über einseitige Approximation durch Polynome, I, Acta Sci. Math. (Szeged) 6 (1955), 12-28.

[6] V. H. Hristov, Best onesided approximations and mean approximations by interpolating polynomials of periodic functions, Math.-Balkanica (N. S.) 3 no. 3-4 (1989), 418-429.

[7] S. O. Kallaev, De la Vallée Poussin means of Fourier-Gegenbauer series, Mat. Zametki 7 (1970), 19-30.

[8] G. G. Lorentz, "Approximation of Functions", Holt, Rinehart, and Winston, 1966.

[9] D. S. Lubinsky, Converse quadrature sum inequalities for polynomials with Freud weights, Acta. Sci. Math. Hung. 60 (1995), 527–557.

[10] D. S. Lubinsky, A. Máté and P. Nevai, Quadrature sums involving p^{th} powers of polynomials, SIAM J. Math. Anal. 18 (1987), 531-544.

[11] D. S. Lubinsky and D. Matjila, Full quadrature sums for pth powers of polynomials with Freud weights, J. Comput. Appl. Math. 60 (1995), 285–296.

[12] D. S. Lubinsky and D. Matjila, Necessary and sufficient conditions for mean convergence of Lagrange interpolation for Freud weights, SIAM J. Math. Anal. 26 (1995), 238–262.

[13] G. Mastroianni and P. Vértesi, Weighted L^p error of Lagrange interpolation, J. Approx. Theory 82 (1995), 321-339.

[14] H. N. Mhaskar, "An introduction to the theory of weighted polynomial approximation", World Scientific, Singapore, 1996.

[15] H. N. Mhaskar and J. Prestin, On bounded quasi-interpolatory polynomial operators, Submitted for publication.

[16] P. Nevai, Einseitige Approximation durch Polynome mit Anwendungen, Acta Math. Acad. Sci. Hungar. 23 (1972), 495-506.

[17] P. Nevai, "Orthogonal polynomials", Mem. Amer. Math. Soc. 18 no. 213 (1979).

[18] J. Prestin and E. Quak, On interpolation and best one-sided approximation by splines in L^p, in: "Approximation Theory," (Ed. G. Anastassiou), pp., 409-420, Lecture Notes in Pure and Applied Mathematics Vol. 138, 1992.

[19] B. Sendov and V. A. Popov, "The Averaged Moduli of Smoothness," Wiley, New York, 1988.

[20] M. Stojanova, The best onesided algebraic approximation in $L^p[-1,1]$ $(1 \leq p \leq \infty)$, Math. Balkanica (N.S.) 2 (1988), 101-113.

[21] A. F. Timan, "Theory of approximation of functions of a real variable", English translation Pergamon Press, 1963.

[22] A. K. Varma, P. Vértesi, Some Erdös-Feldheim type theorems on mean convergence of Lagrange interpolation, J. Math. Anal. Appl. 91 (1983), 68-79.

[23] A. K. Varma, J. Prasad, An analogue of a problem of P. Erdös and E. Feldheim on L_p convergence of interpolatory processes, J. Approx. Theory 56 (1989), 225-240.

[24] Y. Xu, Mean convergence of generalized Jacobi series and interpolating polynomials, I, J. Approx. Theory 72 (1993), 237-251.

[25] Y. Xu, Mean convergence of generalized Jacobi series and interpolating polynomials, II, J. Approx. Theory 76 (1994), 77-92.

[26] A. Zygmund, "Trigonometric Series", Cambridge University Press, Cambridge, 1977.

Extremal Problems for Restricted Polynomial Classes in L^r Norm

Gradimir V. Milovanović
Department of Mathematics, University of Niš
Faculty of Electronic Engineering, P.O. Box 73, 18000 Niš, Yugoslavia

Abstract

We consider extremal problems of Markov-Bernstein type in integral norms, especially for restricted polynomial classes, which were introduced and studied by the late Professor Arun K. Varma. Beside some results on extremal problems of Markov and Bernstein type for the non-restricted polynomial class $\mathcal{P}_n$ in L^r norms, with a special attention to the case $r = 2$, we give a short account of L^2 inequalities of Markov type for curved majorants and on Bernstein inequalities in mixed norms. Also, we consider extremal problems for some classes of nonnegative polynomials on $[0, +\infty)$ and $[-1, 1]$ with respect to the generalized Laguerre and Jacobi measure, respectively.

1 Introduction

There are many results on extremal problems and inequalities of Markov-Bernstein type with algebraic polynomials. The first result of Markov type for polynomials of the second degree was connected with some investigations of the well-known Russian chemist Mendeleev [23]. A general case in the class $\mathcal{P}_n$ of all algebraic polynomials of degree at most n was considered by A. A. Markov [21]. Taking the uniform norm $||f||_\infty = \max_{-1\le t\le 1} |f(t)|$ he solved the extremal problem

$$A_n = \sup_{P\in\mathcal{P}_n} \frac{||P'||_\infty}{||P||_\infty}.$$

The best constant is $A_n = n^2$ and the extremal polynomial $P^*(t) = cT_n(t)$, where T_n is the Chebyshev polynomial of the first kind of degree n and c is an arbitrary constant. The best constant can be expressed also as

$A_n = T_n'(1)$. Thus, the classical Markov's inequality can be expressed in the form

$$\|P'\|_\infty \le T_n'(1)\|P\|_\infty \qquad (P \in \mathcal{P}_n).$$

In 1892, younger brother V. A. Markov [22] found the best possible inequality for k-th derivative ($k \le n$),

$$\|P^{(k)}\|_\infty \le T_n^{(k)}(1)\|P\|_\infty \qquad (P \in \mathcal{P}_n),$$

where the extremal polynomial is also T_n when $k \le n$. The best constant can be expressed in the form

$$T_n^{(k)}(1) = \|T_n^{(k)}\|_\infty = \frac{1}{(2k-1)!!}\prod_{i=0}^{k-1}(n^2 - i^2).$$

A version of this remarkable paper in German was published in 1916.

In 1912 Bernstein [2] considered another type of these inequalities taking $\|f\| = \max_{|z|\le 1}|f(z)|$. He proved the inequality

$$\|P'\| \le n\|P\| \qquad (P \in \mathcal{P}_n),$$

with equality if and only if $P(z) = cz^n$ (c is an arbitrary constant).

There are several different forms of this Bernstein's inequality. A standard form of that can be done as

$$|P'(t)| \le \frac{n}{\sqrt{1-t^2}}\|P\|_\infty, \qquad -1 < t < 1. \tag{1}$$

The equality is attained at the points $t = t_\nu = \cos\frac{(2\nu-1)\pi}{2n}$, $\nu = 1, \ldots, n$, if and only if $P(t) = \gamma T_n(t)$, where $|\gamma| = 1$.

Combining the inequalities of Markov and Bernstein we can state the following result:

Theorem 1 *If $P \in \mathcal{P}_n$ then*

$$|P'(t)| \le \min\Big\{n^2, \frac{n}{\sqrt{1-t^2}}\Big\}\|P\|_\infty, \qquad -1 \le t \le 1.$$

Several monographs and papers have been published in this area (cf. Durand [9], Govil [12], [13], Milovanović [25], [26], Milovanović, Mitrinović and Rassias [29], [30], Mohapatra, O'Hara and Rodriguez [32], Rahman and Schmeisser [34], [35]).

In this survey we consider extremal problems of Markov-Bernstein type in integral norms, especially for restricted polynomial classes, which were introduced and studied by the late Professor Arun K. Varma. The paper is organized as follows. In Section 2 we give some definitions and preliminary results on extremal problems of Markov and Bernstein type for the non-restricted polynomial class $\mathcal{P}_n$ in L^r norms, with a special attention to the case $r = 2$. Sections 3 and 4 are devoted to L^2 inequalities of Markov type for curved majorants and Bernstein inequalities in mixed norms, respectively. In Section 5 we consider extremal problems for the class of nonnegative polynomials on $[0, +\infty)$ with respect to the generalized Laguerre measure. Finally, Section 6 is devoted to the corresponding extremal problems for Lorentz classes of nonnegative polynomials on the interval $(-1, 1)$ with respect to the Jacobi measure.

2 Extremal Problems in L^r Norm on $\mathcal{P}_n$

The classical Markov and Bernstein inequalities and corresponding extremal problems were generalized for various domains, various norms and for various subclasses for polynomials, both algebraic and trigonometric (for details see Chapter 6 in [30]).

Let

$$||f||_r = \Big(\int_{\mathbb{R}} |f(t)|^r \, d\lambda(t)\Big)^{1/r}, \quad r \geq 1, \tag{2}$$

where $d\lambda(t)$ is a given nonnegative measure on the real line $\mathbb{R}$, with compact support or otherwise, for which all moments $\mu_k = \int_{\mathbb{R}} t^k \, d\lambda(t)$, $k = 0, 1, \ldots$, exist and are finite and $\mu_0 > 0$. In a special case $r = 2$, (2) becomes

$$||f||_2 = \Big(\int_{\mathbb{R}} |f(t)|^2 \, d\lambda(t)\Big)^{1/2}. \tag{3}$$

In that case we have an inner product defined by

$$(f, g) = \int_{\mathbb{R}} f(t)\overline{g(t)} \, d\lambda(t)$$

such that $||f||_2 = \sqrt{(f, f)}$. Then also, there exists a unique set of (monic) orthogonal polynomials $\pi_k(\cdot) = \pi_k(\,\cdot\,; d\lambda)$, $k \geq 0$, with respect to $(\cdot, \cdot)$, such that

$$\pi_k(t) = t^k + \text{lower degree terms}, \qquad (\pi_k, \pi_m) = ||\pi_k||_2^2 \, \delta_{km},$$

where δ_{km} is Kronecker's delta. In this paper we deal with the measures of the classical orthogonal polynomials $d\lambda(t) = w(t)\, dt$, where the weight functions $t \mapsto w(t)$ satisfy the differential equation

$$\frac{d}{dt}(A(t)w(t)) = B(t)w(t),$$

where

$$A(t) = \begin{cases} 1-t^2, & \text{if } (a,b) = (-1,1), \\ t, & \text{if } (a,b) = (0,+\infty), \\ 1, & \text{if } (a,b) = (-\infty,+\infty), \end{cases} \tag{4}$$

and $B(t)$ is a polynomial of the first degree. For such classical weights we will write $w \in CW$.

Based on this definition, the classical orthogonal polynomials $\{Q_k\}$ on (a,b) are the *Jacobi polynomials* $P_k^{(\alpha,\beta)}(t)$ $(\alpha,\beta > -1)$ on $(-1,1)$, the *generalized Laguerre polynomials* $L_k^s(t)$ $(s > -1)$ on $(0,+\infty)$, and finally as the *Hermite polynomials* $H_k(t)$ on $(-\infty,+\infty)$. The classical orthogonal polynomial $Q_k(t)$ satisfies a second order linear differential equation of hypergeometric type $A(t)y'' + B(t)y' + \lambda_k y = 0$, where λ_k is a constant. The weight functions, the constants λ_k and the corresponding polynomials $B(t)$ are given in Table 1.

TABLE 1

(a,b)	$w(t)$	$B(t)$	λ_k
$(-1,1)$	$(1-t)^\alpha(1+t)^\beta$	$\beta-\alpha-(\alpha+\beta+2)t$	$k(k+\alpha+\beta+1)$
$(0,+\infty)$	$t^s e^{-t}$	$s+1-t$	k
$(-\infty,+\infty)$	e^{-t^2}	$-2t$	$2k$

The first results on extremal problems in the L^2-norm and corresponding Markov's inequalities

$$||P'||_2 \le A_n ||P||_2 \qquad (P \in \mathcal{P}_n), \tag{5}$$

were given by E. Schmidt [36] and Turán [38]:

Theorem 2 *Let* $||\,.\,||_2$ *be defined by* (3). (a) *If* $(a,b) = (-\infty,+\infty)$ *and* $d\lambda(t) = e^{-t^2}\,dt$ *the best constant in* (5) *is given by* $A_n = \sqrt{2n}$. *An extremal polynomial is Hermite's polynomial* H_n.

(b) *Let* $(a,b) = (0,+\infty)$ *and* $d\lambda(t) = e^{-t}\,dt$. *Then* (5) *holds with*

$$A_n = \left(2\sin\frac{\pi}{4n+2}\right)^{-1}.$$

The extremal polynomial is

$$P(t) = \sum_{\nu=1}^{n} \sin\frac{\nu\pi}{2n+1} L_\nu(t),$$

where L_ν *is the Laguerre polynomial.*

Theorem 2 (b), in this form, was formulated by Turán [38].

An important generalization of A. A. Markov's inequality for algebraic polynomials in an integral norm was given by Hille, Szegő, and Tamarkin [17], who proved the following result:

Theorem 3 *Let $r \geq 1$, $(a,b) = (-1,1)$, $P \in \mathcal{P}_n$, and let $\|\,.\,\|_r$ be given by (2). Then*

$$\|P'\|_r \leq An^2\|P\|_r, \tag{6}$$

where the constant $A = A(n,r)$ is given by

$$A(n,r) = 2(r-1)^{1/r-1}\Big(r+\frac{1}{n}\Big)\Big(1+\frac{r}{nr-r+1}\Big)^{n-1+1/r},$$

for $r > 1$, and

$$A(n,1) = 2\Big(1+\frac{1}{n}\Big)^{n+1}.$$

The factor n^2 in (6) cannot be replaced by any function tending to infinity more slowly. Namely, for each n, there exist polynomials $P(t)$ of degree n such that $\|P'\|_r/\|P\|_r \leq Bn^2$, where B is a constant of the same nature as $A = A(n,r)$.

The constant $A(n,r)$ in Theorem 3 is not the best possible. We can see that $A(n,r) \leq 6\exp(1+1/e)$, for every n and $r \geq 1$. Also,

$$A(n,r) \;\to\; \begin{cases} 2(1+1/(n-1))^{n-1} < 2e & (n \ \text{fixed},\ r \to +\infty), \\ 2e & (r = 1,\ n \to +\infty), \\ 2er(r-1)^{(1/r)-1} & (r > 1 \ \text{fixed},\ n \to +\infty). \end{cases}$$

Some improvements of the constant $A(n,r)$ have recently been obtained by Goetgheluck [11]. He found that

$$A(n,1) = \sqrt{\frac{8}{\pi}}\Big(1+\frac{3}{4n}\Big)^2,$$

as well as a very complicated expression for $A(n,r)$ when $r > 1$.

Recently Guessab and Milovanović [15] have considered weighted L^2-analogues of the Bernstein's inequality (see Theorem 1), which can be stated in the following form:

$$\|\sqrt{1-t^2}\,P'(t)\|_\infty \leq n\|P\|_\infty. \tag{7}$$

Using the norm $\|f\|^2 = (f,f)$, with $w \in CW$, they determined the best constant $C_{n,m}(w)$ $(1 \leq m \leq n)$ in the inequality

$$\|A^{m/2}P^{(m)}\| \leq C_{n,m}(w)\|P\|, \tag{8}$$

where A is defined by (4).

Theorem 4 *For all polynomials $P \in \mathcal{P}_n$ the inequality* (8) *holds, with the best constant*

$$C_{n,m}(w) = \sqrt{\lambda_{n,0}\lambda_{n,1}\cdots\lambda_{n,m-1}},$$

where $\lambda_{n,k} = -(n-k)\left(\frac{1}{2}(n+k-1)A''(0) + B'(0)\right)$.

The equality is attained in (8) *if and only if P is a constant multiple of the classical polynomial $Q_n(t)$ orthogonal with respect to the weight function $w \in CW$.*

We note that $\lambda_{n,0} = \lambda_n$, where λ_n is given in Table 1. In some special cases we have:

(1) Let $w(t) = (1-t)^\alpha(1+t)^\beta$ $(\alpha, \beta > -1)$ on $(-1,1)$ (Jacobi case). Then

$$\|(1-t^2)^{m/2}P^{(m)}\| \le \sqrt{\frac{n!\Gamma(n+\alpha+\beta+m+1)}{(n-m)!\Gamma(n+\alpha+\beta+1)}}\,\|P\|,$$

with equality if and only if $P(t) = cP_n^{(\alpha,\beta)}(t)$.

(2) Let $w(t) = t^s e^{-t}$ $(s > -1)$ on $(0, +\infty)$ (generalized Laguerre case). Then

$$\|t^{m/2}P^{(m)}\| \le \sqrt{n!/(n-m)!}\,\|P\|,$$

with equality if and only if $P(t) = cL_n^s(t)$.

(3) The Hermite case with the weight $w(t) = e^{-t^2}$ on $(-\infty, +\infty)$ is the simplest. Then the best constant is $C_{n,m}(w) = 2^{m/2}\sqrt{n!/(n-m)!}$.

In connection with the previous results is also the following characterization of the classical orthogonal polynomials given by Agarwal and Milovanović [1].

Theorem 5 *For all $P \in \mathcal{P}_n$ the inequality*

$$(2\lambda_n + B'(0))\|\sqrt{A}\,P'\|^2 \le \lambda_n^2\|P\|^2 + \|AP''\|^2 \tag{9}$$

holds, with equality if only if $P(t) = cQ_n(t)$, where Q_n is the classical orthogonal polynomial with respect to the weight function $w \in CW$ and c is an arbitrary real constant.

The Hermite case was considered by Varma [46]. Then, the inequality (9) reduces to

$$\|P'\|^2 \le \frac{1}{2(2n-1)}\,\|P''\|^2 + \frac{2n^2}{2n-1}\,\|P\|^2.$$

In the generalized Laguerre case, the inequality (9) becomes

$$\|\sqrt{t}\,P'\|^2 \le \frac{n^2}{2n-1}\,\|P\|^2 + \frac{1}{2n-1}\,\|tP''\|^2,$$

where $w(t) = t^s e^{-t}$ on $(0, +\infty)$.

In the Jacobi case the inequality (9) reduces to the inequality

$$\begin{aligned}\big((2n-1)(\alpha+\beta) &+ 2(n^2+n-1)\big)\|\sqrt{1-t^2}\,P'\|^2 \\ &\le n^2(n+\alpha+\beta+1)^2\|P\|^2 + \|(1-t^2)P''\|^2.\end{aligned}$$

In the simplest case, when $\alpha = \beta = 0$ (Legendre case), we obtain

$$\|\sqrt{1-t^2}\,P'\|^2 \le \frac{n^2(n+1)^2}{2(n^2+n-1)}\,\|P\|^2 + \frac{1}{2(n^2+n-1)}\,\|(1-t^2)P''\|^2.$$

In the Chebyshev case ($\alpha = \beta = -1/2$), we get

$$\|\sqrt{1-t^2}\,P'\|^2 \le \frac{n^4}{2n^2-1}\,\|P\|^2 + \frac{1}{2n^2-1}\,\|(1-t^2)P''\|^2,$$

where $\|f\|^2 = \int_{-1}^{1}(1-t^2)^{-1/2}f(t)^2\,dt$.

The corresponding result for trigonometric polynomials was obtained by Varma [48].

3 L^2 Inequalities of Markov Type for Curved Majorants

Answering to a question of P. Turán[1], Rahman, Pierre and Rahman, Rahman and Schmeisser, Videnskiĭ and others (see Chapter 6 in [30]) gave several inequalities in the uniform norm on $[-1, 1]$. The first who started with the corresponding inequalities in L^2 norm was Professor Varma. In order to present his results, at first, for polynomials $P \in \mathcal{P}_n$ we define

$$\|P\|_* = \sup_{-1<t<1} \frac{|P(t)|}{\sqrt{1-t^2}}, \tag{10}$$

or generally,

$$\|P\|_\varphi = \sup_{-1<t<1} \frac{|P(t)|}{\varphi(t)},$$

[1]Professor Paul Turán asked this question in 1970 at a conference on *Constructive Function Theory* held in Varna, Bulgaria.

where the majorant $t \mapsto \varphi(t)$ is a nonnegative function on $[-1,1]$. Taking a norm $\|\,.\,\|$ for polynomials on $[-1,1]$, Turán's problem can be stated in the form: *If* $\|P\|_* \le 1$, *or* $\|P\|_\varphi \le 1$, *how large can* $\|P^{(m)}\|$ *be?*

In the case of uniform norm, a general majorant

$$\varphi(t) = (1-t)^{\lambda/2}(1+t)^{\mu/2},$$

where λ, μ are nonnegative integers, was used by Pierre and Rahman [33]. Varma [47], [50] and Varma, Mills and Smith [51] considered L^2 inequalities using the majorants

$$\varphi(t) = \sqrt{1-t^2} \qquad \text{(circular majorant)}$$

and

$$\varphi(t) = 1-t^2 \qquad \text{(parabolic majorant)}.$$

For a circular majorant Varma [47] proved:

Theorem 6 *Let* $P \in \mathcal{P}_{n+1}$ $(n \ge 2)$ *and let* $\|\,.\,\|_*$ *be defined by* (10). *If* $\|P\|_* \le 1$ *then*

$$\int_{-1}^{1} (P'(t))^2 \sqrt{1-t^2}\, dt \le \frac{\pi}{4}(n^2+1),$$

with equality if $P(t) = p_0(t) = (1-t^2)U_{n-1}(t)$, $U_k(t) = \sin(k+1)\theta/\sin\theta$, $t = \cos\theta$.

Under the same conditions, Varma [47] also proved the following inequality

$$\int_{-1}^{1} (P'(t))^2\, dt \le \frac{2n^2(2n^2-1)}{4n^2-1} + 2 + 4\sum_{k=1}^{n} \frac{1}{2k-1},$$

which is at least asymptotically best possible.

Recently, Varma [50] proved:

Theorem 7 *Under same conditions as in the previous theorem, we have*

$$\int_{-1}^{1} (P^{(j)}(t))^2 (1-t^2)^{1/2}\, dt \le \int_{-1}^{1} (p_0^{(j)}(t))^2 (1-t^2)^{1/2}\, dt \qquad (j = 2,3)$$

and

$$\int_{-1}^{1} (P'(t))^2 (1-t^2)^{-1/2}\, dt \le \int_{-1}^{1} (p_0'(t))^2 (1-t^2)^{-1/2}\, dt,$$

with equality if and only if $P(t) = (1-t^2)U_{n-1}(t)$.

In the L^2 norm for real algebraic polynomials of degree $n+2$ that have the parabolic majorant

$$|P(t)| \le 1 - t^2, \qquad -1 \le t \le 1, \tag{11}$$

Varma, Mills and Smith [51] proved the following results:

Theorem 8 *If $P \in \mathcal{P}_{n+2}$ $(n \ge 1)$ and (11) is satisfied then*

$$\int_{-1}^{1} (P'(t))^2 \, dt \le \int_{-1}^{1} (q_0'(t))^2 \, dt, \tag{12}$$

where $q_0(t) = \pm(1-t^2)T_n(t)$, $T_n(t) = \cos n\theta$ and $t = \cos\theta$. Further, equality in (12) occurs if and only if $P(t) = q_0(t)$.

Theorem 9 *If $P \in \mathcal{P}_{n+2}$ $(n \ge 1)$ and (11) is satisfied then*

$$\int_{-1}^{1} (P''(t))^2 \, dt \le \int_{-1}^{1} (q_0''(t))^2 \, dt,$$

with equality if and only if $P(t) = q_0(t)$.

Similarly, Varma [50] proved:

Theorem 10 *Let P be any member of the set of those algebraic polynomials of degree $n+2$ which have only real zeros, all of them in the interval $[-1,1]$, and for which (11) is satisfied. Then*

$$\int_{-1}^{1} (P'(t))^2 (1-t^2)^{-1/2} \, dt \le \int_{-1}^{1} (q_0'(t))^2 (1-t^2)^{-1/2} \, dt$$

and

$$\int_{-1}^{1} (P''(t))^2 (1-t^2)^{1/2} \, dt \le \int_{-1}^{1} (q_0''(t))^2 (1-t^2)^{1/2} \, dt,$$

with equalities if and only if $P(t) = q_0(t) = \pm(1-t^2)T_n(t)$.

Theorem 11 *If $P \in \mathcal{P}_{n+2}$ $(n \ge 1)$ and (11) is satisfied then*

$$\int_{-1}^{1} (P'''(t))^2 (1-t^2)^{1/2} \, dt \le \int_{-1}^{1} (q_0'''(t))^2 (1-t^2)^{1/2} \, dt,$$

with equality if and only if $P(t) = q_0(t) = \pm(1-t^2)T_n(t)$.

It is interesting to remark that the last theorem does not require that the zeros of $P(t)$ are real and lie inside $[-1,1]$. Also, we mention here an interesting auxiliary result proved by Varma [50]:

Theorem 12 *Let $Q \in \mathcal{P}_{n-1}$ and let $|Q(t)| \le (1-t^2)^{-1/2}$, $-1 < t < 1$. Then*

$$\int_{-1}^{1} (Q'(t))^2 (1-t^2)^{3/2}\, dt \le \frac{\pi}{2}(n^2-1),$$

with equality if and only if $Q(t) = \pm \sin n\theta / \sin\theta$, $t = \cos\theta$.

4 Bernstein Inequality in Mixed Norms

In order to find L^2 generalizations of Bernstein's inequality (1), i.e., (7), Varma [49] considered the class H_n of all real polynomials of degree n bounded by 1 on the interval $[-1, 1]$ and proved:

Theorem 13 *If $P \in H_n$ then we have*

$$\int_{-1}^{1} (1-t^2)(P'(t))^2\, dt \le n^2\Big(1 + \frac{1}{4n^2-1}\Big) = \int_{-1}^{1} (1-t^2)(T_n'(t))^2\, dt,$$

with equality only for $P(t) = \pm T_n(t)$.

This result can be interpreted in the form (7) using mixed norms $||\,.\,||_\infty$ and $||\,.\,||_2$, defined by (3) with $d\lambda(t) = dt$ on the interval $(-1, 1)$. Thus, for for all $P \in \mathcal{P}_n$ we have

$$||A^{1/2}P'||_2 \le C_n ||P||_\infty, \tag{13}$$

where $A(t) = 1 - t^2$ and

$$C_n = \Big(\int_{-1}^{1} (1-t^2)(T_n'(t))^2\, dt\Big)^{1/2} = \frac{2n^2}{\sqrt{4n^2-1}}.$$

Defining

$$||f||_{2,\alpha} = \Big(\int_{-1}^{1} f(t)^2 (1-t^2)^\alpha\, dt\Big)^{1/2} \qquad (\alpha > -1),$$

we can consider inequalities of the form

$$||P'||_{2,\alpha} \le C_n(\alpha) ||P||_\infty, \tag{14}$$

Using Varma's method, Shen [37] proved that (14) holds for $\alpha = 3/2$ and $\alpha = 1/2$, with best constants

$$C_n(3/2) = \Big(\int_{-1}^{1} (1-t^2)^{3/2}(T_n'(t))^2\, dt\Big)^{1/2} = \frac{n\sqrt{\pi}}{2}$$

and

$$C_n(1/2) = \left(\int_{-1}^{1} (1-t^2)^{1/2} (T_n'(t))^2 \, dt \right)^{1/2} = n\sqrt{\frac{\pi}{2}},$$

respectively. Furthermore, it was proved that (14) holds also for $\alpha = -1/2$ and $\alpha = 0$. Namely, applying the n-point Gauss-Chebyshev quadrature formula to the integral $\|P'\|_{2,-1/2}^2$ ($P \in \mathcal{P}_n$), we have

$$\int_{-1}^{1} (1-t^2)^{-1/2} (P'(t))^2 \, dt = \frac{\pi}{n} \sum_{\nu=1}^{n} (P'(\tau_\nu))^2, \tag{15}$$

where $\tau_\nu = \cos((2\nu - 1)\pi/(2n))$, $\nu = 1, \ldots, n$, are the zeros of the Chebyshev polynomial T_n. By Bernstein's inequality (1), it follows that

$$|P'(\tau_\nu)| \le n(1-\tau_\nu^2)^{-1/2} \|P\|_\infty, \qquad \nu = 1, \ldots, n.$$

Since $T_n'(\tau_\nu) = n(1-\tau_\nu^2)^{-1/2}(-1)^{\nu-1}$, $\nu = 1, \ldots, n$, we have that $|P'(\tau_\nu)| \le |T_n'(\tau_\nu)| \cdot \|P\|_\infty$. Using (15) we obtain

$$\|P'\|_{2,-1/2}^2 = \frac{\pi}{n} \sum_{\nu=1}^{n} |P'(\tau_\nu)|^2 \le \frac{\pi}{n} \left(\sum_{\nu=1}^{n} |T_n'(\tau_\nu)|^2 \right) \|P\|_\infty^2.$$

Since

$$\frac{1}{n^2} \sum_{\nu=1}^{n} |T_n'(\tau_\nu)|^2 = \sum_{\nu=1}^{n} \frac{1}{1-\tau_\nu^2} = \frac{1}{2} \left(\frac{T_n'(1)}{T_n(1)} - \frac{T_n'(-1)}{T_n(-1)} \right) = n^2$$

we conclude that (14) is valid for $\alpha = -1/2$. Here, $C_n(-1/2) = n\sqrt{n\pi}$. A more general result was obtained by Dimitrov [8].

The case $\alpha = 0$ follows from the following theorem (for $r = 2$) proved by Bojanov [3]:

Theorem 14 *Let $P \in \mathcal{P}_n$ and $r \in [1, +\infty)$. Then*

$$\|P'\|_r \le \|T_n'\|_r \|P\|_\infty.$$

Equality is attained only for $P(t) = cT_n(t)$, where c is an arbitrary constant.

Since

$$\int_{-1}^{1} (T_n'(t))^2 \, dt = 2n^2 \sum_{k=1}^{n} \frac{1}{2k-1}$$

we find that

$$C_n(0) = n \sqrt{\sum_{k=1}^{n} \frac{2}{2k-1}}.$$

Recently, Bojanov [4] proved (14) for $|\alpha| \le 1/2$ with the best constant $C_n(\alpha) = \|T_n'\|_{2,\alpha}$. For $\alpha > 0$ we can give an explicit expression for $\|T_n'\|_{2,\alpha}^2$. After much calculation we obtain

$$\|T_n'\|_{2,\alpha}^2 = \frac{n^2\sqrt{\pi}\,\Gamma(\alpha)}{2\Gamma\left(\alpha+\frac{1}{2}\right)}\left(1 - \prod_{\nu=1}^{n}\frac{\nu-\frac{1}{2}-\alpha}{\nu-\frac{1}{2}+\alpha}\right).$$

Bojanov [4] also considered the corresponding problem for higher derivatives. The inequalities with the second derivative were investigated by Varma [49] and [50] (see Dimitrov [8] too).

Theorem 15 *If $P \in H_n$ then we have*

$$\int_{-1}^{1} (P''(t))^2(1-t^2)^{-1/2}\,dt \le \int_{-1}^{1} (T_n''(t))^2(1-t^2)^{-1/2}\,dt.$$

Theorem 16 *If $P \in H_n$ then we have*

$$\int_{-1}^{1} (P''(t))^2(1-t^2)^{3/2}\,dt \le \int_{-1}^{1} (T_n''(t))^2(1-t^2)^{3/2}\,dt = \frac{\pi}{2}\,n^2(n^2-1).$$

In order to prove the last theorem, Varma [50] used the Bernstein inequality (1) and Theorem 12 with $Q(t) = P'(t)/n$.

Theorem 13 can be also interpreted in terms of trigonometric polynomials t_n of degree n with real coefficients such that

$$\|t_n\|_\infty = \max_{0\le\theta\le\pi} |t_n(\theta)| \le 1.$$

Theorem 17 *Let t_n be a trigonometric polynomial of degree n with real coefficients such that $\|t_n\|_\infty \le 1$. Then*

$$\int_0^\pi (t_n'(\theta))^2 \sin\theta\,d\theta \le n^2\left(1 + \frac{1}{4n^2-1}\right),$$

with equality only for $t_n(t) = \pm\cos n\theta$.

Putting $t_n(\theta) = P(\cos\theta)$, this result reduces to Theorem 13.

Recently, Chen [5] investigated the following quantity:

$$\sup_{\|t_n\|_\infty\le 1} \int_0^\pi (t_n^{(k)}(\theta))^2 w(\theta)\,d\theta, \tag{16}$$

where $w(\theta) = \sin^j\theta$ and j is a positive integer. The case $k = 1$ and $j = 2$ was investigated by Shen [37]. The solution of (16) gives us the best constant in (14) for $\alpha = (j+1)/2$.

At the end of this section we mention that there are several inequalities which reverse the sense of those above (see Labelle [19], Lupaş [20], Daugavet and Rafal'son [7], Konjagin [18], etc.).

5 L^2 Inequalities With Generalized Laguerre Measure for Nonnegative Polynomials

Several results on inequalities of Markov and Turán type in L^2 norm on the restricted polynomial classes were obtained by Professor A. K. Varma [40]–[45] and [52].

In 1981 Varma [44] investigated the problem of determining the best constant $C_n(\alpha)$ in the L^2 inequality

$$\|P'\|^2 \leq C_n(\alpha)\|P\|^2, \tag{17}$$

for polynomials with nonnegative coefficients, with respect to the generalized Laguerre weight function $t \mapsto w(t) = t^\alpha e^{-t}$ $(\alpha > -1)$ on $[0, +\infty)$.

Theorem 18 *Let P_n be an algebraic polynomial of degree exactly equal to n with nonnegative coefficients. Then for $\alpha \geq (\sqrt{5}-1)/2$,*

$$\int_0^\infty (P_n'(t))^2 t^\alpha e^{-t}\, dt \leq \frac{n^2}{(2n+\alpha)(2n+\alpha-1)} \int_0^\infty P_n(t)^2 t^\alpha e^{-t}\, dt. \tag{18}$$

The equality holds for $P_n(t) = t^n$. For $0 \leq \alpha \leq 1/2$,

$$\int_0^\infty (P_n'(t))^2 t^\alpha e^{-t}\, dt \leq \frac{1}{(2+\alpha)(1+\alpha)} \int_0^\infty P_n(t)^2 t^\alpha e^{-t}\, dt. \tag{19}$$

We will briefly review the key points in Varma's proof. At first, we write

$$P_n(t) = a_n t^n + P_{n-1}(t), \qquad P_{n-1}(t) = \sum_{k=0}^{n-1} a_k t^k, \quad a_k \geq 0.$$

Introducing the following inner product and norm by

$$(f, g) = \int_0^{+\infty} f(t)g(t)t^\alpha e^{-t}\, dt \quad \text{and} \quad \|f\| = \sqrt{(f, f)}, \tag{20}$$

respectively, we have

$$\|P_n'\|^2 = \|P_{n-1}'\|^2 + a_n^2 n^2 \Gamma(2n+\alpha-1) + 2na_n(P_{n-1}', t^{n-1})$$

and

$$\|P_n\|^2 = a_n^2 \Gamma(2n+\alpha+1) + \|P_{n-1}\|^2 + 2a_n(P_{n-1}, t^n).$$

Putting

$$b_n = \frac{n^2}{(2n+\alpha)(2n+\alpha-1)} \tag{21}$$

and

$$\lambda_n = 2n(P'_{n-1}, t^{n-1}) - 2b_n(P_{n-1}, t^n),$$

Varma obtained

$$||P'_n||^2 - b_n||P_n||^2 = \lambda_n a_n + ||P'_{n-1}||^2 - b_n||P_{n-1}||^2. \tag{22}$$

Also, he derived that

$$\lambda_n = \frac{2}{n} b_n \sum_{k=0}^{n-1} a_k \mu_{kn} \Gamma(k+n+\alpha-1), \tag{23}$$

where $\mu_{kn} = (k-n)[n(n-k) + (2\alpha-1)n + \alpha(\alpha-1)]$, $0 \le k \le n-1$.

Clearly, for $\alpha \ge 0$ we have

$$\mu_{kn} \le -\alpha(2n+\alpha-1) \le 0, \qquad k = 0, 1, \ldots, n-1. \tag{24}$$

Using (23) and (24), Varma claimed that

$$\lambda_n \le 0, \qquad n = 1, 2, \ldots . \tag{25}$$

Also, he noted that for every $n = 2, 3, \ldots$,

$$b_n \ge b_{n-1} \qquad \text{for} \quad \alpha \ge \frac{\sqrt{5}-1}{2} \tag{26}$$

and

$$b_n < b_{n-1} \qquad \text{for} \quad 0 \le \alpha \le \frac{1}{2}. \tag{27}$$

Using these ideas Varma completed his proof of Theorem 18. Namely, for $\alpha \ge (\sqrt{5}-1)/2$ he obtained from (22), (25) and (26) that

$$\Phi_k \le \Phi_{k-1} \qquad (k = 2, \ldots, n),$$

where we put $\Phi_k = ||P'_k||^2 - b_k||P_k||^2$. From these inequalities Varma concluded that $\Phi_n \le \Phi_1$.

A simple computation shows that for every $P_1(t) = a_1 t + a_0$, $a_1 > 0$, $a_0 \ge 0$, $||P'_1||^2 \le b_1||P_1||^2$, i.e., $\Phi_1 \le 0$, with equality if $P_1(t) = a_1 t$, $a_1 > 0$. So, Varma proved that $\Phi_n \le 0$, i.e., (18).

Using (27) instead of (26), Varma got the following inequalities:

$$\Phi_k \le \Phi_{k-1} + (b_{k-1} - b_k)||P_n||^2 \quad (k = 2, \ldots, n).$$

In a similar way, he concluded that

$$\Phi_n \le \Phi_1 + (b_1 - b_n)||P_n||^2 \le (b_1 - b_n)||P_n||^2,$$

i.e., $||P_n'||^2 \le b_1||P_n||^2$, which is (19).

The case $\alpha = 1$ was considered by Varma [43]. The cases $\alpha \in (-1, 0)$ and $\alpha \in (1/2, (\sqrt{5}-1)/2)$ were not solved in the paper of Varma [44]. Xie [53] tried to solve this problem for $\alpha \in (1/2, (\sqrt{5}-1)/2)$. In fact, he proved the following complicated and crude result:

Theorem 19 *Let $b_n = b_n(\alpha)$ be given by (21) and*

$$\alpha_n = \frac{1 - 2n - 4n^2 + \sqrt{16n^4 + 32n^3 + 20n^2 + 4n + 1}}{2(2n+1)} \qquad (n \ge 1).$$

Then for each polynomial P of degree n with nonnegative coefficients,

$$||P'||^2 \le b_n(\alpha)||P||^2 \qquad (\alpha \ge \alpha_1)$$

and

$$||P'||^2 \le \begin{cases} b_1(\alpha)||P||^2 & (\alpha_\nu \le \alpha < \alpha_{\nu-1},\ n \le \nu), \\ [b_1(\alpha) + b_n(\alpha) - b_\nu(\alpha)]||P||^2 & (\alpha_\nu \le \alpha < \alpha_{\nu-1},\ n > \nu), \end{cases}$$

where $\nu = 2, 3, \ldots$.

In the paper [24], we gave a complete solution to Varma's problem (17) determining

$$C_n(\alpha) = \sup_{P \in W_n} \frac{||P'||^2}{||P||^2}, \tag{28}$$

for all $\alpha \in (-1, +\infty)$, where W_n is defined in the following way:

$$W_n = \left\{ P \mid P(t) = \sum_{\nu=0}^{n} a_\nu t^\nu,\ a_\nu \ge 0\ (\nu = 0, 1, \ldots, n-1),\ a_n > 0 \right\}.$$

We denote by W_n^0 the subset of W_n for which $a_0 = 0$ (i.e., $P(0) = 0$). Note that the supremum in (28) is attained for some $P \in W_n^0$. Indeed,

$$\sup_{P \in W_n} \frac{||P'||}{||P||} = \sup_{\substack{P \in W_n^0 \\ a_0 \ge 0}} \frac{||P'||}{||P + a_0||} = \sup_{P \in W_n^0} \frac{||P'||}{||P||}.$$

Let $(.,.)$ and $||\,.\,||$ be defined by (20). The following theorem (see Milovanović [24]) gives the solution of the extremal problem (17), i.e., (28).

Theorem 20 *The best constant $C_n(\alpha)$ defined in (28) is*

$$C_n(\alpha) = \begin{cases} \dfrac{1}{(2+\alpha)(1+\alpha)} & (-1 < \alpha \le \alpha_n), \\[2ex] \dfrac{n^2}{(2n+\alpha)(2n+\alpha-1)} & (\alpha_n \le \alpha < +\infty), \end{cases} \tag{29}$$

where

$$\alpha_n = \frac{1}{2}(n+1)^{-1}\big((17n^2+2n+1)^{1/2} - 3n + 1\big). \tag{30}$$

In our proof we take that $P \in W_n^0$, i.e., $P(t) = \sum_{\nu=1}^{n} a_\nu t^\nu$, $a_\nu \ge 0$, and put $I_n(\alpha) = ||P||^2$. Then

$$P(t)^2 = \sum_{\nu=2}^{2n} b_\nu t^\nu \qquad (b_\nu \ge 0)$$

and

$$||P||^2 = I_n(\alpha) = \sum_{\nu=2}^{2n} b_\nu \Gamma(\nu+\alpha+1),$$

where Γ is the gamma function.

The inequality

$$t\big(P'(t)^2 - P(t)P''(t)\big) \le P'(t)P(t) \quad (P \in W_n;\ t \ge 0)$$

(see [24]) or [30, Subsection 2.1.5]) and a simple application of integration by parts give us

$$||P'||^2 \le \frac{1}{4}\big\{I_n(\alpha) + (1-2\alpha)I_n(\alpha-1) + (\alpha-1)^2 I_n(\alpha-2)\big\},$$

i.e.,

$$||P'||^2 \le \sum_{\nu=2}^{2n} H_\nu(\alpha) b_\nu \Gamma(\nu+\alpha+1),$$

where

$$H_\nu(\alpha) = \frac{\nu^2}{4(\nu+\alpha)(\nu+\alpha-1)}.$$

Therefore, $||P'||^2 \le \big(\max_{2\le\nu\le2n} H_\nu(\alpha)\big)||P||^2$ and $C_n(\alpha) \le \max_{2\le\nu\le2n} H_\nu(\alpha)$.

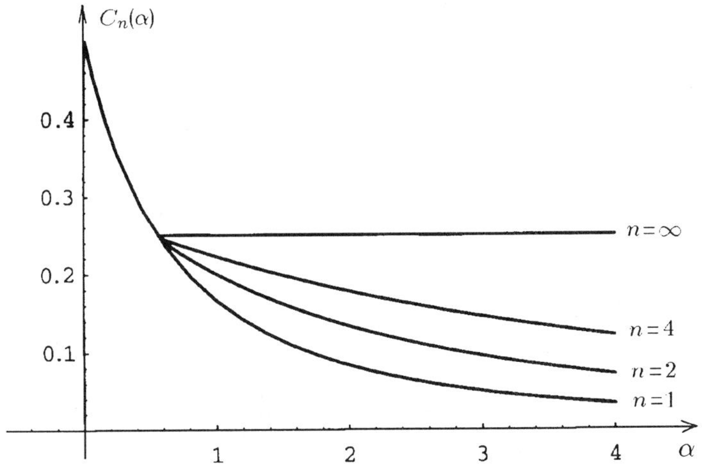

Figure 1: The constant $C_n(\alpha)$ for $n = 1, 2, 4$ and $n = \infty$

Determining the maximum of $f(x) = x^2/\big((x+\alpha)(x+\alpha-1)\big)$ on the interval $[2, 2n]$, we find that

$$\max_{2\le\nu\le 2n} H_\nu(\alpha) = \begin{cases} H_2(\alpha) & \text{if } -1 < \alpha \le \alpha_n, \\ H_{2n}(\alpha) & \text{if } \alpha_n \le \alpha < +\infty, \end{cases}$$

where α_n is given by (30).

We can also show that $C_n(\alpha)$, as it is defined in (29), is the best possible, i.e. that $C_n(\alpha) = \max\limits_{2\le\nu\le 2n} H_\nu(\alpha)$. An extremal polynomial for $\alpha \ge \alpha_n$ is $\tilde{P}(t) = t^n$. If $-1 < \alpha \le \alpha_n$, there exists a sequence of polynomials, for example, $p_{n,k}(t) = t^n + kt$, $k = 1, 2, \ldots$, for which

$$\lim_{k\to+\infty} \frac{||p'_{n,k}||^2}{||p_{n,k}||^2} = C_n(\alpha).$$

The best constant $C_n(\alpha)$ for $n = 1, 2, 3$ and $n = \infty$ as a function of α is displayed in Figure 1. An enlarged nontrivial part of that is given in Figure 2. We can see that:

(a) $C_n(\alpha_n - 0) = C_n(\alpha_n + 0)$;

(b) $C_{n+1}(\alpha) \ge C_n(\alpha)$;

(c) The sequence $\{\alpha_n\}$ is decreasing, i.e.,

$$\alpha_1 > \alpha_2 > \alpha_3 > \cdots > \alpha_\infty,$$

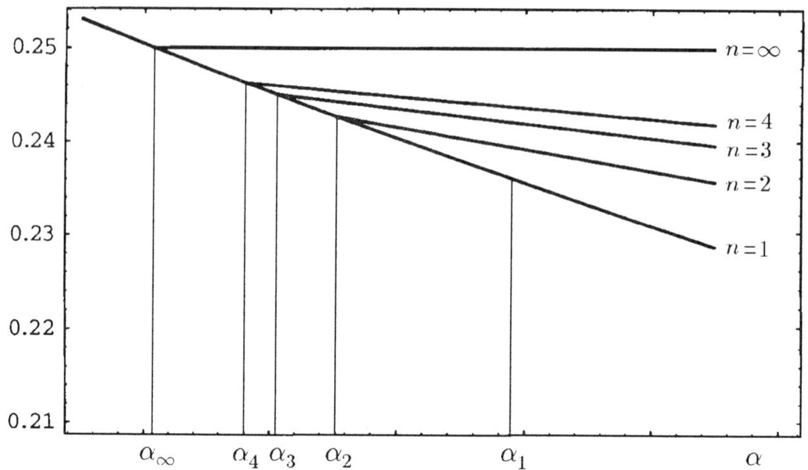

Figure 2: Enlarged nontrivial part in Figure 1

where

$$\alpha_1 = (\sqrt{5}-1)/2, \ \alpha_2 = (\sqrt{73}-5)/6, \ \alpha_3 = (\sqrt{10}-2)/2, \ \text{etc.},$$

and

$$\alpha_\infty = \lim_{n\to\infty} \alpha_n = (\sqrt{17}-3)/2 = 0.561552812\ldots .$$

G. V. Milovanović and I. Ž. Milovanović [28] solved the following extremal problem for higher derivatives

$$C_{n,k}(\alpha) = \sup_{P\in W_n} \frac{\|P^{(k)}\|^2}{\|P\|^2} \qquad (1 \le k \le n). \tag{31}$$

Theorem 21 *The best constant $C_{n,k}(\alpha)$ is given by*

$$C_{n,k}(\alpha) = \begin{cases} \dfrac{(k!)^2}{(\alpha+1)_{2k}}, & -1 < \alpha \le \alpha_{n,k}, \\[2ex] \dfrac{n^2(n-1)^2\cdots(n-k+1)^2}{(2n+\alpha)^{(2k)}}, & \alpha \ge \alpha_{n,k}, \end{cases}$$

where $\alpha_{n,k}$ is the unique positive root of the equation

$$\frac{(2n+\alpha)^{(2k)}}{(2k+\alpha)^{(2k)}} = \binom{n}{k}^2 .$$

Here $(p)_\nu = p(p+1)\cdots(p+\nu-1)$ and $p^{(\nu)} = p(p-1)\cdots(p-\nu+1)$.

In the special case, when $n \to +\infty$, the best constant $C_{n,k}$, defined in Theorem 21, reduces to

$$C_k^*(\alpha) = \lim_{n\to\infty} C_{n,k}(\alpha) = \begin{cases} \dfrac{(k!)^2}{(\alpha+1)_{2k}}, & -1 < \alpha \le \alpha_k^*, \\ \\ \dfrac{1}{4^k}, & \alpha_k^* \le \alpha < +\infty, \end{cases}$$

where α_k^* is the unique positive root of the equation $(\alpha+1)_{2k} = 4^k(k!)^2$. We note that $\alpha_1^* = \alpha_\infty = (\sqrt{17}-3)/2$.

The corresponding extremal problem for polynomials with nonnegative coefficients, with respect to the Freud's weight $t \mapsto w(t) = t^\alpha \exp(-t^s)$, $\alpha > -1$, $s > 0$, on the interval $(0, +\infty)$, was investigated by Milovanović and Djordjević [27]. In this case, using the same method, it was proved that for $P \in W_n^0$

$$\|P\|^2 = (P,P) = \frac{1}{s}\sum_{\nu=2}^{2n} b_\nu \Gamma\Big(\frac{\alpha+\nu+1}{s}\Big)$$

and

$$\|P'\|^2 = (P',P') \le \frac{1}{s}\sum_{\nu=2}^{2n} H_\nu(\alpha;s) b_\nu \Gamma\Big(\frac{\alpha+\nu+1}{s}\Big),$$

where $(f,g) = \int_0^\infty w(t)f(t)g(t)\,dt$ and

$$H_\nu(\alpha;s) = \frac{\nu^2}{2}\cdot\frac{\Gamma\Big(\dfrac{\alpha+\nu-1}{s}\Big)}{\Gamma\Big(\dfrac{\alpha+\nu+1}{s}\Big)}.$$

The corresponding best constant we will denote by $C_n(\alpha;s)$. If $s = 2$ we get the following simple result (see Milovanović and Djordjević [27]):

Theorem 22 *The best constant $C_n(\alpha;2)$ is given by*

$$C_n(\alpha;2) = \begin{cases} \dfrac{2}{\alpha+1}, & -1 < \alpha \le -\dfrac{n-1}{n+1}, \\ \\ \dfrac{2n^2}{2n+\alpha-1}, & -\dfrac{n-1}{n+1} \le \alpha < +\infty. \end{cases}$$

Putting $\alpha = 0$ we obtain the following inequality

$$\int_0^\infty e^{-t^2} P'(t)^2\,dt \le \frac{2n^2}{2n-1}\int_0^\infty e^{-t^2} P(t)^2\,dt$$

for each $P \in W_n$.

The case when s is an arbitrary positive number is more complicated. The following conjecture was stated by Milovanović and Djordjević [27]:

Conjecture 23 *Let $s \geq 1$ and let α_n (> -1) be the unique root of the equation*

$$\frac{\Gamma\left(\frac{\alpha+1}{s}\right)}{\Gamma\left(\frac{\alpha+3}{s}\right)} = n^2 \frac{\Gamma\left(\frac{\alpha+2n-1}{s}\right)}{\Gamma\left(\frac{\alpha+2n+1}{s}\right)}.$$

The best constant $C_n(\alpha; s)$ is given by

$$C_n(\alpha; s) = \begin{cases} H_2(\alpha; s), & -1 < \alpha \leq \alpha_n, \\ H_{2n}(\alpha; s), & \alpha_n \leq \alpha < +\infty. \end{cases}$$

Recently Guessab, Milovanović and Arino [16] considered the extremal problem (31) in L^r-norm,

$$\|P\|_r = \left(\int_0^\infty |P(t)|^r t^\alpha e^{-t}\, dt\right)^{1/r}, \qquad r \geq 1.$$

For every $r \in \mathbb{N}$, using the previous method they determined the best constant in the inequality

$$\|P^{(m)}\|_r^r \leq C_{n,r}^{(m)}(\alpha)\|P\|_r^r \qquad (P \in W_n). \tag{32}$$

Theorem 24 *Let $r \in \mathbb{N}$ and let $\alpha_{n,r,m}$ (> -1) be the unique root of the equation*

$$\frac{\Gamma(\alpha+1)}{\Gamma(mr+\alpha+1)} = \binom{n}{m}^r \frac{\Gamma((n-m)r+\alpha+1)}{\Gamma(nr+\alpha+1)}.$$

Then the best constant $C_{n,r}^{(m)}(\alpha)$ in (32) is given by

$$C_{n,r}^{(m)}(\alpha) = \begin{cases} (m!)^r \dfrac{\Gamma(\alpha+1)}{\Gamma(mr+\alpha+1)}, & -1 < \alpha \leq \alpha_{n,r,m}, \\ \left(n^{(m)}\right)^r \dfrac{\Gamma((n-m)r+\alpha+1)}{\Gamma(nr+\alpha+1)}, & \alpha_{n,r,m} \leq \alpha < +\infty, \end{cases}$$

where $n^{(m)} = n(n-1)\cdots(n-m+1)$.

Our method of proving this theorem works only when r is an integer. We also use the fact that

$$\sup_{P \in W_n} \frac{\|P^{(m)}\|_r}{\|P\|_r} = \sup_{\substack{P \in W_n^0 \\ a_0,\ldots,a_{m-1} \geq 0}} \frac{\|P^{(m)}\|_r}{\|P + Q_{m-1}\|_r} = \sup_{P \in W_n^0} \frac{\|P^{(m)}\|_r}{\|P\|_r},$$

where $Q_{m-1}(t) = \sum\limits_{k=0}^{m-1} a_k t^k$ ($a_k \geq 0$) and W_n^0 is a subset of W_n such that

$$P(0) = P'(0) = \cdots = P^{(m-1)}(0) = 0.$$

The case $r = 3$ and $m = 1$ was considered earlier by Guessab and Milovanović [14]. In that case we have that the best constant $C_{n,3}^{(1)}(\alpha)$ given by

$$C_{n,3}^{(1)}(\alpha) = \begin{cases} \dfrac{1}{(3+\alpha)(2+\alpha)(1+\alpha)} & (-1 < \alpha \leq \alpha_n), \\[2ex] \dfrac{n^3}{(3n+\alpha)(3n+\alpha-1)(3n+\alpha-2)} & (\alpha_n \leq \alpha < +\infty), \end{cases}$$

where α_n is the unique positive root of the equation

$$(n^2+n+1)\alpha^3 + 3(2n^2+2n-1)\alpha^2 + (11n^2-16n+2)\alpha - 3n(7n-2) = 0.$$

In the simplest case ($r = 1$, $m = 1$), we have

$$C_{n,1}^{(1)}(\alpha) = \begin{cases} \dfrac{1}{\alpha+1}, & -1 < \alpha \leq 0, \\[2ex] \dfrac{n}{\alpha+n}, & \alpha \geq 0. \end{cases}$$

Recently, this case was also considered by Chen [6].

For polynomials P ($\not\equiv 0$) with all nonnegative coefficients and degree at most n, and for positive integers r and p ($r \leq p$), Varma [52] proved the L^r inequality

$$\int_0^{+\infty} |P'(t)|^r t^{p-1} e^{-t}\, dt \leq \frac{n^r(nr+p-r-1)!}{(nr+p-1)!} \int_0^{+\infty} |P(t)|^r t^{p-1} e^{-t}\, dt,$$

with equality if and only if $P(t) = ct^n$. In the case $p = 1$, he obtained the best constant in the form $1/r!$, with extremal polynomial $P(t) = ct$.

Evidently, he did not know our more general result given in Theorem 24 (see [16]). We believe that this theorem holds for every real $r \geq 1$.

6 Extremal Problems for Lorentz Classes of Polynomials

In this section we consider the extremal problems of Markov's type for nonnegative algebraic polynomials on $[-1, 1]$ in L^2 metric with Jacobi weight $w(t) = (1-t)^\alpha(1+t)^\beta$ ($\alpha, \beta > -1$). These problems were investigated by Varma [43], Erdős and Varma [10], Milovanović and Petković [31], Chen [6], and Underhill and Varma [39].

Let L_n be the Lorentz class of algebraic polynomials of the form

$$P(t) = \sum_{\nu=0}^{n} b_\nu (1-t)^\nu (1+t)^{n-\nu}, \quad b_\nu \geq 0 \ (\nu = 0, 1, \ldots, n).$$

A subset of the Lorentz class L_n for which $P^{(i-1)}(-1) = P^{(i-1)}(1) = 0$ $(i = 1, \ldots, k)$ will be denoted by $L_n^{(k)}$. Notice that $L_n^{(0)} \supset L_n^{(1)} \supset \cdots$, where $L_n^{(0)} \equiv L_n$. The corresponding representation of a polynomial P from $L_n^{(k)}$ is

$$P(t) = \sum_{\nu=k}^{n-k} b_\nu (1-t)^\nu (1+t)^{n-\nu},$$

where $b_\nu \geq 0$ $(\nu = k, \ldots, n-k)$.

Let $w(t) = (1-t)^\alpha (1+t)^\beta$, $\alpha, \beta > -1$, and $\|f\|^2 = (f, f)$, where

$$(f, g) = \int_{-1}^{1} w(t) f(t) g(t)\, dt \qquad (f, g \in L^2(-1, 1)).$$

For the determination of the best constant

$$C_n^{(k)}(\alpha, \beta) = \sup_{P \in L_n^{(k)} \setminus \{0\}} \frac{\|P'\|^2}{\|P\|^2}, \tag{33}$$

where $k = 0, 1, \ldots, [n/2]$, Milovanović and Petković [31] used the following inequality

$$(1-t^2)(P'(t)^2 - P''(t)P(t)) \leq nP(t)^2 - 2tP(t)P'(t),$$

which holds for every $t \in [-1, 1]$ and $P \in L_n$ (see also [30, Subsection 2.1.5]). They proved:

Theorem 25 *Let $\alpha, \beta \geq 1$, then the best constant $C_n^{(0)}(\alpha, \beta)$, defined in (33), is given by*

$$C_n^{(0)}(\alpha, \beta) = \frac{n^2(2n+\alpha+\beta)(2n+\alpha+\beta+1)}{4(2n+\lambda)(2n+\lambda-1)},$$

where $\lambda = \min(\alpha, \beta)$.

In a special case we obtain:

Corollary 26 *The best constant $C_n^{(0)}(1, 1)$ is given by*

$$C_n^{(0)}(1, 1) = \frac{n(n+1)(2n+3)}{4(2n+1)}.$$

This result was proved earlier by Erdős and Varma [10] (see, also, Varma [43]).

In the same paper [31], Milovanović and Petković proved the following assertion for the class of polynomials $L_n^{(k)}$ $(1 \leq k \leq [n/2])$.

Theorem 27 *Let $1 \le k \le [n/2]$ and $\alpha, \beta > -1$, then*

$$C_n^{(k)}(\alpha,\beta) = \frac{1}{16}(2n+\alpha+\beta)(2n+\alpha+\beta+1)\max\big(H_{2k}(\alpha,\beta), H_{2n-2k}(\alpha,\beta)\big),$$

where $H_\nu(\alpha,\beta) \equiv f(\nu)$ and f is given by

$$\begin{aligned} f(x) &= \frac{(\alpha-1)^2}{(x+\alpha-1)(x+\alpha)} + \frac{(\beta-1)^2}{(2n-x+\beta-1)(2n-x+\beta)} \\ &\quad + \frac{2n+\alpha+\beta-2\alpha\beta}{(x+\alpha)(2n-x+\beta)}. \end{aligned}$$

Especially interesting cases appear when $\alpha = \beta$.

Theorem 28 *Let $k \ge 1$ and $\alpha = \beta > -1$, then*

$$C_n^{(k)}(\alpha,\beta) = \frac{(n+\alpha)(2n+2\alpha+1)q(n,k,\alpha)}{2(2k+\alpha-1)(2k+\alpha)(2n-2k+\alpha-1)(2n-2k+\alpha)},$$

where $q(n,k,\alpha) = \alpha(\alpha-1)n^2 + 2k(n-k)(n-1+3\alpha-2\alpha^2)$.

In the special cases when $\alpha = 0$ (Legendre case), $\alpha = -1/2$ (Chebyshev case), and $\alpha = 1$, we have:

Corollary 29 *Let $k \ge 1$, then*

$$C_n^{(k)}(0,0) = \frac{n(n-1)(2n+1)}{4(2k-1)(2n-2k-1)},$$

$$C_n^{(k)}(-1/2,-1/2) = \frac{2n(2n-1)[3n^2+8k(n-k)(n-3)]}{(4k-3)(4k-1)(4n-4k-3)(4n-4k-1)},$$

$$C_n^{(k)}(1,1) = \frac{n(n+1)(2n+3)}{4(2k+1)(2n-2k+1)}. \tag{34}$$

From Corollary 26 we see that (34) holds for $k = 0$ too.

For $k = 1$, the best constants in Corollary 29 reduce to

$$C_n^{(1)}(0,0) = \frac{n(n-1)(2n+1)}{4(2n-3)}, \tag{35}$$

$$C_n^{(1)}(-1/2,-1/2) = \frac{2n(2n-1)(11n^2-32n+24)}{3(4n-5)(4n-7)},$$

and

$$C_n^{(1)}(1,1) = \frac{n(n+1)(2n+3)}{12(2n-1)}.$$

It is of interest to note that Erdős and Varma [10] proved that the best constant in the Lorentz class L_n $(n \geq 2)$ for $\alpha = \beta = 0$ is the same one as that in (35), i.e. $C_n^{(0)}(0,0) = C_n^{(1)}(0,0)$.

Recently, Underhill and Varma [39] provided a new proof of the ultraspherical case $w(t) = (1-t^2)^\alpha$ $(\alpha > -1)$, without the requirement that $P(\pm 1) = 0$ in the case when $-1 < \alpha < 1$. Namely, they proved:

Theorem 30 *Let $n \geq 2$, $\alpha > -1$, and let α_n be the unique positive solution of the equation*

$$2\alpha^4 + (8n-5)\alpha^3 + (12n^2 - 17n + 4)\alpha^2$$
$$+(8n^3 - 20n^2 + 11n - 1)\alpha - 2n(2n^2 - 5n + 4) = 0.$$

Then for $\alpha_n \leq \alpha \leq 1$ we have

$$C_n^{(0)}(\alpha,\alpha) = \frac{n^2(2n+2\alpha+1)(n+\alpha)}{2(2n+\alpha)(2n+\alpha-1)}, \tag{36}$$

and for $-1 < \alpha \leq \alpha_n$,

$$C_n^{(0)}(\alpha,\alpha) = \frac{(2n+2\alpha+1)(n+\alpha)A(n,\alpha)}{2(\alpha+1)(\alpha+2)(2n+\alpha-2)(2n+\alpha-3)},$$

where $A(n,\alpha) = \alpha(\alpha-1)n^2 + 2(n-1)\big(n-(\alpha-1)(2\alpha-1)\big)$.

Notice that the expression (36) holds for $\alpha \geq 1$ too (see Theorem 25). Underhill and Varma [39] considered also the corresponding problem in L^4 norm with the ultraspherical weight $t \mapsto (1-t^2)^3$ on $(-1,1)$.

At the end we mention a result for polynomials with non-negative coefficients

$$S_n = \Big\{ P \mid P(t) = \sum_{\nu=0}^{n} a_\nu t^\nu,\ a_\nu \geq 0\ (\nu = 0,1,\ldots,n) \Big\},$$

given by Chen [6]:

Theorem 31 *Let $P \in S_n$ and $\alpha > -1$. Then*

$$\int_{-1}^{1} (P'(t))^2 (1-t^2)^\alpha \, dt \leq \frac{2n+2\alpha+1}{2n-1}\, n^2 \int_{-1}^{1} (P(t))^2 (1-t^2)^\alpha \, dt,$$

with equality when $P(t) = t^n$.

Acknowledgment. The author is grateful to Professor Peter Johnson for his careful reading of the paper and useful comments.

References

[1] R. P. Agarwal and G. V. Milovanović, One characterization of the classical orthogonal polynomials, in "Progress in Approximation Theory" (P. Nevai and A. Pinkus, Eds.), pp. 1–4, Academic Press, New York, 1991.

[2] S. N. Bernstein, Sur l'ordre de la meilleure approximation des fonctions continues par des polynômes de degré donné, Mém. Acad. Roy. Belgique (2) **4** (1912), 1–103.

[3] B. D. Bojanov, An extension of the Markov inequality, J. Approx. Theory **35** (1982), 181–190.

[4] B. Bojanov, An inequality of Duffin and Schaeffer type, East J. Approx. **1** (1995), 37–46.

[5] W. Chen, On the L^2 inequalities involving trigonometric polynomials and their derivatives, Trans. Amer. Math. Soc. **347** (1995), 1753–1761.

[6] W. Chen, Some inequalities of algebraic polynomials with nonnegative coefficients, Trans. Amer. Math. Soc. **347** (1995), 2161–2167.

[7] I. K. Daugavet and S. Z. Rafal'son, Some inequalities of Markov-Nikol'skiĭ type for algebraic polynomials, Vestnik Leningrad. Univ. Mat. Mekh. Astronom. 1972, no 1, 15–25 (Russian).

[8] D. K. Dimitrov, Markov inequalities for weight functions of Chebyshev type, J. Approx. Theory **83** (1995), 175–181.

[9] A. Durand, "Quelques aspects de la theorie analytique des polynomes. I et II", Université de Limoges, 1984.

[10] P. Erdős and A. K. Varma, An extremum problem concerning algebraic polynomials, Acta Math. Hung. **47** (1986), 137–143.

[11] P. Goetgheluck, On the Markov inequality in L^p-spaces, J. Approx. Theory **62** (1990), 197–205.

[12] N. K. Govil, Inequalities for the derivative of a polynomial, J. Approx. Theory **63** (1990), 65–71.

[13] N. K. Govil, Some inequalities for derivatives of polynomials, J. Approx. Theory **66** (1991), 29–35.

[14] A. Guessab and G. V. Milovanović, An extremal problem for polynomials with nonnegative coefficients. IV, Math. Balkanica **3** (1989), 142–148.

[15] A. Guessab and G. V. Milovanović, Weighted L^2-analogues of Bernstein's inequality and classical orthogonal polynomials, J. Math. Anal. Appl. **182** (1994), 244–249.

[16] A. Guessab, G. V. Milovanović and O. Arino, Extremal problems for nonnegative polynomials in L^p norm with generalized Laguerre weight, Facta Univ. Ser. Math. Inform. **3** (1988), 1–8.

[17] E. Hille, G. Szegő and J. D. Tamarkin, On some generalizations of a theorem of A. Markoff, Duke Math. J. **3** (1937), 729–739.

[18] S. V. Konjagin, Estimation of the derivatives of polynomials, Dokl. Akad. Nauk SSSR **243** (1978), 1116–1118 (Russian).

[19] G. Labelle, Concerning polynomials on the unit interval, Proc. Amer. Math. Soc. **20** (1969), 321–326.

[20] A. Lupaş, An inequality for polynomials, Univ. Beograd. Publ. Elektrotehn. Fak. Ser. Mat. Fiz. No 461–No 497 (1974), 241–243.

[21] A. A. Markov, On a problem of D.I. Mendeleev, Zap. Imp. Akad. Nauk, St. Petersburg **62** (1889), 1–24 (Russian).

[22] V. A. Markov, On functions deviating least from zero in a given interval, Izdat. Imp. Akad. Nauk, St. Petersburg, 1892 (Russian) [German transl. Math. Ann. **77** (1916), 218–258].

[23] D. Mendeleev, "Investigation of aqueous solutions based on specific gravity", St. Petersburg, 1887 (Russian).

[24] G. V. Milovanović, An extremal problem for polynomials with nonnegative coefficients, Proc. Amer. Math. Soc. **94** (1985), 423–426.

[25] G. V. Milovanović, Various extremal problems of Markov's type for algebraic polynomials, Facta Univ. Ser. Math. Inform. **2** (1987), 7–28.

[26] G. V. Milovanović, Extremal problems for polynomials in L^r-norms: old and new results, in "Open Problems in Approximation Theory" (B. Bojanov, Ed.), pp. 138–155, SCT Publishing, Singapore, 1994.

[27] G. V. Milovanović and R. Ž. Djordjević, An extremal problem for polynomials with nonnegative coefficients. II, Facta Univ. Ser. Math. Inform. **1** (1986), 7–11.

[28] G. V. Milovanović and I. Ž. Milovanović, An extremal problem for polynomials with nonnegative coefficients. III, in "Constructive Theory of Functions '87 (Varna, 1987)" (Bl. Sendov, P. Petrušev, K. Ivanov, and R. Maleev, Eds.), pp. 315–321, Bulgar. Acad. Sci., Sofia, 1988.

[29] G. V. Milovanović, D. S. Mitrinović and Th. M. Rassias, On some extremal problems for algebraic polynomials in L^r norm, in "Generalized Functions and Convergence" (P. Antosik and A. Kamiński, Eds.), pp. 343–354, World Scientific Publishing Company, Singapore, 1990.

[30] G. V. Milovanović, D. S. Mitrinović and Th. M. Rassias, "Topics in Polynomials: Extremal Problems, Inequalities, Zeros", World Scientific, Singapore – New Jersey – London – Hong Kong, 1994.

[31] G. V. Milovanović and M. S. Petković, Extremal problems for Lorentz classes of nonnegative polynomials in L^2 metric with Jacobi weight, Proc. Amer. Math. Soc. **102** (1988), 283–289.

[32] R. N. Mohapatra, P. J. O'Hara and R. S. Rodriguez, Simple proofs of Bernstein-type inequalities, Proc. Amer. Math. Soc. **102** (1988), 629–632.

[33] R. Pierre and Q. I. Rahman, On a problem of Turán about polynomials II, Canad. J. Math. **33** (1981), 701–733.

[34] Q. I. Rahman and G. Schmeisser, "Les inégalités de Markoff et de Bernstein", Presses Univ. Montréal, Montréal, Québec, 1983.

[35] Q. I. Rahman and G. Schmeisser, L^p inequalities for polynomials, J. Approx. Theory **53** (1988), 26–32.

[36] E. Schmidt, Über die nebst ihren Ableitungen orthogonalen Polynomensysteme und das zugehörige Extremum, Math. Ann. **119** (1944), 165–204.

[37] L.-C. Shen, Comments on an L^2 inequality of A.K. Varma involving the first derivative of polynomials, Proc. Amer. Math. Soc. **111** (1991), 955–959.

[38] P. Turán, Remark on a theorem of Erhard Schmidt, Mathematica **2** (25)(1960), 373–378.

[39] B. Underhill and A. K. Varma, An extension of some inequalities of P. Erdős and P. Turán concerning algebraic polynomials, Acta Math. Hung. **73** (1996), 1–28.

[40] A. K. Varma, An analogue of some inequalities of P. Turán concerning algebraic polynomials having all zeros inside $[-1, 1]$, Proc. Amer. Math. Soc. **55** (1976), 305–309.

[41] A. K. Varma, An analogue of some inequalities of P. Erdős and P. Turán concerning algebraic polynomials satisfying certain conditions, in "Fourier Analysis and Approximation Theory", Vol. II, Budapest 1976, Colloq. Math. Soc. János Bolyai **19** (1978), 877–890.

[42] A. K. Varma, An analogue of some inequalities of P. Turán concerning algebraic polynomials having all zeros inside $[-1,1]$. II, Proc. Amer. Math. Soc. **69** (1978), 25–33.

[43] A. K. Varma, Some inequalities of algebraic polynomials having real zeros, Proc. Amer. Math. Soc. **75** (1979), 243–250.

[44] A. K. Varma, Derivatives of polynomials with positive coefficients, Proc. Amer. Math. Soc. **83** (1981), 107–112.

[45] A. K. Varma, Some inequalities of algebraic polynomials having all zeros inside $[-1,1]$, Proc. Amer. Math. Soc. **88** (1983), 227–233.

[46] A. K. Varma, A new characterization of Hermite polynomials, Acta Math. Hung. **49** (1987), 169–17

[47] A. K. Varma, Markoff type inequalities for curved majorants in L_2 norm, in "Approximation Theory", Keckemet (Hungary), 1990, Colloq. Math. Soc. János Bolyai **57** (1991), 689–696.

[48] A. K. Varma, Inequalities for trigonometric polynomials, J. Approx. Theory **65** (1991), 273-278.

[49] A. K. Varma, On some extremal properties of algebraic polynomials, J. Approx. Theory **69** (1992), 48–54.

[50] A. K. Varma, Markoff type inequalities for curved majorants in the weighted L^2 norm, Aequationes Math. **45** (1993), 24–46.

[51] A. K. Varma, T. M. Mills and S. J. Smith, Markoff type inequalities for curved majorants, J. Austral. Math. Soc. Ser. A **58** (1995), 1–14.

[52] A. K. Varma, Some inequalities of algebraic polynomials, Proc. Amer. Math. Soc. **123** (1995), 2041–2048.

[53] D. Xie, A replenishment for the theorem of Varma, A., J. Hangzhou Univ. **10** (1983), 266–270 (Chinese, English summary).

New Developments on Turán's Extremal Problems for Polynomials

Gradimir V. Milovanović
Department of Mathematics, University of Niš
Faculty of Electronic Engineering, P.O. Box 73, 18000 Niš, Yugoslavia

Themistocles M. Rassias
National Technical University of Athens
Department of Mathematics, Zagrafou Campus, 15780 Athens, Greece

Abstract

In this paper we give an account of L^r inequalities of Turán type for algebraic polynomials, mainly initiated and studied by the late Professor Arun K. Varma. Also, this paper is a continuation of our previous survey paper [12].

1 Introduction

Let $\mathcal{P}_n$ be the set of all algebraic polynomials of degree at most n and let W_n be some of its subsets. For a given norm $\|\,.\,\|$ we consider extremal problems

$$B_{n,m} = \inf_{P \in W_n} \frac{\|P^{(m)}\|}{\|P\|} \qquad (1 \leq m \leq n).$$

In comparing with inequalities of Markov's type (cf. Milovanović, Mitrinović, Rassias [13, Chap. 6]), here we have opposite inequalities which are known as *inequalities of Turán type*.

Turán [16] proved the following inequality for polynomials $P \in \mathcal{P}_n$ having all their zeros in $[-1, 1]$,

$$\|P'\|_\infty > \frac{\sqrt{n}}{6}\, \|P\|_\infty\,, \tag{1}$$

taking the uniform norm $\|f\|_\infty = \max\limits_{-1 \leq t \leq 1} |f(t)|$. The constant $\sqrt{n}/6$ is not the best possible.

Turán's inequality (1) has been generalized and extended in several different ways. In this survey we give an account of L^r $(r \geq 1)$ inequalities of Turán type. This kind of inequalities, as well as the inequalities of Markov-Bernstein's type, appear in approximation theory (cf. Dzyadyk [6], Ivanov [8], Lorentz [9], Meinardus [10], Milovanović, Mitrinović, Rassias [13, Chap. 6], Telyakovskiĭ [15]). For instance, Telyakovskiĭ [15] writes: "Among those that are fundamental in approximation theory are the extremal problems connected with inequalities for the derivatives of polynomials. ... The use of inequalities of this kind is a fundamental method in proofs of inverse problems of approximation theory."

Firstly, inequality (1) was sharpened by Erőd [7], who obtained

$$\|P'\|_\infty \geq B_n \|P\|_\infty \,, \tag{2}$$

where $B_2 = 1$, $B_3 = 3/2$, and

$$\begin{aligned} B_{2k} &= \frac{2k}{\sqrt{2k-1}}\left(1 - \frac{1}{2k-1}\right)^{k-1}, \\ B_{2k+1} &= \frac{(2k+1)^2}{2k\sqrt{2k+2}}\left(1 - \frac{\sqrt{2k+2}}{2k}\right)^{k-1}\left(1 + \frac{1}{\sqrt{2k+2}}\right)^{k}, \end{aligned}$$

for $k = 2, 3, \ldots$.

Exactly, equality in (2) is attained for $P(t) = (1-t)^n$, if $n = 1, 2, 3$, and for $P(t) = (1-t)^{n-[n/2]}(1+t)^{[n/2]}$, if $n \geq 4$.

Let W_n be the set of all algebraic polynomials of degree n whose zeros are all real and lie inside $[-1, 1]$. The corresponding inequality for the second derivative of such polynomials was investigated by Babenko and Pichugov [2].

If $P \in W_n$, $n \geq 2$, they proved that

$$\|P''\|_\infty \geq B_{n,2} \|P\|_\infty \,, \tag{3}$$

where $B_{n,2} = \min\{n, (n-1)n/4\}$.

If $n = 2, 3, 4, 5$, then $B_{n,2} = (n-1)n/4$, and equality in (3) is attained only for polynomials of the form $P(t) = C(1 \pm t)^n$, where C is an arbitrary real constant different from zero.

In the case $n \geq 6$, they found that $B_{n,2} = n$, and for $n = 2m$ equality in (3) holds only for polynomials of the form $P(t) = C(1-t^2)^m$, where C is an arbitrary real constant different from zero.

An analogue in L^2 norm for algebraic polynomials was considered first by Professor A. K. Varma [20]. Taking $\|f\|_2^2 = \int_{-1}^{1} f(t)^2\,dt$ he proved:

Theorem 1 *If $P \in W_n$*

$$\|P'\|_2^2 \geq \frac{n}{2}\,\|P\|_2^2. \tag{4}$$

This result is best possible in the sense that there exists a polynomial P_0 of degree n having all zeros inside $[-1,1]$ and for which

$$\|P_0'\|_2^2 = \Big(\frac{n}{2} + \frac{3}{4} + \frac{3}{4(n-1)}\Big)\|P_0\|_2^2, \quad n > 1.$$

The proof of this theorem was based on the following inequality

$$\|\sqrt{1-t^2}P'\|_2^2 \geq \frac{n}{2}\,\|P\|_2^2 \qquad (P \in W_n),$$

which becomes an equality only for $P(t) = C(1+t)^p(1-t)^q$, $p+q=n$, where C is an arbitrary non-zero constant.

2 Turán Type Inequalities in L^2 Norm

In [21] Professor Varma gave a more precise form of (4).

Theorem 2 *Let $\|f\|_2^2 = \int_{-1}^{1} f(t)^2\,dt$, $P \in W_n$ and $P(1) = P(-1) = 0$. Then we have*

$$\|P'\|_2^2 \geq \Big(\frac{n}{2} + \frac{3}{4} + \frac{3}{4(n-1)}\Big)\|P\|_2^2, \tag{5}$$

with equality for $P(t) = (1-t^2)^m$, $n = 2m$.

Taking the norm $\|f\|_2^2 = \int_{-1}^{1}(1-t^2)f(t)^2\,dt$, in 1979 Varma [22] proved the following result:

Theorem 3 *For $P \in W_n$ and $n \geq 2$ we have*

$$\|P'\|_2^2 \geq \Big(\frac{n}{2} + \frac{1}{4} - \frac{1}{4(n+1)}\Big)\|P\|_2^2,$$

with equality for $P(t) = (1-t^2)^m$, $n = 2m$.

Later Varma [23] proved an improvement of one of his earlier results.

Theorem 4 *Let $\|f\|_2^2 = \int_{-1}^{1} f(t)^2\,dt$. If $P \in W_n$ and $n = 2m$; then*

$$\|P'\|_2^2 \geq \Big(\frac{n}{2} + \frac{3}{4} + \frac{3}{4(n-1)}\Big)\|P\|_2^2, \tag{6}$$

where equality holds if and only if $P(t) = (1-t^2)^m$. *Moreover, if* $n = 2m-1$, *then*

$$\|P'\|_2^2 \geq \left(\frac{n}{2} + \frac{3}{4} + \frac{5}{4(n-2)}\right)\|P\|_2^2, \quad n \geq 3, \tag{7}$$

where equality holds if and only if $P(t) = (1-t)^{m-1}(1+t)^m$ *or* $P(t) = (1-t)^m(1+t)^{m-1}$.

This result is an improvement of Theorem 2 in two respects. First, the condition $P(1) = P(-1) = 0$ is not necessary for (5) to hold. Secondly, here there exist precise bounds for n even and also for n odd as mentioned in (6) and (7).

In the same norm, Varma [23] also proved:

Theorem 5 *Let* $P \in W_n$, *subject to the condition* $P(1) = 1$; *then*

$$\|P'\|_2^2 \geq \frac{n}{4} + \frac{1}{8} + \frac{1}{8(2n-1)}, \qquad n \geq 1,$$

where equality holds for $P(t) = ((1+t)/2)^n$.

This inequality is an improvement over $\|P'\|_2^2 > n/4$, given by Szabados and Varma [14]. The corresponding inequality for polynomials $P \in W_n$ in L^r norm, defined on $(-1,1)$ by $\|f\|_r = \left(\int_{-1}^{1} |f(t)|^r\, dt\right)^{1/r}$, was considered by Zhou [25].

Theorem 6 *If* $P \in W_n$, *then for* $1 \leq r \leq +\infty$,

$$\|P'\|_r \geq C\sqrt{n}\,\|P\|_r,$$

where C *is a positive absolute constant.*

A similar result for $0 < r < 1$ was obtained also by Zhou [26]. Recently, Zhou [27] proved the following results:

Theorem 7 *If* $P \in W_n$, *then for* $1 \leq r \leq s \leq +\infty$,

$$\|P'\|_r \geq Cn^{1/2-1/(2r)+1/(2s)}\|P\|_s,$$

where C *is a positive absolute constant.*

The example $P(t) = (1-t^2)^{[n/2]}$ in the previous theorem shows that the order $n^{1/2-1/(2r)+1/(2s)}$ cannot be improved.

Theorem 8 *Let $1 \leq r \leq s \leq +\infty$ and P be an polynomial of degree n with only real zeros. If at most k zeros of P lie outside the interval $[-1, 1]$, then*

$$||P'||_r \geq C_k n^{1/2-1/(2r)+1/(2s)} ||P||_s,$$

where C_k is a positive constant depending only upon k.

After Professor Varma's death, the following result [19] has appeared:

Theorem 9 *Let $||f||^2 = \int_{-1}^{1}(1-t^2)^\alpha f(t)^2\, dt$, $P \in W_n$ $(n \geq 2)$ and $\alpha > 1$ real. Then we have $(n = 2m)$*

$$||P'||^2 \geq \frac{n^2(2n+2\alpha+1)}{4(n+\alpha-1)(n+\alpha)} ||P||^2,$$

with equality if and only if $P(t) = c(1-t^2)^m$. If $P(\pm 1) = 0$, then the previous inequality remains valid for $\alpha > -1$.

This result was proved earlier by Varma for the cases $\alpha = 0$ and $\alpha = 1$. In the same paper [19], Underhill and Varma investigated the corresponding inequality in L^4 norm for $\alpha = 3$:

Theorem 10 *Let $||f||_4^4 = \int_{-1}^{1}(1-t^2)^3 f(t)^4\, dt$ and $P \in W_n$. Then we have $(n = 2m)$*

$$||P'||_4^4 \geq \frac{3n^3(4n+7)(4n+5)}{4(4n+6)(4n+4)(4n+2)} ||P||_4^4,$$

with equality if and only if $P(t) = c(1-t^2)^m$.

Also, they considered the cases when $\alpha = 1$ and $\alpha = 2$, as well as an inequality in L^r norm, when $r \geq 2$ is even. In [24] Varma proved:

Theorem 11 *Let $P \in W_n$, subject to the condition $P(1) = 1$. Then, for $r \geq 1$, we have*

$$\int_{-1}^{1} |P'(t)|^r\, dt \geq \frac{n^r}{2^{r-1}((n-1)r+1)},$$

with equality if and only if $P(t) = ((1+t)/2)^n$.

3 Bojanov's Solution

More general results on Turán type inequalities were obtained by Bojanov [3]. Introducing the notations

$$p_{n,k}(t) = (-1)^{n-k} \frac{n^n}{2^n k^k (n-k)^{n-k}} (t+1)^k (t-1)^{n-k},$$

for $k = 0, 1, \ldots, n$ ($n \in \mathbb{N}$), Bojanov [3] proved the following results:

Theorem 12 *Let $x \mapsto \varphi(x)$ be any continuously differentiable, strictly increasing convex function in $[0, +\infty)$ and let*

$$A_{n,m} = \min_{0 \le k \le n} \left\{ \int_{-1}^{1} \varphi(|p_{n,k}^{(m)}(t)|)\, dt \right\}.$$

Then for every $n \in \mathbb{N}$ and $m \in \{1, \ldots, n\}$, the inequality

$$\int_{-1}^{1} \varphi(|P^{(m)}(t)|)\, dt \ge A_{n,m} \|P\|_\infty \qquad (P \in W_n)$$

holds, where the constant $A_{n,m} > 0$ is the best possible.

Theorem 13 *Let $B_{n,m} = \min_{0 \le k \le n} \{\|p_{n,k}^{(m)}\|_\infty\}$. For any given n and m, the inequality*

$$\|P^{(m)}\|_\infty \ge B_{n,m} \|P\|_\infty \qquad (P \in W_n)$$

holds, where the constant $B_{n,m}$ is the best possible.

Using this theorem one could get the exact previous result of Erőd [7] and Babenko and Pichugov [2], treating the case $m = 1$ and $m = 2$, respectively. In the first case we have that

$$B_{n,1} = \|p'_{n,k}\|_\infty \quad \text{for} \quad k = \left[\frac{n}{2}\right].$$

Combining an idea of Babenko and Pichugov [2] with Theorem 13, Bojanov [3] obtained an explicit value of $B_{n,2}$.

Following Bojanov [3], let $P \in W_n$ and $t_1 \le t_2 \le \cdots \le t_n$ be the zeros of $t \mapsto P(t)$. Then we have

$$P'(t) = P(t)\sigma(t), \quad P''(t) = P'(t)\sigma(t) + P(t)\sigma'(t) \quad (P \in W_n),$$

where

$$\sigma(t) = \sum_{\nu=1}^{n} \frac{1}{t - t_k}.$$

Suppose that $||P||_\infty = |P(\tau)| = 1$ and $\tau \in (-1,1)$. Then $P'(\tau) = 0$ and therefore $\sigma(\tau) = 0$. Thus, $|P''(\tau)| = |\sigma'(\tau)|$. Choose $P = p_{n,k}$, where $k = 1, \dots, n-1$. Then $\tau = b_{n,k} = (2k-n)/n$ and

$$\sigma(t) = \frac{k}{t+1} + \frac{n-k}{t-1}.$$

Therefore,

$$||p''_{n,k}||_\infty \geq |p''_{n,k}(\tau)| = |\sigma'(\tau)| = \frac{n^2}{4}\Big(\frac{1}{k} + \frac{1}{n-k}\Big).$$

But the last expression attains its minimal value for $k = [n/2]$ and this minimal value is n for even n, respectively $n\big(1 + 1/(n^2-1)\big)$, for odd n. Adding the obvious fact that

$$||p''_{n,1}||_\infty = ||p''_{n,n}||_\infty = \frac{1}{4}n(n-1) \geq n \qquad (\text{for } n > 4),$$

we get $B_{n,2} = n$ for even $n \geq 6$, and

$$B_{n,2} \geq n\Big(1 + \frac{1}{n^2-1}\Big) \quad \text{for odd } n \geq 5.$$

Bojanov [3] also proved:

Theorem 14 *Let $x \mapsto \varphi(x)$ be any continuously differentiable, strictly increasing convex function in $[0, +\infty)$. Then for every $n \in \mathbb{N}$ and $m \in \{1, \dots, n\}$,*

$$\int_{-1}^{1} \varphi(|P^{(m)}(t)|)\,dt \geq \int_{-1}^{1} \varphi(|p^{(m)}_{n,n}(t)|)\,dt$$

for every polynomial $P \in W_n$ such that $P(1) = 1$.

If $\varphi(x) = x^r$ $(1 \leq r < +\infty)$ this theorem reduces to the following result:

Corollary 15 *Let $P \in W_n$, $P(1) = 1$, and $1 \leq r < +\infty$. Then*

$$||P^{(m)}||_r \geq \frac{n!}{2^m(n-m)!}\Big(\frac{2}{(n-m)r+1}\Big)^{1/r}.$$

Notice that for $m = 1$ this corollary gives Theorem 11.

Inequalities of Turán type for trigonometric polynomials were investigated by Babenko and Pichugov [1]–[2], Zhou [25], Tyrygin [17]–[18], and Bojanov [3]–[4].

4 A Result of Chen

In this section we mention a recent result of Chen [5], which can be expressed in the same way as the Markov's inequality in [11] (see also [13]).

We consider a general case with a given non-negative measure $d\sigma(t)$ on the real line $\mathbb{R}$, with compact support or otherwise, for which all moments

$$\mu_\nu = \int_{\mathbb{R}} t^\nu \, d\sigma(t), \qquad \nu = 0, 1, \ldots,$$

exist and are finite, and $\mu_0 > 0$. Then there exists a unique set of orthonormal polynomials $\pi_\nu(\cdot) = \pi_\nu(\cdot\,; d\sigma)$, $\nu = 0, 1, \ldots$, defined by

$$\pi_\nu(t) = a_\nu t^\nu + \text{lower degree terms}, \quad a_\nu > 0,$$

and

$$\int_{\mathbb{R}} \pi_\nu(t)\pi_\mu(t)\, d\sigma(t) = \delta_{\nu\mu}, \qquad \nu, \mu \geq 0. \tag{8}$$

For each polynomial $P \in \mathcal{P}_n$, with complex coefficients, we take

$$||P|| = \sqrt{(P, P)} = \Big(\int_{\mathbb{R}} |P(t)|^2 \, d\sigma(t) \Big)^{1/2},$$

where $(f, g) = \int_{\mathbb{R}} f(t)\overline{g(t)}\, d\sigma(t)$.

As a restricted subset of $\mathcal{P}_n$, Chen [5] took

$$W_n = \mathcal{P}_{n,m}(d\sigma) = \big\{ P \in \mathcal{P}_n \mid P \perp \mathcal{P}_{m-1} \big\},$$

i.e., $P \in W_n$ if $P \in \mathcal{P}_n$ and $(P, \pi_\nu) = 0$ for each $\nu = 0, 1, \ldots, m-1$.

Consider now the extremal problem

$$E_{n,m} = E_{n,m}(d\sigma) = \inf_{P \in W_n} \frac{||P^{(m)}||}{||P||} \qquad (1 \leq m \leq n). \tag{9}$$

Theorem 16 *The best constant $E_{n,m}$ defined in* (9) *is given by*

$$E_{n,m} = \big(\lambda_{\min}(G_{n,m})\big)^{1/2}, \tag{10}$$

where $\lambda_{\min}(G_{n,m})$ is the minimal eigenvalue of the matrix

$$G_{n,m} = \big[g_{i,j}^{(m)}\big]_{m \leq i,j \leq n},$$

whose elements are given by

$$g_{i,j}^{(m)} = \int_{\mathbb{R}} \pi_i^{(m)}(t)\pi_j^{(m)}(t)\, d\sigma(t), \qquad m \le i, j \le n. \tag{11}$$

An extremal polynomial is

$$P^*(t) = \sum_{\nu=m}^{n} c_\nu \pi_\nu(t),$$

where $\left[c_k, c_{k+1}, \ldots, c_n\right]^T$ *is an eigenvector of the matrix* $G_{n,m}$ *corresponding to the eigenvalue* $\lambda_{\min}(G_{n,m})$.

Proof. Let $P \in W_n$. Then we can write $P(t) = \sum_{\nu=m}^{n} c_\nu \pi_\nu(t)$ and

$$P^{(m)}(t) = \sum_{\nu=m}^{n} c_\nu \pi_\nu^{(m)}(t), \qquad m \le n,$$

where the coefficients c_ν are uniquely determined. Hence, by (8) and (11), we have

$$\|P\|^2 = \sum_{\nu=m}^{n} |c_\nu|^2 \qquad \text{and} \qquad \|P^{(m)}\|^2 = \sum_{i,j=m}^{n} c_i \bar{c}_j g_{i,j}^{(m)}.$$

Then

$$\frac{\|P^{(m)}\|^2}{\|P\|^2} = \frac{\sum_{i,j=m}^{n} c_i \bar{c}_j g_{i,j}^{(m)}}{\sum_{i=m}^{n} |c_i|^2} = \frac{\langle G_{n,m}\boldsymbol{c}, \boldsymbol{c}\rangle}{\langle \boldsymbol{c}, \boldsymbol{c}\rangle}, \tag{12}$$

where $\langle \cdot, \cdot \rangle$ denotes the standard inner product in an $(n-m+1)$-dimensional space.

The matrix $G_{n,m}$ is evidently positive definite. Since the right side in (12) is not smaller than the minimal eigenvalue of this matrix, we obtain

$$\|P^{(m)}\|^2 \ge \lambda_{\min}(G_{n,m})\|P\|^2. \tag{13}$$

In order to show that $E_{n,m}$, given by (10), is the best possible, we note that (13) reduces to an equality if we put $P(t) = P^*(t) = \sum_{\nu=m}^{n} c_\nu^* \pi_\nu(t)$, where $\left[c_m^*, c_{m+1}^*, \ldots, c_n^*\right]^T$ is an eigenvector of the matrix $G_{n,m}$ corresponding to $\lambda_{\min}(G_{n,m})$. Q.E.D.

An alternative result like Theorem 16 is the following theorem:

Theorem 17 *Let* $Q_{n,m} = [q_{ij}^{(m)}]_{m \le i,j \le n}$ *be an upper triangular matrix of the order* $n - m + 1$*, whose elements* $q_{ij}^{(m)}$ *are given by the following inner product*

$$q_{ij}^{(m)} = (\pi_j^{(m)}, \pi_{i-m}) \qquad (m \le i, j \le n).$$

Then the best constant $E_{n,m}$ *defined in* (9) *is given by*

$$E_{n,m} = \left(\lambda_{\min}(Q_{n,m}Q_{n,m}^T)\right)^{1/2}. \tag{14}$$

Alternatively, (14) *can be expressed in the form*

$$E_{n,m} = \left(\lambda_{\max}(C_{n,m})\right)^{-1/2}, \tag{15}$$

where $C_{n,m} = \left(Q_{n,m}Q_{n,m}^T\right)^{-1}$.

Proof. It is enough to consider only a real polynomial set $\mathcal{P}_n$. Let $P \in W_n$ and $\pi_j^{(m)}(t) = \sum_{i=m}^{j} q_{ij}^{(m)} \pi_{i-m}(t)$, where $q_{ij}^{(m)} = (\pi_j^{(m)}, \pi_{i-m})$. Then

$$P^{(m)}(t) = \sum_{j=m}^{n} c_j \sum_{i=m}^{j} q_{ij}^{(m)} \pi_{i-m}(t) = \sum_{i=m}^{n} \Big(\sum_{j=m}^{n} c_j q_{ij}^{(m)} \Big) \pi_{i-m}(t)$$

and

$$\|P^{(m)}\|^2 = \sum_{i=m}^{n} \Big(\sum_{j=i}^{n} c_j q_{ij}^{(m)} \Big)^2 = \sum_{i=m}^{n} y_i^2,$$

where

$$y_i = \sum_{j=i}^{n} c_j q_{ij}^{(m)}, \qquad i = m, \ldots, n. \tag{16}$$

Let $\boldsymbol{c} = [c_m, \ldots, c_n]^T$, $\boldsymbol{y} = [y_m, \ldots, y_n]^T$, and $Q_{n,m} = [q_{ij}^{(m)}]_{m \le i,j \le n}$. Since $\boldsymbol{y} = Q_{n,m}\boldsymbol{c}$, it follows that

$$\frac{\|P^{(m)}\|^2}{\|P\|^2} = \frac{\langle \boldsymbol{y}, \boldsymbol{y} \rangle}{\langle \boldsymbol{c}, \boldsymbol{c} \rangle} = \frac{\langle \boldsymbol{y}, \boldsymbol{y} \rangle}{\langle (Q_{n,m}Q_{n,m}^T)^{-1}\boldsymbol{y}, \boldsymbol{y} \rangle}.$$

Thus (14) and (15) hold. Q.E.D.

Now, we will consider a few special measures.

1° $d\sigma(t) = e^{-t^2}dt$, $-\infty < t < +\infty$. Here we have

$$\pi_\nu(t) = \hat{H}_\nu(t) = (\sqrt{\pi}\, 2^\nu \nu!)^{-1/2} H_\nu(t),$$

where H_ν is a Hermite polynomial of degree ν. Since

$$H'_\nu(t) = 2\nu H_{\nu-1}(t) \qquad \text{and} \qquad \hat{H}'_\nu(t) = \sqrt{2\nu}\hat{H}_{\nu-1}(t),$$

we have

$$\hat{H}_\nu^{(m)}(t) = \sqrt{2\nu}\sqrt{2(\nu-1)}\cdots\sqrt{2(\nu-m+1)}\hat{H}_{\nu-m}(t),$$

i.e.,

$$\hat{H}_\nu^{(m)}(t) = \sqrt{2^m m! \binom{\nu}{m}}\, \hat{H}_{\nu-m}(t),$$

and

$$g_{ij}^{(m)} = 2^m m! \binom{i}{m} \delta_{ij}, \qquad m \le i, j \le n.$$

Thus, we find $\lambda_{\min}(G_{n,m}) = 2^m m!$ and $E_{n,m} = 2^{m/2}\sqrt{m!}$.

2° $d\sigma(t) = t^s e^{-t} dt$, $0 < t < +\infty$. Here we have the generalized Laguerre case with

$$\pi_\nu(t) = \hat{L}_\nu^s(t) = \sqrt{\nu!/\Gamma(\nu+s+1)} \sum_{i=0}^{\nu} (-1)^{\nu-i} \binom{\nu+s}{\nu-i} \frac{t^i}{i!},$$

where Γ is the gamma function.

First, we consider the simplest case where $m = 1$. Since

$$\frac{d}{dt}\hat{L}_j^s(t) = \sum_{i=1}^{j} q_{ij}^{(1)} \hat{L}_{i-1}^s(t), \qquad q_{ij}^{(1)} = -\sqrt{\frac{j!}{\Gamma(j+s+1)}} \cdot \sqrt{\frac{\Gamma(i+s)}{(i-1)!}},$$

from the equalities (16), it follows that

$$c_i = y_{i+1} - \sqrt{\frac{i+s}{i}}\, y_i, \qquad i = 1, \ldots, n,$$

where we put $y_{n+1} = 0$. The elements $p_{ij}^{(1)}$ of the matrix $P_{n,1} = Q_{n,1}^{-1}$ are

$$p_{ij}^{(1)} = -\sqrt{1 + \frac{s}{i}}, \quad i = 1, \ldots, n; \quad p_{i,i+1}^{(1)} = 1, \quad i = 1, \ldots, n-1;$$

$$p_{ij}^{(1)} = 0, \quad \text{otherwise},$$

so that $C_{n,1} = P_{n,1}^T P_{n,1} = -J_n$, where

$$J_n = \begin{bmatrix} \alpha_0 & \sqrt{\beta_1} & & & \mathbf{O} \\ \sqrt{\beta_1} & \alpha_1 & \sqrt{\beta_2} & & \\ & \sqrt{\beta_2} & \alpha_2 & \ddots & \\ & & \ddots & \ddots & \sqrt{\beta_{n-1}} \\ \mathbf{O} & & & \sqrt{\beta_{n-1}} & \alpha_{n-1} \end{bmatrix}$$

and

$$\alpha_0 = -(1+s), \quad \alpha_\nu = -\Big(2 + \frac{s}{\nu+1}\Big), \quad \beta_\nu = 1 + \frac{s}{\nu}, \quad \nu = 1, \dots, n-1.$$

We see that J_n is the Jacobi matrix for monic orthogonal polynomials $\{Q_\nu\}$, which satisfy the following three-term recurrence relation

$$Q_{\nu+1}(t) = (t - \alpha_\nu)Q_\nu(t) - \beta_\nu Q_{\nu-1}(t), \quad \nu = 0, 1, \dots,$$

with $Q_{-1}(t) = 0$ and $Q_0(t) = 1$. The eigenvalues of $C_{n,1}$ are $\lambda_\nu = -t_\nu$, where $Q_n(t_\nu) = 0$ for $\nu = 1, \dots, n$.

The standard Laguerre case ($s = 0$) can be exactly solved. In fact, for $t = 2(z-1)$ with $-1 \le z \le 1$, we have

$$Q_\nu(t) = \cos(2\nu+1)\frac{\theta}{2} \Big/ \cos\frac{\theta}{2}, \qquad z = \cos\theta.$$

The eigenvalues of the matrix $C_{n,1}$ are

$$\lambda_\nu = -t_\nu = 4\sin^2\frac{(2\nu-1)\pi}{2(2n+1)}, \qquad \nu = 1, \dots, n.$$

Since $\lambda_{\max}(C_{n,1}) = \lambda_n$, we obtain

$$E_{n,1} = \Big(2\cos\frac{\pi}{2n+1}\Big)^{-1}.$$

Now, we consider the case when $m = 2$ and $s = 0$. First, we note that

$$\frac{d^m}{dt^m}\hat{L}_j(t) = (-1)^m \sum_{i=m}^{j} \binom{j-i+m-1}{m-1} \hat{L}_{i-m}(t).$$

The formulae (16), for $m = 2$, become $y_i = \sum_{j=i}^{n} (j - i + 1)c_j$, $i = 2, \dots, n$. Since $\Delta^2 y_i = c_i$ ($y_{n+1} = y_{n+2} = 0$), we find a five-diagonal symmetric matrix of order $n - 1$

$$C_{n,2} = \begin{bmatrix} 1 & -2 & 1 & & & & & \mathbf{O} \\ -2 & 5 & -4 & 1 & & & & \\ 1 & -4 & 6 & -4 & 1 & & & \\ & 1 & -4 & 6 & -4 & 1 & & \\ & & \ddots & \ddots & \ddots & \ddots & \ddots & \\ & & & 1 & -4 & 6 & -4 & 1 \\ & & & & 1 & -4 & 6 & -4 \\ \mathbf{O} & & & & & 1 & -4 & 6 \end{bmatrix}.$$

Thus, using the maximal eigenvalue of this matrix, we obtain the best constant $E_{n,2} = \left(\lambda_{\max}(C_{n,2})\right)^{-1/2}$. In the simplest case when $n = 2$ and $n = 3$ we have $E_{2,2} = 1$ and $E_{3,2} = (3 - 2\sqrt{2})^{1/2}$, respectively.

We conclude this paper with a remark that Varma [22] also studied an extremal problem on $(0, +\infty)$ with respect to the Laguerre measure, i.e., when $\|f\|_2^2 = \int_0^\infty e^{-t} f(t)^2 \, dt$.

Theorem 18 *Let P be an algebraic polynomial of degree n whose zeros τ_ν ($\nu = 1, \dots, n$) all lie in the interval $[0, \infty)$. If $P(0) = 0$ or $\sum_{\nu=1}^{n} \tau_\nu^{-1} \geq 1/2$, then*

$$\|P'\|_2^2 \geq \frac{n}{2(2n-1)} \|P\|_2^2.$$

The equality holds for $P(t) = t^n$.

References

[1] V. F. Babenko and S. A. Pichugov, Exact inequality for the derivative of a trigonometric polynomial having only real zeros, Mat. Zametki **39** (1986), 330–336 (Russian).

[2] V. F. Babenko and S. A. Pichugov, On inequality for the derivative of a polynomial with real zeros, Ukrain. Mat. Zh. **38** (1986), 411–416 (Russian).

[3] B. D. Bojanov, Polynomial inequalities, in "Open Problems in Approximation Theory" (B. Bojanov, Ed.), pp. 25–42, SCT, Singapore, 1993.

[4] B. Bojanov, Turán's inequalities for trigonometric polynomials, J. London. Math. Soc. (2) **53** (1996), 539–550.

[5] W. Chen, On the converse inequality of Markov type in L_2 norm, J. Math. Anal. Appl. **200** (1996), 708–716.

[6] V. K. Dzyadyk, "Introduction to the Theory of Uniform Approximation Functions by Polynomials", Nauka, Moscow, 1977 (Russian).

[7] J. Erőd, Bizonyos polinomok maximumának, Mat. Fiz. Lapok **46** (1939), 58–82.

[8] V. I. Ivanov, Some extremal properties of polynomials and inverse inequalities in approximation theory, Trudy Mat. Inst. Steklov **145** (1979), 79–110 (Russian).

[9] G. G. Lorentz, "Approximation of Functions", Holt, Rinehart and Winston, New York, 1966.

[10] G. Meinardus, "Approximation of Functions: Theory and Numerical Methods" Springer Verlag, Berlin – Heidelberg – New York, 1967.

[11] G. V. Milovanović, Various extremal problems of Markov's type for algebraic polynomials, Facta Univ. Ser. Math. Inform. **2** (1987), 7–28.

[12] G. V. Milovanović, D. S. Mitrinović and Th. M. Rassias, On some Turán's extremal problems for algebraic polynomials, in "Topics in Polynomials of One and Several Variables and Their Applications: A Mathematical Legacy of P. L. Chebyshev (1821–1894)" (Th. M. Rassias, H. M. Srivastava, A. Yanushauskas, Eds.), pp. 403–433, World Scientific, Singapore, 1993.

[13] G. V. Milovanović, D. S. Mitrinović and Th. M. Rassias, "Topics in Polynomials: Extremal Problems, Inequalities, Zeros", World Scientific, Singapore – New Jersey – London – Hong Kong, 1994.

[14] J. Szabados and A. K. Varma, Inequalities for derivatives of polynomial having real zeros, in "Approximation Theory III" (E. W. Cheney, Ed.), pp. 881–887, Academic Press, New York, 1980.

[15] S. A. Telyakovskiĭ, Research in the theory of approximation of functions at the mathematical institute of the academy of sciences, Trudi

Mat. Inst. Steklov. **182** (1988); English transl. in: Proc. Steklov Inst. Math. **1990**, no. 1, 141–197.

[16] P. Turán, Über die Ableitung von Polynomen, Composito Math. **7** (1939), 85–95.

[17] I. Ya. Tyrygin, Inequalities of P. Turán type in mixed integral metrics, Dokl. Akad. Nauk Ukrain. SSR Ser. A **1988**, no. 9, 14–17 (Russian).

[18] I. Ya. Tyrygin, On inequalities of Turán type in some integral metrics, Ukrain. Mat. Zh. **40** (1988), 256–260 (Russian).

[19] B. Underhill and A. K. Varma, An extension of some inequalities of P. Erdős and P. Turán concerning algebraic polynomials, Acta Math. Hung. **73** (1996), 1–28.

[20] A. K. Varma, An analogue of some inequalities of P. Turán concerning algebraic polynomials having all zeros inside $[-1, 1]$, Proc. Amer. Math. Soc. **55** (1976), 305–309.

[21] A. K. Varma, An analogue of some inequalities of P. Turán concerning algebraic polynomials having all zeros inside $[-1, 1]$. II, Proc. Amer. Math. Soc. **69** (1978), 25–33.

[22] A. K. Varma, Some inequalities of algebraic polynomials having real zeros, Proc. Amer. Math. Soc. **75** (1979), 243–250.

[23] A. K. Varma, Some inequalities of algebraic polynomials having all zeros inside $[-1, 1]$, Proc. Amer. Math. Soc. **88** (1983), 227–233.

[24] A. K. Varma, Some inequalities of algebraic polynomials, Proc. Amer. Math. Soc. **123** (1995), 2041–2048.

[25] S. P. Zhou, On Turán inequality in L_p-norm, J. Hangzhou Univ. **11** (1984), 28–33 (Chinese, English summary).

[26] S. P. Zhou, An extension of the Turán inequality in L^p for $0 < p < 1$, J. Math. Res. Exposition **6**, No. 2 (1986), 27–30.

[27] S. P. Zhou, Some remarks on Turán inequality, J. Approx. Theory **68** (1992), 45–48.

Recent Progress in Multivariate Markov Inequality

W. Pleśniak
Jagiellonian University of Cracow, Institute of Mathematics
Reymonta 4, 30-059 Kraków, Poland
plesniak@im.uj.edu.pl

Abstract

In this paper we present a survey on recent research on the multivariate Markov inequality. We illustrate the power of this inequality by giving a number of its applications in the theory of extension and polynomial approximation of $\mathcal{C}^\infty$ functions defined on compact subsets of $\mathbb{R}^n$

1 Markov Inequality

In 1889, A.A. Markov answered a question posed two years earlier by Mendeleev by showing that for every polynomial p in one variable

$$|p'(x)| \leq \deg^2 ||p||_{[-1,1]}, \quad \text{as } x \in [-1,1], \tag{MI_1}$$

where $||p||_I = \sup |p|(I)$. This result is best possible since for the Chebyshev polynomials $T_n(x) = \cos n \arccos x$ ($x \in [-1,1]$), $n = 1, 2, \ldots$, of degree n one has $|T_n'(\pm 1)| = n^2$.

Markov's inequality became soon a fascinating object of investigations. The reason lay with its numerous applications in different domains of mathematics and physics. For an interesting account of classical results in this direction and their refinements in the one-dimensional case, the reader is referred to [40] and [30]. Let us also mention an important contribution to this theory of Arun Kumar Varma discussed in [49].

A corresponding theory in the several variables case is relatively new and until the late 1970's all known extensions of Markov's inequality dealt practically with the case where the line-segment in (MI_1) is replaced by a convex compact subset of $\mathbb{R}^n$ with non-void interior (*see* e.g. [1]). One of the obstacles was the fact that for some cuspidal sets in $\mathbb{R}^n$ no multivariate counterpart of (MI_1) can be proved. A simple example was first given by Zerner [58].

Research partially supported by grant No 2 P03A 057 08 fromKBN (Committee for Scientific Research).

Example 1.1 Let $E = \{(x,y) \in \mathbb{R}^2 : 0 < y \leq \exp(-1/x), 0 < x \leq 1\} \cup \{(0,0)\}$ and let $P_k(x,y) = y(1-x)^k$ for $k = 1,2\ldots$. Then $\deg P_k = k+1$, $\|\partial P_k/\partial y\|_E = 1$, while $\|P_k\|_E < \exp(-\sqrt{k})$ for $k = 1,2,\ldots$, and therefore there are no constants $M > 0$ and $r > 0$ such that for each k,

$$\|\partial P_k/\partial y\|_E \leq M(k+1)^r \|P_k\|_E.$$

In the sequel, a compact subset of $\mathbb{R}^n$ is said to preserve (or admit) *Markov's inequality*, or simply to be *Markov*, if there exist constants $M > 0$ and $r > 0$ such that for each polynomial p in $\mathbb{R}^n$ we have

$$\|\text{grad}\, p\|_E \leq M(\deg p)^r \|p\|_E. \qquad (MI_n)$$

A satisfactory theory of the multivariate Markov inequality was developed in the last decade by W. Pawłucki and W. Pleśniak (see [32], [33], [34], [35], [36], [37], [39]), P. Goetheluck ([18], [19], [20]), M. Baran ([2], [3], [4], [5]), A. Jonsson ([22]), J. Siciak ([46]), A. Zeriahi ([57]), L. Bos and P.D. Milman ([16], [17]), A. Goncharov ([21]) and others. As in the one-dimensional case, such an inequality plays a crucial role in Bernstein-type characterization of $\mathcal{C}^\infty$ functions of several variables (see [32], [34], [37]). On the other hand, it was found by Pawłucki and Pleśniak that Markov's inequality is strictly connected with a classical problem of the existence of a continuous linear operator extending $\mathcal{C}^\infty$ functions from a (sufficiently regular) compact set in $\mathbb{R}^n$ to the whole space $\mathbb{R}^n$ (see [33], [37], [38], [57], [17], [21]), investigated earlier with the aid of Hironaka's theory of the resolution of singularities (*see* [11]) or results of Vogt and Wagner concerning the splitting of exact sequences of nuclear Fréchet spaces (*see* [50]). This illustrates the power of Markov's inequality.

The goal of this paper is to present a state-of-the-art survey of investigations concerning the inequality in question. Although some of the results presented here are valid in the space $\mathcal{C}^n$, we shall restrict ourselves to subsets of the space $\mathbb{R}^n$, since in the real case the theory seems to be more complete. In particular, we shall not discuss here relations between the (complex) Markov inequality and complex dynamics. We refer the reader to papers [23], [54] and [26] for some results in this topic.

We shall start with the following observation due to Goetgheluck ([18]):

Example 1.2 Let $E_k = \{(x,y) \in \mathbb{R}^2 : 0 \leq y \leq x^k, 0 \leq x \leq 1\}$ $(k \geq 1)$. Then the set E is a Markov set with exponent $r = 2k$. Moreover, the exponent $2k$ is best possible.

This example had inspired W. Pawłucki and W. Pleśniak to investigate the Markov property of semianalytic and subanalytic sets, and more general, sets with polynomial cusps. Let us recall that a subset E of $\mathbb{R}^n$ is said to be *semianalytic* if for each point $x \in \mathbb{R}^n$ one can find a neighbourhood U of x and a finite number of real analytic functions f_{ij} and g_{ij} defined in U, such that

$$E \cap U = \bigcup_i \bigcap_j \{f_{ij} > 0, g_{ij} = 0\}.$$

The projection of a semianalytic set need not be semianalytic (*see* [28, p. 133–135]). The class of sets obtained by enlarging that of semianalytic sets to include images under the projections has been called the class of subanalytic sets. More precisely, a subset E of $\mathbb{R}^n$ is said to be *subanalytic* if for each point $x \in \mathbb{R}^n$ there exists an open neighbourhood U of x such that $E \cap U$ is the projection of a bounded semianalytic set A in $\mathbb{R}^{n+m}$, where $m \geq 0$. If $n \geq 3$, the class of subanalytic sets is essentially larger than that of semianalytic sets, the classes being identical if $n \leq 2$ (*see* [28]). The union of a locally finite family and the intersection of a finite family of subanalytic sets is subanalytic. The closure, interior, boundary and complement of a subanalytic set is still subanalytic, the last property being a (non-trivial) theorem of Gabrielov. For further information on the geometry of semianalytic and subanalytic sets, we refer the reader to [13].

It is clear that the set E of Goetgheluck's example is semianalytic, whence subanalytic, while that of Zerner's example is not subanalytic, since it is too flat at the origin. It appears that the family of (fat) subanalytic sets is a subfamily of a family of sets admitting only polynomial-type cusps.

Definition 1.3 A subset E of $\mathbb{R}^n$ is said to be *uniformly polynomially cuspidal* (briefly, *UPC*) if one can choose constants $M > 0, m \geq 1$ and $d \in \mathbb{N}$, and a mapping $h : E \times [0,1] \to \bar{E}$ such that for each $x \in \bar{E}$, $h(x,1) = x$, $h(x,\cdot)$ is a polynomial of degree $\leq d$ and

$$\operatorname{dist}(h(x,t), \mathbb{R}^n \setminus E) \geq M(1-t)^m \quad \text{for} \quad (x,t) \in \bar{E} \times [0,1].$$

By an application of the famous Hironaka *rectilinealization theorem*, it was proved in [32] that

Theorem 1.4 *Every bounded subanalytic subset of* $\mathbb{R}^n$ *with* $\operatorname{int}E$ *dense in* E *is* UPC.

The family of UPC sets is essentially larger than that of subanalytic sets. A simple example of a UPC set which is not subanalytic is given by $[0,1] \times [-1,1] \setminus E$, where E is the set of Zerner's example. For other examples, see [32].

The UPC sets are important from the pluripotential theory point of view, since they admit (pluricomplex) Green functions with nice continuity properties. To explain this, let us suppose that E is a compact subset of $\mathbb{C}^n$. We set

$$V_E(z) = \sup\{u(z) : \ u \in \mathcal{L}(\mathbb{C}^n), \ u|_E \leq 0\}, \ z \in \mathbb{C}^n,$$

where

$$\mathcal{L}(\mathbb{C}^n) = \{u \in PSH(\mathbb{C}^n) : \ \sup_{z \in \mathbb{C}^n} [u(z) - \log(1 + |z|)] < \infty\}$$

is the *Lelong class* of plurisubharmonic functions with minimal growth. The function V_E is called the *(plurisubharmonic) extremal function* associated with E (*see* [45]). Its upper semicontinuous regularization V_E^* is a multidimensional counterpart of the classical *Green function* for $\mathbb{C} \setminus \hat{E}$, where $\hat{E}$ is the polynomial hull of E, since by the pluripotential theory due to E. Bedford and B.A. Taylor it is a solution of the homogeneous complex *Monge-Ampère equation*, which is reduced in the one dimensional case to the *Laplace equation* (for references, *see* [25]). By [45]

$$V_E(z) = \sup\{\tfrac{1}{\deg p}\log|p(z)| : \ p \text{ is a polynomial with } \deg p \geq 1 \text{ and } \|p\|_E \leq 1\}. \tag{1}$$

In other words, $V_E = \log \Phi_E$, where Φ_E is a *(polynomial) extremal function* introduced by Siciak [43]. Now the set E is said to have *Hölder's Continuity Property* (briefly, HCP) if there exist positive constants M and s such that

$$V_E(z) \leq M\delta^s \quad \text{as } \operatorname{dist}(z, E) \leq \delta \leq 1. \tag{HCP}$$

The importance of the class UPC is explained by the following

Proposition 1.5 ([32, Theorem 4.1]) *If E is a compact UPC subset of $\mathbb{R}^n$ with parameter m then E satisfies (HCP) with exponent $s = 1/2[m]$, where $[m] := k$ as $k - 1 < m \leq k$ with $k \in \mathbb{Z}$.*

Here and later on $\mathbb{R}^n$ is treated as a subset of $\mathbb{C}^n$ such that $\mathbb{R}^n = \{(z_1, \ldots, z_n) \in \mathbb{C}^n : \Im z_j = 0,\ j = 1, \ldots, n\}$. Now we can come back to the multivariate Markov inequality. By a simple observation that goes back to Siciak [44],

If E is HCP then it preserves Markov's inequality (MI_n).

Let us recall a simple argument leading to this observation. By Cauchy's Integral Formula, if $a \in E$ and p is a polynomial in n variables of degree at most k then

$$\frac{\partial p}{\partial z_i}(a) = \frac{1}{(2i\pi)^n} \int\limits_{\substack{|z_j - a_j| = \delta \\ j=1,\ldots,n}} \frac{p(z)}{(z_i - a_i)\prod_{j=1}^{n}(z_j - a_j)} dz,$$

for $i = 1, \ldots, n$. Hence, if E is HCP, by (1.1) we get

$$|\frac{\partial p}{\partial z_i}(a)| \leq \frac{1}{\delta}\|p\|_E \exp(Mk\delta^s),$$

whence by putting $\delta = k^{-1/s}$ we complete the proof.

Thus we have yielded a number of examples of sets admitting the multivariate Markov inequality. These are all UPC subsets of $\mathbb{R}^n$. There are, however, sets that are HCP without being UPC. Such (Cantor-type) sets were first constructed by Siciak [46] and Jonsson [22]. For other examples, see [47] and [39]. The problem of whether the classical Cantor ternary set has Markov's property has appeared more difficult and a positive answer was first given by Białas and Volberg [10] who showed that this set is even HCP. It is worth adding that there are also Cantor-type sets which do not preserve Markov's inequality and, at the same time, they are regular with respect to the (classical) Green function (*see* [36], [20], [52]). Up to now, the problem of whether Markov's property of E implies that E is HCP remains open. We know only that the answer is "yes" for a class of one-dimensional Cantor-type sets (*see* [27], [8]). In general, we even do not know whether Markov's property of E implies the continuity of the Green function V_E or else non-pluripolarity of E. We recall that a subset E of $\mathbb{C}^n$ is said to be *pluripolar* if one can find a plurisubharmonic function u on $\mathbb{C}^n$ such that $E \subset \{u = -\infty\}$. However, just recently Białas-Cieź [9] has showed that any plane compact Markov set has a positive logarithmic capacity, whence it is not polar.

2 Polynomial approximation of $\mathcal{C}^\infty$ functions

By the celebrated Bernstein theorem (*see* [7]), a function $f : I = [a, b] \subset \mathbb{R} \to \mathbb{C}$ extends to a $\mathcal{C}^\infty$ function in $\mathbb{R}$ if and only if, for each $s > 0$,

$$\lim_{k\to\infty} k^s \mathrm{dist}_I(f, \mathcal{P}_k) = 0.$$

By a standard argument, this beautiful result can be easily extended to the case of functions defined on (fat) convex compact sets in $\mathbb{R}^n$. In general, the classical proof of Bernstein's theorem does not work, since, contrary to the case of an interval in $\mathbb{R}$, there are compact sets E in $\mathbb{R}^n$ and functions $f : E \to \mathbb{R}$ such that f are $\mathcal{C}^\infty$ in $\operatorname{int} E$ and extend together with all their derivatives to continuous functions in E, but do not admit any $\mathcal{C}^\infty$ extension to an open neighbourhood of E. A standard example is the set $E = E_1 \cup E_2 \subset \mathbb{R}^2$, where $E_1 = \{(x, y) \in \mathbb{R}^2 : 0 \le x \le 1, g(x) \le y \le 1\}$ with $g(x) = \exp(-1/x)$ as $0 < x \le 1$ and $g(0) = 0$, and $E_2 = [0, 1]\times[-1, 0]$, and the function $f(x, y) = g(x)$ if $(x, y) \in E_1$ and $f(x, y) \equiv 0$ if $(x, y) \in E_2$. The problem was solved by Pawłucki and Pleśniak [32] with the aid of special cut-off functions.

Lemma 2.1 ([29], [53]) *There are positive constants C_α ($\alpha \in \mathbb{Z}_+^n$) such that for any compact set E in $\mathbb{R}^n$ and for any $\epsilon > 0$ one can choose a $\mathcal{C}^\infty$ function u_ϵ such that $u_\epsilon = 1$ in a neighbourhood of E, $u_\epsilon(x) = 0$ if dist$(x, E) \ge \epsilon$ and for all $\alpha \in \mathbb{Z}_+^n$ and all $x \in \mathbb{R}^n$, $|D^\alpha u(x)| \le C_\alpha \epsilon^{-|\alpha|}$.*

In the sequel, we shall say that a subset E of $\mathbb{R}^n$ is $\mathcal{C}^\infty$ *determining* if for each function $f \in \mathcal{C}^\infty(\mathbb{R}^n)$, if $f = 0$ on E, then for each $\alpha \in \mathbb{Z}_+^n$, $D^\alpha f = 0$ on E. There is a simple geometric criterion of Glaeser for E to be $\mathcal{C}^\infty$ determining: $E \subset \mathbb{R}^n$ is $\mathcal{C}^\infty$ determining if and only if for each point $x \in E$ and each $\epsilon > 0$, if $E \cap B(x, \epsilon)$ is contained in a $\mathcal{C}^\infty$ submanifold X of $\mathbb{R}^n$ then $\dim X = n$ (cf [24]). It can also be proved (*see* [37, Remark 3.5]) that any compact Markov set in $\mathbb{R}^n$ is $\mathcal{C}^\infty$ determining.

Now, we are able to state a multivariate version of Bernstein's theorem.

Theorem 2.2 ([32, Theorem 5.1], [37, Theorem 3.3]) *If a compact set E in $\mathbb{R}^n$ is $\mathcal{C}^\infty$ determining then the following statements are equivalent:*

(i) E has Markov's property;

(ii) E has the following property: there exist positive constants M and r such that for each polynomial $p \in \mathcal{P}_k$ ($k = 1, 2, \dots$) one has

$$|p(x)| \le M\|p\|_E \quad \textit{if dist}(x, E) \le 1/k^r;$$

(iii) *(Bernstein's Theorem) for every function $f : E \to \mathbb{R}$, if the sequence $\{dist_E(f.\mathcal{P}_k)\}$ is rapidly decreasing, i.e. for each $s > 0$, $k^s dist_E(f, \mathcal{P}_k) \to 0$ as $k \to \infty$, then f extends to a $\mathcal{C}^\infty$ function $\tilde{f}$ in $\mathbb{R}^n$.*

Here $\mathcal{P}_k$ denotes the space of polynomials of degree $\leq k$ and $\mathrm{dist}_E(f, \mathcal{P}_k) := \inf\{\|f - p\|_E : \ p \in \mathcal{P}_k\}$. For the proof of the above theorem we refer the reader to [37]. We only mention here that a $\mathcal{C}^\infty$ extension $\tilde{f}$ of f satisfying the requirements of Theorem 2.2 is given by

$$\tilde{f} := p_0 + \sum_{k=1}^{\infty} u_k(p_k - p_{k-1}),$$

where p_k is a polynomial of degree at most k such that $\|f - p_k\|_E = \mathrm{dist}_E(f, \mathcal{P}_k)$ and u_k is a $\mathcal{C}^\infty$ function of Lemma 2.1 corresponding to $\epsilon = 1/k^r$ with r determined by both (i) and (ii) of Theorem 2.2.

3 Extension of $\mathcal{C}^\infty$ functions from compact sets in $\mathbb{R}^n$

Let E be a compact set in $\mathbb{R}^n$ and let $\mathcal{C}^\infty(E)$ denote the space of all functions $f : E \to \mathbb{C}$ that can be extended to $\mathcal{C}^\infty$ functions in the whole space $\mathbb{R}^n$. We give the space $\mathcal{C}^\infty(E)$ the topology τ_Q endowed with the family of the seminorms

$$q_{K,k}(f) := \inf\{\|g\|_{K,k} : g \in \mathcal{C}^\infty(\mathbb{R}^n), \ g_{|E} = f\}, \tag{3.1}$$

where $k = 0, 1, \ldots$, K is any compact subset of $\mathbb{R}^n$, and

$$\|g\|_{K,k} := \max\{\sup |D^\alpha f(x)| : \ x \in K, |\alpha| \leq k\} \tag{3.2}$$

or, equivalently, the topology endowed with the family of the seminorms

$$q_k(f) := \inf\{\|g\|_{P,k} : g \in \mathcal{C}^\infty(\mathbb{R}^n), \ g_{|E} = f\}, \tag{3.3}$$

where P is a fixed compact cube in $\mathbb{R}^n$ such that $E \subset \mathrm{int}\, P$. Thus τ_Q is the quotient topology of the space $\mathcal{C}^\infty(\mathbb{R}^n)/\mathcal{I}(E)$, where $\mathcal{C}^\infty(\mathbb{R}^n)$ is endowed with the natural topology determined by the seminorms $\|\cdot\|_{K,k}$ and $\mathcal{I}(E) = \{f \in \mathcal{C}^\infty(\mathbb{R}^n) : \ f_{|E} = 0\}$. Since $\mathcal{C}^\infty(\mathbb{R}^n)$ is complete and since $\mathcal{I}(E)$ is a closed subspace of $\mathcal{C}^\infty(\mathbb{R}^n)$, the quotient space $\mathcal{C}^\infty(\mathbb{R}^n)/\mathcal{I}(E)$ is also complete, whence $(\mathcal{C}^\infty(E), \tau_Q)$ is a Fréchet space. If the set E is $\mathcal{C}^\infty$ determining, this space can be identified with the space of Whitney fields on E. Let us recall that a $\mathcal{C}^\infty$ *Whitney field* on E is a vector $F = (F^\alpha)$ $(\alpha \in \mathbb{Z}^n_+)$, where each F^α is a continuous function defined on E, such that

$$|||F|||_{E,k} := \|F\|_{E,k} + \sup\{|(R^k_x F)^\alpha(y)/|x-y|^{k-|\alpha|} : \ x, y \in E, \ x \neq y, \ |\alpha| \leq k\} < \infty,$$

for $k = 0, 1, \ldots$, where

$$\|F\|_{E,k} = \sup\{|F^\alpha(x)| : \ x \in E, \ |\alpha| \leq k\}$$

and

$$(R_x^K F)^\alpha(y) = F^\alpha(y) - \sum_{|\beta|\le k-|\alpha|} (1/\beta!)F^{\alpha+\beta}(x)(y-x)^\beta.$$

Let us denote by $\mathcal{E}(E)$ the space of all $\mathcal{C}^\infty$ Whitney fields on E endowed with the topology τ_W determined by the seminorms $|||\cdot|||_{E,k}$ $(k = 0, 1, \dots)$. It is a Fréchet space. By Whitney's Extension Theorem ([56]), $F \in \mathcal{E}(E)$ if and only if there exists a $\mathcal{C}^\infty$ function f in $\mathbb{R}^n$ such that for all $\alpha \in \mathbb{Z}_+^n$, $D^\alpha f_{|E} = F^\alpha$. In particular, if E is $\mathcal{C}^\infty$ determining, the mapping $J : \mathcal{C}^\infty(E) \ni f \to J(f) = (D^\alpha g_{|E})_{\alpha\in\mathbb{Z}_+^n}$, where $g \in \mathcal{C}^\infty(\mathbb{R}^n)$ and $g_{|E} = f$, is a linear bijection of $\mathcal{C}^\infty(E)$ onto $\mathcal{E}(E)$. Since, for a cube P such that $E \subset \operatorname{int}P$, the seminorms $|||\cdot|||_{P,k}$ and $\|\cdot\|_{P,k}$ are equivalent (*see* [56]), the linear bijection J is a continuous mapping, whence by Banach's theorem, it is a linear isomorphism.

Contrary to the case of $\mathcal{C}^k$ jets, for k finite, Whitney's proof does not yield a continuous linear operator extending jets from $\mathcal{E}(E)$ to functions in $\mathcal{C}^\infty(\mathbb{R}^n)$. Moreover, such an operator does not in general exist, which is e.g. the case when E is a single point (*see* [53, p. 79]). The problem of the existence of such an operator has a long history. Positive examples were first given by Mityagin [31] and Seeley [42] (case of a half-space in $\mathbb{R}^n$). Stein [48] showed that such an operator exists if E is the closure of a domain in $\mathbb{R}^n$ whose boundary is locally of class $Lip\,1$. Bierstone [11] extended this result to the case of $Lip\,\alpha$ domains with $0 < \alpha \le 1$. In [11], Bierstone also proved that an extension operator exists if E is a fat (i.e. $\overline{\operatorname{int} E} \supset E$) closed subanalytic subset of $\mathbb{R}^n$. His method is essentially based on the famous Hironaka Desingularization Theorem. Another method based on results of Vogt and Wagner concerning the splitting of exact sequences of nuclear Fréchet spaces was applied by Tidten [50] to show the existence of an extension operator for closed sets in $\mathbb{R}^n$ admitting some polynomial-type cusps. For a nice introduction to the theory of extension of $\mathcal{C}^\infty$ functions and related problems we refer the reader to [12].

All the above mentioned sets are UPC (whence they are Markov). It appears that some restrictions concerning cuspidality of E are necessary, since Tidten [50] proved that for the set E of Example 1.1 (which is not Markov) there is no continuous linear extension operator from $(\mathcal{C}^\infty(E), \tau_Q)$ to the space $\mathcal{C}^\infty(\mathbb{R}^2)$. However, Pawłucki and Pleśniak ([33], [37]) showed that if E is a Markov compact subset of $\mathbb{R}^n$, then one can easily construct a continuous linear operator extending $\mathcal{C}^\infty$ functions on E to $\mathcal{C}^\infty$ functions in $\mathbb{R}^n$. This result has yielded a new field of applications of Markov's inequality.

In order to present the Pawłucki and Pleśniak construction we have to recall the definition of the multivariate Lagrange interpolation polynomials. Let $\mathcal{P}_k$ be the vector space of all polynomials from $\mathbb{C}^n$ to $\mathbb{C}$ of degree at most k. One can easily check that $m_k := \dim \mathcal{P}_k(\mathbb{C}^n) = \binom{n+k}{k}$. Let

$$\kappa : \{1, 2, \dots\} \ni j \to \kappa(j) = (\kappa_1(j), \dots, \kappa_n(j)) \in \mathbb{Z}_+^n$$

be a one-to-one mapping such that for each j, $|\kappa(j)| \le |\kappa(j+1)|$. Let $e_j(x) := x^{\kappa(j)}$, $j = 1, 2, \dots$. Then the set of monomials $e_1, \dots, e_{m_k}$ is a basis of the space

$\mathcal{P}_k$. (As usual, if $\alpha = (\alpha_1, \ldots, \alpha_n) \in \mathbb{Z}_+^n$, then $|\alpha|$ stands for $\alpha_1 + \cdots + \alpha_n$ and $x^\alpha = x_1^{\alpha_1} \cdots x_n^{\alpha_n}$.)

Let $t^{(l)} = \{t_1, \ldots, t_l\}$ be a system of l points of $\mathbb{C}^n$. Consider the *Vandermonde determinant*

$$V(t^{(l)}) = V(t_1, \ldots, t_l) := \det[e_j(t_i)],$$

where $i, j \in \{1, \ldots, l\}$. If $V(t^{(l)}) \neq 0$, we define, for $j = 1, \ldots, l$,

$$L^{(j)}(x, t^{(l)}) := V(t_1, \ldots, t_{j-1}, x, t_{j+1}, \ldots, t_l)/V(t^{(l)}).$$

Since $L^{(j)}(t_i, t^{(l)}) = \delta_{ij}$ (Kronecker's symbol), we get the following *Lagrange Interpolation Formula* (cf [45, Lemma 2.1]).

(LIF) *If $p \in \mathcal{P}_k$ and $t^{(m_k)}$ is a system of m_k points of $\mathbb{C}^n$ such that $V(t^{(m_k)}) \neq 0$, then $p(x) = \sum_{j=1}^{m_k} p(t_j) L^{(j)}(x, t^{(m_k)})$ for $x \in \mathbb{C}^n$.*

Now let E be a compact subset of $\mathbb{C}^n$. A system $t^{(l)}$ of l points $t_1, \ldots, t_l$ of E is called a *Fekete-Leja system of extremal points* of E of order l if $|V(t^{(l)})| \geq |V(s^{(l)})|$ for all systems $s^{(l)} = \{s_1, \ldots, s_l\} \subset E$. In the sequel, we shall say that a set $E \subset \mathbb{C}^n$ is $\mathcal{P}$ *determining* (or *unisolvent*, cf [45]), if for each $p \in \mathcal{P} = \bigcup_{k=0}^\infty \mathcal{P}_k$, $p = 0$ on E implies $p = 0$ in $\mathbb{C}^n$. It can be easily proved (*see* [45]) that E is $\mathcal{P}$ determining if and only if for each l, $V_l(E) := \sup\{|V(x_1, \ldots, x_l)| : \{x_1, \ldots, x_l\} \subset E\} \neq 0$. In such a case, given a function $f : E \to \mathbb{C}$ and a system $t^{(m_k)}$ of extremal points of E of order m_k, we can define

$$L_k f(x) = \sum_{j=1}^{m_k} f(t_j) L^{(j)}(x, t^{(m_k)})$$

to be a *Lagrange interpolation polynomial* of f of degree k.

Suppose f is continuous on E. Let p_k be any polynomial of degree $\leq k$ such that $\|f - p_k\|_E = \operatorname{dist}_E(f, \mathcal{P}_k)$. Then by (LIF),

$$\|f - L_k f\|_E \leq \|f - p_k\|_E + \|L_k f - L_k p_k\|_E \leq (m_k + 1)\|f - p_k\|_E \leq 4k^n \operatorname{dist}_E(f, \mathcal{P}_k).$$

Hence by a multivariate Jackson theorem (*see* e.g. [51], [38]), if $E \subset \mathbb{R}^n$ is $\mathcal{P}$ determining and $f \in \mathcal{C}^\infty(E)$, the sequence $\{\|f - L_k f\|_E\}$ is rapidly decreasing (*cf.* Theorem 2.2). So the Lagrange interpolation polynomials of f give essentially the same rate of approximation to f on E as those of best uniform approximation. In addition, the operators $f \to L_k f$ are linear.

In order to state the main result of this chapter, we give the space $\mathcal{C}^\infty(E)$ another topology connected with Jackson's theorem. To this end, let us put $d_{-1}(f) := \|f\|_E$, $d_0(f) := \operatorname{dist}_E(f, \mathcal{P}_0) = \inf\{\sup_{x \in E} |f(x) - c| : c \in \mathbb{C}\}$ and, for $k \geq 1$,

$$d_k(f) := \sup_{l \geq 1} l^k \operatorname{dist}_E(f, \mathcal{P}_l).$$

By Jackson's theorem the functionals d_k are seminorms on $\mathcal{C}^\infty(E)$. Let us denote by τ_J the topology of $\mathcal{C}^\infty(E)$ determined by this family of seminorms. In general, it is not Fréchet. We are now in a position to state the following.

Theorem 3.1 ([37, Theorem 3.3]) *Let E be a $\mathcal{C}^\infty$ determining compact subset of $\mathbb{R}^n$. Then the following requirements are equivalent:*

(i) E is Markov;

(iv) the space $(\mathcal{C}^\infty(E), \tau_J)$ is complete;

(v) the topologies τ_J and τ_Q for $\mathcal{C}^\infty(E)$ coincide;

(vi) there exists a continuous linear operator

$$L : (\mathcal{C}^\infty(E), \tau_J) \to \mathcal{C}^\infty(\mathbb{R}^n)$$

such that $Lf_{|E} = f$ for each $f \in \mathcal{C}^\infty(E)$.

Moreover, if E is Markov, such an operator can be defined by

$$Lf = u_1 L_1 f + \sum_{k=1}^{\infty} u_k (L_{k+1} f - L_k f), \qquad (PP)$$

where $L_k f$ is a Lagrange interpolation polynomial of f of degree k and u_k is a cut-off function of Lemma 2.1 corresponding to $\epsilon = k^{-r}$ with r fulfilling both conditions (i) and (ii) of Theorem 2.2.

Theorem 3.1 together with Theorem 2.2 yield different characterizations of Markov's property of a compact set in $\mathbb{R}^n$. For a proof of Theorem 3.1, we refer the reader to [37, Theorem 3.3]. We only mention here that (iv) follows from (i) by Theorem 2.2, and consequently, we get (v) by a Banach theorem. That (i) implies (vi) with the operator L defined by (PP), it follows from Leibniz's rule and the equivalence (i)$\Leftrightarrow$(ii) ensured by Theorem 2.2. Finally, both implications, (v)$\Rightarrow$(i) and (vi)$\Rightarrow$(i), follow by a remark due to Siciak (*see* [37]).

Applying a similar technique to that of [37] Zeriahi [57] showed that if E is Markov then $(\mathcal{C}^\infty(E), \tau_Q)$ is isomorphic with the space $\mathcal{S}$ of rapidly decreasing sequences of positive real numbers, which is a nontrivial extension of a known theorem of Mityagin [31] for a line-segment in $\mathbb{R}$.

By Jackson's theorem, the topology τ_Q is finer than the Jackson topology τ_J. Hence by Theorem 3.1 we get

Corollary 3.2 *If E is a Markov compact subset of $\mathbb{R}^n$ then the assignement (PP) defines a continuous linear extension operator*

$$L : \ (\mathcal{C}^\infty(E), \tau_Q) \to \mathcal{C}^\infty(\mathbb{R}^n).$$

For the topology τ_Q the converse of Corollary 3.2 is not true. For, Goncharov [21] has recently constructed a subset E of $\mathbb{R}$ of type $E = \{0\} \cup \bigcup_{i=1}^{\infty} [a_i, b_i]$ with $b_{i+1} < a_i < b_i$, $i = 1, 2, \ldots$, which is not Markov and at the same time it admits a continuous linear extension operator $L : (\mathcal{C}^\infty(E), \tau_Q) \to \mathcal{C}^\infty(\mathbb{R})$. However, a modification of Pawłucki and Pleśniak's proof of Theorem 3.1 has permitted to Bos and Milman to show, in particular, the following

Theorem 3.3 ([17, Theorem E]) *Let E be a $\mathcal{C}^\infty$ determining compact subset of $\mathbb{R}^n$. Then E is Markov if and only if there exists a continuous linear extension operator*

$$L : (\mathcal{C}^\infty(E), \tau_Q) \to \mathcal{C}^\infty(\mathbb{R}^n)$$

that is bounded in the following sense: there are positive constants c_1, c_2 and c_3, and a positive integer a such that for each $f \in \mathcal{C}^\infty(E)$, the support of Lf is compact and

$$\max_{|\alpha|\le k} \sup_{\mathbb{R}^n} |D^\alpha Lf(x)| \le c_1(c_2 k)^{c_3 k} q_{ak}(f), \quad \textit{for each } k = 1, 2. \dots,$$

where $q_{ak}(\cdot)$ is defined by (3.3).

It is clear that Goncharov's set E does not admit a bounded continuous linear extension of $\mathcal{C}^\infty$ functions on E in the sense of Bos and Milman.

Let us also notice a recent result of Schmets and Valdivia.

Theorem 3.4 ([41]) *Let E be a compact subset of $\mathbb{R}^n$. If there exists a continuous linear extension operator from $\mathcal{E}(E)$ into $\mathcal{C}^\infty(\mathbb{R}^n)$, then there is a continuous linear extension map L_1 such that for every $F \in \mathcal{E}(E)$, all the derivatives of $L_1 F$ are uniformly bounded on $\mathbb{R}^n$ and $L_1 F$ is analytic on $\mathbb{R}^n \setminus E$.*

We close this chapter by an application of Theorem 3.1 to the study of Jackson's property of sets in $\mathbb{R}^n$. Let E be a compact subset of $\mathbb{R}^n$ such that $E = \overline{\text{int} E}$. Let $\mathcal{C}^\infty(\overline{\text{int}\, E})$ denote the space of all $\mathcal{C}^\infty$ functions in $\text{int}\, E$ which can be continuously extended together with all their derivatives to E. We shall say that E has *Jackson's property* (or else that E admits *Jackson's inequality*) if for each $k = 0, 1 \dots$ there exist: a positive constant C_k and a positive integer m_k such that for all $f \in \mathcal{C}^\infty(\overline{\text{int} E})$ and all $n > k$ we have

$$n^k \text{dist}_E(f, \mathcal{P}_n) \le C_k \|f\|_{E, m_k}, \tag{$\mathcal{J}$}$$

where $\|\cdot\|_{E,m_k}$ is defined by (3.2). By a multivariate version of Jackson's theorem (*see* e.g. [51], [38]), if E is a compact cube in $\mathbb{R}^n$ then E admits Jackson's inequality ($\mathcal{J}$) with $m_k = k + 1$. For other examples (or counter-examples) of sets satisfying ($\mathcal{J}$), see [39]. In particular, if E is the above mentioned set $\{(x, y) \in \mathbb{R}^2 : 0 \le x \le 1, g(x) \le y \le 1\} \cup [0, 1] \times [-1, 0]$, where $g(x) = \exp(-x)$ if $0 < x \le 1$ and $g(0) = 0$, then it cannot have Jackson's property. By an application of Theorem 3.1, we can show that

Theorem 3.5 ([39, Theorem 6]) *A set $E = \bar{\Omega}$, where Ω is a bounded open Markov set in $\mathbb{R}^n$, admits Jackson's inequality ($\mathcal{J}$) if and only if it fulfils the following requirement*
($\mathcal{E}$) For each $f \in \mathcal{C}^\infty(\bar{\Omega})$ there exists a function $g \in \mathcal{C}^\infty(\mathbb{R}^n)$ such that $g_{|\Omega} = f$.

A sufficient condition for a compact subset E of $\mathbb{R}^n$ to fulfil the extension requirement ($\mathcal{E}$) is the following *Whitney regularity* (*see* [12]): E is said to be p-regular in the sense of Whitney, where $p \ge 1$, if there exists a constant $A \le 1$ such that for each two points $x, y \in E$ one can find a rectifiable arc σ joining

x and y, that is contained in int E except perhaps for a finite number of points and satisfies

$$|x - y| \geq A|\sigma|^p.$$

E is said to be regular if it is p-regular for some $p \geq 1$. By a Wachta [55] example, there exist compact sets in $\mathbb{R}^2$ which are not Whitney regular and, at the same time, they have the extension property $(\mathcal{E})$. This disproved Bierstone's conjecture [12]. If E is a p-regular compact set in $\mathbb{R}^n$ (with p integer), then by [16] there exist positive constants m_1, m_2 and m_3 such that for each $f \in \mathcal{C}^\infty(E)$ and each $k \in \mathbb{Z}_+$,

$$q_k(f) \leq m_1(m_2k)^{m_3k}||f||_{E,pk}.$$

Hence by Theorem 3.5 we get

Corollary 3.6 ([39]) *Let Ω be a bounded open subset of $\mathbb{R}^n$ such that $\bar{\Omega}$ is Markov and p-regular. Then for $s = 0, 1, \ldots$ one can find a constant $C_s > 0$ such that for each $f \in \mathcal{C}^\infty(\bar{\Omega})$,*

$$k^s\, dist_{\bar{\Omega}}(f, \mathcal{P}_k) \leq C_s||f||_{\Omega,p(s+1)}.$$

4 Markov Exponent

By Theorem 2.2, if E is a Markov compact subset of $\mathbb{R}^n$ and $f : E \to \mathbb{C}$ admits rapid uniform approximation by polynomials on E then f extends to a $\mathcal{C}^\infty$ function in $\mathbb{R}^n$. In general, the extension is done at the cost of a lost of regularity of f. It is seen by the following

Example 4.1 [38] Let $F_p = \{(x, y) \in \mathbb{R}^2 : x^p \leq y \leq 1, 0 < x \leq 1\}$ and let $F = \{(x, y) \in \mathbb{R}^2 : 0 \leq x \leq 1, -1 \leq y \leq 0\}$. Let $f(x, y) = \exp(-1/x)$, if $(x, y) \in F_p$ and $f(x, y) = 0$, if $(x, y) \in F$. Then f is $\mathcal{C}^\infty$ in int E_p where $E_p = F_p \cup F$ and all derivatives of f extend continuously to $\bar{E}_p$. Moreover, they admit the following *Gevrey* type estimates:

$$||D^\alpha f||_{E_p} \leq C^{|\alpha|}|\alpha|^{2|\alpha|}, \quad \text{for } \alpha \in \mathbb{Z}_+^2. \tag{4.1}$$

Since the set $\bar{E}_p$ is p-regular in the sense of Whitney, by [12] f can be extended to a $\mathcal{C}^\infty$ function g on $\mathbb{R}^n$. However, if $p \geq 2$, there is no open neighbourhood U of $\bar{E}_p$ such that the extension g could satisfy estimates (4.1) (with exponent 2) in U, which can be easily seen by the Mean Value Theorem.

It was shown in [38] that if we know the constant r of (MI_n) then we can estimate the loss of regularity of a $\mathcal{C}^\infty$ extension of f. This motivates the following definition of *Markov's exponent* of a compact set E in $\mathbb{R}^n$:

$$\mu(E) := \inf\{r > 0 : E \text{ satisfies } (MI_n) \text{ with } r\}.$$

If E is not a Markov set, we set $\mu(E) = \infty$. By the fact that the Chebyshev polynomials are best possible for (MI_1), one can prove that if E is a compact set in $\mathbb{R}^n$ then $\mu(E) \geq 2$ (*see* [5]). (This is not the case for compact sets in $\mathbb{C}$, since by Bernstein's inequality, for $E = \{|z| \leq 1\}$ we have $r = 1$.) In particular, if E is a fat, convex compact subset of $\mathbb{R}^n$, then by a standard argument based on inequality (MI_1), $\mu(E) = 2$. If E is a UPC compact subset of $\mathbb{R}^n$ with parameter m then by [4], $\mu(E) = 2m$.

It appears that Markov's exponent is invariant under "good" analytic mappings. More precisely, it was proved in [5] that

Theorem 4.2 *If E is a compact subset of $\mathbb{R}^n$ satisfying (MI_n) with an exponent r, and f is an analytic mapping defined in a neighbourhood U of E, with values in $\mathbb{R}^n$, such that $f(E)$ is not pluripolar (in $\mathbb{C}^n$) and $\det d_x f \neq 0$ for each $x \in E$, then $f(E)$ also satisfies (MI_n) with the same exponent r as that of E.*

This result is sharp in the sense that if the assumption $\det d_x f \neq 0$ is not satisfied for all $x \in E$ then the exponent $\mu(f(E))$ may increase (*see* [5]). Moreover, if we knew that Markov's property of E implies that E is not pluripolar, we could remove in the above theorem the assumption for $f(E)$ to be not pluripolar. We notice that a recent result of Białas-Cieź [9] gives such a hope.

5 Final remarks

Another (local) theory of Markov's property of sets in $\mathbb{R}^n$ has been developed by Jonsson and Wallin [24]. It has appeared to be very useful in the study of Lipschitz classes of functions in $\mathbb{R}^n$. For a common point between the two theories see [27] and [8].

It is difficult to survey all ramifications of the multivariate Markov inequality, so we have concentrated only on its uniform norm version. For an L_p version and relations between Markov's inequality and *Sobolev-Gagliardo-Nirenberg* type inequalities we refer the reader to an extensive paper [17].

Finally, let us mention a new topic in the recent research on Markov's inequality. These are Markov and Bernstein-type inequalities on curves or submanifolds in $\mathbb{R}^n$. For references see [14], [15] and [6].

References

[1] M.S. Baouendi and C. Goulaouic, *Approximation of analytic functions on compact sets and Bernstein's inequality,* Trans. Amer. Math. Soc. **189** (1974), 251–261.

[2] M. Baran, *Complex Equilibrium Measure and Bernstein Type Theorems for Compact Sets in* $\mathbb{R}^n$, Proc. Amer. Math. Soc. **123 (2)** (1995), 485–494.

[3] M. Baran, *Bernstein Type Theorems for Compact Sets in* $\mathbb{R}^n$ *Revisited,* J. Approx. Theory **79 (2)** (1994), 190–198.

[4] M. Baran, *Markov inequality on sets with polynomial parametrization,* Ann. Polon. Math. **60 (1)** (1994), 69–79.

[5] M. Baran and W. Pleśniak, *Markov's exponent of compact sets in* $\mathbb{C}^n$, Proc. Amer. Math. Soc. **123** (9) (1995), 2785–2791.

[6] M. Baran and W. Pleśniak, *Bernstein and van der Corput-Schaake type inequalities on semialgebraic curves*, Studia Math. (1997), (to appear).

[7] S.N. Bernstein, *Sur l'ordre de la meilleure approximation des fonctions continues par des polynômes de degré donné*, Mémoires de l'Académie Royale de Belgique **4** (2) (1912), 1–103.

[8] L. Białas-Cież, *Equivalence of Markov's property and Hölder continuity of the Green function for Cantor-type sets*, East Journal on Approximations **1** (2) (1995), 249–253.

[9] L. Białas-Cież, *Markov sets in* $\mathbb{C}$ *are not polar*, Jagiellonian University, Preprint (1996).

[10] L. Białas and A. Volberg, *Markov's property of the Cantor ternary set*, Studia Math. **104** (1993), 259–268.

[11] E. Bierstone, *Extension of Whitney fields from subanalytic sets*, Invent. Math. **46** (1978), 277–300.

[12] E. Bierstone, *Differentiable functions*, Bol. Soc. Bras. Mat. **12** (2) (1980), 139–190.

[13] E. Bierstone and P.D. Milman, *Semianalytic and subanalytic sets*, Inst. Hautes Etudes Sci. Publ. Math. **67** (1988), 5–42.

[14] L. Bos, N. Levenberg and B.A. Taylor, *Characterization of smooth, compact algebraic curves in* $\mathbb{R}^2$, In: "Topics in Complex Analysis", eds. P. Jakóbczak and W. Pleśniak, Banach Center Publications, Institute of Mathematics, Polish Academy of Sciences **31** (1995), 125–134.

[15] L. Bos, N. Levenberg, P. Milman and B.A. Taylor, *Tangential Markov Inequalities Characterize Algebraic Submanifolds of* $\mathbb{R}^N$, Indiana Univ. Math. Journal **44** (1) (1995), 115–138.

[16] L. Bos and P. Milman, *On Markov and Sobolev type inequalities on compact subsets in* $\mathbb{R}^n$, In "Topics in Polynomials in One and Several Variables and Their Applications" (Th. Rassias et al. eds.), World Scientific, Singapore (1992), 81–100.

[17] L. Bos and P. Milman, *Sobolev-Gagliardo-Nirenberg and Markov type inequalities on subanalytic domains*, Geometric and Functional Analysis **5** (6) (1995), 853–923.

[18] P. Goetgheluck, *Inégalité de Markov dans les ensembles efillés*, J. Approx. Theory **30** (1980), 149–154.

[19] P. Goetgheluck, *Polynomial Inequalities on General Subsets of* $\mathbb{R}^N$, Colloq. Math. **57** (1) (1989), 127–136.

[20] P. Goetgheluck and W. Pleśniak, *Counter-examples to Markov and Bernstein Inequalities*, J. Approx. Theory **69** (1992), 318–325.

[21] A. Goncharov, *A compact set without Markov's property but with an extension operator for* C^∞ *functions*, Studia Math. **119** (1996), 27–35.

[22] A. Jonsson, *Markov's inequality on compact sets*, In: "Orthogonal Polynomials and Their Applications" (C. Brezinski, L. Gori and A. Ronveaux, eds.) (1991), 309–313.

[23] A. Jonsson, *Markov's Inequality and Zeros of Orthogonal Polynomials on Fractal Sets*, J. Approx. Theory **78** (1994), 87–97.

[24] A. Jonsson and H. Wallin, *Function Spaces on Subsets of* $\mathbb{R}^n$, Mathematical Reports Vol. 2, Part 1, Harwood Academic, London, 1984.

[25] M. Klimek, *Pluripotential Theory*, Oxford Univ. Press, London, 1991.

[26] M. Kosek, *Hölder Continuity Property of filled-in Julia sets in* $\mathbb{C}^n$, Proc. Amer. Math. Soc. (1997) (to appear).

[27] J. Lithner, *Comparing two versions of Markov's inequality on compact sets*, J. Approx. Theory **77** (1994), 202–211.

[28] S. Łojasiewicz, *Ensembles semianalytiques*, Inst. Hautes Etudes Sci, Bures-sur-Yvette, 1964.

[29] B. Malgrange, *Ideals of Differentiable Functions*, Oxford University Press, Bombay, 1966.

[30] G.V. Milanović and T.M. Rassias, *On Markov-Duffin-Schaeffer Inequalities*, J. Nat. Geometry **5** (1) (1994), 29–41.

[31] B. Mityagin, *Approximate dimension and bases in nuclear spaces*, Russian Math. Surveys **16** (4) (1961), 59–128.

[32] W. Pawłucki and W. Pleśniak, *Markov's inequality and* C^∞ *functions on sets with polynomial cusps*, Math. Ann. **275** (1986), 467–480.

[33] W. Pawłucki and W. Pleśniak, *Extension of* C^∞ *functions from sets with polynomial cusps*, Studia Math. **88** (1988), 279–287.

[34] W. Pawłucki and W. Pleśniak, *Approximation and extension of* C^∞ *functions defined on compact subsets of* $\mathbb{C}^n$, In: "Deformations of Mathematical Structures" (J. Ławrynowicz ed.), Kluwer Academic Publishers (1989), 283–295.

[35] W. Pleśniak, *Compact subsets of* $\mathbb{C}^n$ *preserving Markov's inequality*, Mat. Vesnik **40** (1988), 295–300.

[36] W. Pleśniak, *A Cantor regular set which does not have Markov's property*, Ann. Polon. Math. **51** (1990), 269–274.

[37] W. Pleśniak, *Markov's inequality and the existence of an extension operator for C^∞ functions*, J. Approx. Theory **61** (1990), 106–117.

[38] W. Pleśniak, *Extension and polynomial approximation of ultradifferentiable functions in $\mathbb{R}^n$*, Bull. Soc. Roy. Sci. Liège **63** (5) (1994), 393–402.

[39] W. Pleśniak, *Remarks on Jackson's Theorem in $\mathbb{R}^N$*, East Journal on Approximations **2** (3) (1996), 301–308.

[40] Q.I. Rahman, G. Schmeisser, *Les inégalités de Markoff et de Bernstein*, Presses Univ. Montréal, Montréal, Québec, 1983.

[41] J. Schmets and M. Valdivia, *On the existence of continuous linear analytic extension maps for Whitney jets*, Institut de Mathématique, Université de Liège, Publication n° 95.011.

[42] R. T. Seeley, *Extension of C^∞ functions defined on a half-space*, Proc. Amer. Math. Soc. **15** (1964), 625-626.

[43] J. Siciak, *On some extremal functions and their applications in the theory of analytic functions of several complex variables*, Trans. Amer. Math. Soc. **105** (1962), 322–357.

[44] J. Siciak, *Degree of convergence of some sequences in the conformal mapping theory*, Colloq. Math. **16** (1967), 49–59.

[45] J. Siciak, *Extremal plurisubharmonic functions in $\mathbb{C}^n$*, Ann. Pol. Math. **39** (1981), 175–211.

[46] J. Siciak, *Rapid polynomial approximation on compact sets in $\mathbb{C}^n$*, Univ. Iagello. Acta Math. **30** (1993), 145–154.

[47] J. Siciak, *Wiener's type sufficient conditions in $\mathbb{C}^N$*, ibid **35** (1997), 47-74.

[48] E.M. Stein, *Singular Integrals and Differentiability Properties of Functions*, Princeton University Press, Princeton, 1970.

[49] J. Szabados, *In Memoriam: Arun Kumar Varma (1934-1994)*, J. Approx. Theory **84** (1996), 1–11.

[50] M. Tidten, *Fortsetzungen von C^∞-Funktionen, welche auf einer abgeschlossenen Menge in $\mathbb{R}^n$ definiert sind*, Manuscripta Math. **27** (1979), 291–312.

[51] A.F. Timan, *Theory of Approximation of Functions of a Real Variable*, Pergamon Press, Oxford-London-New York-Paris, 1963.

[52] V. Totik, *Markoff constants for Cantor sets*, Acta Sci. Math. (Szeged) **60** (1995), 715–734.

[53] J. C. Tougeron, *Idéaux de Fonctions Différentiables*, Springer-Verlag, Berlin-Heidelberg-New-York, 1972.

[54] A. Volberg, *An estimate from below for the Markov constant of a Cantor repeller*, In: "Topics in Complex Analysis", eds. P. Jakóbczak and W. Pleśniak, Banach Center Publications, Institute of Mathematics, Polish Academy of Sciences **31** 393–390.

[55] K. Wachta, *Prolongement de fonctions* $\mathcal{C}^\infty$, Bull. Polish Acad. Sci. Math. **31** (1983), 245–248.

[56] H. Whitney, *Analytic extension of differentiable functions defined in closed sets*, Trans. Amer. Math. Soc. **36** (1934), 63–89.

[57] A. Zeriahi, *Inégalités de Markov et développement en série de polynômes orthogonaux des fonctions* $\mathcal{C}^\infty$ *et* $\mathcal{A}^\infty$, in: "Proceedings of the Special Year of Complex Analysis of the Mittag-Leffler Institute 1987-88" (ed. J.F. Fornaess), Princeton Univ. Press, Princeton New Jersey (1993), 693–701.

[58] M. Zerner, *Développement en séries de polynômes orthonormaux des fonctions indéfiniment différentiables*, C. R. Acad. Sci. Paris Sér. I **268** (1969), 218–220.

Orthogonal Expansion and Variations of Sign of Continuous Functions

Gerhard Schmeisser
Mathematisches Institut, Universität Erlangen–Nürnberg
Bismarckstrasse $1\frac{1}{2}$, D–91054 Erlangen, GERMANY

Abstract

Let $f(z) = a_0\phi_0(z) + \cdots + a_n\phi_n(z)$ be an orthogonal expansion of a polynomial of degree n with real coefficients and let $k \leq n$ be a positive integer. We establish a sufficient criterion for f to have at least k variations of sign. It involves only the coefficients $a_0, \ldots, a_k$ and extends to formal orthogonal expansions of continuous functions on an interval of finite length. Although the criterion is not necessary, we also show that for any given admissible function there exists an orthogonal expansion for which the criterion succeeds. These results may be seen as contributions in the spirit of the Landau–Montel problem.

1 Introduction

The late Professor A. K. Varma has enriched Mathematics by numerous fascinating inequalities. The significance of orthogonal polynomials shows in several of them; besides, some of them were inspired by results of the great Hungarian mathematicians, notably of P. Turán. In that respect, the present paper would meet the field of Professor Varma's interests. However, other than he would have done, we have not cared for an explicit calculation or a handy estimate of the constants relevant in our results.

In a keynote speech at the congress of Hungarian mathematicians in 1950, Turán [12] pointed out the importance of studying orthogonal expansions of a polynomial. In a series of papers, he [13, 14] and later Specht [7, 8, 9, 10] as well as Makai and Turán [5] showed that there is a very close relationship between estimating the moduli of the zeros of a polynomial

$$f(z) = c_0 + c_1 z + \cdots + c_n z^n \qquad (c_n \neq 0) \tag{1}$$

in terms of the coefficients $c_0, c_1, \ldots, c_n$ and estimating the imaginary parts of the zeros of the same polynomial in terms of the coefficients $a_0, a_1, \ldots, a_n$ of an expansion

$$f(z) = a_0\phi_0(z) + a_1\phi_1(z) + \cdots + a_n\phi_n(z), \tag{2}$$

where $\{\phi_\nu : \nu \in \mathbb{N}_0\}$ is a system of polynomials orthogonal on the real line. Here we present a result within that scope which is in the spirit of the Landau–Montel problem and shows a striking superiority of the orthogonal expansion (2) as compared to the standard Taylor–Vieta expansion (1).

Let f be a polynomial or, more generally, a holomorphic function

$$f(z) = \sum_{\nu=0}^{N} c_\nu z^\nu \qquad (N \in \mathbb{N} \cup \{\infty\}) \tag{3}$$

such that

$$c_0 = c_1 = \cdots = c_{k-1} = 0 \qquad (k < N). \tag{4}$$

Then, trivially, f has a zero of order at least k at the origin. However, nothing can be said about the location of the zeros of f if instead of (4) we require only that the coefficients $c_0, \ldots, c_{k-1}$ are of sufficiently small modulus as compared to c_k. In fact, prescribing $c_0, \ldots, c_k$ with $c_0 \neq 0$, we can construct a polynomial $\varphi(z) = b_1 z + \cdots + b_k z^k$ such that the entire function $F(z) := c_0 \exp(\varphi(z))$ has expansions

$$F(z) = c_0 \sum_{\nu=0}^{\infty} \frac{(\varphi(z))^\nu}{\nu!} = c_0 + c_1 z + \cdots + c_k z^k + \cdots .$$

To verify this, we observe by equating coefficients that

$$\begin{aligned} c_1 &= c_0 b_1 \\ c_j &= c_0\left(b_j + P_j(b_1, \ldots, b_{j-1})\right) \qquad (j = 2, \ldots, k), \end{aligned}$$

where $P_j(b_1, \ldots, b_{j-1})$ is a polynomial in $b_1, \ldots, b_{j-1}$. Hence the coefficients $b_1, \ldots, b_k$ of the desired polynomial φ can be computed recurrently from $c_0, \ldots, c_k$.

The entire function F does not have zeros at all. Moreover, the polynomials

$$F_n(z) := c_0 \sum_{\nu=0}^{n} \frac{(\varphi(z))^\nu}{\nu!} =: \sum_{j=0}^{nk} c_{n,j} z^j \qquad (n > k)$$

have prescribed initial coefficients

$$c_{n,j} = c_j \qquad (j = 0, \ldots, k)$$

but all their zeros approach infinity as $n \to \infty$. Here the coefficients may be complex, but the situation does not change if we restrict ourselves to real ones.

It is a demanding problem, named after Landau and Montel, to find classes of polynomials of arbitrary degree or entire functions for which a fixed number of coefficients gives a non-trivial information on some of the zeros [3, Chapter IV]. A typical result (see [2]) shows that if f is an entire function with a gap power series

$$f(z) = c_0 + c_1 z + \cdots + c_k z^k + \sum_{j=1}^{\infty} c_{k+j} z^{n_j},$$

where

$$c_k \neq 0, \qquad k < n_1 < n_2 < \cdots, \qquad \textstyle\sum_{j=1}^{\infty} 1/n_j =: M < \infty,$$

then there exists a bound for the moduli of k zeros of f which depends only on $c_0, \ldots, c_k$ and M.

Analogously to (3), we now consider an orthogonal expansion

$$f(z) = \sum_{\nu=0}^{N} a_\nu \phi_\nu(z).$$

If it has *real* coefficients $a_0, \ldots, a_N$ and, analogously to (4),

$$a_0 = a_1 = \cdots = a_{k-1} = 0,$$

then it is known [1, 6] that f has at least k real zeros of odd multiplicities. As we shall show here (see Theorem 1 below), this result remains true if the coefficients $a_0, \ldots, a_{k-1}$ are of sufficiently small modulus as compared to a_k. Moreover, there is an extension to real-valued continuous functions (Theorem 2). As a special case it includes a sufficient criterion for a holomorphic function

$$f(z) = \sum_{\nu=0}^{\infty} a_\nu \phi_\nu(z) \qquad (a_\nu \in \mathbb{R} \text{ for } \nu \in \mathbb{N}_0)$$

to have at least k zeros of odd multiplicities in an interval. It involves $a_0, \ldots, a_k$, but does not need any information about the subsequent coefficients. As such it may be seen as a contribution to the Landau–Montel problem. Finally, we prove a converse result (Theorem 3). It shows that if a function f is continuous and real-valued on an interval $[a, b]$ and has k variations of sign in (a, b), then there exists a formal orthogonal expansion for which the criterion of Theorem 2 applies. For precise statements we need some notations.

2 Notations

Let $\sigma : \mathbb{R} \to \mathbb{R}$ be a non-decreasing bounded function which attains infinitely many distinct values and is such that the moments

$$\int_{-\infty}^{\infty} x^n \, d\sigma(x) \qquad (n = 0, 1, \ldots),$$

understood as improper Stieltjes integrals, exist. The set

$$\operatorname{supp}(d\sigma) := \{x \in \mathbb{R} : \sigma(x+\varepsilon) - \sigma(x-\varepsilon) > 0 \quad \text{for all} \quad \varepsilon > 0\}$$

is called the *support* of $d\sigma$.

As is well known [11], there exists a unique infinite sequence of *monic* polynomials $\phi_0(z), \phi_1(z), \ldots$ with each $\phi_n(z)$ being of exact degree n such that

$$\int_{-\infty}^{\infty} \phi_m(x)\phi_n(x) \, d\sigma(x) = \gamma_n \delta_{mn} \qquad (m, n \in \mathbb{N}_0), \tag{5}$$

where $\gamma_n > 0$ and δ_{mn} is Kronecker's delta. We call $\{\phi_\nu : \nu \in \mathbb{N}_0\}$ the monic orthogonal system associated with the *distribution* σ.

The function σ induces the norm

$$\|f\|_\sigma := \left(\int_{-\infty}^{\infty} |f(x)|^2 \, d\sigma(x)\right)^{1/2}. \tag{6}$$

If $\operatorname{supp}(d\sigma) \subset [a, b]$, then $\int_{-\infty}^{\infty} \ldots d\sigma(x) = \int_a^b \ldots d\sigma(x)$. In this case we say that the polynomials $\phi_0(z), \phi_1(z), \ldots$ are orthogonal on the interval $[a, b]$. It is known that their zeros lie in (a, b).

If σ is absolutely continuous, then $\int \ldots d\sigma(x) = \int \ldots \sigma'(x) \, dx$. Here σ' can be identified with a non-negative function ω called a *weight function.*

If a function f is continuous on an interval which contains $\operatorname{supp}(d\sigma)$, then it has a formal expansion

$$f(z) \sim \sum_{\nu=0}^{\infty} a_\nu \phi_\nu(z) \tag{7}$$

with coefficients

$$a_\nu = \frac{1}{\gamma_\nu} \int_{-\infty}^{\infty} f(x)\phi_\nu(x) \, d\sigma(x) \qquad (\nu = 0, 1, \ldots). \tag{8}$$

The series in (7) need not converge and if it converges, then it need not represent $f(z)$.

Much more can be said about orthogonal expansions of analytic functions for a special class of weight functions, which may be introduced as follows.

Definition 1 *Let $a < b$ be real numbers and*

$$\varphi(t) := \frac{a+b}{2} - t\,\frac{a-b}{2}.$$

Then $\Omega(a,b)$ denotes the class of all non-negative measurable functions $\omega : [a,b] \to \mathbb{R}$ for which the integrals

$$\int_{-\pi}^{\pi} \omega\big(\varphi(\cos\theta)\big)\,|\sin\theta|\,d\theta, \qquad \int_{-\pi}^{\pi} \log\big(\omega\big(\varphi(\cos\theta)\big)\,|\sin\theta|\big)\,d\theta$$

exist with the first one supposed to be positive.

For $\omega \in \Omega(a,b)$ the monic orthogonal system associated with the distribution

$$\sigma(x) := \begin{cases} 0 & \text{if} \quad x < a \\ \int_a^x \omega(t)\,dt & \text{if} \quad x \in [a,b] \\ \sigma(b) & \text{if} \quad x > b \end{cases}$$

is said to be associated with the weight ω. If $\{\phi_\nu : \nu \in \mathbb{N}_0\}$ is such a system and f is holomorphic on $[a,b]$, then the series (7) converges for all complex z inside an appropriate ellipse with foci at a, b and represents $f(z)$; see [11, Theorem 12.7.3].

3 Statement of the Results

Following a common convention, we denote by $\lfloor x \rfloor$ the largest integer not exceeding x.

Theorem 1 *Let $\{\phi_\nu : \nu \in \mathbb{N}_0\}$ be a monic orthogonal system associated with a distribution σ. For integers k and n with $0 \le k \le n$ and $m := \lfloor (n+k+2)/2 \rfloor$ let*

$$C_{m,k} := \frac{1}{\gamma_k} \max\Big\{ \|\psi_k\|_\sigma^2 \;:\; \psi_k \text{ monic divisor of degree } k \text{ of } \phi_m \Big\}. \tag{9}$$

If $f(z) = \sum_{\nu=0}^{n} a_\nu \phi_\nu(z)$ is a polynomial with real coefficients $a_0, \ldots, a_n$ satisfying

$$\gamma_k a_k^2 > (C_{m,k} - 1) \sum_{j=0}^{k-1} \gamma_j a_j^2, \tag{10}$$

then f has at least k real zeros of odd multiplicities lying in the smallest interval spanned by the zeros of ϕ_m.

If f is of degree n and $k = n-1$, then (10) implies that f has n distinct real zeros. In case of the Hermite expansion and $k = n-1$ a related result has been obtained by Turán [14].

If the support of $d\sigma$ is contained in an interval of finite length, then Theorem 1 extends to a considerably wider class of functions.

Theorem 2 *Let $\{\phi_\nu : \nu \in \mathbb{N}_0\}$ be a monic orthogonal system associated with a distribution σ for which* $\operatorname{supp}(d\sigma) \subset [a,b]$. *For $k \in \mathbb{N}$ let*

$$C_k := \frac{1}{\gamma_k}\left(\sup_{a\le x_1\le\cdots\le x_k\le b}\int_a^b \prod_{j=1}^{k}\left(x-x_j\right)^2\,d\sigma(x)\right). \tag{11}$$

If f is a continuous real-valued function on $[a,b]$ with coefficients (8) satisfying

$$\gamma_k a_k^2 > (C_k - 1)\sum_{j=0}^{k-1}\gamma_j a_j^2 \tag{12}$$

for some $k \in \mathbb{N}$, then f has at least k variations of sign in (a,b).

It can be shown and may have been known since a long time that the supremum in (11) is attained when the points $x_1,\ldots,x_k$ coalesce either at a or at b. A simple upper bound for C_k is $(b-a)^{2k}\gamma_0/\gamma_k$.

A different criterion for a function to have variations of sign in an interval was established by Fejér [4]. It is in terms of variations of sign in a sequence of moments.

Theorem 2 allows a lot of flexibility. Of course, we have freedom in the choice of the orthogonal system. Besides, if μ is a continuous function without zeros in (a,b), then f and μf have the same variations of sign in (a,b). Therefore we may apply Theorem 2 to μf. For the coefficients (8) this is like varying the distribution σ but keeping ϕ_ν and γ_ν fixed.

Various interesting examples could be given. We content ourselves with a very simple one.

Example. In case of an expansion by Legendre polynomials, we find that $\gamma_0 = 2$, $\gamma_1 = 2/3$, and $C_1 = 4$. Therefore the criterion of Theorem 2 says for $k = 1$ that if

$$\left|\int_{-1}^{1} x f(x)\,dx\right| > \left|\int_{-1}^{1} f(x)\,dx\right|,$$

then f changes sign in $(-1,1)$. Of course, this is also obtained as a simple observation. Besides, it can be deduced from the before-mentioned result of Fejér.

Finally, we turn to a converse of Theorem 2.

Theorem 3 *Let f be a continuous real-valued function on $[a, b]$. If f has k variations of sign in (a, b), then there exists an absolutely continuous distribution σ, defined by a weight $\omega \in \Omega(a, b)$, with an associated monic orthogonal system $\{\phi_\nu : \nu \in \mathbb{N}_0\}$ such that the coefficients (8) satisfy (12).*

Theorem 3 implies that if a function f is holomorphic and real-valued on an interval $[a, b]$ and changes sign k times inside, then there exists an orthogonal expansion which represents f inside an ellipse with foci at a, b and is such that the criterion of Theorem 2 detects k variations of sign.

4 Proofs of the Theorems

Proof of Theorem 1
Let $x_1, \ldots, x_m$ be the zeros of ϕ_m arranged in increasing order. The conclusion of the theorem is obviously true if the sequence

$$f(x_1),\ f(x_2),\ \ldots\ ,\ f(x_m), \tag{13}$$

where vanishing elements are ignored, has at least k variations of sign.

Let us now assume that the sequence (13) has less than k variations of sign. Then a short reflection shows that there exists a monic divisor ψ_k of ϕ_m which is of degree k and achieves that

$$\frac{1}{a_k}\,\psi_k(x_\mu) f(x_\mu) \le 0 \qquad \text{for} \quad \mu = 1, \ldots, m. \tag{14}$$

Next, let

$$\int_{-\infty}^{\infty} F(x)\, d\sigma(x) \;=\; \sum_{\mu=1}^{m} H_\mu F(x_\mu)$$

be the Gaussian quadrature formula associated with ϕ_m. It has positive coefficients $H_1, \ldots, H_m$ and holds for all polynomials F up to degree $2m-1$. In particular, it holds for the polynomial $\psi_k f$. In conjunction with (14), this allows us to conclude that

$$\begin{aligned}
0 \;\ge\; \frac{1}{a_k}\sum_{\mu=1}^{m} H_\mu \psi_k(x_\mu) f(x_\mu) \;&=\; \frac{1}{a_k}\int_{-\infty}^{\infty} \psi_k(x) f(x)\, d\sigma(x) \\
=\; \sum_{\nu=0}^{n} \frac{a_\nu}{a_k}\int_{-\infty}^{\infty} \psi_k(x)\phi_\nu(x)\, d\sigma(x)& \\
=\; \sum_{\nu=0}^{k-1} \frac{a_\nu}{a_k}\int_{-\infty}^{\infty} \psi_k(x)\phi_\nu(x)\, d\sigma(x) \;&+\; \int_{-\infty}^{\infty} \psi_k(x)\phi_k(x)\, d\sigma(x).
\end{aligned}$$

Since the polynomial $\psi_k(x)$ has an expansion of the form

$$\phi_k(x) + \sum_{j=0}^{k-1} c_j \phi_j(x),$$

we infer that $\int_{-\infty}^{\infty} \psi_k(x)\phi_k(x)\,d\sigma(x) = \gamma_k > 0$. Using this and the Cauchy–Schwarz inequality, we find that

$$\begin{aligned}
\gamma_k &\leq -\sum_{\nu=0}^{k-1} \frac{a_\nu}{a_k} \int_{-\infty}^{\infty} \psi_k(x)\phi_\nu(x)\,d\sigma(x) \\
&\leq \left(\sum_{\nu=0}^{k-1} \gamma_\nu \left(\frac{a_\nu}{a_k}\right)^2\right)^{1/2} \cdot \left(\sum_{\nu=0}^{k-1} \frac{1}{\gamma_\nu}\left(\int_{-\infty}^{\infty} \psi_k(x)\phi_\nu(x)\,d\sigma(x)\right)^2\right)^{1/2} \\
&= \left(\sum_{\nu=0}^{k-1} \gamma_\nu \left(\frac{a_\nu}{a_k}\right)^2\right)^{1/2} \cdot \left(\|\psi_k\|_\sigma^2 - \gamma_k\right)^{1/2}.
\end{aligned}$$

Finally, noting that $\|\psi_k\|_\sigma^2 \leq \gamma_k C_{m,k}$, we readily find a contradiction to (10). This completes the proof of Theorem 1.

Proof of Theorem 2

Obviously, there is nothing to prove if f has an infinite number of variations of sign. Therefore let $t_1, \ldots, t_\ell \in (a, b)$ be all the points where f changes sign arranged as

$$a =: t_0 < t_1 < \cdots < t_\ell < t_{\ell+1} := b.$$

With $\ell = 0$ the following considerations allow an extension to the simpler case that f is assumed to have no variation of sign.

Given $\varepsilon_1 > 0$, there exist a positive $\delta < \min_{0 \leq j \leq \ell} (t_{j+1} - t_j)/2$ and linear functions of the form

$$L_j(x) := \alpha_j(x - t_j) \qquad (\alpha_j \neq 0)$$

with the same change of sign as f at t_j such that

$$\left|f(x) - L_j(x)\right| < \varepsilon_1 \qquad \text{for} \quad |x - t_j| \leq \delta, \ j = 1, \ldots, \ell.$$

Let

$$a_0 := t_0, \quad a_j := t_j + \delta, \quad b_{j-1} := t_j - \delta \ \ (j = 1, \ldots, \ell), \quad b_\ell = t_{\ell+1}.$$

Clearly, on the intervals $[a_j, b_j]$ there exist continuous functions F_j having no zeros and satisfying

$$F_j(a_j) = L_j(a_j), \quad F_{j-1}(b_{j-1}) = L_j(b_{j-1}) \quad (j = 1, \dots, \ell)$$

and

$$\big|f(x) - F_j(x)\big| < \varepsilon_1 \qquad \text{for } x \in [a_j, b_j] \quad (j = 0, \dots, \ell).$$

The functions

$$F_0,\ L_1,\ F_1,\ \dots,\ L_\ell,\ F_\ell$$

constitute a piecewise defined continuous function $F : [a,b] \to \mathbb{R}$ which has no other zeros on $[a,b]$ than $t_1, \dots, t_\ell$ and satisfies

$$|f(x) - F(x)| < \varepsilon_1 \qquad \text{for} \quad x \in [a,b]. \tag{15}$$

Obviously, for every $x \notin \{t_1, \dots, t_\ell\}$, the function

$$H(x) := \frac{F(x)}{(x - t_1) \cdots (x - t_\ell)} \tag{16}$$

is continuous and different from zero. For each $x = t_j$ $(j = 1, \dots, \ell)$ it has a continuous extension such that $H(t_j) \neq 0$. Hence it may be defined on the whole of $[a,b]$ as a continuous function without zeros. Therefore, given $\varepsilon_2 > 0$, there exists, by Weierstrass' theorem, a real polynomial $P(x)$ without zeros in $[a,b]$ such that

$$|H(x) - P(x)| < \varepsilon_2 \qquad \text{for} \quad x \in [a,b]. \tag{17}$$

Combining (15)–(17), we conclude that for any given $\varepsilon > 0$ there exists a real polynomial $Q(x)$ having in $[a,b]$ no other zeros than $t_1, \dots, t_\ell$, each with multiplicity 1, such that

$$|f(x) - Q(x)| < \varepsilon \qquad \text{for} \quad x \in [a,b]. \tag{18}$$

Now our aim is to apply Theorem 1 to Q. Since the orthogonal system of Theorem 2 is associated with a distribution σ such that $\operatorname{supp}(d\sigma) \subset [a,b]$, we know that the polynomials ϕ_ν have all their zeros in (a,b). Therefore, a comparison of the numbers (9) and (11) shows that

$$C_{m,k} \leq C_k \qquad \text{for} \quad m \geq k.$$

Consequently (12) and (18) imply that for sufficiently small $\varepsilon > 0$ the coefficients

$$b_j := \frac{1}{\gamma_j} \int_a^b Q(x)\phi_j(x)\, d\sigma(x) \qquad (j = 0, \dots, k)$$

satisfy

$$\gamma_k b_k^2 > (C_{m,k} - 1) \sum_{j=0}^{k-1} \gamma_j b_j^2$$

with $m = \lfloor (n+k+2)/2 \rfloor$ and n being the degree of Q. Therefore Theorem 1 assures that Q has at least k zeros of odd multiplicities in (a, b) and so $\ell \geq k$. This completes the proof of Theorem 2.

Proof of Theorem 3
Under the hypothesis of Theorem 3, there exist $k + 1$ points

$$a < \xi_0 < \cdots < \xi_k < b$$

such that

$$f(\xi_{j-1})\, f(\xi_j) < 0 \qquad \text{for} \quad j = 1, \ldots, k.$$

Consider now the linear system

$$\sum_{j=0}^{k} w_j\, x_{\ell j}^{\ell} f(x_{\ell j}) = \delta_{\ell k} f(\xi_k) \qquad (\ell = 0, \ldots, k) \tag{19}$$

with given $x_{\ell j} \in (a, b)$ and unknowns $w_0, \ldots, w_k$. When

$$x_{\ell j} = \xi_j \qquad (\ell = 0, \ldots, k;\ j = 0, \ldots, k),$$

then, introducing $L(x) := \prod_{j=0}^{k} (x - \xi_j)$, we find by Cramer's rule that

$$w_j = \frac{f(\xi_k)}{f(\xi_j)} \cdot \frac{1}{L'(\xi_j)} > 0 \qquad (j = 0, \ldots, k).$$

By continuity, the system (19) has also a positive solution when the numbers $x_{\ell j}$ $(\ell = 0, \ldots, k)$ are sufficiently close to ξ_j for $j = 0, \ldots, k$. Consequently, for all sufficiently small $\varepsilon_1 > 0$ the system

$$\sum_{j=0}^{k} w_j \frac{1}{\varepsilon_1} \int_{\xi_j - \varepsilon_1/2}^{\xi_j + \varepsilon_1/2} x^{\ell} f(x)\, dx = \delta_{\ell k} f(\xi_k) \quad (\ell = 0, \ldots, k)$$

has a positive solution. Again by continuity, it has also a positive solution when we replace the right hand side by

$$\delta_{\ell k} f(\xi_k) - \varepsilon_2 \int_a^b x^{\ell} f(x)\, dx$$

with a sufficiently small $\varepsilon_2 > 0$. Hence there exist positive numbers ε_1, ε_2 and $w_0, \dots, w_k$ defining disjoint subintervals

$$\mathcal{I}_j := [\xi_j - \varepsilon_1/2\,,\, \xi_j + \varepsilon_1/2] \qquad (j = 0, \dots, k)$$

of (a, b) and a weight function

$$\omega(x) := \begin{cases} \varepsilon_2 + w_j/\varepsilon_1 & \text{if } x \in \mathcal{I}_j \quad (j = 0, \dots, k) \\ \varepsilon_2 & \text{for all other } x \in [a, b] \end{cases}$$

which obviously belongs to $\Omega(a, b)$ and yields that

$$\int_a^b x^\ell f(x)\omega(x)\,dx = \delta_{\ell k} f(\xi_k) \qquad (\ell = 0, \dots, k). \tag{20}$$

Because of (20), the monic orthogonal system $\{\phi_\nu : \nu \in \mathbb{N}_0\}$ associated with that weight ω provides coefficients (8) so that

$$a_0 = \cdots = a_{k-1} = 0, \quad a_k \neq 0.$$

Hence (12) is satisfied. This completes the proof of Theorem 3.

References

[1] T. Anghelutza, Sur une équation algébrique; application à la formule de Taylor, Mathematica (Cluj) 6 (1932), 140–145.

[2] M. Biernacki, Sur les zéros des polynômes et sur les fonctions entières dont le développement taylorien présente des lacunes, Bull. Sci. Math. (2) 69 (1945), 197–203.

[3] J. Dieudonné, "La théorie analytique des polynômes d'une variable (à coefficients quelconques)", Mémorial des Sciences Mathématiques No. 93, Gauthier–Villars, Paris, 1938.

[4] L. Fejér, Nombre des changements de signe d'une fonction dans un intervalle et ses moments, Comptes Rendus (Paris) 158 (1914), 1328–1331.

[5] E. Makai and P. Turán, Hermite expansion and distribution of zeros of polynomials, Publ. Math. Inst. Hung. Acad. Sci., Ser. A, 8 (1963), 157–163.

[6] N. Obreschkoff, Sur les zéros réels des polynômes, Mathematica (Cluj) 10 (1935), 132–136.

[7] W. Specht, Die Lage der Nullstellen eines Polynoms, Math. Nachr. 15 (1956), 353–374.

[8] W. Specht, Die Lage der Nullstellen eines Polynoms II, Math. Nachr. 16 (1957), 257–263.

[9] W. Specht, Die Lage der Nullstellen eines Polynoms III, Math. Nachr. 16 (1957), 369–389.

[10] W. Specht, Die Lage der Nullstellen eines Polynoms IV, Math. Nachr. 21 (1960), 201–222.

[11] G. Szegö, "Orthogonal Polynomials", Amer. Math. Soc. Colloq. Publ. 23, 4th edn., Providence, R. I., 1975.

[12] P. Turán, Sur l'algèbre fonctionelle, in "Comptes Rendus du Premier Congrès des Mathématiciens Hongrois 1950", pp. 279–290, Akadémiai Kiadó, Budapest, 1952.

[13] P. Turán, Hermite-expansion and strips for zeros of polynomials, Archiv Math. 5 (1954), 148–152.

[14] P. Turán, To the analytic theory of algebraic equations, Bulgar. Akad. Nauk., Otdel. Mat. Fiz. Nauk., Izvestija Mat. Institut 3 (1959), 123–137.

Convolution Properties of Two Classes of Starlike Functions Defined by Differential Inequalities

Vikramaditya Singh
3A/95 Azad Nagar
Kanpur - 208002, INDIA

Abstract

Let $\mathcal{H}$ be the space of analytic functions in the unit disc U, $|z| < 1$, with the topology of local uniform convergence and $A \subset \mathcal{H}$ be the set of analytic functions $f(z)$ with normalization $f(0) = f'(0) - 1 = 0$. Further, let $M_{1,\lambda} = \{f \in A \mid \text{Re } zf''(z) > -\lambda,\ z \in U\}$ and $M_{2,\mu} = \{f \in A \mid |zf''(z)| < \mu,\ z \in U\}$. Mocanu [4] had posed the problems to determine (i) $\sup\{\lambda \mid M_{1,\lambda} \subset S^*\}$ and (ii) $\sup\{\mu \mid M_{2,\mu} \in S^*\}$. It had been shown [1] that for $M_{1,\lambda}$, $0 \le \lambda \le \frac{1}{\log 4}$ is sharp. But the sharp bound on μ for $M_{2,\mu} \subset S^*$ is not known. In the present paper we show that (i) $M_{2,\mu} \in K$ (convex) if and only if $0 \le \mu \le \frac{1}{2}$ and (ii) $M_{2,\mu} \in S^*$ if and only if $0 \le \mu \le 1$. We further investigate the starlikeness and convexity of $h(z) = f(z) * g(z)$ when f and g belong to one or both of the sets $M_{1,\lambda}$ and $M_{2,\mu}$.

1 Introduction

Let U be the unit disc $|z| < 1$, and let $\mathcal{H}$ be the space of analytic functions in U with the topology of local uniform convergence. The subclasses A and A_0 of $\mathcal{H}$, respectively consist of functions $f \in \mathcal{H}$ such that $f(0) = f'(0) - 1 = 0$ and $f(0) = 1$. By S, S^* and K we denote, respectively, the well known subsets of A which are univalent, starlike (with respect to the origin) and convex. We also need the classes of functions defined below:

$$B = \{f \in \mathcal{H} : f(0) = 0,\ |f(z)| < 1, z \in U\}. \tag{1}$$

$$P_\alpha = \{f \in A_0 : \text{Re } f(z) > \alpha, z \in U\}, \quad \alpha \le 1,\ P_0 = P. \tag{2}$$

$$R_1(\beta) = \{f \in A : |f'(z) - 1| < \beta, \ z \in U\}. \tag{3}$$

$$R(\alpha, \lambda) = \{f \in A : f'(z) + \alpha z f''(z) \prec 1 + \lambda z, z \in U\}. \tag{4}$$

$$M_{1,\lambda} = \{f \in A : \text{Re}\ z f''(z) \geq -\lambda, \ z \in U\}, \quad \lambda > 0, . \tag{5}$$

$$M_{2,\mu} = \{f \in A : |z f''(z)| \leq \mu, \ z \in U\}. \tag{6}$$

Here $\prec$ denotes the subordination. An analytic function $f \in \mathcal{H}$ is said to be subordinate to a univalent function $g \in \mathcal{H}$ if $f(z) = g(\omega(z))$, $\omega \in B$.

The convolution or the Hadamard product of two functions, $f, g \in \mathcal{H}$ of the form

$$f(z) = a_0 + \sum_{n=1}^{\infty} a_n z^n, \quad g(z) = b_0 + \sum_{n=1}^{\infty} b_n z^n,$$

is defined as

$$h(z) = (f * g)(z) = a_0 b_0 + \sum_{n=1}^{\infty} a_n b_n z^n. \tag{7}$$

Mocanu [4] posed the following two problems:

$$\text{determine } \sup \lambda \text{ such that } M_{1,\lambda} \subset S^*, \tag{8}$$

$$\text{determine } \sup \mu \text{ such that } M_{2,\mu} \subset S^*. \tag{9}$$

In [1] the solution of (8) was established in the following form:

Theorem A *$M_{1,\lambda} \subset S^*$ for $0 \leq \lambda \leq \frac{1}{\log 4}$, and this bound is the best possible.*

This theorem is a special case of a more general theorem and uses the technic of duality in convolution [7] and results in [3].

However, for $M_{2,\mu}$ the best known result so far is [6]

$$M_{2,\mu} \subset S^* \text{ for } 0 \leq \mu \leq \frac{2}{\sqrt{5}}.$$

Further, it is known [8] that if $g \in K$ and $f \in S^*$ or $f \in K$, then $g * f \in S^*$, K respectively. Moreover, $(f * g)(z)$, f, $g \in S^*$ need not even be univalent. It is thus, of interest to generalize the convolution properties of the classes $M_{1,\lambda}$ and $M_{2,\mu}$ where they yield stronger results than in [8].

In order to prove the results in this paper we use the following:

Theorem B [9] *If* $f, g \in \mathcal{H}, F, G \in K$ *such that* $f \prec F$, $g \prec G$, *then* $f * g \prec G * F$.

Theorem C [11] *For* $\alpha \leq 1$, $\beta \leq 1$

$$P_\alpha * P_\beta = P_\delta, \quad \delta = 1 - 2(1-\alpha)(1-\beta).$$

Theorem D [3] *If*

$$\tilde{R}(\alpha, \beta) = \{f \in A \mid \text{Re}\,[f'(z) + \alpha z f''(z)] > \beta, \ \ z \in U\}, \beta < 1,$$

then

(i) $\tilde{R}(1, \beta) \subset S^*$ *if* $\beta > \beta_0 = \frac{1-2\log 2}{2-2\log 2} = -0.629$,

(ii) $\tilde{R}(\frac{1}{2}, \beta) \subset S^*$ *if* $\beta > \beta_1 = \frac{3-4\log 2}{2-4\log 2} = -0.294$.

The bounds β_0 *and* β_1 *are the best possible.*

Theorem E [12] *If* $f \in K$, $g(z) = \sum_{k=1}^{\infty} b_k z^k$ *and* $zg'(z) \prec zf'(z)$ *then* $g \prec f$.

2 Statement of results

In the present paper we prove the following theorems:

Theorem 1

(i) $M_{2,\mu} \subset K$ *if and only if* $0 \leq \mu \leq \frac{1}{2}$,

(ii) $M_{2,\mu} \subset S^*$ *if and only if* $0 \leq \mu \leq 1$.

As

$$f_1(z) = z + \mu \frac{z^2}{2}, \tag{10}$$

satisfies the case of equality in (6) and $f_1 \in K$ if and only if $0 \leq \mu \leq \frac{1}{2}$, $f_1 \in S^*$ if and only if $0 \leq \mu \leq 1$, one part of the above theorem follows.

Theorem 2

(i) *If* $f, g \in M_{1,\lambda}$ *and* $h(z) = (f * g)(z)$ *then* $h \in S^*$ *if* $\lambda^2 \leq \frac{3}{2(1-\log 2)\pi^2}$. *This value of* λ *is the best possible.*

(ii) *If* $f \in M_{1,\lambda}$ *and* $g \in A$ *satisfies* $\operatorname{Re} \frac{g(z)}{z} > \frac{1}{2}$

then

$$(f * g)(z) \subset M_{1,\lambda}.$$

Theorem 3 *If* $f, g \in M_{2,\mu}$ *and* $h(z) = (f * g)(z)$ *then*

(i) $zh'(z) \in K$ *if* $0 \le \mu \le \frac{1}{\sqrt{2}}$,

(ii) $h \in K$ *if* $0 \le \mu \le 1$,

and

(iii) $h \in S^*$ *if* $0 \le \mu \le \frac{2e}{\sqrt{1+2e^2}} = 1.368\ldots$.

The bounds on μ in (i) and (ii) are sharp as is shown by taking $f(z) = g(z) = z + \mu\frac{z^2}{2}$.

Theorem 4 *If* $f \in M_{1,\lambda}$ *and* $g \in M_{2,\mu}$ *and* $h(z) = (f * g)(z)$ *then*

(i) $zh'(z) \in K$ *if* $0 \le \lambda\mu \le \frac{1}{4}$,

(ii) $h \in K$ *if* $0 \le \lambda\mu \le \frac{1}{2}$,

and

(iii) $h \in S^*$ *if* $0 \le \lambda\mu \le \frac{2e^2}{1+2e^2} = .9366$.

The bounds on $\lambda\mu$ *in (i) and (ii) are sharp, the extremal functions being*

$$f(z) = z + 2\lambda \sum_{n=1}^{\infty} \frac{z^{n+1}}{n(n+1)}, \tag{11}$$

and

$$g(z) = z + \frac{\mu}{2} z^2.$$

In order to prove (iii) in Theorems 3 and 4 we use the following:

Lemma *If* $h \in A$ *satisfies*

$$h'(z) + zh''(z) \prec 1 + \nu z, \tag{12}$$

then

(i) $h \in K$ *if and only if* $0 \le \nu \le \frac{2}{\sqrt{5}}$,

and

(*ii*) $h \in S^*$ *if* $0 \le \nu \le \frac{4e^2}{1+2e^2} = 1.87324\ldots.$

It should be noticed that (12) implies

$$\text{Re}\ (h'(z) + zh''(z)) > 1 - \nu, \tag{13}$$

and therefore Theorem D (i) yields that $h \in S^*$ if $0 \le \nu \le 1.629$. But under the hypothesis (12) (ii) of the above lemma gives much stronger result. Further,

$$h_1(z) = z + \frac{\nu}{4}z^2, \tag{14}$$

satisfies the case of equality in (12) and shows that $h_1(z) \in S^*$, for $0 \le \nu \le 2$. However, (14) does not seem to be extremal as is shown by (i) which is sharp.

Further, although (i) of the above lemma is sharp we do not use it to establish (i) and (ii) and Theorem 3 and 4. This is because the condition (12) does not use the full force of the hypotheses in these theorems.

3 Proof of Theorem 1

Notice that $f \in M_{2,\mu}$ if and only if

$$zf''(z) = \mu\, \omega(z), \quad \omega \in B, \quad z \in U. \tag{15}$$

Then

$$f'(z) = 1 + \mu \int_0^1 \frac{\omega(tz)}{t} dt = 1 + \mu\, \omega_1(z), \tag{16}$$

and

$$\frac{f(z)}{z} = 1 + \mu \int_0^1 \frac{\omega(tz)}{t}(1-t)dt = 1 + \mu\, \omega_2(z). \tag{17}$$

Thus

$$\phi(z) = 1 + z\frac{f''(z)}{f'(z)} = \frac{1 + \mu(\omega(z) + \omega_1(z))}{1 + \mu\, \omega_1(z)} \tag{18}$$

and

$$\psi(z) = z\frac{f'(z)}{f(z)} = \frac{1 + \mu\,\omega_1(z)}{1 + \mu\,\omega_2(z)}. \tag{19}$$

Since the proof of (i) and (ii) is on similar lines we shall prove (ii) only.

We observe that

$$\text{Re}\ \psi(z) > 0 \text{ if and only if } \psi(z) \neq -iT,\ T \in \mathbb{R}. \tag{20}$$

This leads to

$$\frac{\mu}{2}[\int_0^1 \frac{\omega(tz)}{t}(2-t)dt + \frac{1-iT}{1+iT}\int_0^1 \omega(tz)dt] \neq -1. \tag{21}$$

Let

$$M = \sup_{\substack{T\in\mathbb{R}\\ z\in U}} \left|\int_0^1 \frac{\omega(tz)}{t}(2-t)dt + \frac{1-iT}{1+iT}\int_0^1 \omega(tz)dt\right|. \tag{22}$$

In view of the fact that B is rotation invariant (21) will hold if

$$\mu \leq \frac{2}{M}. \tag{23}$$

But

$$|\int_0^1 \frac{\omega(tz)}{t}(2-t)dt| \leq \int_0^1 \frac{|\omega(tz)|}{t}(2-t)dt \leq 3/2$$

and

$$|\int_0^1 \omega(tx)dt| \leq \int_0^1 tdt = \frac{1}{2}.$$

Thus $M \leq 2$ and by (23) $\mu \leq 1$.

This completes the proof.

4 Proof of Theorem 2

We notice that $f \in M_{1,\lambda}$ if and only if for some $\phi \in S^*$

$$f'(z) = 1 + \lambda \log\frac{\phi(z)}{z}. \tag{24}$$

Thus Theorem A states that if $\phi \in S^*$ and $f \in A$ is defined by (24), where principal value of log is taken, then $f \in S^*$ if and only if $0 \leq \lambda \leq \frac{1}{\log 4}$.

In view of well known Herglotz formula, (24) yields

$$f'(z) \prec 1 + 2\lambda \log \frac{1}{1-z}. \tag{25}$$

If $f, g \in M_{1,\lambda}$ and $h(z) = f(z) * g(z)$, then we obtain

$$h'(z) + zh''(z) = f'(z) * g'(z), \tag{26}$$

and because $\log \frac{1}{1-z}$ is a convex function, Theorem B gives that

$$\begin{aligned} h'(z) + zh''(z) &\prec (1 - 2\lambda \log(1-z)) * (1 - 2\lambda \log(1-z)) \\ &= 1 - 4\lambda^2 \textstyle\int_0^1 \log(1-tz)\frac{dt}{t}. \end{aligned} \tag{27}$$

Hence

$$\text{Re}\ \{h'(z) + zh''(z)\} \geq 1 - 4\lambda^2 \int_0^1 \log(1+t)\frac{dt}{t} = 1 - \frac{\lambda^2\pi^2}{3}. \tag{28}$$

and use of Theorem D (i) now establishes the theorem. Since (28) is sharp and so is Theorem D it follows that (i) is sharp.

In order to prove (ii) we observe that, if

$$h(z) = f(z) * g(z), \quad f \in M_{1,\lambda},\ \text{Re}\ \frac{g(z)}{z} > \frac{1}{2}$$

then

$$zh''(z) = \frac{g(z)}{z} * zf''(z)$$

and use of Theorem C then gives the result.

5 Proof of the Lemma

We note that (12) is equivalent to

$$\frac{d}{dz}(zh'(z)) = 1 + \nu\omega(z), \quad \omega \in B \tag{29}$$

or

$$|\frac{d}{dz}(zh'(z)) - 1| \le \nu,\ z \in U. \tag{30}$$

(i) of the Lemma now follows from [2, 10]. In order to prove (ii) we observe that (29) yields

$$h'(z) = 1 + \nu \int_0^1 \omega(tz)dt, \tag{31}$$

and

$$\frac{h(z)}{z} = 1 + \nu \int_0^1 \omega(tz)(-\log t)dt. \tag{32}$$

Proceeding as in the proof of Theorem 1 we obtain

$$\text{Re } (z\frac{h'(z)}{h(z)}) > 0 \tag{33}$$

if and only if

$$\frac{\nu}{2}\left[\int_0^1 \omega(tz)(1-\log t)dt + \frac{1-iT}{1+iT}\int_0^1 \omega(tz)(1+\log t)dt\right] \ne -1. \tag{34}$$

Once again refering to proof of Theorem 1 we find that (33) holds for

$$\nu \le \frac{2}{M} \tag{35}$$

where

$$M = \sup_{\substack{T\in\mathbb{R}\\ \omega\in B\\ z\in U}} \left|[\int_0^1 \omega(tz)(1-\log t)dt + \frac{1-iT}{1+iT}\int_0^1 \omega(tz)(1+\log t)dt]\right|. \tag{36}$$

But it is easily seen that

$$M \le 1 + \frac{1}{2e^2} \tag{37}$$

and hence (33) holds for $\nu \le \frac{4e^2}{1+2e^2} = 1.87324$.

This bound on ν is not likely to be sharp but what we know is that $\nu \le 2$. In a private communication Ruscheweyh informed that he has obtained sharp bound on ν for h to be starlike of negative order.

6 Proof of Theorem 3

As $f, g \in M_{2,\mu}$ we have

$$zf''(z) \prec \mu z, \quad zg''(z) \prec \mu z, \tag{38}$$

and by Theorem E

$$f'(z) \prec 1 + \mu z, \quad g'(z) \prec 1 + \mu z. \tag{39}$$

Further, if $h(z) = (f * g)(z)$, then

$$(zh')' = h'(z) + zh''(z) = f'(z) * g'(z) \prec 1 + \mu^2 z, \tag{40}$$

by Theorem B. Moreover, from (40) we obtain

$$z(z(zh')'')' = zf''(z) * zg''(z) \prec \mu^2 z \tag{41}$$

by Theorem B.

Use of Theorem E now gives

$$z(zh'(z))'' \prec \mu^2 z.$$

By Theorem 1 we obtain (i) $zh'(z) \in K$ if and only if $0 \le \mu \le \frac{1}{\sqrt{2}}$ and (ii) $h \in K$ or $zh' \in S^*$ for $0 \le \mu \le 1$.

In order to prove (iii) we use (40) and the Lemma and the result follows

7 Proof of Theorem 4

In this case

$$zf''(z) \prec \frac{2\lambda z}{1-z}, \quad zg'' \prec \mu z, \tag{42}$$

and

$$f'(z) \prec 1 + 2\lambda \log \frac{1}{1-z}, \qquad g'(z) \prec 1 + \mu z. \tag{43}$$

If $h(z) = (f * g)(z)$ then

$$h'(z) + zh''(z) = (zh'(z))' = f'(z) * g'(z) \prec 1 + 2\lambda\mu z, \tag{44}$$

by Theorem B. If $t(z) = zh'(z)$ then from (44) we obtain,

$$z(zt''(z))' = zf''(z) * zg''(z) \prec 2\lambda\mu z, \tag{45}$$

by (42) and Theorem B. By Theorem E we obtain

$$zt''(z) \prec 2\lambda\mu z, \tag{46}$$

and (i) and (ii) follow from Theorem 1.

In order to obtain (iii) we use (44) and Lemma (ii). This completes the proof.

References

[1] Rosihan H. Ali, S. Ponnusamy and V. Singh, Starlikeness of functions satisfying a differential inequality, Annales Polonici Mathematici, 61 (1995), 135-140.

[2] R. Fournier, On integrals of bounded analytic functions in the closed disc, Complex Variables Theory and Appl. (2) 11 (1989), 125-133.

[3] R. Fournier and St. Ruscheweyh, On two extremal problems related to univalent functions. Rocky Mountain J. Math. 24 (2) (1994), 529-538.

[4] P. T. Mocanu, Two simple sufficient conditions for starlikeness, Mathematica (CLUJ) 34 (57) (1992), 175-181.

[5] S. Ponnusamy and V. Singh, Convolution properties of some classes of analytic functions, Journal of Mathematical Sciences (Plenum Publication Corporation).

[6] S. Ponnusamy and V. Singh, Criteria for strongly starlike functions, to appear in Complex Variables Theory and Appl.

[7] St. Ruscheweyh, Convolutions in geometric function theory, Sem. Math. Superieurs, Presses Univ. Montreal, 1982.

[8] St. Ruscheweyh and T. Sheil-Small, Hadamard product of schlicht functions and the Pólya-Schoenberg conjecture, Comment. Math. Helv. 48 (1973), 119-135.

[9] St. Ruscheweyh and J. Stankiewicz, Subordination and convex univalent functions, Bull. Pol. Acad. Sci. Math. 33 (1985) 499-502.

[10] V. Singh, Univalent functions with bounded derivative in the unit disc, Indian J. Pure and Appl. Math. 8 (1977), 1370-1377.

[11] J. Stankiewicz and Z. Stankiewicz, Some applications of Hadamard convolutions in the theory of functions, Ann. Univ. Mariae Curie-Sklodowska, 40 (1989), 251-265.

[12] T. J. Suffridge, Some remarks on convex maps of the unit disc, Duke Math. J. 37 (1970), 775-777.

Weighted Lagrange Interpolation on Generalized Jacobi Nodes

P. Vértesi [1]

Mathematical Institute, Hungarian Academy of Sciences

H-1364 Budapest, P.O. 127, Hungary

To the memory of Professor Arun Kumar Varma

Abstract

The aim of this paper is to give some examples using generalized Jacobi nodes when the corresponding weighted Lebesgue constants have the best possible order.

1 Introduction. Notations. Preliminary results

1.1. The investigation of weighted Lagrange interpolation using Jacobi nodes is relatively new. In 1978 Barbara Háy and P. Vértesi [4] proved that if $X = X^{(\alpha,\beta)}$ (= the interpolatory matrix containing the roots of the Jacobi polynomials with parameters α, β) and $w(x) = (1-x)^{\alpha/2+1/4}(1+x)^{\beta/2+1/4}$ then

$$\lambda_n(w, X) = \begin{cases} (1 + o(1))\frac{2}{\pi}\log n & \text{if } \ -\frac{1}{2} \leq \alpha, \beta \leq \frac{1}{2}, \\ (1 + o(1))\delta(\eta)\log n & \text{if } \ \eta := \max(\alpha, \beta) > \frac{1}{2}, \end{cases}$$

where $\delta(\eta) > 2/\pi$ (cf. I. Melinder [4; [2]]). Very recently, in paper [1] G. Mastroianni and M. G. Russo constructed many pointsystems X where the weighted Lebesgue constant

$$\Lambda_n(w, X) \sim \log n \quad \text{for} \quad w(x) = (1-x)^{\alpha}(1+x)^{\beta},$$

$\alpha, \beta > -1$ ($A_n \sim B_n$ iff $c_1 \leq A_n/B_n \leq c_2 \quad n \geq n_0$ where $0 < c_1 \leq c_2$ are proper constants).

[1]Supported by the Hungarian National Science Foundation Grant Ns T7570, T22943 and T17425

1.2. In this paper we give two examples which give the same optimal order using *generalized Jacobi weights* (cf. (1.14)).

First we give the definition (cf. G. Mastroianni, P. Vértesi [2]; the present paper uses only a special case of [2; Definition 1.1]).

In what follows, $L^p[a,b]$ denotes the set of functions F such that

$$\begin{cases} \|F\|_{L^p[a,b]} := \left\{ \int_a^b |F(t)|^p dt \right\}^{1/p} & \text{if} \quad 0 < p < \infty, \\ \|F\|_\infty := \text{ess } \sup_{a \le t \le b} |F(t)| & \text{if} \quad p = \infty \end{cases}$$

is finite. If $p \ge 1$ it is a norm; for $0 < p < 1$ its pth power defines a metric in $L^p[a,b]$.

By a *modulus of continuity* we mean a nondecreasing, continuous semi-additive function $\omega(\delta)$ on $[0,\infty)$ with $\omega(0) = 0$. If, in addition,

$$\omega(\delta) + \omega(\eta) \le 2\omega(\delta/2 + \eta/2) \text{ for any } \delta, \eta \ge 0,$$

then $\omega(\delta)$ is a *concave* modulus of continuity. In the later case $\delta/\omega(\delta)$ is nondecreasing for $\delta \ge 0$. We define $\omega(f,\delta)_p = \sup_{|\lambda| \le \delta} \|f(\lambda + \cdot) - f(\cdot)\|_p$, the *modulus of continuity of* f in L^p (where L^p stands for $L^p[0, 2\pi]$).

For a fixed $m \ge 0$ let

$$-1 \equiv u_{m+1} < u_m < \ldots < u_1 < u_0 \equiv 1$$

and with $\ell_r \in N$ $(r = 0, 1, \ldots, m+1)$

$$w_r(\delta) := \prod_{s=1}^{\ell_r} \{\omega_{rs}(\delta)\}^{\alpha(r,s)},$$

where $\omega_{rs}(\delta)$ are concave moduli of continuity with $\alpha(r,s) > 0$ $(s = 1, 2, \ldots, \ell_r;\ r = 0, 1, \ldots, m+1)$.

Further, let $H(x)$ be a *positive continuous* function on $[-1,1]$ such that for $h(\vartheta) := H(\cos\vartheta)$

$$\omega(h,\delta)_\infty \, \delta^{-1} \in L^1[0,1] \quad \text{or} \quad \omega(h,\delta)_2 = 0(\sqrt{\delta}), \quad \delta \to 0.$$

Then we give the following

Definition. *The function*

(1.1)

$$w(x) = H(x) w_0(\sqrt{1-x}) w_{m+1}(\sqrt{1+x}) \prod_{r=1}^{m} w_r(|x - u_r|), \quad -1 \le x \le 1,$$

is a generalized Jacobi weight ($w \in GJ$, *shortly*), *with singularities* u_r ($0 \leq r \leq m+1$).

Remark. By $\omega_{rs}(\tau) \leq \omega_{rs}(\delta)$ $(0 \leq \tau \leq \delta)$,

$$\int_0^\delta w_r(\tau)d\tau \leq \delta w_r(\delta); \tag{1.2}$$

in the original [2, Definition 1.10] where $\alpha(r,s)$ might be negative, we had to *suppose* this important inequality (cf. [2, (1.12)]). Actually, by (1.2) and [2, (1.24)] we get

$$\int_0^\delta w_r(\tau)d\tau \sim \delta w_r(\delta), \qquad r = 0, 1, \ldots, m+1. \tag{1.3}$$

1.3. Let $\mathcal{R} = \mathcal{R}(w) = \{u_r; r = 0, 1, \ldots m+1\}$ denote the set containing the singularities of $w \in GJ$.

The matrix $X = \{x_{kn}; 1 \leq k \leq n, n \in N\} \subset I = I(w) := [-1,1] \setminus \mathcal{R}(w)$ is an *interpolatory* matrix iff

$$-1 \leq x_{nn} < x_{n-1,n} < \ldots < x_{2n} < x_{1n} \leq 1, \qquad n \in \mathbf{N}; \tag{1.4}$$

for $f \in C(w, I)$ where $w \in GJ$ and

$$C(w,I) :=$$

$$\{f;\ f \text{ is continuous on } I \text{ and } \lim_{x \to u_r} f(x)w(x) = 0,\ r = 0, 1, \ldots, m+1\},$$

we investigate the *weighted Lagrange interpolation* defined by

$$L_n(f, w, X, x) = \sum_{k=1}^n f(x_{kn})w(x_{kn})t_{kn}(w, X, x), \qquad n \in \mathbf{N}, \tag{1.5}$$

where

$$t_k(x) = t_{kn}(w, X, x) = \frac{w(x)}{w(x_{kn})}\ell_{kn}(X, x), \qquad 1 \leq k \leq n, \tag{1.6}$$

$$\ell_k(x) = \ell_{kn}(X, x) = \frac{\omega_n(X, x)}{\omega_n'(X, x_{kn})(x - x_{kn})}, \qquad 1 \leq k \leq n, \tag{1.7}$$

and

$$(1.8) \qquad \omega_n(x) = \omega_n(X,x) = c_n \prod_{k=1}^{n} (x - x_{kn}), \qquad n \in N.$$

The polynomials ℓ_k of degree exactly $n-1$ are the fundamental functions of the (usual) Lagrange interpolation while the functions t_k are the fundamental functions of the weighted Lagrange interpolation.

The classical Lebesgue estimation now has the form

$$(1.9) \qquad |L_n(f,w,X,x) - f(x)w(x)| \le \{\lambda_n(w,X,x) + 1\} E_{n-1}(f,w)$$

where the (weighted) Lebesgue function is

$$(1.10) \qquad \lambda_n(w,X,x) := \sum_{k=1}^{n} |t_{kn}(w,X,x)|, \qquad x \in I, \ n \in N$$

and

$$(1.11) \qquad E_{n-1}(f,w) := \inf_{p \in \mathcal{P}_{n-1}} \|(f-p)w\|, \qquad n \in \mathbf{N}.$$

Here $\|\ldots\|$ is the sup norm on I. If $w \in GJ$ then as it is well-known $E_{n-1}(f,w) \to 0$ if $n \to \infty$ and $f \in C(w,I)$.

Relation (1.9) and its immediate consequence

$$(1.12) \qquad \|L_n(f,w,X) - fw\| \le \{\Lambda_n(w,X) + 1\} E_{n-1}(f,w),$$

where

$$(1.13) \qquad \Lambda_n(w,X) := \|\lambda_n(w,X,x)\|$$

show that the investigation of $\lambda_n(w,X,x)$ and $\Lambda_n(w,X)$, the weighted Lebesgue constant, are fundamental.

In paper [3, Theorem 2.2] we proved *that for arbitrary fixed* $w \in GJ$ *and* $0 < \varepsilon < 1$

$$(1.14) \qquad \lambda_n(w,X,x) > \eta(\varepsilon,w)\log n \quad \text{if} \quad x \in I \setminus H_n, \quad n \ge n_1,$$

whatsoever interpolatory matrix X *in* I *is given. Here* $H_n = H_n(w,\varepsilon,X)$ *with* $|H_n| \le \varepsilon$.

That means, that the order $\log n$ obtained in [1] and [4] are *optimal.* Our forthcoming results give the same order for $w \in GJ$.

2 Results

2.1. For $w \in GJ$ let $p_n(w,s) = \gamma_n \prod_{k=1}^{n} (x - x_{kn}(w))$ be the corresponding orthonormal polynomials with respect to w, i.e $p_n(w) \in \mathcal{P}_n \setminus \mathcal{P}_{n-1}$ and $\int_{-1}^{1} p_n(w)p_m(w)w = \delta_{nm}$, $n, m \in \mathbf{N}$.

As it is well-known

(2.1)
$$x_{n+1,n}(w) \equiv -1 < x_{nn}(w) < x_{n-1,n}(w) < \ldots < x_{1n}(w) < 1 \equiv x_{0n}(w).$$

Generally $X = \{x_{kn}(w)\} \not\subset I(w)$ however a simple modification can do the job.

Let with $x_{kn} = x_{kn}(w^2)$ (sic!)

$$\left|x_{t(r,n),n}(w^2) - u_r\right| = \min_{1 \le k \le n} \left|x_{kn}(w^2) - u_r\right|, \qquad 1 \le r \le m. \tag{2.2}$$

If $|x_{tn} - u_r| > \delta/n$ ($\delta = \delta(w) > 0$ will be given later), then $v_{tn} := x_{tn}$; if $|x_{tn} - u_r| \le \delta/n$ then let $v_{tn} := x_{tn} + 2\delta/n$ $(1 \le r \le m)$. Further, if $k \ne t(r,n)$, $(1 \le r \le m)$, let $v_{kn} = x_{kn}$. Finally let $v_{0n} := \frac{1+x_{1n}}{2}$ and $v_{n+1,n} := \frac{x_{nn}-1}{2}$.

Let $U = U(w) := \{v_{kn}; 1 \le k \le n, n \in N\}$ and $V = V(w) := \{v_{kn}; 0 \le k \le n+1, n \in N\}$. We state

Theorem 2.1. *Let $w \in GJ$. For the interpolatory matrices $U(w)$ and $V(w)$*

$$\Lambda_n(w, U(w)) \sim n^{1/2}, \tag{2.3}$$

$$\Lambda_n(w, V(w)) \sim \log n. \tag{2.4}$$

Remarks 1. If w has no inner singularities, then $U = \{x_{kn}(w^2)\}$ and the results correspond to the ones with exponential weights on $(-1,1)$ if we replace $T(a_n)$ with n^2 (n^2 turns out to be the equivalent of $T(a_n)$ if we consider GJ-type weights).

2. Estimations (2.3) and (2.4) hold using singularities with ,,negative exponents" (see the original [2, Definition 1.1]).

3. Similar result was obtained using a different method by J. Szabados (oral communication).

2.2. The proof of Theorem 2.1 suggests another solution namely a matrix *without* additional nodes with weighted Lebesgue constants of optimal order. For this aim let $W = w/\sqrt{\varphi}$ and whenever $W^2 \in GJ$, we can construct the matrix $S = S(w) := \{s_{kn} \equiv \cos\sigma_{kn}; 1 \le k \le n, n \in N\}$ by shifting the roots $\{x_{kn}(W^2)\}$ (according to $u_r = u_r(w), 1 \le r \le m$) as we did before to get U from $x_{kn}(w)$. We state

Theorem 2.2. *Let $w, W^2 \in GJ$. For the interpolatory matrix $S(w)$*

$$\Lambda_n(w, S(w)) \sim \log n. \tag{2.5}$$

3 Proofs

3.1. *Proof of Theorem 2.1.* First we quote some relations. If $w^2 \in GJ$ then

(3.1)

$$|p_n'(w^2, x_{kn}(w^2))| \sim \frac{n}{(\varphi(n, x_{kn}(w^2)))^{3/2} w(n, x_{kn}(w^2))}, \qquad 0 \le k \le n+1,$$

$$|p_n(w^2, x)| \sim |p_n'(w^2, x_{jn}(w^2))||x - x_{jn}(w^2)|, \quad |x| \le 1, \tag{3.2}$$

$$\vartheta_{k+1,n}(w^2) - \vartheta_{kn}(w^2) \sim \frac{1}{n}, \qquad 0 \le k \le n. \tag{3.3}$$

Here $\varphi(x) = \sqrt{1-x^2}$, $\varphi(n, x) = \varphi(x) + \frac{1}{n}$;

$$w(n, x) = w_0(\sqrt{1-x} + \frac{1}{n}) w_{m+1}(\sqrt{1+x} + \frac{1}{n}) \prod_{r=1}^{m} w_r(|x - u_r| + \frac{1}{n});$$

$x_{jn}(w^2)$ is the closest root(s) of $p_n(w^2)$ to x (by definition $j = j(n, x)$ and $1 \le j \le n$); finally we applied notations $x_{kn}(w^2) = \cos\vartheta_{kn}(w^2)$, $0 \le k \le n+1$ (cf. [2, Theorem 3.3]).

Let $\omega_n(x) = \gamma_n(w^2) \prod_{k=1}^{n} (x - v_{kn})$. With $v_{kn} = \cos\eta_{kn}$ and $|x - v_{jn}| = \min_{1 \le k \le n} |x - v_{kn}|$ we state

$$|\omega_n'(v_{kn})| \sim \frac{n}{(\varphi(n, v_{kn}))^{3/2}} \frac{1}{w(n, v_{kn})}, \qquad 0 \le k \le n+1, \tag{3.4}$$

$$|\omega_n(x)| \sim |\omega_n'(v_{jn})||x - v_{jn}|, \qquad |x| \le 1, \tag{3.5}$$

$$\eta_{k+1,n} - \eta_{kn} \sim \frac{1}{n}, \qquad 0 \le k \le n. \tag{3.6}$$

Indeed, (3.6) immediately comes from the definition of v_{kn}. Next, we prove (with obvious short notations)

$$0 < c_1 \le \frac{\omega'(v_{kn})}{p_n'(x_{kn})} \le c_2, \qquad 0 \le k \le n, \quad n \in N, \tag{3.7}$$

if $\delta > 0$ is small enough. To get (3.7) (and (3.5)), for sake of simplicity, let $m = 1$, $u_1 = 0$, $n = 2N + 1$ and $w(x)$ be even, whence $x_{Nn} = 0$ and $v_{Nn} = \frac{2\delta}{nX}$. Then by definition $\omega_n(x) = p_n(x)\frac{x-v_{Nn}}{x-x_{Nn}} = p_n(x)\left\{1 - \frac{2\delta}{nx}\right\}$. Then

$$\omega_n'(x) = p_n'(x)\left\{1 - \frac{2\delta}{nx}\right\} + p_n(x)\frac{2\delta}{nx^2}. \tag{3.8}$$

If $k \neq N$, then $v_{kn} = x_{kn}$, so $\omega_n'(v_{kn}) = p_n'(x_{kn})\left\{1 - \frac{2\delta}{nx_{kn}}\right\}$. By (3.3), $n|x_{kn}| \ge c(w) > 0$ $(k \neq N)$, whence $\frac{3}{4} \le \{\ldots\} \le \frac{5}{4}$ with a proper $\delta > 0$. On the other hand, if $k = N$, then by (3.8) and (3.2) $\omega_m'(v_{Nn}) = \omega_n'(\frac{2\delta}{n}) = p_n\left(\frac{2\delta}{n}\right)\frac{n}{2\delta} \sim p'(x_{Nn})\frac{2\delta}{n} \cdot \frac{n}{2\delta} = p_n'(x_{Nn})$, which completes the proof of (3.7). Then, by $w(n, x_{kn}) \sim w(n, v_{kn})$, $\varphi(n, x_{kn}) \sim \varphi(n, v_{kn})$ and (3.7), we obtain (3.4).

Finally, we prove relation (3.5). First let $|x| \ge \frac{4\delta}{n}$. Then $\frac{1}{2} \le \{\ldots\} \le \frac{3}{2}$, so $\frac{3}{2} \le \frac{\omega_n(x)}{p_n(x)} \le 2$, which by (3.3), (3.4) and (3.7) verify (3.5).

If $0 < |x| < \frac{4\delta}{n}$ and $n = 2N + 1$, by (3.2) and (3.7)

$$|\omega_n(x)| = \left|p_n(x)\frac{x - v_N}{x - x_N}\right| \sim \left|(p_n'(x_N)|x - x_N|)\frac{x - v_N}{x - x_N}\right| =$$

$$= |p_n'(x_N)||x - v_N|| \sim |\omega_n'(v_N)||x - v_N|,$$

as it was stated. The other cases are similar.

3.2. By definition,

$$\lambda_n(w, U, x) = \sum_{k=1}^{n} |t_{kn}(w, U, x)| = \sum_{k=1}^{n} \left|\frac{w(x)\omega_n(x)}{w(v_k)\omega_n'(v_k)(x - v_k)}\right| \le \tag{3.9}$$

$$\le c\sum_{k=1}^{n} \left|\frac{w(n, x)\omega_n(x)}{w(n, v_k)\omega_n'(v_k)(x - v_k)}\right| = c\sum_{k=j} + c\sum_{k\neq j} := S_1 + S_2,$$

where we used relation $w(v_k) \sim w(n, v_k)$, which *generally does not hold for* x_k.

By (3.4)-(3-6), $S_1 \sim 1$, further again by (3.4)-(3.6), we get for $x \geq 0$, say,

$$S_2 \sim \sum_{k \neq j} \frac{n|x - v_j|(\sin \eta_k)^{3/2}}{(\sin \eta_j)^{3/2} n|x - v_k|} \leq c \sum_{k \neq j} \frac{(\sin \eta_k)^{3/2}}{(\sin \eta_j)^{1/2} n|x - v_k|} \sim$$

$$\text{(3.10)} \qquad \sim \sum_{k=1}^{3n/4} + \sum_{k=3n/4}^{n} \sim \sum_{k=1}^{n} \frac{k^{3/2}}{j^{1/2}} \frac{1}{(k+j)(|k-j|+1)} \sim$$

$$\sim \sum_{k=1}^{j/2} + \sum_{k=j/2}^{2j} + \sum_{k=2j}^{n} \sim 1 + \log 2j + \left(\frac{n}{j}\right)^{1/2} \leq c n^{1/2},$$

as it was stated. The case $x < 0$ is similar. That means, relation (2.3) has been proved.

3.3. Here we prove (2.4). By definition

$$\text{(3.11)} \quad \lambda_{n+2}(w, V, x) = \sum_{k=0}^{n+1} |t_{k,n+2}(w, V, x)| = \sum_{k=0,n+1} + \sum_{k=1}^{n} := T_1 + T_2.$$

Again, by definition

$$T_2 = \sum_{k=1}^{n} \left| \frac{v_{0n}^2 - x^2}{v_{0n}^2 - v_{kn}^2} \right| |t_{kn}(w, U, x)|.$$

As above, let $x \geq 0$, say. If $v_{0n} - n^{-2} \leq x \leq 1$, by $v_{0n}^2 - v_{kn}^2 \sim \sin^2 \eta_{kn}$ (cf. (3.6)) we have using the considerations of (3.10) (now $j = 1$)

$$T_2 \leq \frac{c}{n^2} \sum_{k=1}^{n} \frac{n^2}{k^{5/2}} \sim 1; \quad \text{if} \quad 0 \leq x \leq v_{0n} - n^{-2}, \quad \text{then}$$

$$v_{0n}^2 - x^2 \sim \sin \eta_j^2, \quad \text{i.e.} \quad T_2 \leq c \sum_{k=1}^{n} \frac{j^{3/2}}{k^{1/2}} \frac{1}{(k+j)(|k-j|+r)} \sim$$

$$\sim \sum_{k=1}^{j/2} + \sum_{k=j/2}^{2j} + \sum_{k=2j}^{n} \sim 1 + \log 2j + 1 \sim \log 2j \leq c \log n.$$

The estimation $T_1 \leq c$ is left to the reader.

The proof of Theorem 2.1 is complete.

3.4. *Proof of Theorem 2.2.* As before we can prove relations corresponding to (3.4)-(3.6). Namely, if $\Omega_n(x) = \gamma_n(W^2) \prod_{k=1}^{n} (x - s_{kn})$ then

$$|\Omega_n'(s_{kn})| \sim \frac{n}{(\varphi(n, s_{kn}))^{3/2}} \frac{1}{W(n, s_{kn})}, \qquad 1 \le k \le n, \tag{3.12}$$

$$|\Omega_n(x)| \sim |\Omega_n'(s_{jn})||x - s_{jn}|, \qquad |x| \le 1, \tag{3.13}$$

$$\sigma_{k+1,n} - \sigma_{kn} \sim \frac{1}{n}, \qquad 0 \le k \le n. \tag{3.14}$$

Here s_{jn} is the closest node(s) to x ($1 \le j \le n$); $s_{0n} = \cos\sigma_{0n} \equiv 1$, $s_{n+1,n} = \cos\sigma_{n+1,n} = -1$. Using these relations, $w(s_k) \sim w(n, s_k)$, and $w = W\sqrt{\varphi}$, one has if $x \ge 0$, say,

$$\lambda_n(w, S, x) = \sum_{k=1}^{n} |t_{kn}(w, S, x)| \le$$

$$\le c \sum_{k=1}^{n} \left| \frac{W(n, x)\sqrt{\varphi(n, x)}\, \Omega_n(x)}{W(n, s_k)\sqrt{\varphi(n, s_k)}\, \Omega_n'(s_k)(x - s_k)} \right| \sim$$

$$\sim \sum_{k=1}^{n} \frac{k}{(k + j)(|k - j| + 1)} \sim \log n.$$

(for other details, see Part 3.1).

Acknowledgement. The author thanks Professor Ted Kilgore for his suggestions and comments.

References

[1] G. Mastroianni, M. G. Russo, Lagrange interpolation in some weighted uniform spaces (to appear).

[2] G. Mastroianni, P. Vértesi, Some applications of generalized Jacobi weights, Acta Math. Hungar. (to appear).

[3] P. Vértesi, On the Lebesgue function of weighted Lagrange interpolation. II., Bull. of the Aust. Math. Soc. (to appear).

[4] Barbara Háy, P. Vértesi, Interpolation in spaces of weighted maximum norm, Studia Sci. Math. Hung., 14(1979), 1-9.

Relative Differentiation, Descartes' Rule of Signs, and the Budan-Fourier Theorem for Markov Systems

R. A. Zalik
Department of Mathematics, Auburn University
Auburn, Alabama 36849–5310

Abstract

Using appropriate generalized differentiation operators, we introduce the concept of zero of arbitrary order. These operators are expressed in terms of functions in the linear span of any "Generalized Extended Tchebycheff System". We then prove versions of the Budan-Fourier theorem and of Descartes' rule of signs valid for these systems. As a corollary, we obtain an interpolation theorem involving values of the function and its generalized derivatives at arbitrary points. These results are illustrated by means of examples.

1 Introduction and Statement of Results

Given a function $f(x)$, let $Z^*_{(a,b)}(f)$ denote the number of zeros of $f(x)$ in (a, b), counting multiplicities. Let $S^+[a_1, \ldots, a_n]$ denote the number of weak sign changes in the sequence $a_1, \ldots, a_n$, i.e. the number of sign changes, where the zeros are regarded as $+1$ or -1, whichever makes the count largest; $S^-[a_1, \ldots, a_n]$ will denote the number of strong sign changes, i.e. the number of sign changes in $a_1, \ldots, a_n$, where zero terms are ignored.

The Budan-Fourier theorem for polynomials states that if $p(x)$ is a polynomial of order n, then

$$\begin{aligned} Z^*_{(a,b)}(p) \quad \leq \quad & n - S^+[p(a), -p'(a), \ldots, (-1)^n p^{(n)}(a)] \\ & -S^+[p(b), p'(b), \ldots, p^{(n)}(b)]. \end{aligned}$$

From the inequality

$$S^+[a_1, -a_2, \ldots, (-1)^n a_{n+1}] + S^-[a_1, a_2, \ldots, a_{n+1}] \geq n, \tag{1}$$

(cf. [5, (2.48)]), it is also easily seen that if $p(x)$ is a nontrivial polynomial of order at most n, then

$$Z^*_{(a,b)}(p) \leq S^-[p(a), p'(a), \ldots, p^{(n)}(a)] - S^-[p(b), p'(b), \ldots, p^{(n)}(b)].$$

Descartes' rule of signs states that if $p(x) = \sum_{k=0}^{n} c_k x^k$, and not all the c_k are 0, then $Z^*_{(0,\infty)}(p) \leq S^-(c_0, \ldots, c_n)$.

For a discussion of these results, we refer the reader to [5].

L. L. Schumaker has generalized the Budan-Fourier theorem in the manner we describe forthwith.

Following [5, pp. 407-410], let $q_0(x)$ be a bounded strictly positive function on the interval $[a, b]$ and suppose that the functions $w_k(x)$, $k = 1, \ldots, n$ are bounded and strictly increasing on $[a, b]$, and right-continuous on $[a, b)$. For x in $[a, b]$, let

$$\begin{aligned} q_1(x) &:= q_0(x) \int_a^x dw_1(t)dt, \\ q_2(x) &:= q_0(x) \int_a^x \int_a^{t_1} dw_2(t_2)dw_1(t_1) \\ &\vdots \\ q_n(x) &:= q_0(x) \int_a^x \int_a^{t_1} \cdots \int_a^{t_{n-1}} dw_n(t_n)dw_{n-1}(t_{n-1}) \ldots dw_1(t_1). \end{aligned} \tag{2}$$

We call $Q_n := \{q_0, \ldots, q_n\}$ a Canonical Complete Tchebycheff (CCT) System on $[a, b]$.

Given a function g defined on $[a, b]$, we define $D_0 g(t) := g(t)/q_0(t)$, $D_0^+ g := D_0 g$, and

$$D_k g(t) := \lim_{\delta \to 0} \frac{D_{k-1}g(t+\delta) - D_{k-1}g(t)}{w_k(t+\delta) - w_k(t)}, k = 1, \ldots, n,$$

provided that these limits exist. The operators $D_k^+, k = 1, \ldots, n$, are similarly defined in terms of one sided limits (from the right). Note that the D_k^+ are undefined at b (except for D_0^+), and that $D_k g(a) = D_k^+ g(a)$. Moreover, if the functions $w_k(x)$, $k = 1, \ldots, n$ are continuous on $[a, b]$, (2) implies that $D_k^+ q(x)$, $k = 0, \ldots, n$, is defined for any linear combination $q(x)$ of the elements of Q_n, and any $x \in [a, b)$. (Since $w_k(x+\delta) - w_k(x) = \int_x^{x+\delta} dw_k(t)$, this follows by repeated application of [9, Lemma 3]). The operator D_k is called the *generalized differentiation operator with respect to* $\{q_0, w_1, ..., w_n\}$. It is readily seen that $f(x)$ has an nth order derivative in the ordinary sense if and only if it is differentiable with respect to $\{1, w_1, ..., w_n\}$, where $w_k(t) := t, k = 1, ..., n$.

If, for example, D denotes the differentiation operator with respect to $\{1, t^3\}$ and P the differentiation operator with respect to $\{1, t^5\}$, applying L'Hôpital's rule we deduce that $Dx^4 = (4/3)x, Px^4 = (4/5)x^{-1}$, and x^4 is not differentiable with respect to $\{1,\ t^5\}$ at $x = 0$.

Given a set U of functions, let $S(U)$ denote the linear span of U. If $q \in S(Q_n)$, where Q_n is a CCT system on $[a,b]$ and $1 \leq k \leq n$, we say that $q(x)$ has a zero of multiplicity (or order) k at $x \in [a,b)$, if $q(x) = D_1^+ q(x) = \cdots = D_{k-1}^+ q(x) = 0$, and $D_k^+ q(x) \neq 0$. If $q(x) = D_1^+ q(x) = \cdots D_n^+ q(x) = 0$, we say that $q(x)$ has a zero of multiplicity $n+1$ at x. With this definition of multiple zero, Schumaker's generalization of the Budan-Fourier theorem ([5, p. 411]) may be stated as follows:

Theorem A *Let $Q_n := \{q_0, \ldots, q_n\}$ be a CCT system on $[a,b]$, $c \in (a,b)$, and assume that $q(x) \in S(Q_n) \setminus S(Q_{n-1})$. Then*

$$\begin{aligned} Z^*_{(a,c)}(q) \quad &\leq \quad n - S^+[q(a), -D_1^+ q(a), \ldots, (-1)^n D_n^+ q(a)] \\ &\qquad -S^+[q(c), D_1^+ q(c), \ldots, D_n^+ q(c)]. \end{aligned}$$

The use of a point $c \in (a,b)$ is required because, as already mentioned, the operators $D_k^+, k = 1, \ldots, n$, are undefined at b. This is no major restriction: since the functions $q_0(x)$ and $w_k(x)$ are bounded, Q_n can be extended to any interval of the form $[a, b']$, $b < b'$ by setting, for example, $q_0(x) = q(b)$, and $w_k(x) = w_k(b) + x - b$, $k = 1, \ldots, n$, for $x \in (b, b']$.

One of the purposes of this paper is to prove versions of the Budan-Fourier theorem and Descartes' rule of signs. But we first need to introduce some additional definitions.

Let $A \subset (-\infty, \infty)$, let $I(A)$ denote the convex hull of A (for example, if $A := (-2,3) \cup \{4\} \cup (7,8]$, then $I(A) = (-2,8]$), and let $F(A)$ denote the set of real valued functions defined on A. A is said to satisfy *property (B)* if between any two distinct points of A there is another point of A. If in addition, $\inf A \notin A$ and $\sup A \notin A$, then A is said to satisfy *property (D)*. The numbers $\inf A$ and $\sup A$ are called the *endpoints* of A, with the caveat that, since A is not necessarily bounded, the endpoints could be $\pm\infty$.

If A satisfies property (B), $a := \inf A$, and $b := \sup A$, then the complementary set of A in $[a,b] \bigcap A$ is either empty or can be represented in the form $\bigcup_{i \in J} I(a_n, b_n)$, where J is a subset of the natural numbers, $a_i < b_i$, and $\{I(a_i, b_i), i \in J\}$ is a collection of disjoint intervals. For each i, $I(a_i, b_i)$ is an open or semiopen interval having a_i and b_i as its endpoints (i. e. $I(a_i, b_i)$ is one of the intervals (a_i, b_i), $[a_i, b_i)$, or $(a_i, b_i]$). Thus,

$$\varphi(x) := x - \sum_{b_i < x} (b_i - a_i)$$

is a strictly increasing function that maps A onto a dense subset of the interval (a, c), where $c := \sup A - \sum_{b_i \in J}(b_i - a_i)$.

If A satisfies property (B), $f \in F(A)$, and a, c, and φ are defined as in the preceding paragraph, we say that f is *D–continuous* if $f \circ \varphi^{-1}$ is continuous in the relative topology of $\varphi(A)$. Let $A^1 := A$ and, for $n > 1$, let A^n denote the n–fold cartesian product of A with itself. Assume that $g \in F(A^n)$, and let $\mathbf{x} = (x_1, \dots, x_n) \in A^n$, $\mathbf{e} = (e_1, \dots, e_n) \in A^n$, $\mathbf{y} = (y_1, \dots, y_n) \in [\varphi(A)]^n$, and $\mathbf{\Phi}(\mathbf{e}) := (\varphi(e_1), \dots, \varphi(e_n))$. Then the *D–limit* is defined as follows:

$$\text{D–lim}_{\mathbf{x} \to \mathbf{e}} g(\mathbf{x}) := \lim_{\mathbf{y} \to \mathbf{\Phi}(\mathbf{e})} g(\mathbf{\Phi}^{-1}(\mathbf{y})).$$

If $\mathbf{x}_0 := (x, \dots, x) \in A^n$, then

$$\text{D–lim}_{x_1, \dots x_n \to x} g(\mathbf{x}) := \text{D–lim}_{\mathbf{x} \to \mathbf{x}_0} g(\mathbf{x}).$$

When we say that a set of points $\{x_1, \dots, x_n\} \in A$ coalesce, we mean that the points $\varphi(x_1), \dots, \varphi(x_n)$ coalesce.

Here is an equivalent definition of D–limit: let $\mathbf{e} \in A^n$, $\alpha \in A$, $m(\alpha) := \sup\{x \in A : x < \alpha\}$, $M(\alpha) := \inf\{x \in A : x > \alpha\}$, $\mathbf{m}(\mathbf{e}) := (m(e_1), \dots, m(e_n))$, $\mathbf{M}(\mathbf{e}) := (M(e_1), \dots, M(e_n))$, and $g(\mathbf{x}) \in F(A^n)$. Assume

$$\lim_{\mathbf{x} \to \mathbf{m}(\mathbf{e})^-} g(\mathbf{x}) = \lim_{\mathbf{x} \to \mathbf{M}(\mathbf{e})^+} g(\mathbf{x}) = \ell,$$

where the limits are in the relative topology of A^n. Then

$$\text{D–lim}_{\mathbf{x} \to \mathbf{e}} g(\mathbf{x}) = \ell.$$

If A has at least $n + 2$ points, a sequence of functions $Z_n := \{z_0, \dots, z_n\} \subset F(A)$ is called a Tchebycheff system (T system) on A if for all points $x_0 < x_1 < \dots < x_n$ in A, $\det[z_i(x_j)]_{i,j=0}^n > 0$. If Z_k is a T system for $k = 0, \dots, n$, we say that Z_n is a Markov (or Complete Tchebycheff) system. Note that in this case $z_0(x) > 0$ on A. It follows from [5, Theorem 9.53] that a CCT system is a Markov system.

We shall use the following notation:

$$\bigcup\begin{pmatrix} z_0, \dots, z_n \\ x_0, \dots, x_n \end{pmatrix} := \det\{z_i(t_j)\}_{i,j=0}^n. \tag{3}$$

Let Z_n be a T system on A, let $x_0, \dots, x_n$ be distinct points of A, and let $f \in F(A)$. The divided difference of $f(x)$ with respect to Z_n at $x_0, \dots, x_n$ is defined by

$$\begin{bmatrix} z_0, \dots, z_n \\ x_0, \dots, x_n \end{bmatrix} f := \frac{\bigcup\begin{pmatrix} z_0, \dots, z_{n-1}, f \\ x_0, \dots, x_n \end{pmatrix}}{\bigcup\begin{pmatrix} z_0, \dots, z_n \\ x_0, \dots, x_n \end{pmatrix}}$$

(for $n = 0$ this reduces to $\left[\begin{smallmatrix} z_0 \\ x_0 \end{smallmatrix}\right] f = f(x_0)/z_0(x_0)$). We also define the generalized differentiation operators: $H_0 f(x) = f(x)/z_0(x)$, and

$$H_k f(x) := \text{D–lim}_{x_0,\ldots,x_k \to x} \begin{bmatrix} z_0, \ldots, z_k \\ x_0, \ldots, x_k \end{bmatrix} f, \; k = 1, \ldots, n,$$

where the points x_i are assumed all distinct. We say that $p(x)$ has a zero of order k ($1 \le k \le n$) at t with respect to Z_n, if $H_r p(t) = 0$, $r = 0, \ldots, k-1$, and $H_k p(t) \neq 0$. If $H_r p(t) = 0$, $r = 0, \ldots, n$, we say that $p(x)$ has a zero of order $n+1$ at t with respect to Z_n. Proceeding as in the proof of Theorem 4 below, it is easy to see that if Z_n is a CCT system, then $H_k p(x) = D_k p(x)$, $k = 0, \ldots, n$, at every point x where the functions w_k, $k = 1, \ldots, n$, are continuous.

We say that $Z_n \subset F(A)$ is representable if *for some* $c \in A$ there is a sequence $U_n := \{u_0, \ldots, u_n\} \subset F(A)$, obtained from Z_n by a triangular linear transformation (i.e. $u_0(x) = z_0(x)$, $u_k(x) - z_k(x) \in S(U_{k-1}), k = 1, \ldots, n$), a strictly increasing function $h(x) \subset F(A)$ (an "embedding function"), with $h(c) = c$, and a sequence $W_n := \{w_1, \ldots, w_n\}$ of continuous increasing functions defined on $I(h(A))$, such that, if

$$\begin{aligned} p_1(x) &:= \textstyle\int_c^x dw_1(t) \\ p_2(x) &:= \textstyle\int_c^x \int_c^{t_1} dw_2(t_2) dw_1(t_1) \\ &\vdots \\ p_n(x) &:= \textstyle\int_c^x \int_c^{t_1} \cdots \int_c^{t_{n-1}} dw_n(t_n) \ldots dw_1(t_1), \end{aligned} \tag{4}$$

then, for every $x \in A$, $u_k(x) = u_0(x) p_k(h(x))$, $k = 1, \ldots, n$. In this case, we say that (h, c, W_n, U_n) is a *representation* of Z_n([3, 4, 5]).

Remark 1 From [2, Lemma 4.9] we know that if c and d are points of A, (g, c, R_n, U_n) is a representation of Z_n, $h(x) := g(x) - g(d) + d$, $w_i(t) := r_i(t + g(d) - d)$, $p_i(x)$ is given by (4) with c replaced by d, and $v_i(x) := u_0(x) p_i(x)$, $1 \le i \le n$, then also (h, d, W_n, V_n) is a representation of Z_n. Clearly, if the functions $r_i(x)$ are strictly increasing and continuous, so are the functions $w_i(x)$. We also observe that if a representation of Z_n exists for any *one* point c of A, then a representation exists for *every* point c of A.

If Z_n has a representation (h, c, W_n, U_n) such that the elements of W_n are strictly increasing on $I(h(A))$, we say that Z_n is a *Generalized Extended Complete Tchebycheff System*, or GECT system.

Remark 2 From e. g. [5, Theorem 9.53] and Remark 1 we readily see that a GECT system is a Markov system.

In the sequel, $a := \inf A$, $b := \sup A$, and $A_1 := A \setminus \{b\}$. If $a > -\infty$, then $A_0 := A \cup \{a\}$. Note that if A has property (D), then A_0 has property (B).

Remark 3 If $Z_n \subset F(A)$ is a GECT system, $a \in A$, and $B := (-\infty, a) \cup A$, then there is a GECT system $Y_n \in F(B)$ such that $y_i(x) = z_i(x)$ on A, for $i = 0, \ldots, n$. Indeed, if (g, c, R_n, U_n) is a representation of Z_n such that the functions $r_i(x)$ are strictly increasing and continuous on $I(h(A))$, it suffices to define $h(x) := g(x)$ on A, $h(x) := x - a + g(a)$ on $(-\infty, a)$, $w_i(x) := r_i(x)$ on A, and $w_i(x) := x - g(a) + r_i(g(a))$ on $(-\infty, g(a))$. If $b \in A$, a similar argument can be used to extend Z_n beyond b. (The problem of extending the domain of definition of a Markov or T system is discussed in detail in [2] and references thereof.) From the definition of a CCT system Q_n defined on an interval $[a, b]$, we also see that there is a CCT system Q'_n, defined on an interval $[a', b']$, with $a' < a$, $b < b'$, such that Q_n is the restriction of Q'_n to $[a, b]$. Our version of Theorem A requires a condition of this kind:

Theorem 1 *Let A have property (D), assume that $a > -\infty$, and let $Z_n \subset F(A_0)$ be a GECT system such that z_1/z_0 is D–continuous on A_0. Let $d \in A$, $a < d < b$, and $B := (a, d) \cap A$. Then, for every $p(x) \in S(Z_n)$ and $k = 0, \ldots, n$, $H_k p(x)$ is defined for every $x \in A_0$, and, if $p(x) \not\equiv 0$,*

$$\begin{aligned} Z_B^*(p) \le n \quad &- \quad S^+[H_0 p(a), -H_1 p(a), \ldots, (-1)^n H_n p(a)] \\ &- \quad S^+[H_0 p(d), H_1 p(d), \ldots, H_n p(d)]. \end{aligned}$$

If, in particular, $p(x) \in S(Z_n) \setminus S(Z_{n-1})$, then

$$Z_B^*(p) \le S^-[H_0 p(a), \ldots, H_n p(a)] - S^+[H_0 p(b), \ldots, H_n p(b)].$$

An immediate consequence of Theorem 1 is:

Theorem 2 *Let A have property (D), and let $Z_n \subset F(A_0)$ be a GECT system such that z_1/z_0 is D–continuous on A. Then, for any nontrivial $p(x) \in S(Z_n)$, $Z_A^*(p) \le n$.*

As a consequence of Theorem 1 we will prove the following generalization of Descartes' rule of signs:

Theorem 3 *Let A have property (D), assume that $a > -\infty$, let $A_{0,1} := \{a\} \cup A_1$, and let $Z_n \subset F(A_{0,1})$ be a GECT system such that z_1/z_0 is D–continuous on $A_{0,1}$. Let (h, a, W_n, V_n) be a representation of Z_n such that all the elements of $W_n(x)$ are strictly increasing on $I(h(A_{0,1}))$ (note that we have set $c = a$), and let $p(x) = \sum_{k=0}^{n} c_k u_k(x)$ be such that not all the c_k vanish. Then $Z_A^*(p) \le S^-(c_0, \ldots, c_n)$.*

The following proposition, of independent interest, will be used in the proof of Theorem 1:

Theorem 4 *Let A have property (B), and let $Z_n \subset F(A_0)$ be a GECT system such that z_1/z_0 is D–continuous on A. Let (h, c, W_n, U_n) be a representation of Z_n such that the elements of W_n are strictly increasing. Let $1 \leq k \leq n$, $f \in F(A)$, and $x \in A$. If $D_k(f \circ h^{-1})$ is defined at $h(x)$, then $H_k f$ is defined at x, and $H_k f(x) = D_k(f \circ h^{-1})(h(x))$.*

Theorem 4 generalizes results of de la Vallée–Poussin and Franklin (cf. [3]).

Let A be an interval that contains the origin, and $Q_n := \{1, x, x^2, ..., x^n\}$. Then Q_n has a representation of the form (4) with $w_k(t) := kt, k = 1, ..., n$ and $c = 0$, and Theorem 4 implies that a function $f(x)$ has a zero of order n at x_0 in the ordinary sense if and only if it has a zero of order n at x_0 with respect to Q_{n+1}.

We now extend the definition in (3) to allow for equalities among the x_j : if for instance $x_j < x_{j+1} = x_{j+2} = \ldots = x_{j+k} < x_{j+k+1}$, then $\bigcup^* \binom{z_0, \ldots, z_n}{x_0, \ldots, x_n}$ is defined as the determinant in (3), where, for $1 \leq r \leq k-1$, the $j+r+1^{th}$ column vector is replaced by the column vector $(H_r z_i(x_j), 0 \leq j \leq n)$. With this definition we have:

Theorem 5 *Let A have property (B), and let $Z_n \subset F(A)$ be a GECT–system such that $z_1(x)/z_0(x)$ is D–continuous on A. Then, for $1 \leq k \leq n$ and any $x_0 \leq x_1 \leq \ldots \leq x_k$ in A, $\bigcup^* \binom{z_0, \ldots, z_n}{x_0, \ldots, x_n} > 0$.*

If H_k is the ordinary k^{th} order differentiation operator, then Z_n is an Extended Complete Tchebycheff System ([4, 5]).

In view of Theorem 5 we can generalize the definition of divided difference to allow for coalescing points:

$$\begin{bmatrix} z_0, \ldots, z_n \\ x_0, \ldots, x_n \end{bmatrix} f := \frac{\bigcup^* \binom{z_0, \ldots, z_{n-1}, f}{x_0, \ldots, x_n}}{\bigcup^* \binom{z_0, \ldots, z_n}{x_0, \ldots, x_n}}.$$

With this definition we have:

Theorem 6 *Let A have property (B), and let $Z_n \subset F(A)$ be a GECT system such that $z_1(x)/z_0(x)$ is D–continuous on A. Let $\{c_i\}_{i=0}^r$ be a sequence of distinct elements of A, $0 \leq r \leq n$. Let $\{k_i\}_{i=0}^r$ be a sequence of strictly positive integers such that $\sum_{i=0}^r k_i = n + 1$. Finally, let $\{d_{i,1}, d_{i,2}, \ldots, d_{i,k_i}\}_{i=0}^r \subset R$. Then there is a unique $z \in S(Z_n)$ such that $H_{j-1} z(c_i) = d_{i,j}, j = 1, \ldots, k_i, i = 0, \ldots, r$.*

2 Proofs

Proof of Theorem 4. Without essential loss of generality we may assume that $z_0(x) = 1$. Let (h, c, W_n, U_n) be a representation of Z_n such that the elements of W_n are strictly increasing on $I(h(A))$. Since the value of a divided difference is the same for any permutation of the sequence $\{x_k\}_{k=0}^n$, there is no essential loss of generality in assuming that $x_0 < x_1 < \ldots < x_n$. Since U_n is obtained from Z_n by a triangular linear transformation,

$$\begin{bmatrix} z_0, \ldots, z_k \\ x_0, \ldots, x_k \end{bmatrix} f = \begin{bmatrix} u_0, \ldots, u_k \\ x_0, \ldots, x_k \end{bmatrix} f = \begin{bmatrix} p_0, \ldots, p_k \\ h(x_0), \ldots, h(x_k) \end{bmatrix} (f \circ h^{-1}),$$

where $p_0(x) = 1$, and the functions $p_k(x)$, $k = 1, \ldots, n$, are given by (4). From Remark 2 we know that P_n is a Markov system on $I(h(A))$, and therefore the divided differences in the preceding displayed formula are well defined.

Since $h(x_0) < h(x_1) < \ldots < h(x_k)$, from [9, Theorem 2] we see that

$$\begin{bmatrix} p_0, \ldots, p_k \\ h(x_0), \ldots, h(x_k) \end{bmatrix} (f \circ h^{-1}) = D_k(f \circ h^{-1})(\xi),$$

where $h(x_0) < \xi < h(x_k)$. (We can apply [9, Theorem 2] even if the interval of definition of the functions w_i contains one or both endpoints, see Remark 3 above.) Since p_1 is strictly increasing and continuous, the inverse function p_1^{-1} is strictly increasing. It must also be continuous, otherwise the domain of p_1 could not be an interval. Since the hypotheses imply that $u_1(x) = p_1[h(x)]$, it is clear that $h(x) = p_1^{-1}[u_1(x)]$. Since u_1 is D–continuous, this implies that h is D–continuous as well. Passing to the limit, the conclusion follows.

Proof of Theorem 1. Without essential loss of generality we may assume that $z_0(x) = 1$. From Remark 1 we know there is a representation (h, a, W_n, U_n) of Z_n such that the elements of W_n are strictly increasing on $I(h(A_0))$. Let e be any point in $(d, \infty) \cap A$. Since for $1 \leq i \leq n$, $u_i(x) = p_i[h(x)]$ and $S(Z_i) = S(U_i)$, and, moreover, P_n is a CCT system on $[h(a), h(e]$, the conclusion follows from Theorem A, Theorem 4, and (1).

Proof of Theorem 3. The hypotheses imply that A does not contain its endpoints. Let m be the largest of the indices i such that $c_i \neq 0$. Then $p \in S(Z_m) \setminus S(Z_{m-1})$. (If $m = 0$, $p \in S(Z_0)$). Let $d \in \cap A_1$, $d > a$, and let $B := (a, d) \cap A$. Applying Theorem 1 we see that

$$Z_B^*(p) \leq S^-[H_0 p(a), \ldots, H_m p(a)] - S^+[H_0 p(b), \ldots, H_m p(b)].$$

From Theorem 4 and [9, Theorem 2] we deduce that $H_r z_k(a) = 0$ for $k = 1, \ldots, n$ and $0 \leq r < k$, and that $H_k z_k(a) = 1, k = 0, \ldots, n$; thus

$c_k = H_k z_k(a)$, and therefore $S^-[H_0 p(a), \ldots, H_m p(a)] = S^-[c_0, \cdots, c_m] = S^-[c_0, \cdots, c_n]$. Since $S^+[H_0 p(b), \ldots, H_m p(b)] \geq 0$ and d is arbitrary, the conclusion follows.

Proof of Theorem 5. Without essential loss of generality we may assume that $z_0(x) = 1$. Let $x_0 \leq x_1 \leq \ldots \leq x_n$ be in A, and let (h, c, W_n, U_n) be a representation of Z_n. Since U_n is obtained from Z_n by a triangular linear transformation,

$$\bigcup^* \begin{pmatrix} u_0, \ldots, u_n \\ x_0, \ldots, x_n \end{pmatrix} = \bigcup^* \begin{pmatrix} z_0, \ldots, z_n \\ x_0, \ldots, x_n \end{pmatrix}.$$

If $t_j = h^{-1}(x_j), p_0(x) := 1$, and the functions $p_k(x)$, $k = 1, \ldots, n$ are given by (4), then $p_k(t_j) = H_0 u_k(x_j)$. Applying the Mean Value Theorem for Riemann-Stieltjes integrals (cf., e.g. [1]), we readily see that for any k and j, $1 \leq k, j \leq n$, $D_k p_j$ is defined for every $t \in h(A)$. Therefore $D_k(p_j \circ h^{-1})$ is defined at $h(x)$ for every $x \in A$, and from Theorem 4 we deduce that

$$\bigcup^* \begin{pmatrix} u_0, \ldots, u_n \\ x_0, \ldots, x_n \end{pmatrix} = c \bigcup^* \begin{pmatrix} p_0, \ldots, p_n \\ t_0, \ldots, t_n \end{pmatrix},$$

where $c = \prod_{r=0}^{m} u_0(x_{k_r})$, $\{x_{k_r}\}_{r=0}^{m}$ is the set of distinct points in the sequence $\{x_0, \ldots, x_n\}$, and the columns $\{p_i(t_{k+j})\}_{i=0}^{n}$ corresponding to coalescing points, say $t_{k-1} < t_k = t_{k+1} = \ldots t_{k+r}$, are replaced by $\{H_j p_i(t_k)\}_{i=0}^{n} = \{D_j p_i(t_k)\}_{i=0}^{n}$, $j = 1, \ldots, r$.

Let $W[p_0, \ldots, p_n](t) := \bigcup^* \begin{pmatrix} p_0, \ldots, p_n \\ t, t, \ldots, t \end{pmatrix}$. From the Mean Value Theorem for Riemann-Stieltjes integrals we readily deduce that $D_k p_k(t) = 1$, $k = 1, 2, \ldots, n$, and therefore that $W[p_0, \ldots, p_n](t) = 1$. Similarly, if $n > 0$, $\tilde{p}_1(x) := 1$,

$$\begin{aligned} \tilde{p}_2(x) &:= \int_c^x dw_2(t) \\ \tilde{p}_3(x) &:= \int_c^x \int_c^{t_1} dw_3(t_2) dw_2(t_1) \\ &\vdots \\ \tilde{p}_n(x) &:= \int_c^x \int_c^{t_1} \cdots \int_c^{t_{n-2}} dw_n(t_{n-1}) \ldots dw_2(t_1), \end{aligned} \tag{5}$$

then $W[\tilde{p}_1, \ldots, \tilde{p}_n](t) = 1$. The remainder of the proof follows exactly as in the proof of [4, Chapter XI, Theorem 1.1], with $\{p_0, \ldots, p_n\}$ and $\{\tilde{p}_1, \ldots, \tilde{p}_n\}$ taking the place of $\{u_0, \ldots, u_n\}$ and $\{v_0, \ldots, v_{n-1}\}$ respectively, and using [9, Lemma 1] instead of the mean value theorem. The details will be omitted.

Proof of Theorem 6. We may assume without any loss of generality that $c_0 < c_1 < \ldots < c_r$. Let $z = \sum_{j=0}^{n} c_j z_j$. Then the interpolation conditions are equivalent to

$$\sum_{j=0}^{r} a_j H_k z_j(c_i) = d_{k,i}; k = 0, \ldots, w_i, i = 0, \ldots, r. \tag{6}$$

Since $\sum_{i=0}^{r} w_i = n + 1$ by hypothesis, this is a system of $n + 1$ equations in $n + 1$ unknowns a_j. Let $s_{-1} = 0$, $s_m = \sum_{k=0}^{m-1} w_k$, $x_{s_{m-1}+i} = c_r$; $i = 0, \ldots, w_m, m = 0, \ldots, r$. Then the determinant of the matrix of coefficients of (6) is $\mathsf{U}^*\left(\begin{smallmatrix} z_0, \ldots, z_n \\ x_0, \ldots, x_n \end{smallmatrix}\right)$, and the assertion follows from Theorem 5.

3 Examples

Example 1 Let $A := [-3, -1) \bigcup [0, 1)$, and let $Z_3 := \{z_0, z_1, z_2, z_3\}$ be defined on A in the following way: $z_0(x) := 1$,

$$z_1(x) := \begin{cases} x^3, \ -3 \le x \le -2 \\ (x+1)^3, \ -2 < x < -1, \\ (x+2)^3, \ 0 \le x < 1, \end{cases}$$

$$z_2(x) := \begin{cases} (x-1)^2 + 1\}e^x - x^3/3 - 2, \ -3 \le x \le -2 \\ \{x^2 + 1\}e^{x+1} - (x+1)^3/3 - 2, \ -2 < x < -1, \\ \{(x+1)^2 + 1\}e^{x+2} - (x+2)^3/3 - 2, \ 0 \le x < 1, \end{cases}$$

$$z_3(x) := \begin{cases} 3(x-1)e^x\{\sin x - x\cos x\} + x^3, \ -3 \le x \le -2 \\ 3xe^{x+1}\{\sin(x+1) - (x+1)\cos(x+1)\} + (x+1)^3, \\ \hfill -2 < x < -1, \\ 3(x+1)e^{x+2}\{\sin(x+2) - (x+2)\cos(x+2)\} + (x+2)^3, \\ \hfill 0 \le x < 1. \end{cases}$$

We shall first show that Z_3 is a GECT system on A by exhibiting an integral representation (h, c, W_n, U_n) such that the functions in W_n are strictly increasing.

We note that the functions $z_i(x)$ can be written in the form $z_i(x) = u_i[h(x)]$, where

$$h(x) = \begin{cases} x, & -3 \le x \le -2 \\ x+1, & -2 < x < -1, \\ x+2, & 0 \le x < 1, \end{cases}$$

and the functions $u_i(x)$ are defined on $[-3, 3) = I(h(A))$ as follows: $u_0(x) := 1$, $u_1(x) := x^3$, $u_2(x) := \{(x-1)^2 + 1\}e^x - (1/3)x^3 - 2$, and $u_3(x) :=$

$3(x-1)e^x\{\sin x - x\cos x\} + x^3$. Since $h(0) = u_1(0) = u_2(0) = u_3(0) = 0$, let us look for a representation with $c = 0$.

Since $u_1'(x) = 3x^2$, we have: $u_1(x) = \int_0^x dw_1(t)$, where $w_1(x) = u_1(x)$. Moreover, $u_2'(x) = x^2(e^x - 1)$ and $u_3'(x) = 3x^2\{e^x(\sin x - \cos x) + 1\}$. Thus, by a simple application of L'Hôpital's rule, we see that $D_1u_2(x) = (1/3)(e^x - 1)$, and $D_1u_3(x) = e^x(\sin x - \cos x) + 1$. Thus, $u_2(x) = \int_0^x D_1u_2(t_1)dw_1(t_1) = \int_0^x \int_0^{t_1} dw_2(t_2)dw_1(t_1)$, where $w_2(x) = D_1u_2(x)$. Moreover, since $(D_1u_3(x))' = 2e^x \sin x$, and $u_2'(x) = (1/3)e^x$, we have $D_2u_3(x) = 6\sin x$. Thus

$$\begin{aligned} u_3(x) &= \int_0^x u_3'(t_1)dt_1 = \int_0^x D_1u_3(t_1)dw_1(t_1) \\ &= \int_0^x \int_0^{t_1} [D_1u_3(t_1)]'dw_1(t_1) = \int_0^x \int_0^{t_1} D_2u_3(t_2)dw_2(t_2)dw_1(t_1) \\ &= \int_0^x \int_0^{t_1} \int_0^{t_2} dw_3(t_3)dw_2(t_2)dw_1(t_1), \end{aligned}$$

where $w_3(x) = 6\sin x$. Note that all the functions $w_i(x)$ are strictly increasing and continuous on $[-3, 3]$.

Let $z(x) := 3z_2(x) - 4z_3(x)$. Then $z(x) = u[h(x)]$, where $u(x) = 3u_2(x) - 4u_3(x) = 3\{x^2 - 2x + 2\}e^x - 12(x-1)e^x\{\sin x - x\cos x\} - 5x^3 - 6$. Let $B := A \cap (-2, 0.5)$. Then $I(h(B)) = (-2, 2.5)$. We now use Theorem 1 to estimate the zeros of $z(x)$ in B and of $u(x)$ in $(-2, 2.5)$. We have: $D_0u(x) = u(x)$, $D_1u(x) = 3D_1u_2(x) - 4D_1u_3(x) = e^x - 1 - 4\{e^x(\sin x - \cos x) + 1\}$, $D_2u(x) = 3D_2u_2(x) - 4D_2u_3(x) = 3 - 24\sin x$, and $D_3u(x) = -4$. Thus,

$$\begin{aligned} & S^-[H_0z(-2), H_1z(-2), H_2z(-2), H_3z(-2)] \\ & \qquad -S^-[H_0z(0.5), H_1z(0.5), H_2z(0.5), H_3z(0.5)] \\ &= S^-[u(-2), D_1u(-2), D_2u(-2), D_3u(-2)] \\ & \qquad -S^-[u(2.5), D_1u(2.5), D_2u(2.5), D_3u(2.5)] \\ &= S^-[29.57, -4.59, 24.82, -4] - S^-[-608.87, -61.02, -11.36, -4] \\ &= 3 - 0 = 3. \end{aligned}$$

Thus, from Theorem 1, $Z_B^*(z) \le 3$, and $Z_{(-2,2.5)}^*(u) \le 3$.

Note that $u(x)$ has a double zero at 0. Moreover, since $u(-1.5) = 8.57$ and $u(-1) = -7.68$, $u(x)$ vanishes somewhere in $(-1.5, -1)$. Since $\{u_0, u_1, u_2, u_3\}$ is a GECT system on $(-3, 3)$, we know from Theorem 2 that $Z_{(-3,3)}^*(u) \le 3$. Thus, $Z_{(-3,3)}^*(u) = 3$. On the other hand, $0 \in h(A)$, but $(-1.5, -1) \cap h(A) = \phi$. Thus, $Z_A^*(u) = 2$.

Example 2 Let $A = (-\infty, \infty), h(x) := [x]$, (where $[x]$ is the integral part of x), $w_1(x) := x^3, w_2(x) := x$, $w_3(x) := e^x$,

$$
\begin{aligned}
u_1(x) &:= \int_0^x dw_1(t) = x^3 \\
u_2(x) &:= \int_0^x \int_0^{t_1} dw_2(t_2) dw_1(t_1) = \\
&\quad 3\int_0^x \int_0^{t_1} t_2 t_1^2 dt_2 dt_1 = (3/4)x^4
\end{aligned}
$$

and

$$
\begin{aligned}
u_3(x) &:= \int_0^t \int_0^{t_1} \int_0^{t_2} dw_3(t_3) dw_2(t_2) dw_1(t_1) = \\
&\quad 3\int_0^x \int_0^{t_1} \int_0^{t_2} e^{t_3} t_1^2 dt_3 dt_2 dt = (3/4)x^4 - x^3 - 6
\end{aligned}
$$

Since u_3 has a zero of multiplicity 3 at the origin, applying Theorem 3 we conclude that neither $u_3(x)$ nor $z_3(x) := u_3[h(x)]$ vanish on $(0, \infty)$.

Example 3 Let Z_3 be defined as in Example 1, let $c_0 = -2$, $c_1 = 0$, $c_2 = 0.5$, $w_0 = w_2 = 1$, $w_2 = 2$, $d_{0,1} = 1$, $d_{1,1} = -1$, $d_{1,2} = 3$, $d_{2,1} = 0.3$. We find the unique $z \in S(Z_3)$ such that $H_0 z(c_0) = d_{0,1}$, $H_0 z(c_1) = d_{1,1}$, $H_1 z(c_1) = d_{1,2}$, and $H_0 z(c_2) = d_{2,1}$, whose existence is guaranteed by Theorem 6.

Let $z(x) = a_0 z_0(x) + a_1 z_1(x) + a_2 z_2(x) + a_3 z_3(x)$, and $u(x) := a_0 u_0(x) + a_1 u_z(x) + a_2 u_2(x) + a_3 u_3(x)$. Since $h(-2) = -2$, $h(0) = 1$, and $h(0.5) = 2.5$, we need to solve the following system of equations:

$$
\begin{aligned}
&a_0 + u_1(-2)a_1 + u_2(-2)a_2 + u_3(-2)a_3 = 1 \\
&a_0 + u_1(1)a_1 + u_2(1)a_2 + u_3(1)a_3 = -1 \\
&D_1 u_1(1)a_1 + D_1 u_2(1)a_2 + D_1 u_3(1)a_3 = -3 \\
&a_0 + u_1(2.5)a_1 + u_2(2.5)a_2 + u_3(2.5)a_3 = 0.3,
\end{aligned}
$$

i.e.

$$
\begin{aligned}
&a_0 - 8a_1 + 2.02a_2 - 5.878a_3 = 1 \\
&a_0 + a_1 + 0.385a_2 + a_3 = -1 \\
&a_1 + 0.5728a_2 + 0.819a_3 = -3 \\
&a_0 + 15.625a_1 + 32.385a_2 + 158.233a_3 = 0.3,
\end{aligned}
$$

which yields: $a_0 = 1.129$, $a_1 = -1.607$, $a_2 = 3.75$, and $a_3 = 0.921$.

References

[1] T. A. Apostol, "Mathematical Analysis", 2^d Ed., Addison-Wesley, New York, 1974.

[2] J. M. Carnicer, J. M. Peña and R. A. Zalik, Strictly totally positive systems, to appear in J. Approx. Theory.

[3] P. Franklin, Derivatives of higher order as single limits, Bull. Amer. Math. Soc. **41** (1935), 573–582.

[4] S. Karlin and W. J. Studden, "Tchebycheff Systems: With Applications in Analysis and Statistics", Interscience, New York, 1966.

[5] L. L. Schumaker, "Spline Functions: Basic Theory", John Wiley & Sons, Inc., New York, 1981.

[6] F. Schwenker, Integral representation of normalized weak Markov systems, J. Approx. Theory **68** (1992), 1-24.

[7] R. A. Zalik, Integral Representation and Embedding of Weak Markov Systems, J. Approx. Theory **58** (1989), 1-11.

[8] R. A. Zalik, Integral representation of Markov systems and the existence of adjoined functions for Haar spaces, J. Approx. Theory **65** (1991), 22-31.

[9] R. A. Zalik and D. Zwick, Some properties of Markov systems, J. Approx. Theory **65** (1991), 32-45.

[10] R. A. Zalik, Nondegeneracy and integral representation of weak Markov systems, J. Approx. Theory **68** (1992), 25-32.

Index

S

T

U

V

W

Z